ENVIRONMENTAL SCIENCE, ENGINEERING AND TECHNOLOGY

FLY ASH

SOURCES, APPLICATIONS AND POTENTIAL ENVIRONMENTAL IMPACTS

ENVIRONMENTAL SCIENCE, ENGINEERING AND TECHNOLOGY

Additional books in this series can be found on Nova's website under the Series tab.

Additional e-books in this series can be found on Nova's website under the e-book tab.

ENVIRONMENTAL SCIENCE, ENGINEERING AND TECHNOLOGY

FLY ASH

SOURCES, APPLICATIONS AND POTENTIAL ENVIRONMENTAL IMPACTS

PRABIR KUMAR SARKER, PH.D.
EDITOR

nova publishers
New York

For permission to use material from this book please contact us:
Telephone 631-231-7269; Fax 631-231-8175
Web Site: http://www.novapublishers.com

Library of Congress Cataloging-in-Publication Data

Fly ash : sources, applications, and potential environments impacts / editors, Prabir Kumar Sarker (Curtin University of Technology, Australia).
pages cm
Includes index.
ISBN 978-1-62948-044-2 (hardcover)
1. Fly ash--Industrial applications. 2. Fly ash--Environmental aspects. I. Sarker, Prabir Kumar, editor of compilation.
TA455.F55F588 2013
363.72'88--dc23

2013033673

Published by Nova Science Publishers, Inc. † New York

CONTENTS

PREFACE

Fly ash is mainly generated as a by-product of the combustion of coal and other waste materials. It has been a major subject of research since it was recognised as a pozzolanic material in the 1930s. The characteristics of fly ash vary in a considerable range depending on the source material and the burning process. In the recent years, there has been intensive worldwide research on various utilizations of fly ash. This is because of the increasing demand for recycling of industrial by-product materials in the manufacturing processes and infrastructures. Research has been carried out to increase the utilization of fly ash in various applications and maximize its economic and environmental benefits.

This book presents some of the latest developments on the generation, characterization, utilization and environmental impacts of fly ash. The book has collected the recent research works on the methods of collection and analysis of fly ash, and novel applications such as in geopolymers and recovery of resources from fly ash. Each chapter ends with an extensive list of references for the readers to consult with beyond the scope of the book. It is intended as a book for the professionals, researchers, academics and students interested on materials, recycling and sustainability.

The book consists of 17 chapters contributed by 42 authors. The chapters have been grouped in the following 3 parts:

Part 1: Various sources, methods of collection, factors influencing collection efficiency, flow and hydrodynamic behaviours, methods of characterization, physical, chemical, leaching and radioactive properties, and extraction of metals from fly ash.

Part 2: Different technological initiatives on utilization, applications in construction industry, use as a supplementary cementitious material, alkali-activated fly ash binder, polypropylene composite materials, and behaviour of fly ash geopolymer concrete.

Part 3: Health and environmental issues such as the effects on lung and heart, bioleaching for detoxification, ash from incinerated wastes, and lifecycle assessment of Civil structures using fly ash.

I feel myself privileged to have the opportunity to work with all the authors. I thank the authors for their contributions in the book and their help in peer-review of the chapters.

I wish to thank Professor Vijaya Rangan and Professor Hamid Nikraz of the Civil Engineering Department of Curtin University, Australia, for their supports in completion of this task.

I am grateful to Ms Carra Feagaiga of the acquisitions department of Nova Science Publishers Inc., New York, for her prompt responses to all my queries.

Foremost, I would like to thank my wife Subarna and children Prasun, Promiti and Joyoti for their patience and support to complete the book.

Prabir K. Sarker

Department of Civil Engineering
Curtin University
Perth, Western Australia

July 2013

PART A: SOURCES, COLLECTION AND CHARACTERIZATION OF FLY ASH

In: Fly Ash
Editor: Prabir Kumar Sarker

ISBN: 978-1-62948-044-2

Chapter 1

GENERATION, CHARACTERIZATION AND EXTRACTING OF SILICON AND ALUMINUM FROM COAL FLY ASH

Zhi-tong Yao*

College of Materials Science and Environmental Engineering,
Hangzhou Dianzi University, Hangzhou, China

ABSTRACT

Coal fly ash is a bulky waste and by-product derived from coal combustion in thermal power plant. It is one of the most complex anthropogenic materials that can be characterized and may well be the biggest and yet most unseen hazard to our environment and health. This waste must be reduced prior to disposal. Coal fly ash is rich in SiO_2 and Al_2O_3, and has been considered as a potential resource of silicon and aluminum. This chapter presents the work related to extraction of silicon and aluminium from coal fly ash. In this work, the generation, characterizations and hazards of coal fly ash were firstly investigated. Thereafter, the raw fly ash was pretreated by dry magnetic separation and foam floatation removing the iron-rich component and unburnt carbon. Lastly, the silicon and aluminium were extracted to prepare silica white and Al_2O_3, respectively. The various factors on iron and carbon removal rate were investigated and the optimum conditions were determined. The prepared silica white and γ-Al_2O_3 were characterized by means of X-ray diffraction (XRD), Scanning electron microscopy (SEM) and N_2 adsorption isotherms. The results showed that silicon extraction rate displayed an uptrend with acid concentrations, reaction time and stirring rates increasing. The BET surface area of prepared silica white was calculated as large as 364.8 m^2/g. Based on the Thermo-gravimetric/differential thermal analyzer (TG-DTA) analysis, the $Al(OH)_3$ precursor obtained was calcined forming γ-Al_2O_3.

Keywords: Coal fly ash; chemical component; hazard; silicon and aluminium extraction

* Email: sxyzt@126.com; ztyao@hdu.edu.cn (Z.T. YAO).

1. Introduction

Coal fly ash is a by-product derived from coal combustion in thermal power plant. It could cause air, water and soil pollution, disrupt ecological cycles and set off environmental hazards if not properly treated. This numerous waste must be reduced prior to disposal. Recently, the conduct of fly ash recycling has underwent aggressively (Ilic, et al., 2003; Temimi, et al., 1995). For example, around 20% of fly ash has been used in concrete production. It has also been used in road base construction, soil amendment, sewage sludge treatment, and as filler in plastic materials (Cho, et al., 2005). However, these applications are not sufficient for the complete utilization of fly ash. The remainder finds no applications and has to be discharged into ash ponds or landfills. In the future, lacking of landfill spaces and escalating disposal costs are anticipated. There is a constant need for exploring new recycling method for coal fly ash.

2. Generation, Characterizations and Hazards

2.1. Generation

Coal fly ash is generated from the combustion of pulverized coal in coal-fired power stations. It accounts for 5~20% of the feed coal weight during coal combustion and typically in the form of coarse bottom ash and fine fly ash, which occupy 70~85 wt.% and 15~30 wt.%, respectively. The coal combustion process and by-product collection are generally accepted as follows (see Figure 1). The ground coal is introduced into the furnace together with a part of combustion air (primary air). The remaining air may be added as secondary-, tertiary-, over burner-, and over fire air through the burners or in ports further downstream to ensure complete burnout (Pedersen, et al., 2008). The coal particles are transported in suspension and combust in the form of fine particles or coal cluster at temperature range of 1200~1700 °C. The residence time is approximately 1-2 s before reaching the superheater. At high temperatures it achieves melting. Most organics in the coal are burnt off and the unburnt carbon and inorganic matter will be carried away by the high-temperature flue gas. These particles go through superheater, economiser and air preheater successively at the exhausting of draught fan. In addition, their temperature reduces significantly, causing the internal gas condensed and making the ash to transform into hollow spherical particles. Most particles surface become smooth at the effect of surface tension, however, some particles collide each other forming coarse honeycomb fragments. These dust-like particles are captured and collected lastly by electrostatic or mechanical precipitation from the flue gas. Fly ash is disposed of either by dry or wet methods. In the dry disposal, fly ash is dumped in landfills or fly ash basins. In the wet method, it is washed out with water into artificial lagoons and is called pond ash. The bottom ash is a coarse and solid mineral residue, which is formed when ash adheres as hot particles to the boiler walls, agglomerates and then falls to the base of the furnace at a temperature around 1200 °C (Kim, et al., 2005). The individual particles are much larger than fly ash particles and fall down through the airflow to the bottom of the boiler. It is mechanically removed from the bottom of boilers (Jala and Goyal, 2006). The fly ash formation can be generally divided into three steps: 1) Feed coal transforms into porous carbon particle. During coal combustion, volatiles firstly escape from the gap between minerals and fixed carbon, thereafter the coal transforms into porous carbon particles. Its shape maintains irregular and clastic, however, its surface area becomes larger as the porous characteristic. 2) Porous carbon transforms into vitreous matter. With the organics in porous carbon burnt out and boiler temperature increasing, dehydration, decomposition and oxidizing of minerals (such as kaolin, illite, calcite and pyrite) occurred. It transforms into inorganics (such as mullite, calcium oxide and hemetite) and becomes porous vitreous matter. Its shape is roughly the same as that of porous carbon, but the surface area decreases significantly. 3) Porous vitreous matter transforms into spherical glass bead. With the temperature increasing, porous vitreous matter gradually shrinks and forms dense glass bead at the effect of surface tension. At this stage, its porosity decreases, roundness improves and particle size decreases (Sarkar, et al., 2005).

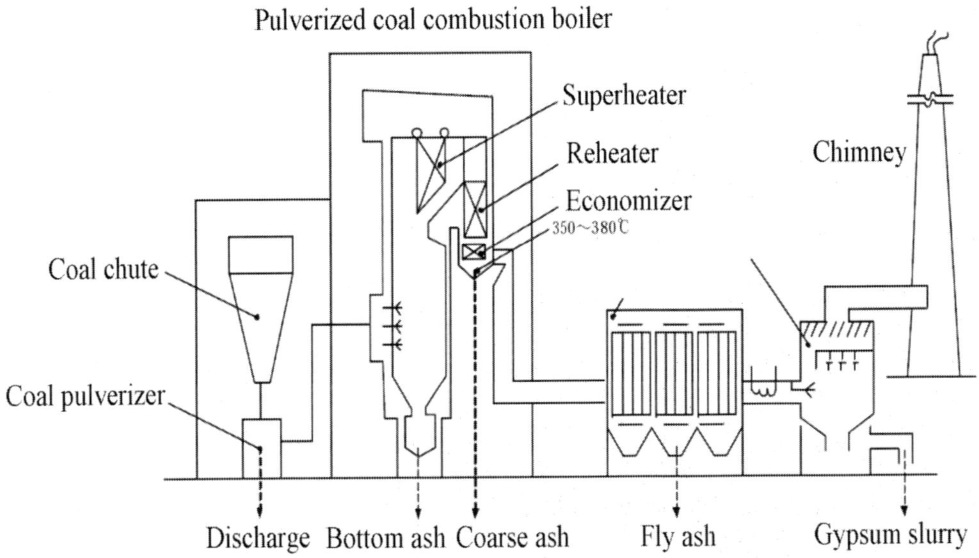

Figure 1. The collection of by-products from coal combustion (Chen, 2003).

Some researchers have put the coal fly ash generation amount in the region of 500 million tons annually in the world (Cho, et al., 2005; Molina and Poole, 2004; Querol, et al., 2001; Wang, et al., 2008). However, these estimates are based on data several years old and need be updated. Coal consumption has increased significantly over the intervening period (International Energy Statistics, 2011). Taking this into account, a more up to date estimate 750 million tons of fly ash is generated each year (Blissett and Rowson, 2012). Installation of the same amount of power generation capacity as that installed in the 20th century was predicted. Coal continues to fuel the largest share of worldwide electric power production. In 2006, coal-fired generation accounted for 41% of the world electricity supply and its share is anticipated to be 43% by 2030. Sustained high prices for oil and natural gas make coal-fired generation more attractive economically, particularly in nations that are rich in coal resources, such as China, India, and the United States (Lior, 2010). The proportion of coal in primary energy consumption reached approximate 68.6% in 2011 for China and 51.07% in 2008 for India. Figure 2 displays the generation and utilization of coal fly ash in China. The annual generation amount gradually increases and is anticipated to reach 580 million tons by 2015 (National Development and Reform Commission of China, 2011.). In India, about 118 million tons is generated at present and is projected to reach 170 million tons by 2012, 225 million tons by 2017 and 440 million tons by 2030 (Ram, et al., 2008). The amount in the US and EU is estimated in around 115 million tons per annum.

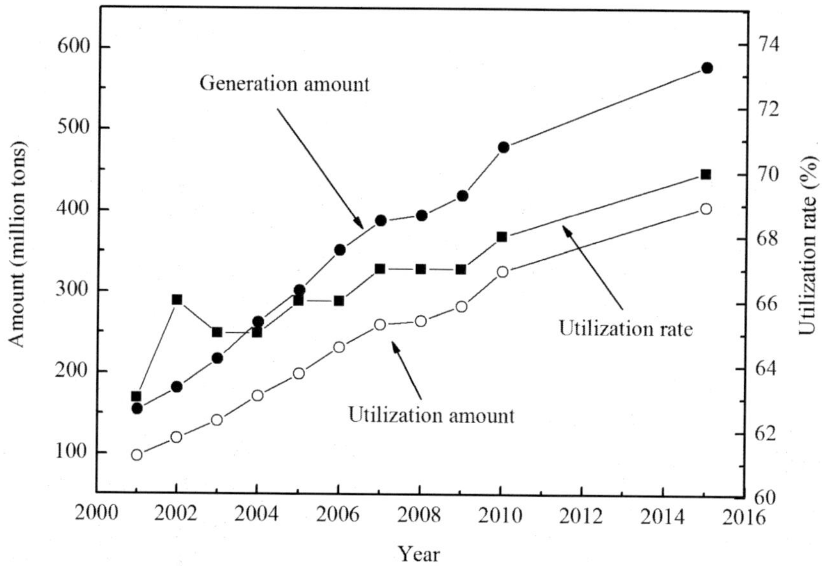

Figure 2. Fly ash generation and utilization in China.

2.2. Characterizations

Understanding of the physical, chemical and mineralogical properties of fly ash is important as these properties influence the use and disposal of fly ash. The properties mainly depend on parent coal, combustion method and conditions, collector setup, etc. Physically, fly ash occurs as fine particles with an average diameter of <10 μm and has low to medium bulk density, high surface area and light texture. Generally, the mechanical collector ash is coarser, with sand-sized particles predominating; electrostatic precipitators also remove finer, silt-sized particles. Ash from bituminous coal is usually finer than that produced by the burning of lignite (Tolle, et al., 1982). The specific gravity ranges from 2.1 to 2.6 g/cm^3. Bulk density varies from 1 to 1.8 g/cm^3. The iron content influences their apparent color which ranges from water-white to yellow orange to deep red or brown to opaque. The pH value varies from 4.5 to 12.0 depending largely on the sulphur content of the parent coal (Plank and Martens, 1974). The morphology of fly ash particles is controlled primarily by the combustion temperature and subsequent cooling rate. SEM observation (see Figure 3) revealed that fly ash samples are predominantly spherical in shape and consist of solid spheres, hollow spheres (cenospheres) and irregular unburned carbon. The cenospheres are collected in a sink/float process, all particles less dense than water, whether they are spherical or non-spherical, porous or non-porous, are considered to be ash cenosphere products (Blissett and Rowson,

2012). The magnetic particles are extracted using magnetic separation that exerts a medium intensity magnetic force over fly ash slurry and lifts the magnetic fraction out. The surface of the magnetic sphere is rougher than that of the cenospheres. Vassilev et al. (2004) described these magnetic spheres as containing iron crystals set in a lattice of amorphous alumina and silica. The carbon enriched product possesses an entirely different morphology to the other ash fractions and is a larger porous structure exhibiting a more irregular shape. The carbon particles result from an incomplete oxidation of coal. Because of the increase in porosity, the carbon has a higher surface area as compared to the inorganic matter in the fly ash (Blissett and Rowson, 2012). In the circulating fluidized bed (CFB) ash, spherical particles are hardly observed and the particles mainly present irregular shapes. The reason is that most minerals in the coal soften and not melt under the low temperature of 850~900 °C for CFB boiler. The irregular fragments mainly consist of unburnt carbon, anhydrate and calcite.

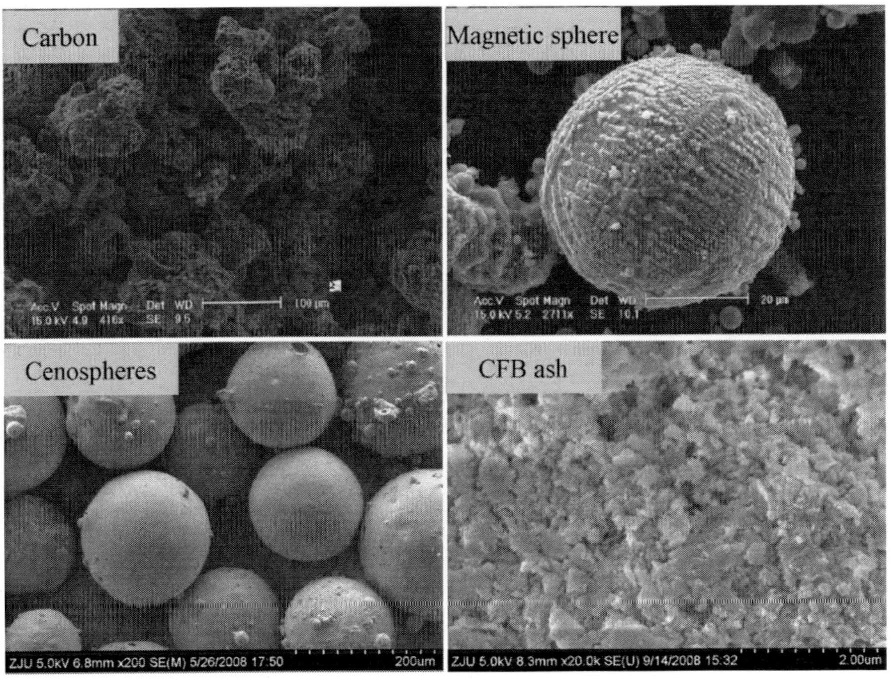

Figure 3. SEM images of fly ash.

Coal fly ash is one of the most complex anthropogenic materials that can be characterized. Approximately, 316 individual minerals and 188 mineral groups have been identified in different ashes (Vassilev and Vassileva, 2005). Its chemical compositions include major (>1%), minor (0.1-1%) and trace (<0.1%) elements. The major components are metallic oxides with varying contents of carbon as measured by a loss on ignition (LOI) test (Ahmaruzzaman, 2010). Table 1 presents the major compositions of fly ash in different regions. It mainly consists of oxides in the order $SiO_2 > Al_2O_3 > Fe_2O_3 > CaO > MgO > K_2O$. However, there are distinctions in composition, not only between regions, but also within the regions themselves. In order of decreasing amounts, the other elements are normally H, Na, Ti, N, P and occasionally F and Cl (Vassilev, et al., 2003; Vassilev, et al., 2005; Vassilev and Vassileva, 1997).

Table 1. Major compositions of coal fly ash in different regions

Components (wt.%)	EU (Akar, et al., 2012; Koukouzas, et al., 2011; Moreno, et al., 2005; Silva, et al., 2012; Vassilev and Vassileva, 2007)	US (Diaz, et al., 2010; Hower, et al., 1996; Kim and Prezzi, 2008)	China (Bai, et al., 2010; Bai, et al., 2011; Cao, et al., 2008; Liu, et al., 2004; Qi and Yuan, 2011; Wu, et al., 2012; Yan, et al., 2012)	India (Dutta, et al., 2009; Mishra, et al., 2010)	Australia (Jankowski et al., 2006; Iyer, 2002)
SiO_2	20.79~60.70	37.8~58.5	35.6~57.2	50.2~61.61	48.8~66.0
Al_2O_3	12.5~35.6	19.1~28.6	18.8~55.0	14.0~32.42	16.25~27.8
Fe_2O_3	2.08~21.2	6.8~25.5	1.86~19.3	2.70~14.4	1.1~22.31
CaO	0.5~38.24	1.4~22.4	0.9~7.0	0.6~2.6	2.9~18
MgO	0.6~4.66	0.7~4.8	0.03~4.8	0.10~2.1	0.3~2.0
Na_2O	0.1~4.07	0.3~1.8	0.6~1.3	0.5~1.2	0.2~1.3
K_2O	0.39~4	0.9~2.6	0.8~0.9	0.74~4.7	1.1~2.9
P_2O_5	0.07~2.49	0.1~0.3	1.1~1.5	0.09~0.64	0.2~3.9
TiO_2	0.41~2.6	1.1~1.6	0.2~0.7	1.0~2.7	1.09~3.7
SO_3	0.1~24.58	0.1~2.1	1.0~2.9		0.1~0.6
LOI	0.8~32.8	0.2~11.0		0.47~5.0	

The chemical components of fly ash are determined by parent coal type (see Table 2). Fly ash from lignite and subbituminous coals is generally characterized by higher CaO, MgO, SO_3 and lower SiO_2, Al_2O_3 relative to the high rank bituminous and anthracite coals. It shows pozzolanic properties and possesses cementious properties due to the high content of lime. Fly ash from bituminous and lignite coals contains <10% CaO and mainly consists of aluminosilicate glass. Ash containing >15% CaO is composed of calcium aluminosilicate glass in addition to substantial proportions of crystalline calcium compounds including C_3A, C_4A_3S, CS, and CaO (Manz, 1999).

Table 2. Normal range of chemical compositions of fly ash by coal types (Ahmaruzzaman, 2010)

Components (wt.%)	Sub-bituminous	Bituminous	Lignite
SiO_2	40~60	20~60	15~45
Al_2O_3	20~30	5~35	10~25
Fe_2O_3	4~10	10~40	4~15
CaO	5~30	1~12	15~40
MgO	1~6	0~5	3~10
Na_2O	0~2	0~4	0~6
K_2O	0~4	0~3	0~4
SO_3	0~2	0~4	0~10
LOI	0~3	0~15	0~5

The major chemical compositions have traditionally been the basis for evaluating the suitability of fly ash for use as a cement replacement material. The American Society for Testing and Materials (ASTM) groups coal fly ash into three classes: C, F and N (Table 3).

Table 3. Standards for coal fly ash used in cement and concrete

Components	US ASTM C618-12			BS EN 450-1: 2005+A1:2007			Japan JIS A6201-2008	China GBT1596-2005			
								Fly ash used for cement and concrete		Fly ash used for reactivity cement admixture	
	C	N	F	A	B	C		F	C	F	C
SiO_2				< 22.0			> 45				
SiO_2+Al_2O_3+Fe_2O_3	> 50.0	> 70.0	> 70.0	> 65.0			> 70.0				
CaO				Free CaO < 2.6 Reactive CaO < 11.0			< 6.0	< 1.0	< 4.0	< 4.0	
MgO	< 5.0		< 5.0	< 4.5			< 5.0				
SO_3	< 5.0	< 4.0	< 5.0	< 3.0			< 5.0	< 3.0		< 3.5	
Na_2O	< 1.5		< 1.5	< 5.5%							
LOI	< 6.0	< 10.0	< 12.0	< 7.0	< 9.0	< 11.0	< 10.0	<5.0 (Class I) <8.0 (Class II) < 15.0 (Class III)		< 8.0	

Table 4. Trace elements in fly ash in different regions

Elements (ppm)	EU (Moreno, et al., 2005)	US (Meawad, et al., 2010)	Australian (Jankowski, et al., 2006)
As	22~162	0.0003~391	6.58~22.3
B	24~534	2.98~2,050	11~89
Ba	311~3,134	0.02~10,850	393~3,520
Be	3~34	0.2~2,105	
Cd	1~6	0.01~76	0.25~1.34
Co	20~112	4.9~79	5.6~100
Cr	47~281	3.6~437	18~130
Cu	39~254	0.2~655	28.1~93.8
Ge	1~61		
Hg	<0.01~1.4	0.013~49.5	
Li	36~377		
Mn		24.5~750	87.5~1,630
Mo	5~22		4.9~21
Ni	49~377	0.1~1,270	10.5~242
P			648~7,360
Pb	40~1,075	0.02~273	48~80
Se	3~30	0.0003~49.5	2.26~5.15
Rb	22~202		
Sb	1~120	0.2~205	
Sn	4~15		
Sr	131~4,406	30~3,885	
Th	17~65		
U	5~29		
V	154~514	43.5~5,015	
Zn	70~924	0.28~2,200	
F		0.04~320	
Ag		0.01~49.5	
Ti		0.15~85	

Class F ash shows a combined SiO_2, Al_2O_3, and Fe_2O_3 content of >70% compared to >50% for Class C. It is for the reason sometime held that class C ash is derived from lignite and sub-bituminous coals and class F ash is derived from bituminous and anthracite coals (Wang, 2008).

The chemical properties of fly ash have been well studied and they contain many trace elements, some of which are of environmental concern (Bhangare, et al., 2011; Izquierdo and Querol, 2012; Vassilev and Vassileva, 2007).

Table 4 shows the concentrations of trace elements in fly ash in different regions. It is noticeable that elements such as Cr, Pb, Ni, Ba, Sr, V and Zn are in significant quantity. Because of this, mobility studies have been conducted to assess the likelihood of these elements escaping to the environment. The leachability of the elements is closely related to the phase with which they are associated and the leaching conditions such as pH that they are exposed to (Jankowski, 2006).

2.3. Hazards

Coal fly ash disposal methods ultimately lead to the dumping of fly ash on open land. In China, the coal fly ash is classified as a non-hazardous waste and its treatment is not subject to hazardous waste procedures. According to the standards for pollution control on the storage and disposal site for general industrial solid wastes (GB 18599-2001), the fly ash dumping site is grouped into Class II-general industrial solid waste dumping site. Irregular accumulation and inappropriate disposal of fly ash will lead to the occupancy of vast lands, degrade the soil and endanger human health and the environment. Fly ash particles are small enough to escape emission control devices, easily get suspended in the air and have become a major source of pollution to the atmosphere. Repeated exposure to fly ash can cause irritation in eyes, skin, nose, throat and respiratory tract and result in arsenic poisoning. The fly ash even reaches the sub-soil and ultimately causes siltation, clogs the natural drainage system and contaminates the ground water with heavy metals. These ash particles contain toxic elements and the concentration increases with decreasing particle size. Because of these toxic elements, disposal of fly ash e.g. in a landfill may be a major source of contamination. The water soluble trace elements contaminate the soil, surface and ground water (Choi, et al., 2002; Sushil and Batra, 2006). It was reported that the metal contaminants like V, Cr, Ni, Cd and Pb are potential hazards to the environment even at low concentration levels. Analysis of water extracts (Elseewi, et al., 1980; Menon, 1980) showed that the principal cations in water extracts are calcium and sodium, whereas anions are dominated by OH^- and CO_3^{2-} with aqueous extracts of ash saturated with $Ca(OH)_2$. Leachates have markedly different compositions, depending on the fly ash surface, flue gas process conditions and whether lime or lime stone injection processes were implemented for desulfurization. The alkalinity and acidity control the extractability of elements, such as As, B, Be, Cd, Cr, Cu, F, Mo, Se, V and Zn. Usually the trace metals are present as a relatively small fraction in ashes but special attention should be given to them because of their long lives, and high toxicity to humans, plants and animals. While the leaching toxic contaminates into our environment is a slow process, fly ash landfill accidents have spread large amounts of contaminated sludge into large waterways that have contaminated the surrounding communities. On 22 December, 2008 in Roane County the TVA Kingston Fossil Power Plant experienced an ash dike rupture at their 84 acre solid waste containment area. The resulting spill measure 1.1 billion US gallons of coal ash sludge into the surrounding Emory River and Clinch River. The spill covered 300 acres of surrounding land and was up to 6 feet deep in this area.

Assessment of the radiation exposure from fly ash is critically dependent on the concentration of radioactive elements in coal and in the fly ash that remains after combustion. During coal combustion, most of U, Th and their decay products are released from the original coal matrix and are distributed between the gas phase and the solid combustion products. Mathur et al (2008) investigated the Rn exhalation rates in fly ash samples in India and observed that the Rn exhalation rate was higher than from coal samples and that the activity concentration of radionuclides was enhanced after coal combustion. Fly ash contains the radiochemical pollution of U and Th series along with other radioactive contaminants like ^{222}Ru and ^{220}Ru. Though several reports regarding the presence of radionuclides in fly ash are available, studies on their impact are lacking. The results of radioactivity analyses revealed that the activity levels of gamma emitting radionuclides ^{40}K, ^{226}Ra were within the permissible limits and that mixing of fly ash with soil at 24% (v/v) was of no consequence.

Kant et al (2010) analysed the activity concentration of ^{238}U, ^{232}Th, ^{40}K and radon activity and exhalation rates for fly ash in northern India. The activity concentrations of ^{238}U, ^{232}Th and ^{40}K varied from 99 to 203 Bq/kg, 145~288 Bq/kg and 355~516 Bq/kg, respectively in various fly ash samples. The radium equivalent activity was varied from 317 to 614 Bq/kg, radon activity varied from 214 to 590 Bq/m^3, radon exhalation rate varied from 7.8 to 21.6 $mBqkg^{-1}h^{-1}$ for mass exhalation rate and from 138 to 381 $mBqm^{-2}h^{-1}$ for surface exhalation rate. The presence of natural radioactivity levels in the fly ash was well below the acceptable limits. The activity concentration, absorbed dose and annual effective dose were marginally higher as compared to the other building materials. However, these materials satisfied the universal standards limiting the radioactivity within the safe limits. Papastefanou (2010) examined the radioactivity of fly ashes in Greece and found that the activity concentrations ranged from 263 to 950 Bq/kg for ^{238}U, 142 ~605 Bq/kg for ^{226}Ra, 133~428 Bq/kg for ^{210}Pb, 27~68 Bq/kg for ^{228}Ra and 204~382 Bq/kg for ^{40}K. In Southwest China, some fly ash produced also has relative high contents of radionuclides, such as ^{226}Ra 135.5~699 Bq/kg, ^{238}U 305.3~2170 Bq/kg and ^{232}Th 4.2~74.9 Bq/kg (Xiong, et al., 2007).

2.4. Utilization and Recycling

The use of coal fly ash has important economic and environmental implications. The utilization as a by-product has received a great deal of attentions over the past two decades as more sustainable solutions to waste problems have been sought. The current fly ash utilization rates are 39% and 47% in the US and EU (European Coal Combustion Products Association, 2008; The American Coal Ash Association, 2010.), respectively. The global average is estimated to be nearly 25% (Wang, 2008). In China, this rate increases annually and keeps at nearly 67%. It is expected to reach 70% by 2015. However, there is a contradiction and the Greenpeace reported that the practical utilization rate is about 30% (Greenpeace, 2010). It is clear that a significant proportion of fly ash left untreated must be disposed of and there is a constant need for developing new recycling method for coal fly ash.

2.4.1. Construction Industry

Fly ash can be referred to as either cementitious or pozzolanic. The silica in fly ash can react with calcium hydroxide released by the hydration of calcium silicate to produce calcium silicate hydrate (Nonavinakere and Reed, 1995). It is when the silicates have an amorphous structure, rather than crystalline, that materials tend to be pozzolanic and contribute to the formation of hydration products (Iyer and Scott, 2001). The cementitious and pozzolanic properties make pulverized fly ash useful for cement replacement in concrete and other building applications (González, et al., 2009), such as in producing high quality and cost-effective bricks, interlocking pavers, kerbstones and mosaic tiles, where several benefits arise from this such as improved workability of the freshly mixed concrete and lower heat evolution during hydration. The hardened fly ash concrete shows increased strength together with a lower permeability, where the latter leads to a higher resistance toward aggressive admixtures (Maroto-Valer, et al., 2001; Taylor, 1997). In addition, partial replacement of cement with fly ash reduces the production costs of concrete due to the lower price of fly ash compared to cement (Manz, 1999). In general, the additive amount of fly ash for typical durable concrete ranges from 15 to 35 wt.% and can be up to 70 wt.% for concrete in constructions like pavement, walls and parking lots and 80 wt.% in autoclaved aerated concrete (Dilmore and Neufeld, 2001). In the EU, 47% of fly ash generated in 2003 (44 million tons) was used and 71% of which was used in the cement and concrete manufacture (European Coal Combustion Products Association, 2004.). In the US, approximately 40% of the ash produced in 2004 (71 million tons) was used and 59% of which was utilized in the cement and concrete industry (The American Coal Ash Association, 2005.). McCarthy and Dhir (2005) examined the use of high levels (beyond 30% by mass commonly adopted) of low-lime fly ash as a cement component in concrete. The results indicated that fly ash leveled up to 45% can be combined with Portland cement (PC, C_1) to produce the range of practical concrete design strengths, although early strength can be reduced compared to PC. The study progressed to consider the use of a rapid hardening Portland cement (C_2) and low energy clinker (C_3) combined with fly ash at 45%, as a means of overcoming these early strength shortfalls. Both were found to be effective in matching early strength behaviour of PC concrete. Tests covering fresh, engineering and durability properties of these concretes were then carried out. The results indicated that in almost all cases, either similar or enhanced performance was achieved with the high volume fly ash concrete. A beneficial use has recently been demonstrated by JEA to recycle both the bottom ash and fly ash from two CFB boilers in pavement construction applications (Jackson, 2009). The results of laboratory and field testing demonstrated that fly ash was beneficial in pavement and roadway construction applications. The JEA hydrated ash had an stiffness equal to or greater than that of limerock when used in both subgrade stabilization and base course applications. Horiuchi et al. (2000) studied coal fly ash slurry use for civil engineering materials and reported its potential for underwater fills, light weight backfills, light weight structural fills, through laboratory tests and construction works. From the data obtained, it can be concluded that fly ash slurry can be effectively utilized in civil engineering projects. However, concerns have been raised about using fly ash as a cement substitute in concrete. Kikuchi (1999) highlighted the problem that both construction industry using fly ash in cement and the coal fired power plants producing fly ash were seasonal. The different activities peak at different times; the construction industry needs most of the cement in the summer months when building conditions are optimum, whereas most of the ash is generated during the winter. In the summer, coal generators tend to be restricted to double shifting e.g. operating during breakfast and evening meal times to satisfy peak demand and in doing so the LOI content of the ash is increased (Sears, 2011). The increase in LOI is an indication of the increase of unburnt carbon. It changes the adsorption properties of ash in terms of the air entrainment admixture that are added in cement mixes. According to ASTM C 618 and EN 450-1, an increase in LOI reduces the quality of fly ash. High carbon content can also lead to discolouration of concrete and mixture segregation (Freeman, et al., 1997).

Brick is an extremely old building material. Cultrone and Sebastián (2009) evaluated the technical quality of solid bricks with added fly ash for use in building restoration. The sample bricks were made using two types of clayey raw materials and were then fired at different temperatures. The textures of the bricks with fly ash were very similar to the textures of those without it, except that the samples with the additive contained spherical fly ash particles with diameters ranging from 0.1 to 10 μm. These particles led to a reduction in the density of bricks and a substantial improvement in their durability, with less decay being caused by salt crystallization in the pores. Use of this additive could have practical implications as a means of recycling and for achieving cost savings in brick production. Cicek and Tanriverdi (2007) also observed the positive effect of the addition of fly ash, sand and hydrated lime in the compressive strength of bricks. An optimum raw material composition was found to be a mixture of 68% fly ash, 20% sand and 12% hydrated lime. The optimum brick forming pressure, autoclaving time and pressure were 20 MPa, 6 h and 1.5 MPa, respectively. The compressive strength, unit volume weight, water absorption and thermal conductivity of the fly ash-sand-lime bricks obtained are 10.25 MPa, 1.14 g/cm^3, 40.5% and 0.34W $m^{-1}K^{-1}$, respectively.

Beyond incorporation in construction materials, fly ash is also used in geotechnical applications. In many countries regulations require all hazardous waste disposal facilities to be lined with suitable impermeable barriers to protect against contamination. Fly ash generally consists of silt size particles and consequently possesses high permeability. However, pozzolanic fly ash with lime exhibits low permeability on curing because of the formation of gelatinous compounds which block the pores. Thus, pozzolanic fly ash appears to be promising for construction of liners to contain alkaline leachate. Prashanth et al. (2001) examined the potential of pozzolanic fly ash as a hydraulic barrier in landfill. Geotechnical properties, such as shrinkage, compaction, permeability, consolidation and strength characteristics were needed to evaluate. The results indicated that fly ash possessed low shrinkage and hence did not crack. Compacted fly ash underwent very little volume changes. They also found that fly ash developed good strength properties with time. Cokca and Yilmaz (2004) evaluated the feasibility of utilizing fly ash, rubber and bentonite as a low hydraulic conductivity liner material. To investigate the properties of rubber and bentonite added fly ash, hydraulic conductivity, leachate analysis, unconfined compression, split tensile strength, one-dimensional consolidation, swell and freeze/thaw cycle tests were performed. The overall evaluation of results revealed that rubber and bentonite added fly ash showed good promise and a candidate for construction of a liner. Fly ash has been used as stabilizer for soil because of the beneficial properties it imparts. It is found that the addition of fly ash to soils tends to decrease the propensity for the soil to absorb water and thus less swelling of the soil results. For soils with high montmorillonites content, the swelling problems can be severe. Such soils expand when they are wetted and shrink when dried. This movement exerts pressure to crack sidewalks, basement floors, driveways, pipelines and foundations. The damages due to expansive soils are sometimes minor maintenance but often they are much worse, causing major structural distress. It is thought that adding fly ash to the soil changes the mineralogy due to pozzolanic reaction. The soil becomes more granular and holds less water which results in a decrease of swelling associated with water absorption (Zha, et al., 2008). Nalbantoglu (2004) evaluated the changes in engineering properties of fly ash treated soils. The fly ash treatment was effective in improving the plasticity of Degirmenlik and Tuzla soils. The reduction in swell pressure values of both soils indicated that the swelling of soils was prevented under smaller pressure values. Thus, high swell potential values were not expected beneath the smaller foundation pressures. The reduced CEC values indicated that fly ash treatment caused changes in the mineralogy of the treated soils and produced new secondary reaction minerals. The formation of these new pozzolanic reaction products caused the soils to become more granular and resulted in less water absorption potential. Dermatas and Meng (2003) used fly ash along with quicklime to immobilize Pb^{2+}, Cr^{3+} and Cr^{6+} present in artificially contaminated clayey sand soils. The degree of heavy metal immobilization was evaluated using the Toxicity Characteristic Leaching Procedure (TCLP) as well as controlled extraction experiments. The results suggested that the controlling mechanism for both Pb^{2+}

and Cr^{6+} immobilization was surface adsorption, whereas for Cr^{3+} it is hydroxide precipitation. Addition of quicklime and fly ash to the contaminated soils effectively reduced heavy metal leachability well below the nonhazardous regulatory limits. Overall, fly ash addition increased the immobilization pH region for all heavy metals tested, and significantly improved the stress-strain properties of the treated solids, thus allowing their reuse as readily available construction materials. Bin-Shafiquea et al. (2010) investigated the long-term performance of fly ash stabilized two fine-grained soil subbases. Specimens prepared from low plasticity clay and high plasticity expansive soil stabilized with Class C fly ash were subjected to 12 cycles of wet-dry cycles and freeze-thaw cycles in a controlled laboratory environment and the change in geotechnical properties of the weathered fly ash stabilized soils was evaluated to understand the long-term performance. The fly ash stabilization increased the unconfined compressive strength of fine-grained soils 3~4 times and decreased the plasticity and swell potential by approximately 50% and 75%, respectively. Even after losing strength due to freeze-thaw cycles, the strength of stabilized soils was still at least 3 times higher than that of the unstabilized soils. Fly ash grout injection is currently being considered for use at a closed underground mining site. The injection process would reduce acid mine drainage (AMD) through a two-fold approach, by neutralizing AMD and by preventing contact between water and pyritic materials. The technical feasibility of disposing fly ash in underground mines has been proven and the selection of this disposal alternative will be decided based primarily on cost and regulatory compliance issues.

Overall, a relatively small percentage of fly ash finds application as an ingredient in cement and other construction products, but the vast majority is held in ash dams or similar dumps. This unproductive use of land and the associated long-term financial burden of maintenance have led to realization that alternative uses for fly ash as a value-added product are needed (Iyer, 2001).

2.4.2. Soil Amendment

Fly ash, acidic or alkaline depending on the source, can be used for buffering the soil pH. The addition of fly ash to soil neutralizes pH to a level suitable for agriculture depending on the initial pH of the soil (Moliner and Street, 1982). Fly ash also contains some useful nutrients, such as Mg, P, S, Ca, Mn, Fe, Cu and Zn, which are beneficial for plant growth (Roy and Joy, 2011; Sharma and Kalra, 2006; Yunusa, et al., 2006). The addition of alkaline fly ash to acidic soil can raise the pH, which may be attributed to the release of Ca, Na, Al, and OH ions (Jala and Goyal, 2006). Studies has demonstrated the ability of raising acidic soil pH using class C fly ash, i.e., ash with high CaO contents ($\geq$15%), and were applied at excessively high rates. Most of class F ashes with low CaO content have limited potential to ameliorate soil acidity with the exception of ashes that contain relatively high calcium contents (Seoane and Leirós, 2001; Taylor, et al., 1988). Researchers have noted other beneficial effects of the application of fly ash soil systems: it improves the soil texture, aeration, percolation, and water retention in the treated zone; it reduces soil bulk density, the consumption of other soil amelioration agents such as fertilizers or lime; it serves as an insecticide and decrease the mobility and availability of metal in the soil (Pandey and Singh, 2010). Manoharan et al. (2010) applied two acidic and two alkaline Class F ashes to the top layer (0-10 cm) of 100 cm long intact cores of acidic sandy clay and clay loam soils. The results showed that it was the CEC rather than pH that determined the liming value of fly ash. Of the four ashes, only Alk-FAII substantially increased the soil pH of the top layer in

Belangalo soil. These increases in pH were 2.0 and 2.3 units with additions of ash at 36 and 108 Mg/ha, respectively. The other ashes had smaller effects on soil pH. Riehl et al (2010) studied the physical, chemical and physico-chemical modifications in a fluvisol (calcaric) after integration of a silico-aluminous (SiAl) and a sulfocalcic (SCa) fly ash. The results showed that SCa ash seemed less well adapted to the integration in soils, mainly because of the basicity of soil solution which it induced. The SiAl ash seemed better adapted for amendment from an agricultural and farming point of view as it allowed an improvement in CEC, thus furthering the reserve of nutrients and the adsorption of water as a useful reserve for plants. However, it lowered the structural stability of the soil. Patra et al (2012) applied fly ash with 200 t/hectare in the crop lands. The increase in yield of maize grains was 28% and 34% over the control values at Malud and Dhenkanal respectively whereas for rice grains it was 40% and 13%. The trace elements like Mn, Fe, Cu, Zn, As, Se, Mo and Pb were detected at a trace level in the rice grains produced in ash treated soil. The concentration of trace elements was in the normal range and well below the critical levels in the plants and hence the grains were safe for consumption. Lee et al (2006) evaluated rice (*Oryza sativa*) productivity in silt loam and loamy sand soils to which fly ash were added with 2 Mg/ha Si as a control. Fly ash increased the soil pH (<0.5 units) and available Si and P contents of both soils. The rice plants did not show B toxicity effects. The highest rice yields were achieved following the addition of around 90 Mg/ha fly ash. The application of fly ash increased Si, P and K uptake by the rice plants, but did not result in an excessive uptake of heavy metals in the submerged paddy soil.

However, in fly ash most elements are associated with the mineral phase. Thus, harmful effects such as the low availability of some nutrients elements resulting from application of fly ash have been reported. Other undesirable effects include excess salinity and a high content of phytotoxic elements, potential toxic side effects of heavy metals and slow rate of degradation of fly ash in soils (Bhattacharya, et al., 2012; Pandey, et al., 2009; Pandey and Singh, 2010). Manoharan et al. (2010) reported that high salt content of class F ash may prevent its widespread application as a liming material in agricultural soils. More than half of the ash-derived Mo and Se in the ashes with the highest concentrations of these elements remained in the near-surface layers of the soil into which the ash was incorporated. Reductions in plant growth and yield with some of the ash treatments are probably due to interactive effects of salinity and other elemental toxicity. Douay et al (2002) tested the effect of fly ash integration on the bioavailability of toxic elements in calcium carbonate rich soils but noticed a reduced emergence of the seedlings and delayed growth with respect to the control soil. They explained it by a rise in pH and a poor integration of ashes with the soil.

2.4.3. Sintered Material

Coal fly ash contains appreciable amounts of SiO_2, Al_2O_3, CaO and Fe_2O_3, among other oxides. These oxides have been mainly considered as a low cost material for ceramic industry. In addition, the fine powder form makes it suitable to be directly incorporated into ceramic pastes with almost no pretreatment (Erol, et al., 2008). Recent research has investigated the production of glass, glass-ceramic and sintered materials prepared from coal fly ash (Barbieri, et al., 1999; Olgun, et al., 2005; Zhao, et al., 2010). Due to their low or negative coefficient of thermal expansion, glass-ceramics materials based on Li_2O-Al_2O_3-SiO_2 ternary system have important industrial applications. Yao et al. (2011a) prepared $Li_2Al_2Si_3O_{10}$ using fly ash and lithium hydroxide monohydrate ($LiOH{\cdot}H_2O$) as precursors.

The results showed that the optimum additive amount of $LiOH{\cdot}H_2O$ was about 20%. The onset of calcining temperature and time was identified as 980 °C and 1 h, respectively. The obtained $Li_2Al_2Si_3O_{10}$ phase was well crystallized with small grain size. Indialite ceramic was also prepared using fly ash and magnesium carbonate powder as precursors (Yao, et al., 2011b). X-ray diffraction results indicated that the formation of indialite was achieved by solid-state sintering reactions at 1200 °C for 4 h. Zhao et al (2010) investigated the effect of alkali on the preparation process of foam material. The experimental results showed that homogenous microstructures of interconnected pores could be obtained by adding 13 wt.% foaming agent at 1050 °C, leading to foams presenting water absorption, apparent density and compressive strength values of about 126.5%, 0.414 g/cm^3, 6.76 MPa, respectively. Tan (2011) prepared mullite whiskers in sodium sulfate flux using fly ash. The mullite whisker was obtained under sintering at 1000 °C; with an aspect ratio of >25 and about 0.06-0.3 μm in diameter. Erol et al. (2008) prepared glass, glass–ceramic and ceramic materials from thermal power plant fly ash without any additives. TCLP results indicated that the produced samples could be taken as non-hazardous materials. Produced samples showed high resistance to alkali solutions in contrast to acidic solutions. Microstructural, physical, chemical and mechanical properties of the produced glass-ceramic samples were better than those of the produced glass and ceramic samples. Jing et al. (2012) prepared ceramic granules using fly ash, clay and diatomite and employed them in the trickling filters. Its start-up characteristics, influence of hydraulic loading rates, carbon/nitrogen ratio and filter depth on pollutants removal were investigated. The results indicated that the biofilter was started quickly in 16 days with river sediment as inoculum. This double-layer biofilter was efficient in treating highly polluted river water.

2.4.4. Geopolymers

Geopolymers exhibit many excellent properties such as high compressive strength, good acid resistance, low shrinkage and low creep. They are generally synthesized by mixing aluminosilicate-reactive material with strong alkali solutions. Alkali activating solution is important for the dissolution of Si and Al to form geopolymer precursors and finally aluminosilicate material. The commonly used alkali solutions are NaOH and KOH (Somna, et al., 2011; Swanepoel and Strydom, 2002; Xu, et al., 2002). Under a strong alkali solution, aluminosilicate-reactive materials dissolve and form free SiO_4 and AlO_4 tetrahedral units. With the progress of reaction, water is gradually removed, and the SiO_4 and AlO_4 tetrahedral clusters are linked to yield polymeric precursors through the sharing of all oxygen atoms between two tetrahedral units, thereby forming amorphous geopolymers (Davidovits, 2008). In the last decade fly ash based geopolymer has emerged as a promising new cement alternative in the field of building and construction materials. Normally, good high-strength geopolymers can be made from class F fly ash (Schmucker and MacKenzie, 2004). Fly ash geopolymers have been prepared as geopolymer pastes, mortars and concretes. Temuujin et al (2010) prepared geopolymer mortars with varying levels of sand aggregate and investigated their physical and mechanical properties. The geopolymer binder to sand aggregate weight ratio was varied from 9 to 1. Compressive strength and Young's modulus of the fly ash based geopolymer paste were 60 MPa and 2.27 GPa and these values did not change significantly with addition of up to 50 wt.% sand aggregate. Geopolymer binder exhibited strong bonding to the sand aggregate. Joseph and Mathew (2012) studied the engineering properties of geopolymer concrete made from alkali activated fly ash. The alkalis used were sodium silicate and sodium hydroxide. A geopolymer concrete with total aggregate content of 70% by volume, ratio of fine aggregate to total aggregate of 0.35, NaOH molarity 10, Na_2SiO_3/NaOH ratio of 2.5 and alkali to fly ash ratio of 0.55 when cured for 24 h at 100 °C gave an average cube compressive strength of 52 MPa after temperature curing (56 MPa after 28th day). This geopolymer concrete gave a higher value of Poisons ratio and modulus of elasticity compared to ordinary cement concrete having almost same cube compressive strength as that of geopolymer concrete. Xu et al. (2010) prepared geopolymer using low-reactive circulating fluidized bed combustion (CFBC) fly ashes. An alkali fusion process was employed to promote the dissolution of Si and Al species from the fly ash, and thus to enhanced the reactivity of the ashes. The results indicated that via enhancing the reactivity by alkali fusion and balancing the Na/Al ratio by additional aluminosilicate source, low-reactive fly ash could be recycled as an alternative source material for geopolymer production.

2.4.5. Catalysts and Catalyst Supports

Fly ash can also be used as low-cost catalyst and catalyst support as it contains high silica content, which helps in catalysis of various reactions and provides support to different catalytic species on the surface. The catalytic role of activated fly ash for different reactions such as oxidation (Kastner, et al., 2003; Mallik and Chaudhuri, 1999), chlorination (de Leer, et al., 1989) and condensation of short chain olefins (Jarmohamed and Mulder, 1994) is well documented in the literatures. Such reactions were promoted by metal oxides constituting the fly ash, although the yield of reaction product was quite low. Fly ash supported calcium oxide has been employed as a recyclable solid base catalyst for Knoevenagel condensation reaction (Jain, et al., 2010). Chakraborty et al (2010) prepared a fly ash supported CaO catalyst (derived from waste eggshell) for transesterification of soybean oil. It exhibited high catalytic performance in converting soybean oil into fuel grade biodiesel with a higher fatty acid methyl ester content (96.97%) compared to minimum EN 14214, 2003 requirement of 96.5 wt.%. Fly ash was also used as solid support in the synthesis of 2-mercaptobenzothiazole derivatives under microwave irradiation (Narkhede, et al., 2007). Saputra et al (2012) employed red mud (RM) and fly ash (FA) as supports for synthesis of Co-oxide based catalysts. Their catalytic activity in oxone activation for phenol degradation for wastewater treatment was tested. It was found that Co oxide on the catalysts was present in the Co_3O_4 phase and the oxide showed better dispersion on RM. Co/FA and Co/RM could degrade phenol with the presence of oxone and Co/RM exhibited much higher activity than Co/FA. Fly ash has been used as a support material for nickel in the application of CO_2 methane reforming. It was shown that treatment of the fly ash with CaO prior to Ni loading could produce catalysts capable of high conversion and stability with activities close to the well reported Ni/Al_2O_3 and Ni/SiO_2 systems (Wang and Lu, 2007). Wang et al. (2011) synthesized Fe^{3+}-doped TiO_2 film deposited on fly ash cenosphere (Fe-TiO_2/FAC) by the sol-gel method. The photocatalytic activity and kinetics of Fe-TiO_2/FAC with varying the iron content and the calcination temperatures were investigated by measuring the photodegradation of methyl blue during visible light irradiation. Compared with TiO_2/FAC and Fe^{3+}-doped TiO_2 powder (Fe-TiO_2), the degradation rate using Fe-TiO_2/FAC increased by 33% and 30%, respectively.

Khatri and Rani (2008) synthesized a nano-crystalline activated fly ash catalyst (AFAC) by chemical and thermal treatment of fly ash. The variation of surface and physico-chemical properties of the fly ash by activation methods resulted in improved acidity and therefore catalytic activity for acid catalyzed reactions. The catalyst was found to be highly active solid acid catalyst for liquid phase esterification of salicylic acid with acetic anhydride and methanol giving acetylsalicylic acid and methyl salicylate respectively. A maximum yield of 97% with high purity of acetylsalicylic acid and a very high conversion 87% of salicylic acid to methyl salicylate was obtained with AFAC. Jain et al (2012) prepared solid base catalysts by functionalization of thermally activated F-type fly ash with 3-amino propyltrimethoxysilane. The prepared NH_2-FA catalyst was found to possess significant basicity due to attachment of aminopropyl group on the fly ash surface, which catalyzed the Knoevenagel condensation of cyclohexanone with ethyl cyanoacetate giving high yield (92%) of the desired product in solvent freecondition. Khatri et al (2010) synthesized highly active nano-crystalline, thermally stabilized solid acid catalyst by loading different weight fractions of sulfated zirconia on chemically activated fly ash through two step sol-gel technique. Liquid phase benzylation of benzene and toluene with benzyl chloride was studied as test reaction for catalytic activity. A very high conversion of benzene (87%) and toluene (93%) were observed, which is attributed to significant amount of acid site on the catalyst surface. The catalyst with 12 wt.% zirconia was regenerated and reused up to four reaction cycles with equal efficiency as in the first run. Zhang et al (2012) used nitric-acid-activated fly ash (AFA) as heterogeneous fenton-like catalyst for p-nitrophenol removal from water. The highest removal rate (98%) was observed at pH 2.0 when H_2O_2 dosage 166.5 mg/L, AFA loaded 10 g/L and temperature 298 K. Xuan et al. (2003) investigated the selective catalytic reduction (SCR) of NO with NH_3 using fly ash catalyst. The results indicated that fly ash, after pre-treatment and loading with transition metal elements, can be reasonably used as the SCR catalyst support to remove NO from flue gas. Cu gave the highest catalytic activity and NO conversion, compared with Fe, Ni and V. In the pretreatment process, the nitric acid treatment and drying temperatures for the fly ash particles had little effect on the NO conversion. However, the calcination temperature had an important effect on the catalyst preparation process.

2.4.6. Adsorbents

Fly ash is a strong alkali material, which exhibits pH of 10-13 when added to water, and its surface is negatively charged at high pH's. Hence, it can be expected that metal ions can be removed from aqueous solutions by precipitation or electrostatic adsorption. Besides, it contains a certain volume of unburned carbon, which has a large adsorption capacity. Fly ash has been investigated for its direct use as an adsorbent in both gaseous and aqueous applications. As early as 1984, fly ash was considered as a potential adsorbent to remove Cu^{2+} from industrial waste waters (Panday, et al., 1985). More recently the use of mixed fly ash with different chemical compositions has been evaluated for the removal of B and a variety of different metal ions from waste waters, such as Cu, Pb, Zn, Mn, Cd, Cr, As and Ni (Al-Zboon, et al., 2011; Mohan and Gandhimathi, 2009; Papandreou, et al., 2007; Papandreou, et al., 2011; Wang, et al., 2007). Polat et al (2004) conducted a series of column and batch experiments to investigate the efficiency of boron removal from seawater and desalinated seawater using fly ash. The results showed that the selected fly ash was very effective in removing boron. A considerable amount of boron (>90%) could be removed under optimum conditions (e.g., pH=9, L/S=l/10, reaction time> 6 h). The boron removal was associated with Mg depletion and Ca enrichment in the residual seawater and conversely with relative Mg enrichment and Ca depletion in the residual fly ash. Li et al (2009) used high iron containing fly ash to remove arsenic from aqueous system. The batch experiments showed that it was effective for arsenic removal. About 40 g was sufficient to remove 50 mg arsenic from 1 L aqueous solution to meet the standard value for the industrial wastewater discharge. Cho et al (2005) conducted batch experiments to evaluate the heavy metals (Zn, Pb, Cd and Cu) removal from aqueous solutions by fly ash. These heavy metals were adsorbed onto the fly ash very rapidly (within the first 20 min), while equilibrium was reached in 2 h for Zn and 3 h for the other heavy metals. The heavy metal ions removal increased as the pH increased. Most of the metal ions had a 95% removal rate when the pH was above 8. A recent review on the use of low cost adsorbents for the removal of heavy metals from industrial waste water concluded that fly ash has great potential for use in the treatment of wastewater; this potential was limited by the variability in chemical composition of fly ash and the large volumes that might be required for it to be effective (Ahmaruzzaman, 2011). Fly ash was also used for the removal of organic pollutants from aqueous solution. Matheswaran and Karunanithi (2007) investigated the removal of Chrysoidine R from the aqueous solution using fly ash as adsorbent. The results showed that as the amount of the adsorbent was increased, the dye removal increased accordingly; higher adsorption percentage was observed at lower concentration of chrysoidine. The adsorption was found to obey pseudo-first order kinetics. Singh (2009) studied the sorption potential of fly ash for removal of metribuzin, metolachlor and atrazine from water. The amount of herbicides sorbed increased with increase in the amount of fly ash in the suspension. The maximum capacity of fly ash to adsorb metribuzin, metolachlor and atrazine was found to be 0.20, 0.28 and 0.38 mg/g by Freundlich equation and 0.56, 1.0 and 3.33 mg/g by Langmuir equation. Aksu and Yener (2001) investigated the potential for fly ash to be used in place of activated carbon for the adsorption of phenol. They found the adsorption capacity of fly ash to be 98.7 mg/g and 118.6 mg/g for 2-chlorophenol and 4-chlorophenol respectively. 2, 4-dichlorophenol and pentachlorophenol have been removed from waste water in a fixed bed column operating continuously; removal efficiencies of >99% were achieved for an initial concentration of 10 mg/L of chlorophenols (Estevinho, et al., 2007). Fly ash has also been found to be effective for the removal of polychlorinated biphenyls (Nollet, et al., 2003).

Adsorption is considered to be one of the promising technologies for capturing CO_2, SO_2 and NOx from flue gases. Lu and Do (1991) used fly ash as an adsorbent for *NOx* removal from flue gases. It was found that the unburnt carbon remaining in the fly ash particles contributed the main surface area to fly ash, and the carbon can be activated to further improving the adsorption performance of fly ash. The experimental results on activation of coarser fly ash particles showed that the adsorption capacity can be increased through controlled gasification of the unburnt carbon. Izquierdo and Rubio (2008) evaluated the carbon-enriched fly ash as a low-cost adsorbent for SO_2 removal from stack gases. A good SO_2 removal capacity was shown by some of the studied samples that can be related to their textural properties. Cycles of SO_2 adsorption/regeneration were carried out in order to evaluate the possibility of thermal regeneration and re-use of these carbons. After each cycle, the SO_2 removal capacity of the sample decreased. Sarmah et al. (2013) investigated the CO_2 capturing capacity of two untreated mixed amine/fly ash solid composites (FAA13and FAA23). Both composites were thermally stable up-to 250 °C. The capturing of CO_2 was due to the chemical adsorption process. Results showed the maximum adsorptions of CO_2 were 6.89 and 5.99 mmol/g on FAA13 and FAA23 respectively. A comparison of material cost and adsorption capacity showed that they were cheap and efficient for CO_2 capturing.

2.4.7. Deep Separation

Recovery of cenospheres, unburnt carbon and magnetic sphere is one of the coal fly ash beneficiations, providing economic as well as environmental benefits. Cenospheres are formed from coal combustion in a molten state. Flowing with the combustion gas stream, the molten particles are rapidly cooled to form a spherical shape. Meanwhile, gases within the molten particles are trapped inside the spheres. Bubbles may occur in multiple forms within the cenospheres, in both single or concentric forms. It has been found that the thickness of a cenosphere wall is relatively thin, <10% of the particle diameter. The resultant densities are typically less than 1 g/cm^3 (Raask, 1968). Enospheres are commonly used as fillers in cement to produce low density concrete (Blanco, et al., 2000; Vassilev, et al., 2004). Recently, some researchers have studied the advantages of using cenospheres in composite materials, mixing them with metals and polymers in order to produce lightweight composite materials with higher strengths. Recently, technologies to recover cenospheres from fly ash are related to density by using water as the separating fluid (wet-separation processes; lagoon, pond, multigravity separator, etc.). As Borm (1997) has stated, some materials in fly ash are prone to dissolution upon contact with water. Some of these dissolved materials are toxic; thus, their accumulation in water needs to be controlled, which is a disadvantage of the wet-separation processes. In addition, the availability of land is a problem for applying wet separation processes in densely populated countries, in addition to the technical disadvantages related to the process (drying, variable particle size, etc.). As wet-separation processes have some disadvantages, it is important to study alternative techniques. An air classifier is one of the alternative techniques and a basic theoretical study of comparing the efficiencies of dry separation with that of wet processes has been conducted (Hirajima, et al., 2010). The estimated separation efficiency was as high as that of the wet separation. The estimated optimum Newton's efficiency was about 0.54, slightly lower than that of the wet separation process of 0.6. It was also found that the cenospheres recoveries of two methods can be expected to be similar at about 70%. Petrus et al. (2011) investigated the separation efficiency of two types of air classifiers, namely a closed-type pneumatic separator and a micron separator. It was found that the micron separator has the potential to be applied in cenospheres recovery from fly ash, giving a Newton's efficiency of 0.44, as compared to a value of about 0.26 for the closed-type pneumatic separator. The cenospheres recoveries of both pieces of equipment at their optimum Newton's efficiencies were similar at over 60 wt.%. In order to further confirm the applicability of this technique, the micron separator was deployed in a cenospheres recovery unit prior to the use of a wet-separation process. About 80% of the cenospheres was recovered, with an 87.8% reduction in the total mass of fly ash to be separated in the float and sink tank. Consequently, much less water was needed for the process of cenospheres recovery.

The recovered carbon from fly ash could serve as a fuel or as a source of carbon-derived materials. Niewiadomski et al (1999) used wet flotation to recovery coke fines. The flotation tests were carried out in a 1-L Denver-type flotation cell and in a 2-inch diameter air-sparged hydrocyclone. The flotation concentrate from the air-sparged hydrocyclone contained 35% coke by weight on a dry basis, with a coke recovery to the concentrate of 54%. Yang (2000) used froth flotation to separate carbonaceous portion and low-carbon fly ash. The carbon was separateness as single grains in fly ash and the low-carbon fly ash can fit the role of pozzolanic activity index. Besides, by gathering several times of flotation, the grade of unburnt carbon reached 75% and the heat value of carbonaceous portion was 6000 kcal/kg to use as fuel. Maroto-Valer et al. (2001) recycled unburnt carbon using nitric acid, hydrochloric acid and hydrofluoric acid to dissolve fly ash. The grade of carbon was above 90% and can be used as adsorbents after being activated and as coke filler after being pelleted. Maroto-Valer (2002) investigated two potential routes for the generation of premium carbon products from the unburned carbon present in fly ash. The first route focused on the use of fly ash carbons as precursors for activated carbons by steam activation at 850 °C, while the second route concentrated on the utilization of fly ash carbons as a replacement for calcined petroleum coke in the production of carbon artifacts.

The removal of magnetic concentrate from fly ash is also a significant opportunity to create added value. The amount of magnetic material varies depending on the coal source and operating conditions of the coal boiler, but it can vary from between 0.5% and 18% (Zyryanov, et al, 2011).

The basic principle behind magnetic separation is straightforward; it relies on the fact that materials with different magnetic moments experience varying forces in the presence of magnetic field gradients. In this way, an externally applied magnetic field can lift out those materials with similar magnetic properties (Yavuz, et al., 2009). Groppo and Honaker (2009) conducted a pilot scale investigation into the potential for recovering magnetic fraction from bottom ash.

The process involved feeding the ash slurry through concentrating spirals which separated the ash into heavy and light fractions. From here the heavy ash fraction was fed to a continuous high gradient magnetic separator (HGMS). In a HGMS, a large static magnetic field was applied to a column containing a matrix such as steel wool. The irregular surface of the matrix generated high magnetic gradients which generatd enough force to capture weakly magnetic particles (Gomes, et al., 1999). The processing scheme was capable of producing magnetic material with an iron content of almost 75%.

2.4.8. Zeolite Synthesis

Synthesis of zeolites is noticed as one of the effective usages for coal fly ash (Lin and His, 1995). It is possible due to the similar compositions of fly ash to some volcanic material, precursor of natural zeolites. Although this potential application may consume only a small proportion of fly ash, the obtained products may reach a relatively higher added value. The synthesized products can be used in various engineering and agriculture applications for water purification, gas cleaning, soil improvement, and so on. This could potentially eliminate the ash disposal problem, but also turn an otherwise useless waste into marketable products (Querol, et al., 1995; Shih and Chang, 1996).

Since the pioneering work of Holler and Wirsching in 1985 (Holler and Wirsching, 1985; Tanaka, et al., 2002), extensive studies have been undertaken and different types of zeolites were prepared, such as zeolite X, Na-A, ZSM-5, Na-P, K-L, K-W, faujasite, phillipsite and Li-ABW. Table 5 shows the zeolite types that have been synthesized from coal fly ash.

The zeolite synthesis is conventionally developed by hydrothermal crystallization under alkaline conditions. However, mullite and quartz in ash is considered to be inert and difficult to be dissolved. In some cases, zeolite formation is assessed to be fairly unsuccessful, taking into account that the resultant product usually co-crystallized of zeolites with original crystalline phases (quartz and mullite). The zeolitic aggregates form like an egg white, covering the central core of fly ash particles. The zeolite yield and transformation efficiency is low (Shigemoto and Hayshi, 1993).

Table 5. Different zeolites synthesized from coal fly ash

Zeolites	Chemical formula	JCPDS card number
Na-P1	$Na_6Al_6Si_{10}O_{32}·12H_2O$	39-0219
Na-A	$(NaAlSiO_4)_{12}·27H_2O$	11-590
Na-X	$NaAlSi_{1.18}O_{4.36}·0.22H_2O$	39-0218
K-chabazite	$K_2Al_2SiO_6·H_2O$	12-0194
K-L	$KAlSi_3O_8·H_2O$	16-0385
K-W	$K_9Al_9Si_{23}O_{64}·8H_2O$	86-2338
K-Phillipsite	$KAlSi_2O_6$	31-0967
Kalsilite	$KAlSiO_4$	33-0988
Herschelite	$Na_{1.08}Al_2Si_{1.68}O_{7.44}·1.8H_2O$	31-1271
Faujasite	$Na_2Al_2Si_{3.3}O_{10.6}·7H_2O$	12-0228
Zeolite Y	$NaAlSi_{2.43}O_{6.86}·4.46H_2O$	38-0239
Analcime	$NaAlSi_2O_6·H_2O$	19-1180
Hydroxysodalite	$Na_6Al_6Si_6O_{24}·8H_2O$	40-0102
Hydroxycancrinite	$Na_{14}Al_{12}Si_{13}O_{51}·6H_2O$	28-1036
Sodalite	$Na_8Al_6Si_6O_{24}Cl_2$	73-1733
Tobermorite	$Ca_5(OH)_2Si_6O_{16}·4H_2O$	19-1364
Perlialite	$K_9NaCaAl_{12}Si_{24}O_{72}·15H_2O$	38-0395
Zeolite ABW, (Li)	$LiAlSiO_4·2H_2O$	27-1211
Zeolite F	$Li_{5.04}Si_5Al_5O_{20}(H_2O)_{7.92}$	79-1893

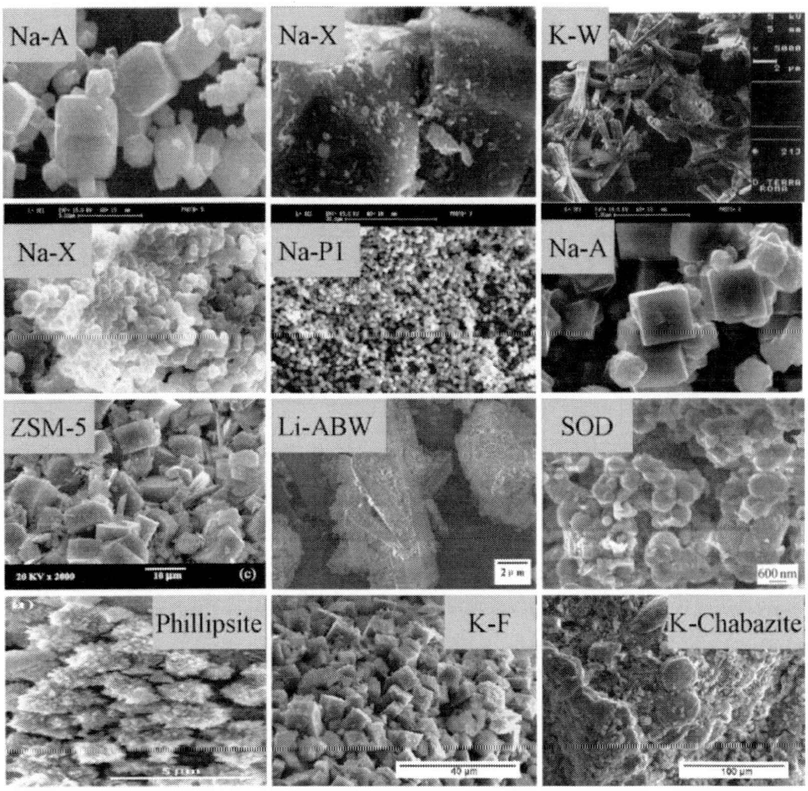

Figure 4. SEM images of zeolites synthesized from coal fly ash (Belardi, et al., 1998; Chareonpanich, et al., 2004; Tanaka, et al., 2006; Ye, et al., 2008; Yao, et al., 2009).

Recently, the conventional method has been improved by using more sophisticated treatments, including an alkaline fusion step followed by hydrothermal treatment (Rayalu, et al., 2000; Rayalu, et al., 2001) facilitated the formation of highly active aluminosilicate readily soluble in water and promoted zeolite formation, the application of microwave-assisted zeolite synthesis (Querol, et al., 1997) resulted in a drastic reduction of the reaction time, and a method for synthesizing zeolite under molten conditions without any addition of water (Park, et al., 2000a; Park, et al., 2000b), although complete zeolitization has not been obtained.

A two-step synthesis process (Hollman, et al., 1999; Moreno, et al., 2004) has been developed consisting of an initial Si-extraction from fly ash by using a light alkaline treatment followed by a synthesis of pure zeolites by adding an Al-bearing residue solution, such as Na_2SiO_3 and $NaAlO_2$ solution. This method could significantly improve the zeolitization process and thus obtain high purity and crystalline zeolites.

2.4.9. Valuable Metals Extraction

Germanium is a valuable element used in the manufacture of semiconductors and novel, high technological industrial applications, so it is attractive to process raw materials with low Ge contents and considerable quantities of other elements (Arroyo Torralvo and Fernández-Pereira, 2011). Around 30% of the Ge used in the world is obtained from recycling materials, particularly electronic devices (Butterman, et al., 2005). Sometimes Ge is concentrated in certain coals basically joined to the organic matter. After coal is burned, Ge is concentrated in combustion and gasification by-products, especially in the fly ash, where a content of 10 times greater than in coal (Font, et al., 2005). Therefore, one attractive source of Ge comes from the leaching of coal fly ash. Many studies have focused on the Ge recovery from fly ash (Jandova and Vu, 2001; Jandova, et al., 2002). The first step is almost the leaching of fly ash, followed by a process separating Ge from other elements, such as As, Mo, Ni, Sb, V or Zn. The methods of separating Ge from other elements include precipitation with tannin (Moskalyk, 2004), distillation of $GeCl_4$ (Jandova and Vu, 2001), flotation (Hernandez-Exposito, et al., 2006), adsorption onto activated carbon (Marco-Lozar, et al., 2006; Marco-Lozar, et al., 2007), precipitation with cetyl trimethyl ammonium bromide (Arroyo and Fernández-Pereira, 2009), solvent extraction (Arroyo, 2007) and adsorption onto chelating exchange resins (Arroyo Torralvo, et al., 2010). Torralvo and Fernández-Pereira (2011) studied Ge recovery from fly ash leachates using ion-exchange procedures. The procedure was based on Ge complexation with catechol (CAT) in an aqueous solution followed by the retention of the Ge-CAT complex onto a conventional strongly basic anionic resin. Optimum conditions were determined as CAT/Ge molar ratio of 9 and resin/Ge dosage of 3, obtaining 96.1% Ge retention. The maximum amount of retained Ge was calculated to be 215.5 mg/g resin. The elution of the germanium complex was achieved with HCl in 50% of ethanol. Font et al (2007) evaluated the Ga and V recovery from Puertollano Integrated Coal Gasification Combined Cycle (IGCC) fly ash. A series of NaOH based extraction tests were performed with different temperatures, NaOH/fly ash ratios, NaOH concentrations and extraction times. The optimum extraction conditions were temperature 25 °C, NaOH concentration 0.7-1 M, NaOH/fly ash ratio of 5 L/kg, attaining Ga extraction yields of 60–86%, equivalent to 197–275 mg of Ga/kg of fly ash. Carbonation of concentrated Ga leachate demonstrated that 99% of the bulk Ga content in the leachate precipitated at pH 7.4. A second carbonation of the remaining solution (at pH 7.5) recovered 98.8% of the bulk Ga. Re-dissolution (at pH 0) of the precipitate increased Ga purity from 7 to 30%, this being a suitable Ga end product for further purification by electrolysis. This method produced higher recovery efficiency than currently applied for Ga on an industrial scale. In contrast, low V extraction yields (< 64%) were obtained even when using extreme alkaline extraction conditions, which given the current marked price of this element, limits considerably the feasibility of V recovery from IGCC fly ash. Shabtai and Mukmenev (1996) described a novel biomagnetic titanium-aluminium extraction process and used concentrated sulphuric acid to leach both valuable and toxic elements from the fly ash. The pH was lowered until titanium precipitated from solution. This was carried out concurrently with a biomagnetic adsorption process. This biomagnetic separation procedure was carried out twice to yield a titanium dioxide rich product with ferrous, silicate, and sulphate impurities.

Coal fly ash produced in the southern Inner Mongolia, northern Shanxi and Shaanxi province of China is characterized by high aluminum content. The average content is about 40% and in some cases it reaches as high as 51% (Cao, et al., 2008). At present, bauxite, the main aluminum source for various industrial applications, is mainly dependent on import. Therefore, extraction of aluminum from coal fly ash is significant for developing new aluminum source, and has attracted extensive attention recently. The high aluminium content of fly ash could meet the demand of Bayer process for alumina production (above 30%) and makes it a potential substitute for bauxite (Authier-Martin, et al., 2001; Iyer, 2002). Alkali sintering process involves calcination of the mixture of coal fly ash and calcium oxide (or calcium carbonate or sodium carbonate) at high temperature to produce aluminates which can dissolve in sodium carbonate molten salt. Eventually, Al_2O_3 was attained through self-decomposition, alkali dissolution and desilication treatment, Bayer cycle process (Grzymek, et al., 1979; Peng, et al., 2005; Qin 2007). However, attempts have failed to directly use the fly ash as a substitute for bauxite either in Bayer process or lime-soda sintering process. The problem results from the substantial SiO_2 content, which creates a problem in separating alumina from silica and other silicon compounds. SiO_2 is soluble under the alkali condition designed for dissolving aluminum oxide, which is contributed to a significant amount of alkali consumption in the process that produces a large volume of silicate residue. It is estimated that 7~10 t calcium silicates slag will be generated for 1 t alumina produced using lime soda sintering process. Therefore, there is a strict requirement on the Al/Si ratio of fly ash for Bayer process above 8 and for lime-soda sintering process around 2. However, the Al/Si ratio of fly ash is in the range of 0.8 to 1.0, thus enriching aluminum content by removing silicon seems to be the logic choice. Bai et al (2010) enriched aluminum in fly ash by alkali desilication and carbonation to transform amorphous silicon dioxide into an amorphous nano particles product. The desilicated coal fly ash residue was alumina enriched from 42 to 49% with Al/Si ratio improved from 0.86 to 1.63. These nano silicon dioxide particles were in size of 50 nm with a purity of 96%. However, alkali sintering process requires high energy-consumption and produces more solid wastes, which have restricted the large-scale application of alkali sintering process in extracting aluminum from coal fly ash. Recently, gas phase extraction and mineral acid leach processes for aluminium extraction has been reported. However, one major shortcoming is that the aluminum extraction efficiency by direct acid leaching method is obviously lower than that by alkali sinter process (Kelmers, et al., 1982; Seidel, et al., 1999; Seidel and Zemmles, 1998). Shemi et al (2012) employed two methods of extracting aluminium from fly ash. In the first technique, fly ash was optimally leached with 6 M sulphuric acid at slurry solid to liquid ratio of 1: 4 and temperature of 75 °C for 8 h 45 min. The leaching experiment yielded an aluminium extraction efficiency of 23.5%. In the second technique, fly ash was extracted with acetylacetone in gas phase at 250°C and flow rate of 6 mL/min after a reaction time of 6 h and an extraction efficiency of 17.9% was obtained. A higher extraction efficiency of 46.7% was achieved from synthetic fly ash with the same reaction conditions. The extent of extraction with acetylacetone was comparatively lower than that of acid leaching but had advantages of reduced extraction times and the possibility of recycling the unreacted acetylacetone. Acid leaching on the other hand had longer leaching times but relatively higher extraction efficiencies at lower temperatures. Nayak and Panda (2010) revealed that aluminum extraction from fly ash at a low acid concentration and ambient temperature was not suitable for high recovery of aluminum. In order to improve the aluminum extraction efficiency, various researches have been developed. Li et al (2011) provided the optimum conditions of acid leaching process with leaching temperature of 200-210 °C, leaching time of 80 min, and volumetric ratio of acid to coal fly ash of 5: 1, and the extraction efficiency of Al_2O_3 can reach 87%. Bai et al. (2010) and Matjie et al. (2005) represented that the aluminum extraction efficiency by acid leaching method can be improved through adding calcium oxide (or calcium sulfate or lime) in coal fly ash and calcining the mixture at high temperature before acid-leaching reaction. Bai et al (2011) also boosted the alumina extraction efficiency by sintering the mixture of coal fly ash and concentrated sulfuric acid at 300°C to transform most of aluminum in coal fly ash into aluminum sulfate, which can be extracted out from coal fly ash by hot water. Wu et al. (2012) leached out aluminum from coal fly ash by pressure acid-leaching method. The effects of fly ash particle size, sulfuric acid concentration, reaction time and temperature on extraction efficiency were investigated. The optimum conditions were determined as fly ash size 74 μm, sulfuric acid concentration 50%, reaction temperature and time 180°C for 4 h. Under optimum conditions, the extraction efficiency of aluminum can reach 82.4%.

The silica white is usually prepared by chemical vapor deposition method using silicon tetrachloride as raw material (Gun'ko, et al., 2005) and acid precipitation method using water glass solution and acid as precursors (Einarsrud, et al., 1998). In order to reduce the cost of production, many researches were undertaken to prepare silica white from silicon-rich non-metallic minerals. Thus, in this study, we attempt to extract silicon and aluminium to prepare silica white and Al_2O_3 respectively, using alkaline fusion-acid leaching joint method.

3. Magnetic Separation of Coal Fly Ash

The high iron contents will influence the characteristics of synthesized zeolites (Molina and Poole, 2004), glass-ceramic (Kniess, et al., 2007) and rubber filled with fly ash. It is also not helpful for extracting valuable metals from fly ash. Thus, it is necessary to remove iron components from fly ash before further utilization. The recent iron removal methods include acid pretreatment and magnetic separation. Wang et al (2008) used 10% hydrochloric acid to treat fly ash with S/L ratio of 1: 25 and reported that Fe_2O_3 content decreased by 76%; SiO_2 and Al_2O_3 contents increased to some extent. Molina and Poole (2004) pretreated the fly ash by magnetic separation performed on a fly ash-water slurry of 15% solids at 9000 G field intensity using an Eriez wet high intensity magnetic separator. The results showed a reduction of about 65% of iron components and increases of nearly 34% and 3.5% in silicon and aluminium compounds, respectively. As compared with acid pretreatment and wet magnetic separation, the dry magnetic separation will not produce secondary environmental pollution and is helpful for magnetic materials recycling. Thus, in this work, the dry magnetic separation was selected to pretreat fly ash.

3.1. Experimental

3.1.1. Materials

The raw coal fly ash was supplied by a thermal power plant in Zhejiang province, China. Its chemical composition consists of following oxides: SiO_2 50.5%, Al_2O_3 35.9%, Fe_2O_3 5.3%, CaO 4.6%, TiO_2 1.5%, K_2O 1.3%, and other compounds 0.9%. It has $\geq$ 85% of SiO_2+Al_2O_3 and low contents of impurities and thus groups in class F.

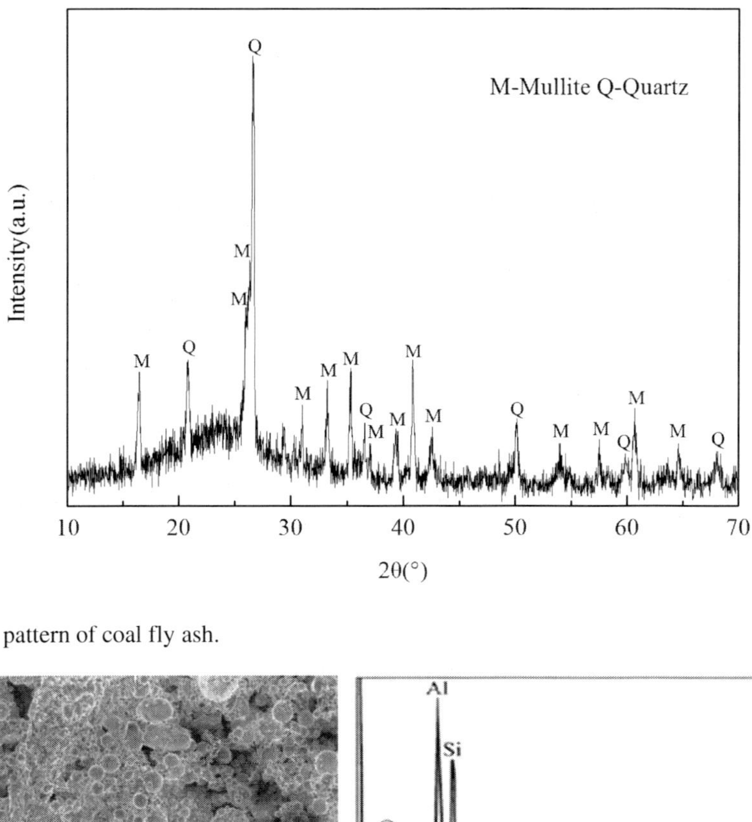

Figure 5. XRD pattern of coal fly ash.

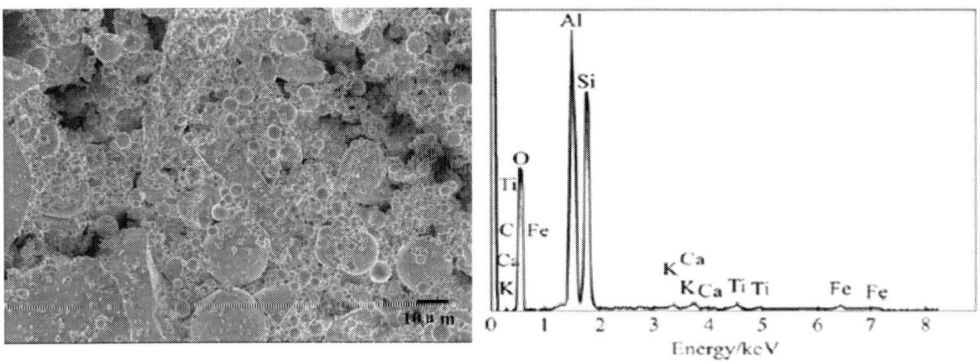

Figur 6. SEM-EDS analysis of coal fly ash.

XRD analysis in Figure 5 revealed that the major crystalline phases were quartz (SiO_2, JPCDS card no. 79-1910) and mullite ($3Al_2O_3{\cdot}2SiO_2$, JPCDS card no. 5-490). The morphology was also observed (see Figure 6.). It presented mainly spherically and irregularly round shapes. The EDS analysis confirmed that the chemical elements consist of Si, Al, O, Ca, Fe, Ti and K.

3.1.2. Magnetic Separation

The process was performed on WCF-2 multi-purpose magnetic analytical meter (Beijing Geological Instrument Factory, Beijing, China). The effects of magnetic current, separation times, dip angle and separation process on separation efficiency were investigated. The steps were listed as follows: Adjust the magnetic current (0~5 A) and open the oscillator. A quantity of fly ash was fed into the hopper and separation was performed. At the end of separation, magnetic particles and fly ash pretreated were collected respectively. The iron contents were analysed and iron removal rate was calculated.

3.1.3. Separation Mechanism

After fed into the hopper, the coal fly ash is affected by the magnetic force and mechanical forces, including gravity force, centrifugal force, fluid resistence, etc. The particles with different magnetic characteristics are affected by different magnetic forces and mechanical forces, which incur them go through the magnetic field via different routes. Thus, the coal fly ash is seperated obtaining magnetic particles and fly ash pretreated. The schematic diagram is displayed in Figure 7.

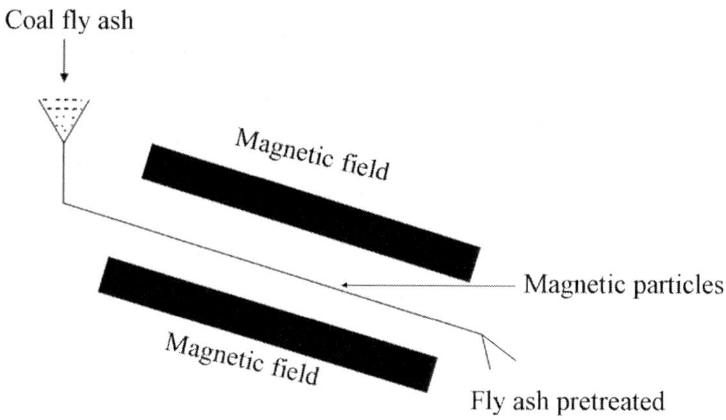

Figure 7. The schematic diagram of magnetic separation.

3.2. Results and Discussion

3.2.1. Effect of Magnetic Current

The dip angle of magnetic analytical meter was kept at 30° in the series experiments. The effect of magnetic current was investigated through adjusting the current to change the magnetic field intensities. Figure 8 revealed that the iron removal rate increased with an increase of current. The optimum current was determined as 2.5 A with the highest removal rate of 80.74%. An increase of currency meant magnetic field intensity increasing, thus obtaining more magnetic materials and higher iron removal rate. The larger current could increase the iron grade of magnetic components, however, the yield will reduce, which resulted in the decrease of iron removal rate. Thus, the current was selected as 2.5 A for further studies.

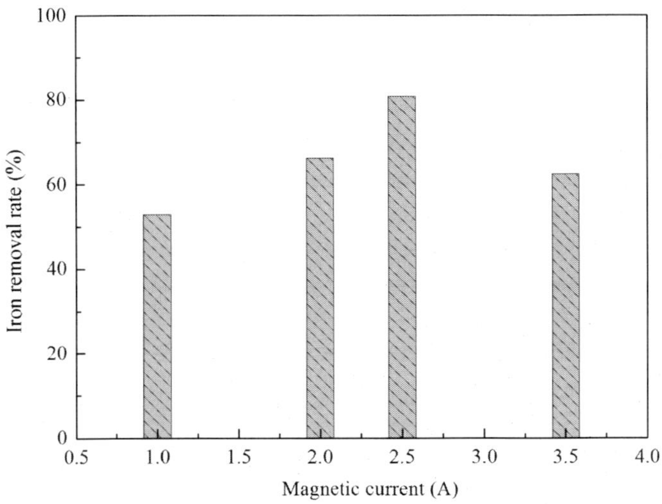

Figure 8. Effect of magnetic current on separation efficiency.

3.2.2. Effect of Dip Angle

The magnetic current was kept at 2.5 A in the series experiments. The effect of dip current was investigated through adjusting dip angle to change the forces on magnetic and non-magnetic particles. The results showed that the iron removal rate increased with an increase of dig angle (see Figure 9).

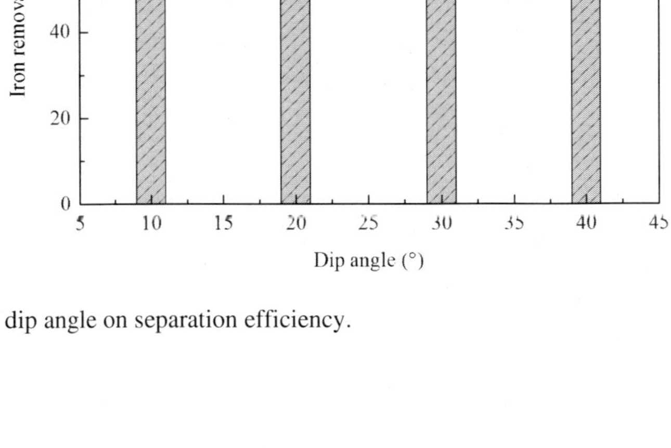

Figure 9. Effect of dip angle on separation efficiency.

The optimum dip angle was determined as 30° with the highest removal rate of 80.19%. However, the rate decreased when the angle was larger than 30°. Increasing dip angle could increase the gravity force component and thus making the fly ash go through magnetic field quickly and distributed uniformly for separation. However, the larger dip angle would cause the gravity component become larger than the magnetic force. The too large flow speed was not helpful for separation and resulted in the decrease of iron removal rate. Thus, the dip angle was selected as 30° for further studies.

3.2.3. Effect of Separation Times

The series experiments were conducted with a constant current and dip angle of 2.5 A and 30°, respectively. The results showed that the iron removal rate did not increase significantly with an increase of separation times. The removal rate almost maintained at about 80% (see Figure 10).

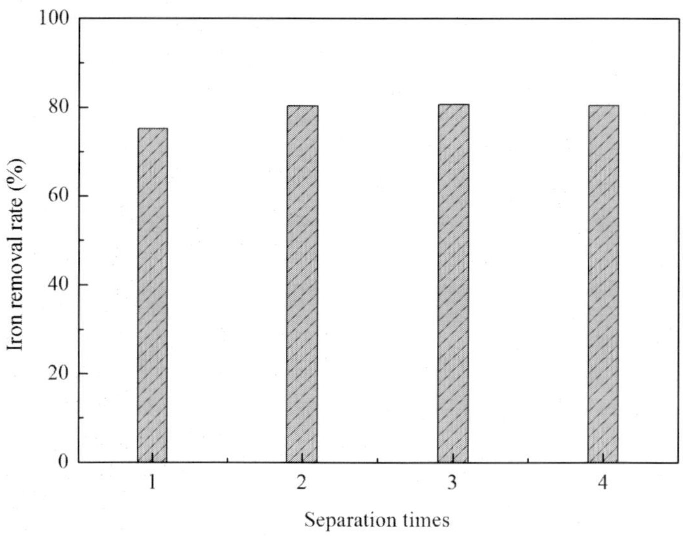

Figure 10. Effect of separation times on separation efficiency.

3.2.4. Effect of Separation Process

The series experiments were conducted with a constant current and dip angle of 2.5 A and 30°, respectively. The results showed that the iron removal rate was 69.83% with a primary separation (see Figure 11). The fly ash also contained higher iron contents, thus a scavanger separation and cleaning concentration were conducted. Comparatively analysing the three different separation processes, it was found that the iron removal rate was the highest as 81.06% using primary separation-scavanger separation. The elevation of removal rate was not significant using primary separation-cleaning concentration.

3.2.5. XRD Analysis of Magnetic Particle

XRD pattern of magnetic particle separated is displayed in Figure 12. Using a built in computer search program, the crystalline phases were identified by comparing the diffraction peak positions and intensities with those in the JCPDS data files. After analysing, these diffraction peaks were well indexed to quartz, mullite and maghemite ($Fe_{21.16}O_{31.92}$, JPCDS card no. 89-5892).

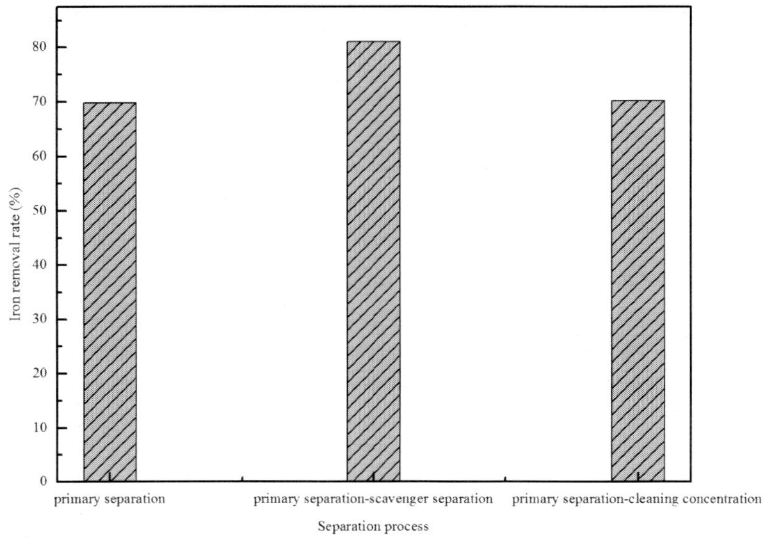

Figure 11. Effect of separation process on separation efficiency.

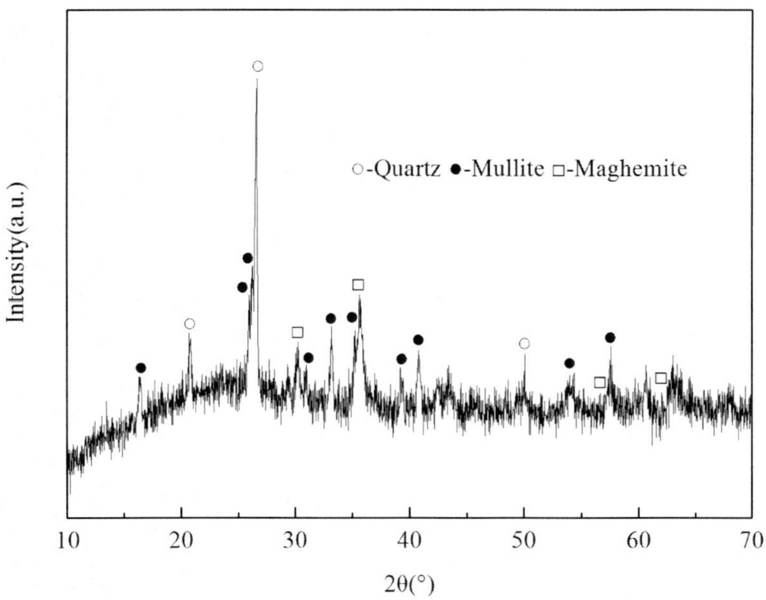

Figure 12. XRD pattern of magnetic particle.

3.2.6. SEM Observation of Magnetic Particle

SEM image of magnetic particle in Figure 13 revealed that the particle mainly represented round shapes with rough surface. They were generated interlocking or wrapping by vitreous material and silicates. Flake and needle crystals were observed on the surface.

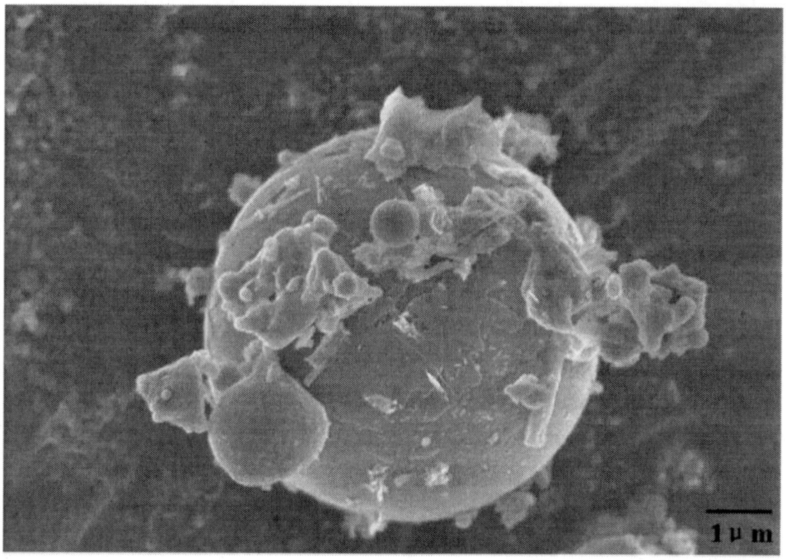

Figure 13. SEM image of magnetic particle.

4. Wet Floatation of Coal Fly Ash

The complex composition has proved to its bulk utilization in many fields. Each component may be useful or inert in a particle application but actively detrimental in another. For example, the feed coal will not burn off using anthracite coal and bitumite coal as fuels, which causes the high carbon content in coal fly ash. The effect of residual carbon in fly ash concrete, besides poor air entrainment behavior, reduces product strength, freeze-thaw stability and anti-permeability, increases water requirement to obtain normal consistency in the concrete paste (Pedersen, et al., 2008; Wesche, 1991), mixture segregation, and discoloration (Freeman, et al., 1997). The latter is aesthetically unacceptable in certain applications.

Therefore, it is necessary to reduce the carbon content meeting the need of cement and concrete products. Besides, the recycled carbon can be used as fuel, rubber filler, raw material of carbon black and activated carbon. The carbon removal methods can be divided into two types: chemical and physical method.

Physical separations, such as drum electric separation, fluidization bed separation and friction electrostatic separation, are mainly based on the specific resistance difference between carbon and ash.

The carbon and ash particles have positive and negative charge through contactting and rubbing between particles, then move to magnetic field poles and separate at the effect of high-voltage electrostatic field. This method offers advantages of avoiding hydration and flotation agent residue, however, the separation efficiency is small as the influence of ash characteristics especially particles size. Wet floatation is based on the wettability difference between ash and carbon. It is usually conducted through adding collectors and frothers, and using floatation machine to separate. This method is helpful for obtaining high purity carbon (> 70%), thus is selected in this work.

4.1. Experimental

4.1.1. Materials

The coal fly ash was used as the same as that in Section *3.1.1*.

4.1.2. Carbon removal

A quantity of coal fly ash and water was fed into a small-scale flotation machine forming 30% w/v slurry. During severe agitating, different volumes of collectors (kerosine oil and light diesel oil) and frothers (2-octanol, n-butyl alcohol and no. 2 oil) were added. At the end, the filtration was conducted. The solid was dried at 70 $^{\circ}C$ overnight and carbon content was analysed.

4.1.3. Separation Mechanism

The unburned carbon in fly ash is hydrophobic with surface wettability and floatability similar with that of coal. Its contact angle is about 60° as compared with that of 10° for other particles in ash. During the foam flotation, the carbon particles could adhere to the bubbles and surface from the ash slurry. This wettability distinction could increase significantly at the help of flotation agents, making the carbon separate from other particles in ash.

4.2. Results and Discussion

4.2.1. Effect of Collector Type

The collectors mainly include two types namely tar and petroleum products. The tar is helpful for floatation, but limited in use as containing phenol. Thus the petroleum products kerosine oil and light diesel oil were used in this study. The 2-octanol was used as frother with an additive amount of 600 g/t. The additive amount of collector was kept at 800 g/t in the series experiments. The results showed that carbon removal rate was higher as 83.41% using light diesel oil as collector (see Figure 14). Besides, the cost of kerosene oil is higher than light diesel oil. Thus, light diesel oil was selected as collector for further studies.

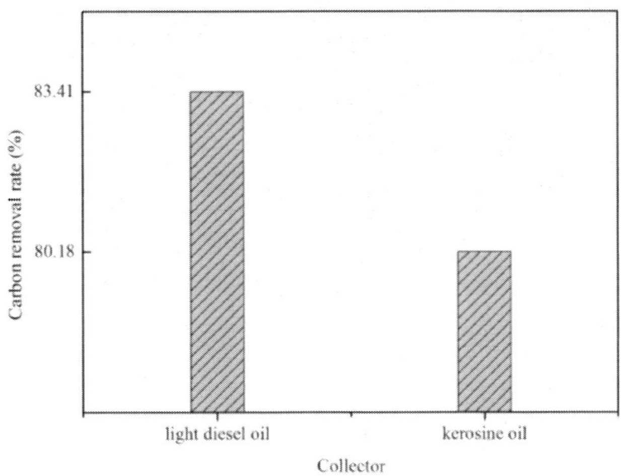

Figure 14. Effect of collector on separation efficiency.

4.2.2. *Effect of Collector Volume*

In the series experiments, the light diesel oil was used as collector and other conditions were kept. The results showed that the carbon removal rate increased with an increase of collector volume (see Figure 15). It reached 82.53% with a volume of 800 g/t. However, this uptrend became slow when the volume was above 800 g/t. Thus, the volume was determined as 800 g/t and selected for further studies.

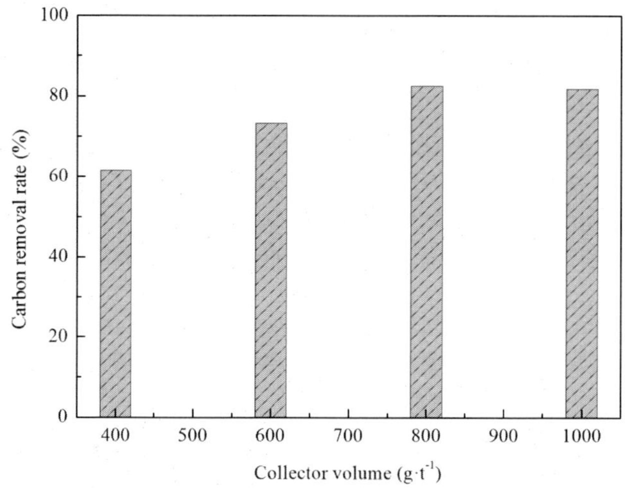

Figure 15. Effect of collector volume on separation efficiency.

4.2.3. *Effect of Frother Type*

Frothers are surface-active agents widely used in flotation to aid generation of small bubbles and to stabilize the froth phase. In the series of experiments, the light diesel oil was used as a collector and other conditions were kept the same. The results showed that the carbon removal rate was as high as 83.82% using 2-octanol as frother (see Figure 16). Thus, 2-octanol was selected as frother for further studies.

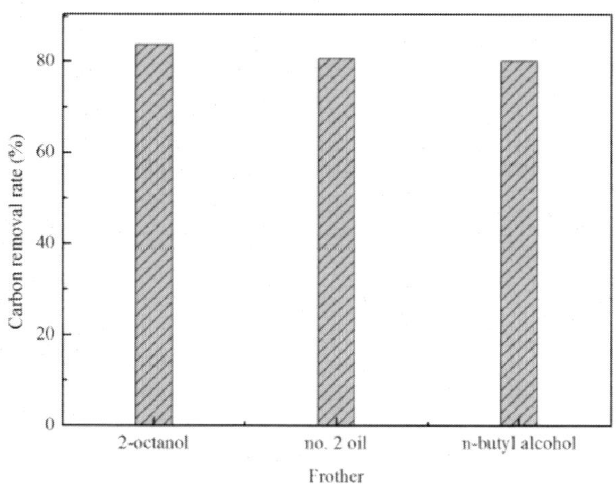

Figure 16. Effect of frother on separation efficiency.

4.2.4. Effect of Frother Volume

In the series of experiments, the 2-octanol was used as frother and other conditions were kept the same. The results showed that the carbon removal rate increased with frother volume increasing (see Figure 17). It reached 80.03% with a volume of 600 g/t. However, this uptrend became slow when the volume was greater than 600 g/t. Thus, this volume was selected for further studies.

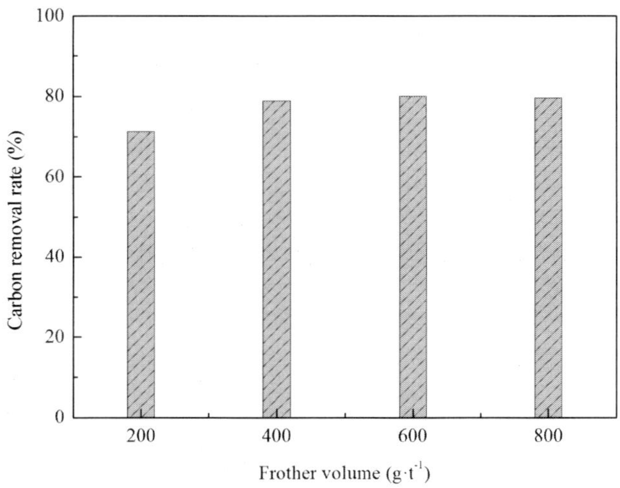

Figure 17. Effect of frother volume on separation efficiency.

4.2.5 XRD Analysis of Unburnt Carbon

The XRD pattern of unburnt carbon is displayed in Figure 18. It revealved a broad bump in the region of 20~30°, indicating the presence of amorphous phase. After comparing the diffraction peak positions and intensities with those in the JCPDS data files, these peaks were indexed to carbon. Besides, a wide range was observed in the peaks.

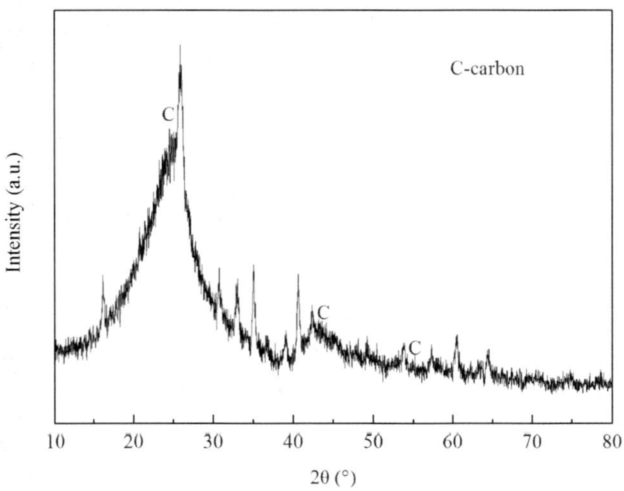

Figure 18. XRD pattern of unburnt carbon.

An asymmetrical peak located near to the peak with the highest intensity ($2\theta \approx 24^{\circ}$) was observed, which was mainly formed by several peaks. After analysing, there was mullite phase present, which was attributed to that the unburnt carbon adhered to vitreous material during the coal combustion.

5. Extracting of Silicon

5.1. Experimental

5.1.1. Materials

Sodium carbonate anhydrous and hydrochloric acid were of analytical grade from Sinopharm Chemical Reagent Co. Ltd. The fly ash was pretreated by magnetic separation and froth floatation removing the unburnt carbon and iron-rich component at the optimum conditions.

5.1.2. Extracting Experiment

(1) Fly ash activation

Based on a series of experiments, the optimum conditions for fly ash activation was determined. The activation process was as follows: The ash was firstly mixed with sodium carbonate anhydrous with a weight ratio of 1: 0.5 and milled for 2 hours. The mixture was then calcined at 950 °C for 1.5 hrs. Finally, the fused product was naturally cooled and crushed into powder obtaining the activated product.

(2) Silicon extraction

The activated product was added into a three-neck flask fitted with an impeller stirrer. The hydrochloric acid with different concentrations (4.5, 5.0, 5.5, 6.0, 6.5 and 7.0 mol/L) were transferred into flask at different S/L ratios (1: 4, 1: 4.5, 1: 5, 1: 5.5, 1: 6 and 1: 7). The effects of reaction time (30, 60, 90, 120 and 150 min) and stirring rates (100, 150, 200, 250, 300 and 400 r/min) on silicon extraction rate were investigated. At the end, the solid was separated by filtration. The filtrate was collected and underwent sol-gel transform in water bath. After transformation, the silica precipitated in the form of $SiO_2 \cdot nH_2O$. It was separated by filtration and washed with deionized water until no Cl^- ion was detected in the supernatant. Thereafter, the precipitate was dried overnight and characterized.

(3) Extracting mechanism

The XRD pattern of activated fly ash is displayed in Fig. 19. The diffraction peaks were well indexed to nephelite ($NaAlSiO_4$, JPCDS card no. 2-625). Its morphology was also observed (see Figure 20.). It can be seen that the round-shaped particles in raw fly ash was almost reduced. Simultaneously, irregular particles substantially interlocked together forming agglomeration. Higher-magnification images revealed that the glass shell of spherical particles had become rough and gradually fall out.

The following equations may represent these transformations:

$$3Al_2O_3 \cdot 2SiO_2 + 4SiO_2 + 3Na_2CO_3 \rightarrow 6NaAlSiO_4 + 3CO_2 \quad (1)$$

$$NaAlSiO_4 + 4HCl \rightarrow H_4SiO_4 + AlCl_3 + NaCl \quad (2)$$

$$H_4SiO_4 \rightarrow H_2SiO_3 + H_2O \quad (3)$$

$$mH_2SiO_3 \rightarrow mSiO_2 \cdot nH_2O + (m-n)H_2O \quad (4)$$

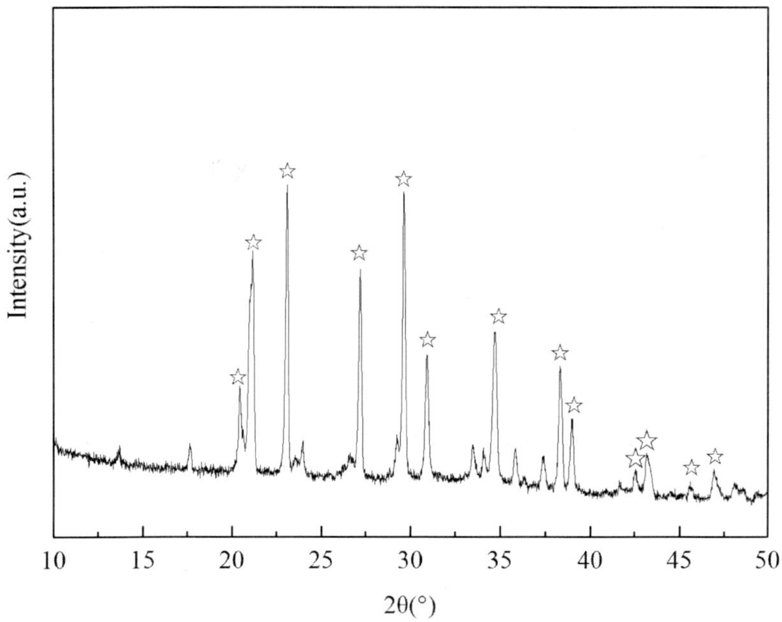

Figure 19. XRD pattern of activated fly ash.

Figure 20. SEM image of activated fly ash.

5.1.3. Characterization and Testing

The crystalline phases present was carried out by Rigaku D-Max/IIB diffractometer with Cu Kα radiation operated at 40 kV and 30 mA. The detector was scanned at a step scan of 0.02° and a scan speed of 4 °/min. SEM investigations were conducted in a Hitachi S-4800 scanning electron microscope to observe the microstructure of samples. Transmission IR spectra was recorded by a Nicolet Nexus-670 Fourier transform infrared (FT-IR) spectrometer with a resolution of 2 cm^{-1} using a KBr disc method. The thermal behavior of $Al(OH)_3$ precursor was observed through TG-DTA analysis and conducted by a ZRY-2P simultaneous thermal analyzer. N_2 adsorption measurement was carried out to evaluate the specific surface area and pore volumes of silica white. N_2 adsorption-desorption isotherms were recorded using a Micromeritics model ASAP 2020 adsorption analyzer. The pore size distribution was evaluated from the desorption branch of isotherm using Barret-Joyner-Halenda method.

5.2. Results and Discussion

5.2.1. Effect of Hydrochloric Acid Concentrations

In order to probe into the effect of hydrochloric acid concentrations on silicon extracting efficiency, a series of experiments were conducted. From Figure 21 it can be seen that the silicon extraction rate increased gradually with the hydrochloric acid concentration increasing. It increased significantly from 37 to 76% as the acid concentration evaluated from 4.5 to 6.0 mol/L. However, a slowdown in the uptrend was found when the acid concentration increased to 6.0 mol/L. This may be attributed to that a greater acid concentration promoted reaction and thus silica gel was formed. The form of silicon precipitate will cover on the surface of reactants and incumber its reaction with acid (Wang, et al., 2006). It was not helpful for filtration as well.

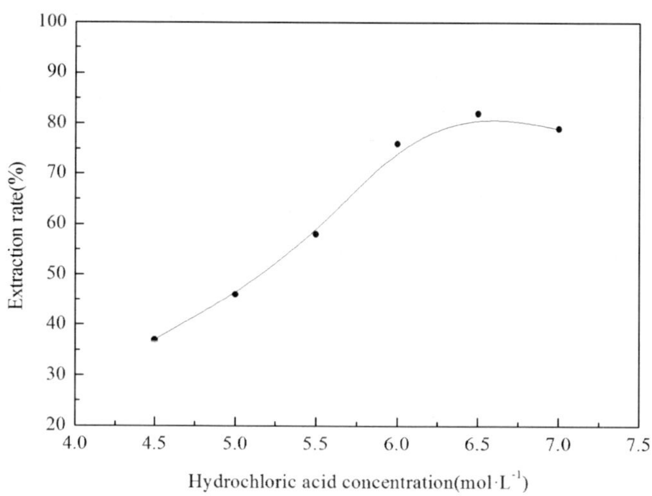

Figure 21. Effect of hydrochloric acid concentrations on silicon extraction rate (Reaction time--90 min; S/L ratio--1: 6; stirring rate--300 r/min).

5.2.2. Effect of S/L Ratios

Figure 22 illustrates the effect of S/L ratios on silicon extracting efficiency. On the premise of acid concentration keeping constant, larger S/L ratio could reduce the consumption of acid. However, it means the reduction of acid in reaction system and will incur incomplete reaction. Besides, it can result in slurry viscidity increase and go against the reaction. It was clear that the extraction efficiency increased with S/L ratio decreasing. The smaller S/L ratio could reduce the slurry viscidity and increase the matter diffusion, and thus promote the reaction.

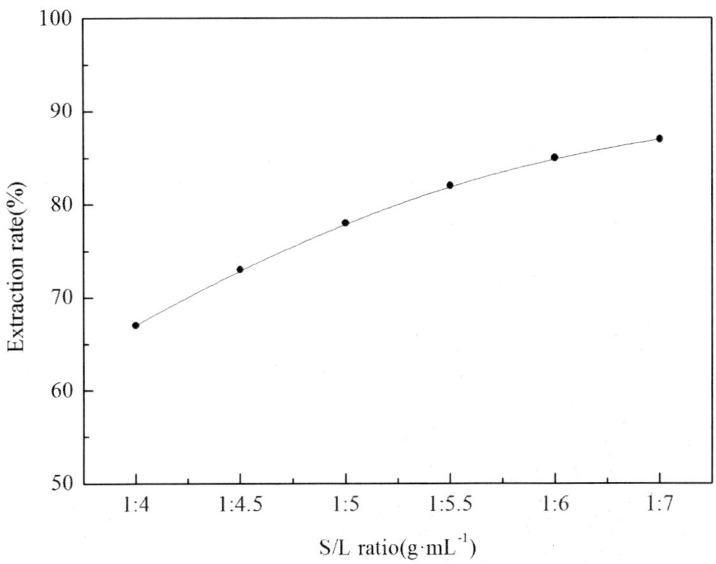

Figure 22. Effect of S/L ratios on silicon extraction rate (HCl concentration--6.5 mol/L; reaction time--90 min, stirring rate--300 r/min).

After calculating according to the reaction formula, the acid addition with S/L ratio of 1: 5 was enough for complete reaction with an acid concentration of 6.5 mol/L. Thus, the optimum S/L ratio was determined as 1: 6, which was selected for further studies.

5.2.3. Effect of Stirring Rates

Figure 23 displays the effect of stirring rates on silicon extraction rate. It revealed that the extraction rate increased at higher stirring rates. Higher stirring rates could create turbulence, speed the diffuse, enhance mass transfer and then promote the reaction. However, the extraction rate increased slowly when the stirring rate was greater than 300 r/min.

According to the principle of hydrometallurgical process, the stirring rate has a significant effect on the reaction and usually increases the reaction rate by 40~65%, when the dominant procedure is liquid film diffuse. However, the effect of stirring rates on extracting rate was not significant in this study. It indicated that liquid film diffuse was not the dominant procedure, but the internal diffusion on the solid product layer or interface reaction.

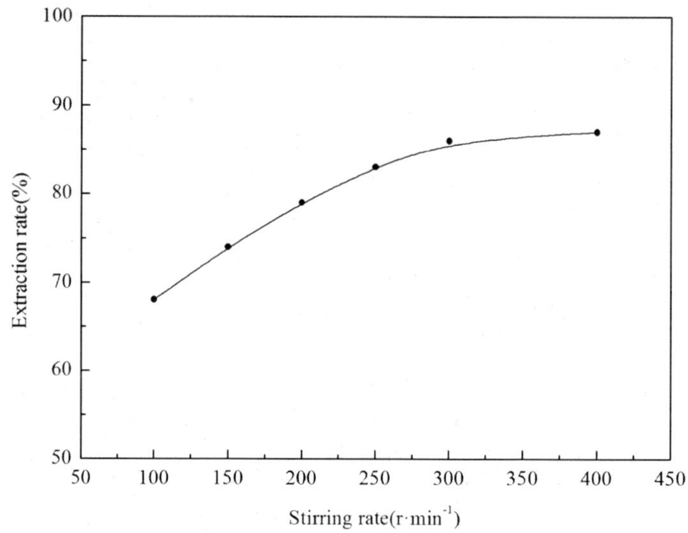

Figure 23. Effect of stirring rates on silicon extraction rate (HCl concentration--6.5 mol/L; S/L ratio--1: 6; reaction time--90 min).

5.2.4. Effect of Reaction Time

The effect of reaction time on silicon extraction rate is also investigated and the results are showed in Figure 24. It was clear that the extracting rate increased with extending of time. However, it increased slowly as reaction time prolonging to 90 min. This was due to that the acid concentration was larger at the initial of reaction and the active sites of could efficiently contact with acid. With the proceeding of reaction, the internal diffusion rate reduced and the acid concentration decreased, thus the extracting rate increased slowly.

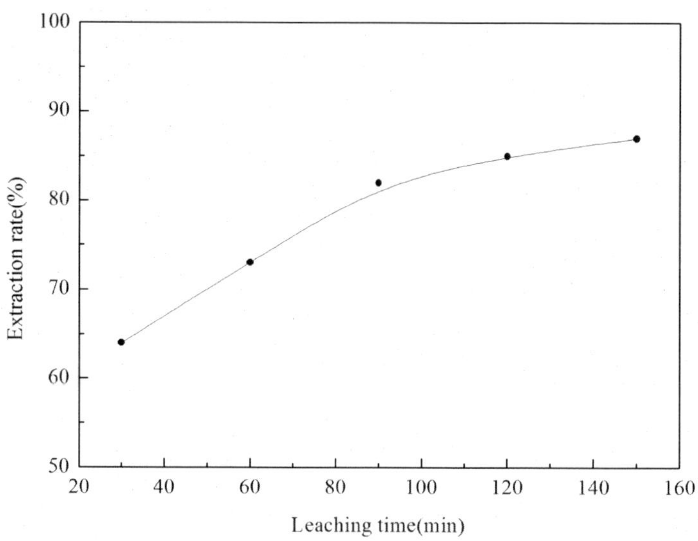

Figure 24. Effect of reaction time on silicon extraction rate (HCl concentration--6.5 mol/L; S/L ratio--1: 6; stirring rate--300 r/min).

5.2.5. XRD Analysis

The XRD pattern of silica precipitate is displayed in Figure 25. It was well indexed to silica white. There was a broad hump in the range of 15~35° (*2θ*), which indicated that the prepared silica white was amorphous.

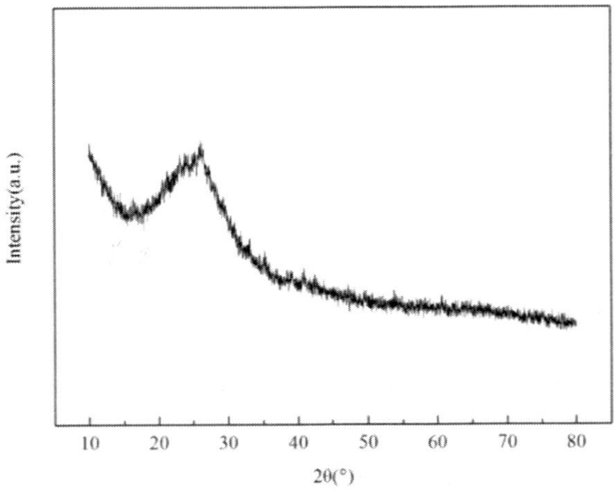

Figure 25. XRD pattern of silica white.

5.2.6. N_2 Adsorption-Desorption and Pore Size Distribution

Figure 26 illustrates the N_2 adsorption-desorption plots for silica white. The pattern was corresponded to the type IV of the BDDT classification (Gregg, et al., 1991). There was a steep increase in the amount of N_2 adsorbed when the relative pressure (P/P_0) was higher than 0.6. The existence of a hysteresis loop indicated the presence of mesopores (Janssen, et al., 2002; Morin, et al., 1998;), which was evidenced in Figure 26 (inset). The BET surface area was calculated as 364.8 m^2/g and the average pore size was determined as 6.02 nm.

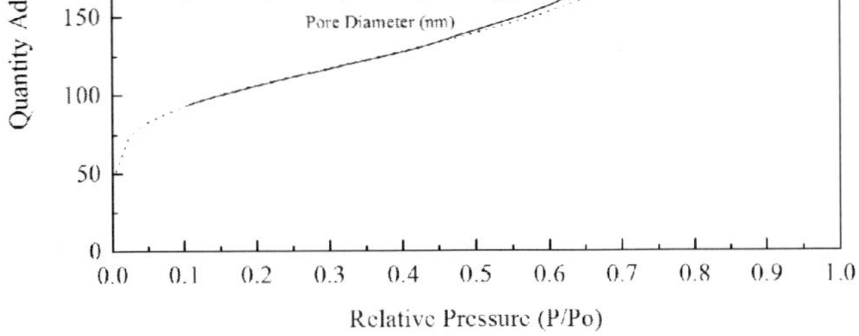

Figure 26. N_2 adsorption-desorption isotherms of silica white and corresponding pore diameter distribution pattern (inset).

6. Extracting of Aluminium

6.1. Experimental

6.1.1. Material

Ammonia solution was of analytical grade from Jiangsu Yonghua Fine Chemical Co., Ltd. The fly ash, sodium carbonate anhydrous and hydrochloric acid were used as the same as that in Section *5.1.1*.

6.1.2. Extracting Experiment

The activated product abovementioned in Section *5.1.2* was added into a three-neck flask fitted with an impeller stirrer. The hydrochloric acid with a concentration of 6.5 mol/L was transferred at an S/L ratio of 1: 6. Reaction time and stirring rate were fixed as 90 min and 300 r/min. At the end of the reaction, the solid was separated by filtration. The filtrate was collected and underwent sol-gel transform in water bath. After transformation, the silicon precipitate was separated and the filtrate was collected as the source of aluminium.

Ammonia solution with a concentration of 2.0 mol/L was added slowly into the filtrate containing aluminium with continuous vigorous stirring. The dosage of $NH_3{\cdot}H_2O$ solution was determined according to the desired pH value. After reaction, the gel was aged for 2 h and separated by filtration. The wet gel was dried forming $Al(OH)_3$ precursor and calcined at optimum temperature to obtain Al_2O_3 based on the analysis of TG-DTA.

6.2. Results and Discussion

6.2.1. XRD Analysis of Precursor

The XRD pattern of precursor is presented in Figure 27. Comparing the diffraction peaks with those in the JCPDS (Joint Committee on Powder Diffraction Standards) data files, the XRD pattern was accorded well with that of $Al(OH)_3$ (JPCDS card no. 26-25).

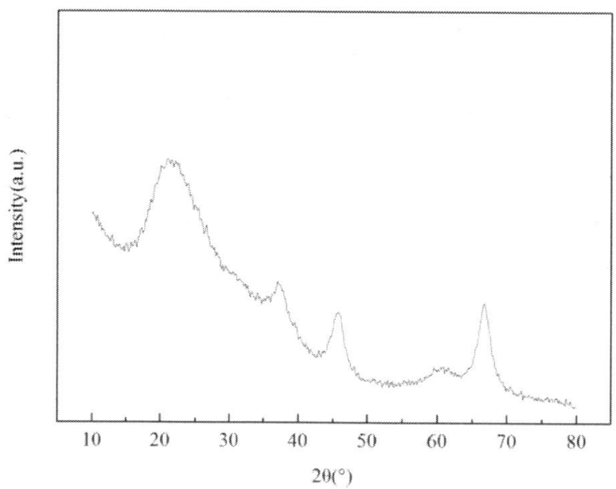

Figure 27. The XRD pattern of precursor.

In addition, the peaks were broadening, which indicated that the precursor crystallized imperfectly and the particle size may be small (Li and Shao, 2008).

6.2.2. FT-IR Analysis of Precursor

FT-IR transmittance spectrum of the precursor is displayed in Figure 28. The peaks located at approximate 3450 cm^{-1} and 1650 cm^{-1} were attributed to the stretching and bending vibration of OH group, respectively. The bands at range of 1410~584 cm^{-1} were assigned to the in-plane bending vibration, bending vibration of OH group and stretching vibration of Al-O bound, respectively.

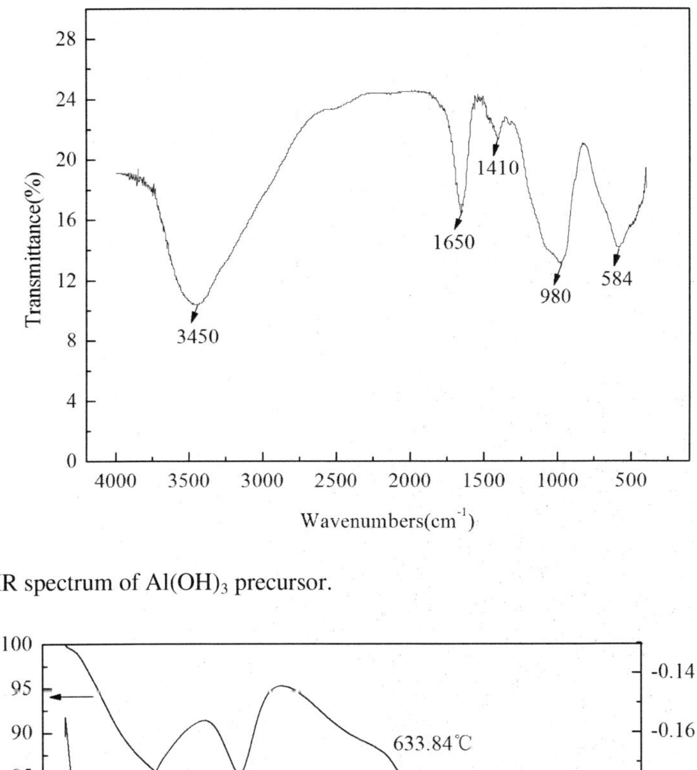

Figure 28. FT-IR spectrum of $Al(OH)_3$ precursor.

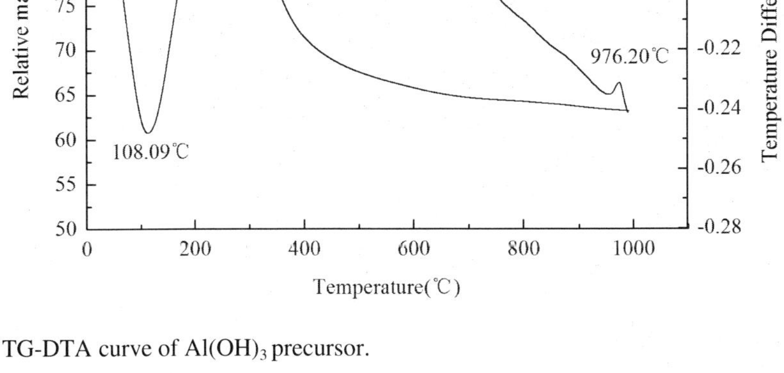

Figure 29. TG-DTA curve of $Al(OH)_3$ precursor.

6.2.3. TG-DTA Analysis of Precursor

TG-DTA analysis is conducted to determine the optimum calcining temperature. From the TG curve in Figure 29, it can be seen that the relative mass decreased continuously. The decreasing trend became slow when the temperature reached 800°C. DTA curve presented two endothermic peaks at 108.09 °C and 358.14 °C, which attributed to the desorption of physical adsorbed water on solid surface and chemical structure water in crystal, respectively. The two weak exothermic peaks at 633.84 °C and 976.20 °C belonged to the thermal decomposition of precursor.

6.2.4. XRD Patterns Of Calcined Precursor

The $Al(OH)_3$ precursor was calcined on the basis of TG-DTA analysis. The calcined product was characterized by XRD analysis (see Figure 30). The diffraction peaks for calcined product at 700 °C for 2 h were well accorded with that of $Al_{2.144}O_{3.2}$ (JPCDS card no. 79-1557). The diffraction peaks intensities for sample calcined at 980 °C increased significantly. The d values and 2θ matched well with that of γ-Al_2O_3 (JPCDS card no. 10-425).

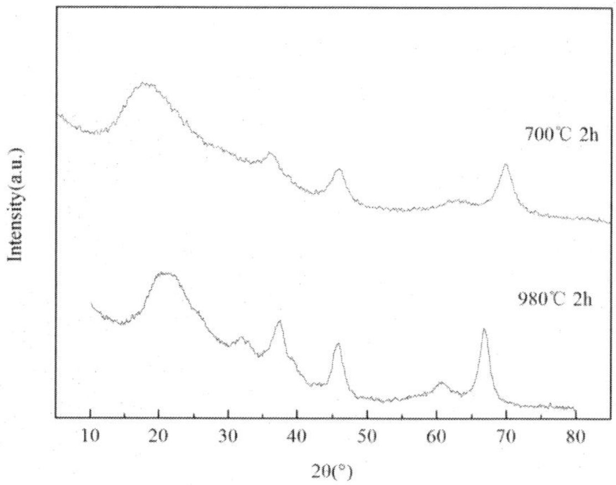

Figure 30. XRD patterns of calcined precursor.

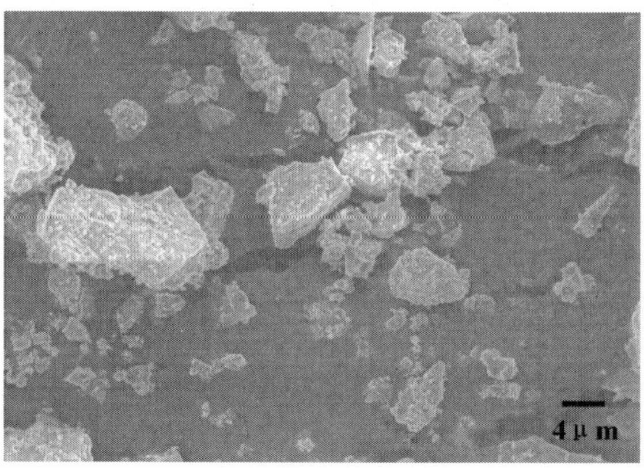

Figure 31. SEM image of γ-Al_2O_3.

6.2.5. SEM image of γ-Al_2O_3

SEM analysis is conducted to observe the morphology of obtained γ-Al_2O_3. It presented irregular particles with non-uniform size distribution (see Figure 31).

Conclusion

Fly ash has been proved to have significant value-added potential as an ingredient in cementitious and other construction process. However, in many cases these markets are close to saturation, plus they are vulnerable to economic down-turn in the building industry. In most countries at least 75% the fly ash generated annually is dumped with no subsequent reutilization, which poses a significant economic and environmental burden. As a consequence, there exists significant interest in developing alternative recycling method for the bulky waste. Coal fly ash is rich in SiO_2 and Al_2O_3, and has been considered as a potential resource of aluminum and silicon. In this work, silicon and aluminium were extracted to prepare silica white and Al_2O_3 respectively, using sintering-acid leaching method. The silicon extraction rate displayed an uptrend with the increase of acid concentration, reaction time and stirring rate. The optimum conditions, such as acid concentration, reaction time, stirring rate and S/L ratio were determined as 6.5 mol/L, 90 min, 300 r/min and 1: 6, respectively. The prepared silica white was amorphous and its specific surface area was determined as large as 364.8 m^2/g. Based on the TG-DTA analysis, $Al(OH)_3$ precursor was calcined at 980 °C forming γ-Al_2O_3.

Acknowledgments

We gratefully acknowledge financial support from the Scientific Research Foundation of Hangzhou Dianzi University (Grant no. KYS205612029), Zhejiang Provincial Natural Science Foundation of China (Grant no. Q13B070018), Chinese National 863 High Technology (Grant no. 2007AA06Z128) and Key Laboratory for Solid Waste Management and Environment Safety Open Fund (Grant no. SWMES 2011-07).

References

Ahmaruzzaman, M. (2010). A review on the utilization of fly ash. *Prog. Energy Combust.* Sci. 36, 327-363.

Ahmaruzzaman, M. (2011). Industrial wastes as low-cost potential adsorbents for the treatment of wastewater laden with heavy metals. *Adv. Colloid Interfac.* 166, 36-59.

Akar, G., Polat, M., Galecki, G., Ipekoglu, U. (2012). Leaching behavior of selected trace elements in coal fly ash samples from Yenikoy coal-fired power plants. *Fuel Process. Technol.* 104, 50-56.

Aksu, Z., Yener, J. (2001). A comparative adsorption/biosorption study of monochlorinated phenols onto various sorbents. *Waste Manage.* 21, 695-702.

Al-Zboon, K., Al-Harahsheh, M.S., Hani, F.B. (2011). Fly ash-based geopolymer for Pb removal from aqueous solution. *J. Hazard. Mater*. 188, 414-421.

Arroyo Torralvo, F., Fernández-Pereira, C. (2011). Recovery of germanium from real fly ash leachates by ion-exchange extraction. *Miner. Eng*. 24, 35-41.

Arroyo Torralvo, F., Fernández-Pereira, C., Campaña, M.C. (2010). Recovery of germanium from aqueous solutions by ion-exchange extraction of its catechol complex. *Ind. Eng. Chem. Res*. 49, 4817-4823.

Arroyo, F. Recuperación del Germanio Presente en Una Central GICC Que Emplea Carbones de la Cuenca de ENCASUR en Puertollano, PhD. dissertation, University of Seville, 2007.

Arroyo, F., Fernández-Pereira, C. Method for the recovery of germanium in solution by means of complexing and use of ion-exchange resins. Patent WO2009/106660, 2009.

Authier-Martin, M., Forte, G., Ostap, S., See, J. (2001). The mineralogy of bauxite for producing smelter-grade alumina. *JOM*, 53, 36-40.

Bai, G.H., Qiao, Y.H., Shen, B., Chen, S.L. (2011). Thermal decomposition of coal fly ash by concentrated sulfuric acid and alumina extraction process based on it. *Fuel Process. Technol*. 92, 1213-1219.

Bai, G.H., Teng, W., Wang, X.G., Qin, J.G., Xu, P., Li, P.C. (2010). Alkali desilicated coal fly ash as substitute of bauxite in lime-soda sintering process for aluminum production. Trans. *Nonferrous Met. Soc. China* 20, s169-s175.

Bai, G.H., Teng, W., Wang, X.G., Zhang, H., Xu, P. (2010). Processing and kinetics studies on the alumina enrichment of coal fly ash by fractionating silicon dioxide as nano particles. *Fuel Process. Technol*. 91, 175-184.

Barbieri, L., Lancellotti, I., Manfredini, T., Queralt, I., Rincon, J.M., Romero, M. (1999). Design obtainment and properties of glasses and sintered materials from coal fly ash. *Fuel* 78, 271-276.

Belardi, G., Massimilla, S., Piga, L. (1998). Crystallization of K-L and K-W zeolites from fly-ash. *Resour. Conserv. Recy*. 24, 167-181.

Bhangare, R.C., Ajmal, P.Y., Sahu, S.K., Pandit, G.G., Puranik, V.D. (2011). Distribution of trace elements in coal and combustion residues from five thermal power plants in India. *Int. J. Coal Geol*. 86, 349-356.

Bhattacharya, S.S., Iftikar, W., Sahariah, B., Chattopadhyay, G.N. (2012). Vermicomposting converts fly ash to enrich soil fertility and sustain crop growth in red and lateritic soils. *Resour. Conserv. Recy*. 65, 100-106.

Bin-Shafiquea, S., Rahmanb, K., Yaykirana, M., Azfara, I. (2010). The long-term performance of two fly ash stabilized fine-grained soil subbases. *Resour. Conserv. Recy*. 54, 666-672.

Blanco, F., Gareis, P., Materos, P., Ayala, J. (2000). Characteristics and properties of lightweight concrete manufactured with cenospheres. *Cement Concrete Res*. 30, 1715-1722.

Blissett, R.S., Rowson, N.A., (2012). A review of the multi-component utilization of coal fly ash. *Fuel* 97, 1-23.

Borm, P.J.A. (1997). Toxicity and occupational health hazards of fly ash (CFA): a review of data and comparison to coal mine dust. *Ann. Occup. Hyg*. 41, 659-676.

Butterman, W.C., Jogerson, J.D. (2005). *Germanium. In: Mineral Commodity Profiles*. US Department of the Interior, US Geological Survey, Reston, Virginia.

Cao, D.Z., Selic, E., Herbell, J.D. (2008). Utilization of fly ash from coal-fired power plants in China. *J. Zhejiang Univ-Sc.* A 9, 681-687.

Chakraborty, R., Bepari, S., Banerjee, A. (2010). Transesterification of soybean oil catalyzed by fly ash and egg shell derived solid catalysts. *Chem. Eng. J.* 165, 798-805.

Chareonpanich, M., Namto, T., Kongkachuichay, P., Limtrakul, J. (2004). Synthesis of ZSM-5 zeolite from lignite fly ash and rice husk ash. *Fuel Process. Technol.* 85, 1623-1634.

Chen, T.P. Crystallization of glass ceramic prepared from bottom ash and magnesium carbonate. Master Dissertation, Cheng Kung University, Taiwan, 2003.

Cho, H., Oh, D., Kim K. (2005). A study on removal characteristics of heavy metals from aqueous solution by fly ash. *J. Hazard. Mater.* 127, 187-195.

Choi, S.K., Lee, S., Song, Y.K., Moon, H.S. (2002). Leaching characteristics of selected Korean fly ashes and its implication for the ground water composition near the ash disposalmound. *Fuel* 81, 1083-1090.

Cicek, T., Tanriverdi, M. (2007). Lime based steam autoclaved fly ash bricks. *Constr. Build. Mater.* 21, 1295-300.

Cokca, E., Yilmaz, Z. (2004). Use of rubber and bentonite added fly ash as a liner material. *Waste Manage*. 24, 153-164.

Cultrone, G., Sebastián, E. (2009). Fly ash addition in clayey materials to improve the quality of solid bricks. *Constr. Build. Mater.* 23, 1178-1184.

Davidovits, J. (2008). *Geopolymer chemistry and applicatio*n, 3ed Ed. Geopolymer Institute, Saint-Quentin, France.

de Leer, E.W.B., Lexmond R.J., de Zeeuw, M.A. (1989). ''De novo"-synthesis of chlorinated biphenyls, dibenzofurans and dibenzo-p-dioxins in the fly ash catalyzed reaction of toluene with hydrochloric acid. *Chemosphere* 19, 1141-1152.

Dermatas, D., Meng, X.G. (2003). Utilization of fly ash for stabilization/solidification of heavy metal contaminated soils. *Engineering Geology* 70, 377-394.

Diaz, E., Allouche, E., Eklund, S. (2010). Factors affecting the suitability of fly ash as source material for geopolymers. *Fuel* 89, 992-996.

Dilmore, R.M., Neufeld, R.D. (2001). Autoclaved aerated concrete produced with low-NOx burner/selective catalytic reduction fly ash. *J. Energy Eng.* 127, 37-50.

Douay, F., Dubourguier, H. C., Pruvot, C., Warin, A. (2002). Evaluation de lintérêt des cendres LFC pour le phytomanagement des sites contaminés par des métaux lourds. Rapport intermédiaire. Pôle de Compétences sur les Sites et Sédiments Pollués, Metaleurop, ISA, Lille 1, EDF, Surschiste, Apinor, p. 50.

Dutta, B.K., Khanra, S., Mallick, D. (2009). Leaching of elements from coal fly ash: assessment of its potential for use in filling abandoned coal mines. *Fuel* 88, 1314-1323.

Einarsrud, M.A., Kirkedelen, M.B., Nilsen, E., Mortensen, K., Samseth, J. (1998). Structural development of silica gels aged in TEOS. *J. Non-Cryst. Solids* 231, 10-16.

Elseewi, A.A., Page, A.L., Grimm, S.R. (1980). Chemical characterization of fly ash aqueous system. *J. Environ. Qual.* 9, 224.

Erol, M., Küçükbayrak, S., Ersoy-Meriçboyu, A. (2008). Characterization of sintered coal fly ashes. *Fuel* 87, 1334-1340.

Erol, M., Küçükbayrak, S., Ersoy-Meriçboyu, A. (2008). Comparison of the properties of glass, glass-ceramic and ceramic materials produced from coal fly ash. *J. Hazard. Mater.* 153, 418-425.

Estevinho, B.N., Martins, I., Ratola, N., Alves, A., Santos, L. (2007). Removal of 2,4-dichlorophenol and pentachlorophenol from waters by sorption using coal fly ash from a portuguese thermal power plant. *J. Hazard. Mater.* 143, 535-540.

European Coal Combustion Products Association (ECOBA). *Production and utilisation of CCPs in 2003 in Europe*, 2004, http://www.ecoba.org/index.html.

European Coal Combustion Products Association. Production and Utilisation of CCPs in 2008 in Europe, 2008, http://www.ecoba.com/evjm, media/ccps/Ecoba_Stat_2008_EU15. pdf>.

Font, O., Querol, X., Juan, R., Casado, R., Ruiz, C.R., López-Soler, Á., Coca, P., Peña, F.G. (2007). Recovery of gallium and vanadium from gasification fly ash. *J. Hazard. Mater.* 139, 413-423.

Font, O., Querol, X., Lopez-Soler, A., Chimenos, J.M., Fernandez, A.I., Burgos, S., Garcia Pena, F. (2005). Ge extraction from gasification fly ash. *Fuel* 84, 1384-1392.

Freeman, E., Gao, Y.M., Hurt, R., Suuberg, E. (1997). Interactions of carbon-containing fly ash with commercial air-entraining admixtures for concrete. *Fuel* 76, 761-765.

Freeman, E., Gao, Y.M., Hurt, R., Suuberg, E. (1997). Interactions of carboncontaining fly ash with commercial air-entraining admixtures for concrete. *Fuel* 76, 761-765.

Gomes, S., François, M., Abdelmoula, M., Refait, P., Pellissier, C., Evrard, O. (1999). Characterization of magnetite in silico-aluminous fly ash by SEM, TEM, XRD, magnetic susceptibility, and Mössbauer spectroscopy. *Cem. Concr. Res.* 29, 1705-1711.

González, A., Navia, R., Moreno, N. (2009). Fly ashes from coal and petroleum coke combustion: current and innovative potential applications. *Waste Manage. Res.* 27, 976-987.

Greenpeace. The true cost of coal—an investigation into coal ash in China, 2010, http://www.docin.com/p-86430929.html.

Gregg, S. J., Sing, K.S.W. (1991). *Adsorption Surface Area and Porosity,* Academic Press, London.

Groppo, J., Honaker, R. (2009). Economical recovery of fly ash-derived magnetics and evaluation for coal cleaning. 2009 World of Coal Ash (WOCA) Conference, Lexington, KY, USA.

Grzymek, J., Warsaw, Derdaka, A. *Method for obtaining aluminum oxide.* US patent 4149898, 1979-04-17.

Gun'ko, V.M., Turov, V.V., Bogatyrev, V.M., Zarko, V.I., Leboda, R., Goncharuk, E.V., Novza, A.A., Turov, A.V., Chuiko, A.A. (2005). Unusual properties of water at hydrophilic/hydrophobic interfaces. *Adv. Colloid Interfac.* 118, 125-172.

Hernandez-Exposito, A., Chimenos, J.M., Fernandez, A.I., Font, O., Querol, X., Coca, P., Garcia Pena, F. (2006). Ion flotation of germanium from fly ash aqueous leachates. *Chem. Eng. J.* 118, 69-75.

Hirajima, T., Petrus, H.T.B.M., Oosako, Y., Nonaka, M., Sasaki, K., Ando, T. (2010). Recovery of cenospheres from fly ash using dry separation process: Separation estimation and potential application. *Int. J. Miner. Process.* 95, 18-24.

Holler, H., Wirsching, U. (1985). Zeolite formation from fly ash. *Forsch. Miner.* 63, 21-43.

Hollman, G.G., Steenbruggen, G., Janssen-Jurkovičová M. (1999). A two-step process for the synthesis of zeolites from coal fly ash. *Fuel* 78, 1225-1230.

Horiuchi, S., Kawaguchi, M., Yasuhara, K. (2000). Effective use of fly ash slurry as fill material. *J. Hazard. Mater.* 76, 301-337.

Hower, J.C., Robertson, J., Thomas, G.A., Wong, A.S., Schram, W.H., Graham, U.M., Rathbone, R.F., Robi, T.L. (1996). Characterization of fly ash from Kentucky power plants. *Fuel* 75, 403-411.

Ilic, M., Cheeseman, C., Sollars, C., Knight, J. (2003). Mineralogy and microstructure of sintered lignite coal fly ash. *Fuel* 82, 331-336.

International Energy Statistics, 2011. US Energy Information Administration, http://www.eia.gov/cfapps/ipdbproject/iedindex3.cfm.

Iyer, R. (2002). The surface chemistry of leaching coal fly ash. *J. Hazard. Mater.* B 93, 321-329.

Iyer, R.S., Scott, J.A. (2001). Power station fly ash—a review of value-added utilization outside of the construction industry. *Resour. Conserv. Recy.* 31, 217-228.

Izquierdo, M., Querol, X. (2012). Leaching behaviour of elements from coal combustion fly ash: an overview. *Int. J. Coal Geol.* 94, 54-66.

Izquierdo, M.T., Rubio, B. (2008). Carbon-enriched coal fly ash as a precursor of activated carbons for SO_2 removal. *J. Hazard. Mater.* 155, 199-205.

Jackson, N.M., Schultz, S., Sander, P., Schopp, L. (2009). Beneficial use of CFB ash in pavement construction applications. *Fuel* 88, 1210-1215.

Jain, D., Khatri, C., Rani, A. (2010). Fly ash supported calcium oxide as recyclable solid base catalyst for Knoevenagel condensation reaction. *Fuel Process. Technol.* 91, 1015-1021.

Jain, D., Mishra, M., Rani, A. (2012). Synthesis and characterization of novel aminopropylated fly ash catalyst and its beneficial application in base catalyzed Knoevenagel condensation reaction. *Fuel Process. Technol.* 95 (2012) 119-126.

Jala, S., Goyal, D. (2006). Fly ash as a soil ameliorant for improving crop production—a review. *Bioresource Technol.* 97, 1136-1147.

Jandova, J., Vu, H. (2001). Processing of germanium-bearing fly ash. Metallurgy, *Refractories and Environment*, 107-112.

Jandova, J., Vu, H., Fecko, P. (2002). Recovery of germanium from Czech brown-coal fly ash. In: *Proceedings of Annual International Pittsburgh Coal Conference*, vol. 19, p. 1075-1080.

Jankowski, J., Ward, C.R., French, D., Groves, S. (2006). Mobility of trace elements from selected Australian fly ashes and its potential impact on aquatic ecosystems. *Fuel* 85, 243-256.

Janssen, A.H., Koster, A.J., de Jong K.P. (2002). On the shape of the mesoporous in zeolite Y: a three-dimensional transmission electron microscopy study combined with texture analysis. *J. Phys. Chem.* B 106, 11905-11909.

Jarmohamed, W., Mulder, P. (1994). Oxychlorination and combustion of propene on fly ash formation of chlorinated benzenes, dibenzodioxines and mono- and dibenzofurans. *Chemosphere,* 29, 1911-1917.

Jing, Z.Q., Li, Y.Y., Cao, S.W., Liu, Y.Y. (2012). Performance of double-layer biofilter packed with coal fly ash ceramic granules in treating highly polluted river water. *Bioresource Technol.* 120, 212-217.

Joseph, B., Mathew, G. (2012). Influence of aggregate content on the behavior of fly ash based geopolymer concrete. *Sci. Iran.* A 19, 1188-1194.

Kant, K., Rashmi, Kuriakose, S., Sonkawade R.G., Chauhan, R.P., Chakarvarti, S.K., Sharma, G.S. (2010). Radon activity and exhalation rates in Indian fly ash samples. *Indian J. Pure Ap. Phy.* 48, 457-462.

Kastner J.R., Das, K.C., Buquoi, Q., Melear, N.D. (2003). Low temperature catalytic oxidation of hydrogen sulfide and methanethiol using wood and coal fly ash. *Environ. Sci. Technol.* 37, 2568-2574.

Kelmers, A.D., Canon, R.M, Egan, B.Z., Felker, L.K., Gilliam, T.M., Jones, G., Owen, G.D., Seeley, F,G., Watson, J.S. (1982). Chemistry of the direct acid leach, calsinter, and pressure digestion-acid leach methods for recovery of aluminum from fly-ash. *Resources and Conservation* 9, 271-279.

Khatri, C., Mishra, M.K., Rani, A. (2010). Synthesis and characterization of fly ash supported sulfated zirconia catalyst for benzylation reactions. *Fuel Process. Technol.* 91, 1288-1295.

Khatri, C., Rani, A. (2008). Synthesis of a nano-crystalline solid acid catalyst from fly ash and its catalytic performance. *Fuel* 87, 2886-2892.

Kikuchi, R. (1999). Application of coal ash to environmental improvement: transformation into zeolite, potassium fertilizer, and FGD absorbent. *Resour. Conserv. Recycl.* 27, 333-346.

Kim, B., Prezzi, M. (2008). Evaluation of the mechanical properties of class-F fly ash. *Waste Manage.* 28, 649-659.

Kim, B.J., Prezzi, M., Salgado, S. (2005). Geotechnical properties of fly and bottom ash mixtures for use in highway embankments. *J. Geotech. Geoenviron.* 131, 914-924.

Kniess, C.T., de Lima, J.C., Prates, P.B., Kuhnen, N.C., Riella, H.G. (2007). Dilithium dialuminium trisilicate phase obtained using coal bottom ash. *J. Non-Cryst. Solids* 353, 4819-4822.

Koukouzas, N., Ketikidis, C., Itskos, G. (2011). Heavy metal characterization of CFB-derived coal fly ash. *Fuel Process. Technol.* 92, 441-446.

Lee, H., Ha, H.S., Lee, C.H., Lee, Y.B., Kim, P.J. (2006). Fly ash effect on improving soil properties and rice productivity in Korean paddy soils. *Bioresource Technol.* 97, 1490-1497.

Li, B., Shao, L.L. (2008). Appraisal of alumina and aluminium hydroxide by XRD. *Inoganic Chemicals Industry* 40, 54-57.

Li, L.S., Wu, Y.S., Liu, Y.Y., Zhai, Y.C. (2011). Extraction of alumina from coal fly ash with sulfuric acid leaching method. *The Chinese Journal of Process Engineering* 11, 254-258.

Li, Y., Zhang, F.S., Xiu, F.R. (2009). Arsenic (V) removal from aqueous system using adsorbent developed from a high iron-containing fly ash. Sci. Total Environ. 407, 5780-5786.

Lin, C.F., His, H.C. (1995). Resource recovery of waste fly ash: synthesis of zeolite-like materials, *Environ. Sci. Technol.* 29, 1109-1117.

Lior, N. (2010). Sustainable energy development: the present (2009) situation and possible paths to the future. *Energy* 35, 3976-3994.

Liu, G., Zhang, H., Gao, L., Zheng, L., Peng, Z. (2004). Petrological and mineralogical characterizations and chemical composition of coal ashes from power plants in Yanzhou mining district, China. *Fuel Process Technol.* 85, 1635-1646.

Lu, G.Q., Do, D.D. (1991). Adsorption properties of fly ash particles for NOx removal from flue gases. *Fuel Process. Technol.* 27, 95-107.

Mallik, D., Chaudhuri, S.K. (1999). Air oxidation of aqueous sodium sulfide solutions with coal fly ash. *Wat. Res.* 33, 585-590.

Manoharan, V., Yunusa, I.A.M., Loganathan, P., Lawrie, R., Skilbeck, C.G., Burchett, M.D., Murray, B.R., Eamus, D. (2010). Assessments of Class F fly ashes for amelioration of soil acidity and their influence on growth and uptake of Mo and Se by canola. *Fuel* 89, 3498-3504.

Manz, O.E. (1999). Coal fly ash: a retrospective and future look. *Fuel* 78, 133-136.

Marco-Lozar, J.P., Cazorla-Amorós, D., Linares-Solano, A. (2007). A new strategy for germanium adsorption on activated carbon by complex formation. *Carbon* 45, 2519-2528.

Marco-Lozar, J.P., Cazorla-Amorós, D., Linares-Solano, A. Procedimiento parala Recuperación de Germanio en Disolución Mediante Carbón Activo. Patent ES2257181, 2006.

Maroto-Valer, M., Taulbee, D., Hower, J. (2001). Characterization of differing forms of unburned carbon present in fly ash separated by density gradient centrifugation. *Fuel* 80, 795-800.

Maroto-Valer, M.M. (2002). Recovery and utilization of fly ash carbon. *Fuel and Energy Abstracts* 43, 290.

Maroto-Valer, M.M., Zhang, Y., Andrésen, J.M., Jones, A., Morrison, J.L. (2001). Routes for the development of value added products from high carbon fly ashes. *ACS. Fuel Chemistry*, Preprints, 46, 316-317.

Matheswaran, M., Karunanithi, T. (2007). Adsorption of Chrysoidine R by using fly ash in batch process. *J. Hazard. Mater.* 145, 154-161.

Mathur, A.K., Kumar, R., Mishra, M., Sengupta, D., Prasad, R., (2008). An investigation of radon exhalation rate and estimation of radiation doses in coal and fly ash samples. *Appl. Radiat. Isotopes* 66, 401-406.

Matjie, R.H., Bunt, J.R., Van, Heerder, J.H.P. (2005). Extraction of alumina from coal fly ash generated from a selected low rank bituminous South African coal. *Miner. Eng.* 18, 299-310.

McCarthy, M.J., Dhir, R.K. (2005). Development of high volume fly ash cements for use in concrete construction. *Fuel* 84, 1423-1432.

Meawad, A.S., Bojinova, D.Y., Pelovski, Y.G. (2010). An overview of metals recovery from thermal power plant solid wastes. *Waste Manage.* 30, 2548-2559.

Menon, M.P., Ghuman, G.S., James, J., Chandra, K., Adriano, D.C. (1980). Physico-chemical characteristics of water extracts of different coal ashes and fly ash amended composts. *Water Air Soil Pollut.* 50, 343-353.

Mishra, D.P., Das, S.K. (2010). A study of physico-chemical and mineralogical properties of Talcher coal fly ash for stowing in underground coal mines. *Mater. Charact.* 61, 1252-1259.

Mohan, S., Gandhimathi, R. (2009). Removal of heavy metal ions from municipal solid waste leachate using coal fly ash as an adsorbent. *J. Hazard. Mater.* 169, 351-359.

Molina, A., Poole, C. (2004). A comparative study using two methods to produce zeolite from fly ash. *Miner. Eng.* 17, 167-173.

Moliner, A.M., Street, J.J. (1982). Effect of fly ash and lime on growth and composition of corn on acid sandy soils. *Soil Crop Sci. Soc. Fla. Proc.* 41, 217-220.

Moreno, N., Querol, X., Andrés, JM, Stanton, K., Towler, M., Nugteren, H., Janssen-Jurkovicová M., Jones R. (2005). Physico-chemical characteristics of European pulverized coal combustion fly ashes. *Fuel* 84, 1351-1363.

Moreno, N., Querol, X., López-Soler, A., Andres, J.M., Janssen, M., Nugteren, H., Towler M., Stanton, K. (2004). Determining suitability of a fly ash for silica extraction and zeolite synthesis. *J. Chem. Technol. Biot.* 79, 1009-1018.

Morin, S., Ayrault, P., Gnep, N.S., Guisnet, M. (1998). Influence of the framework composition of commerical HFAU zeolites on their activity and selectivity in m-xylene transformation. *Appl. Catal. A: Gen.* 166, 281-292.

Moskalyk, R.R. (2004). Review of germanium processing worldwide. *Miner. Eng.* 17, 393-402.

Nalbantoglu, Z. (2004). Effectiveness of class C fly ash as an expansive soil stabilizer. *Constr. Build. Mater.* 18, 377-381.

Narkhede, H.P., More, U.B., Dalas, D.S., Pawar, N.S., More, D.H., Mahulikar, P.P. (2007). Fly-ashsupported synthesis of 2-mercaptobenzothiazole derivatives under microwave irradiation. *Synt. Comm.* 37, 573-577.

National Development and Reform Commission of China, 2011. Implementation of comprehensive utilization of bulky waste. http://www.sdpc.gov.cn/zcfb/zcfbtz/2011tz /t20111229_453571.htm.

Nayak, N., Panada, C.R. (2010). Aluminium extraction and leaching characteristics of Talcher thermal power station fly ash with sulphuric acid. *Fuel* 89, 53-58.

Niewiadomski, M., Hupka, J., Bokotko, R., Miller, J.D. (1999). Recovery of coke fines from fly ash by air sparged hydrocyclone flotation. *Fuel* 78, 161-168.

Nollet, H., Roels, M., Lutgen, P., Van der, M.P., Verstraete, W. (2003). Removal of PCBs from wastewater using fly ash. *Chemosphere* 53, 655-665.

Nonavinakere, S., Reed, B.E. (1995). Fly ash enhanced metal removal process. In: Sengupta, A.K, editor. Hazardous and Industrial Wastes. *Proceedings of the 27th Mid-Atlantic Industrial Waste Conference*. Technomart Publishing, 1995, p. 588-594.

Olgun, A., Erdogan, Y., Ayhan, Y., Zeybek, B. (2005). Development of ceramic tiles from coal fly ash and tincal ore waste. *Ceram. Int.* 31, 153-158.

Panday, K.K., Prasad, G., Singh, V.N. (1985). Copper(II) removal from aqueous solutions by fly ash. *Water Res.* 19, 869-873.

Pandey, V.C., Abhilash, P.C., Singh, N. (2009). The Indian perspective of utilizing fly ash in phytoremediation, phytomanagement and biomass production. *J. Environ. Manage.* 90, 2943-2958.

Pandey, V.C., Singh, N. (2010). Impact of fly ash incorporation in soil systems. *Agric. Ecosyst. Environ.* 136, 16-27.

Papandreou, A., Stournaras, C.J., Panias, D. (2007). Copper and cadmium adsorption on pellets made from fired coal fly ash. *J. Hazard. Mater.* 148, 538-547.

Papandreou, A.D., Stournaras, C.J., Panias, D., Paspaliaris, I. (2011). Adsorption of Pb (II), Zn (II) and Cr (III) on coal fly ash porous pellets. *Miner. Eng.* 24, 1495-1501.

Papastefanou, C. (2010). Escaping radioactivity from coal-fired power plants (CPPs) due to coal burning and the associated hazards: a review. *Journal of Environmental Radioactivity*, 101, 191-200.

Park, M., Choi, C.L., Lim, W.T., Kim, M.C., Choi, J., Heo, N.H. (2000). Molten-salt method for the synthesis of zeolitic materials: I. Zeolite formation in alkaline molten-salt system. *Micropor. Mesopor. Mater.* 37, 81-89.

Park, M., Choi, C.L., Lim, W.T., Kim, M.C., Choi, J., Heo, N.H. (2000). Molten-salt method for the synthesis of zeolitic materials: II. Characterization of zeolitic materials. *Micropor. Mesopor. Mater.* 37, 91-98.

Patra, K.C., Rautray, T.R., Nayak, P. (2012). Analysis of grains grown on fly ash treated soils. *Appl. Radiat. Isotopes* 70, 1797-1802.

Pedersen, K.H., Jensen, A.D., Skjøth-Rasmussen, M.S., Dam-Johansen, K. (2008). A review of the interference of carbon containing fly ash with air entrainment in concrete. *Prog. Energ. Combust.* 34, 135-154.

Peng, F., Liang K.M., Hu, A.M. (2005). Nano-crystal glass-ceramics obtained from high alumina coal fly Ash. *Fuel* 2005, 84, 341-343.

Petrus, H.T.B.M., Hirajima, T., Oosako, Y., Nonaka, M., Sasaki, K., Ando, T. (2011). Performance of dry-separation processes in the recovery of cenospheres from fly ash and their implementation in a recovery unit. *Int. J. Miner. Process.* 98, 15-23.

Plank, C.O., Martens, D.C. (1974). Boron availability as influenced by application of fly ash to soil. *Soil Sci. Soc. Am. Proc.* 38, 974-977.

Polat, H., Vengosh, A., Pankratov, I., Polat, M. (2004). A new methodology for removal of boron from water by coal and fly ash. *Desalination* 164, 173-188.

Prashanth, J.P., Sivapullaiah, P.V., Sridharan, A. (2001). Pozzolanic fly ash as hydraulic barrier in land fills. *Eng. Geol.* 60, 245-252.

Qi, L., Yuan, Y. (2011). Characteristics and the behavior in electrostatic precipitators of high-alumina coal fly ash from the Jungar power plant, Inner Mongolia, China. *J. Hazard. Mater.* 192, 222-225.

Qin, J.G. A technique of preparing silicon oxide and alumina with fly ash. Chinese patent 200710061662.X, 2007-04-03.

Querol, X., Alastuey, A., Fernández-Turiel, J., López-Soler, A., (1995). Synthesis of zeolites by alkaline activation of ferro-aluminous fly ash, *Fuel*, 74, 1226-1231.

Querol, X., Alastuey, A., Lopez-Soler, A., Plana, F. (1997). A fast method of recycling fly ash: microwave assisted zeolite synthesis. *Environ. Sci. Technol.* 31, 2527-2533.

Querol, X., Umana, J., Alastuey, A., Lopez-Soler, A., Plana, F. (2001). Extraction of soluble major and trace elements from fly ash in open and closed leaching system. *Fuel* 80, 801-813.

Raask, E. (1968). Cenospheres in pulverized-fuel ash. *Journal of the Institute of Fuel* 9, 339-344.

Ram, L.C., Jha, S.K., Tripathi, R.C., Masto, R.E., Selvi, V.A. (2008). Remediation of fly ash landfills through plantation. *Remediation* 18, 71-90.

Rayalu, S., Meshram, S.U., Hasan, M.Z. (2000). Highly crystalline faujasitic zeolites from fly ash. *J. Hazard. Mater.* 77, 123-131.

Rayalu, S., Udhoji, J.S., Munshi, K.N., Hasan, M.Z. (2001). Highly crystalline zeolite A from fly ash of bituminous and lignite coal combustion. *J. Hazard. Mater.* 88, 107-121.

Riehl, A., Elsass, F., Duplay, J., Huber, F., Trautmann, M. (2010). Changes in soil properties in a fluvisol (calcaric) amended with coal fly ash. *Geoderma,* 155, 67-74.

Roy, G., Joy, V.C. (2011). Dose-related effect of fly ash on on edaphic properties in laterite cropland soil. *Ecot. And Environ. Safety* 74, 769-775.

Saputra, E., Muhammad, S., Sun, H.Q., Anga, H.M., Tadéa, M.O., Wang, S.B. (2012). Red mud and fly ash supported Co catalysts for phenol oxidation. *Catal. Today* 190, 68-72.

Sarkar, A., Rano, R., Mishra, K., Sinha, I. (2005). Particle size distribution profile of some indian fly ash-a comparative study to assess their possible uses. *Fuel Process Technol.* 86, 1221-1238.

Sarmah, M., Baruah, B.P., Khare, P. (2013). A comparison between CO_2 capturing capacities of fly ash based composites of MEA/DMA and DEA/DMA. *Fuel Process. Technol.* 106, 490-497.

Schmucker, M., MacKenzie, K.J.D. (2004). Microstructure of sodium polysialate siloxo geopolymer. *Ceram. Int.* 31, 433-437.

Sears, L. Future trends for PFA in cementitious systems. In: *1st Future cement conference and exhibition.* London Chamber of Commerce and Industry, 2011.

Seidel, A., Sluszny, A., Shelef, G., Zimmels, Y. (1999). Self inhibition of aluminum leaching from coal fly ash by sulfuric acid. *Chem. Eng. J.* 72, 195-207.

Seidel, A., Zemmles, Y. (1998). Mechanism and kinetics of aluminum and iron leaching from coal fly ash by sulfuric acid. *Chem. Eng. Sci.* 53, 3835-3852.

Seoane, S., Leirós, M.C. (2001). Acidification-neutralization processes in a lignite mine spoil amended with fly ash or limestone. *J. Environ. Qual.* 30, 1420-1431.

Shabtai, Y., Mukmenev, I. (1996). A combined chemical-biotechnological treatment of coal fly ash (CFA). *J. Biotechnol.* 51, 209-217.

Sharma, K.S., Kalra, N. (2006). Effect of fly ash incorporation on soil properties and productivity of crops: A review. *J. Scient. Industr. Res.* 65, 383-390.

Shemi, A., Mpana, R.N., Ndlovu, S., van Dyk, L.D., Sibanda, V., Seepe, L. (2012). Alternative techniques for extracting alumina from coal fly ash. *Minerals Engineering* 34, 30-37.

Shigemoto, N., Hayshi, H. (1993). Selective formation of Na-X zeolite from coal fly ash by fusion with sodium prior to hydrothermal reaction, *J. Mater. Sci.* 28, 4781-4786.

Shih, W.H., Chang, H.L. (1996). Conversion of fly ash into zeolites for ion-exchange applications, *Mater. Lett.* 28, 263-268.

Silva, L.F.O., DaBoit, K., Sampaio, CH., Jasper, A., Andrade, M.L., Kostova, I.J., Waanders, F.B., Henke, K.R., Hower, J.C. (2012). The occurrence of hazardous volatile elements and nanoparticles in Bulgarian coal fly ashes and the effect on human health exposure. *Sci. Total Environ.* 416, 513-526.

Singh, N. (2009). Adsorption of herbicides on coal fly ash from aqueous solutions. *J. Hazard. Mater.* 168, 233-237.

Somna, K., Jaturapitakkul, C., Kajitvichyanukul, P., Chindaprasirt, P. (2011). NaOH-activated ground fly ash geopolymer cured at ambient temperature. *Fuel* 90, 2118-2124.

Survey_Goss_FINAL_110310.pdf.

Sushil, S., Batra, V.S. (2006). Analysis of fly ash heavy metal content and disposal in three thermal power plants in India. *Fuel* 85, 2676-2679.

Swanepoel, J.C., Strydom C.A. (2002). Utilisation of fly ash in a geopolymeric material. *Appl. Geochem.* 17, 1143-1148.

Tan, H.B. (2011). Preparation of mullite whiskers from coal fly ash using sodium sulfate flux. *Int. J. Miner. Process.* 100, 188-189.

Tanaka, H., Eguchi, H., Fujimoto, S., Hino, R. (2006). Two-step process for synthesis of a single phase Na-A zeolite from coal fly ash by dialysis. *Fuel* 85, 1329-1334.

Tanaka, H., Sakai, Y., Hino, R. (2002). Formation of Na-A and Na-X zeolites from waste solutions in conversion of coal fly ash to zeolites. *Mater. Res. Bull.* 37, 1873-1884.

Taylor, E.M., Schuman, E.E. (1988). Fly ash and lime amendment of acidic coal spoil to aid vegetation. *J. Environ. Qual.* 17, 120-124.

Taylor, H.F.W. *Cement chemistry*. 2nd ed. Thomas Telford, London, 1997, p. 261-262, 272-280.

Temimi, M., Camps, J.P., Laquerbe, M. (1995). Valorization of fly ash in the cold stabilization of clay materials. *Resour. Conserv. Recy.* 15, 219-234.

Temuujin, J., van Riessen, A., MacKenzie, K.J.D., (2010). Preparation and characterisation of fly ash based geopolymer mortars. *Constr. Build. Mater.* 24, 1906-1910.

The American Coal Ash Association. Coal combustion product (CCP)—production and use survey, 2005, http://www.acaa-usa.org/PDF/2004_CCP_Survey(9-9-05).pdf.

The American Coal Ash Association. CORRECTED 2009 Coal Combustion Product (CCP) Production & Use Survey, 2010, http://www.acaa-usa.org/associations/8003/files/2009_ Production_and_Use_

Tolle, D.A., Arthur, M.F., Pomeroy, S.E. Fly ash use for agriculture and land reclamation: a critical literature review and identification of additional research needs. RP-1224-5. Battelle Columbus Laboratories, Columbus, Ohio, 1982.

Vassilev, S.V., Menendez, R., Alvarez, D., Diaz-Somoano, M., Martinez-Tarazona, M.R. (2003). Phase-mineral and chemical composition of coal fly ashes as a basis for their multicomponent utilization. 1. Characterization of feed coals and fly ashes. *Fuel* 82, 1793-1811.

Vassilev, S.V., Menendez, R., Borrego, A.G., Diaz-Somoano, M., Martinez-Tarazona, M.R. (2004). Phase-mineral and chemical composition of coal fly ashes as a basis for their multicomponent utilization. 3. Characterization of magnetic and char concentrates. *Fuel* 83, 1563-1583.

Vassilev, S.V., Menendez, R., Diaz-Somoano, M., Martinez-Tarazona, R. (2004). Phasemineral and chemical composition of coal fly ashes as a basis for their multicomponent utilization. 2. Characterization of ceramic cenosphere and salt concentrates. *Fuel* 83, 585-603.

Vassilev, S.V., Vassileva, C.G. (1997). Geochemistry of coals, coal ashes and combustion wastes from coal-fired power stations. *Fuel Process. Technol.* 51, 19-45.

Vassilev, S.V., Vassileva, C.G. (2005). Methods for characterization of composition of fly ashes from coal-fired power stations: a critical overview. *Energ. Fuel.* 19, 1084-1098.

Vassilev, S.V., Vassileva, C.G. (2007). A new approach for the classification of coal fly ashes based on their origin, composition, properties, and behaviour. *Fuel* 86, 1490-1512.

Vassilev, S.V., Vassileva, C.G., Karayigit, A.I., Bulut, Y., Alastuey, A., Querol, X. (2005). Phase-mineral and chemical composition of composite samples from feed coals, bottom ashes and fly ashes at the Soma Power Station, Turkey. *Int. J. Coal Geol.* 61, 35-63.

Wang, B., Li, Q., Wang, W., Li, Y., Zhai, J.P. (2011). Preparation and characterization of Fe^{3+}-doped TiO_2 on fly ash cenospheres for photocatalytic application. *Appl. Surf. Sci.* 257, 3473-3479.

Wang, C.F., Li, J.S., Wang, L.J., Sun, X.Y. (2008) Influence of NaOH concentrations on synthesis of pure-form zeolite A from fly ash using two-stage method. *J. Hazard. Mater.* 155, 58 64.

Wang, C.F., Li, J.S., Wang, L.J., Sun, X.Y. (2008). Study on adsorption of Cr (VI) using single-phase zeolites synthesized from fly ash. *Chinese Journal of Environmental Engineering* 2, 1121-1126.

Wang, L., Ma, H.W., Nie, Y.M., Li, H.X. (2006). Experimental research on preparation of high specific surface area SiO_2 from fly ash. *Bulletin of the Chinese Ceramic Society*. 25, 3-7.

Wang, S. (2008). Application of solid ash based catalysts in heterogeneous catalysis. *Environ. Sci. Technol.* 42, 7055-7063.

Wang, S., Lu, G.Q. (2007). Effect of chemical treatment on Ni/fly-ash catalysts in methane reforming with carbon dioxide. In: Noronha, F.B., Schmal, M., Sousa-Aguiar, E.F., editors. *Studies in surface science and catalysis*, vol. 167. Elsevier, 2007, p. 275-280.

Wang, S.B., Li, L., Zhu, Z.H. (2007). Solid-state conversion of fly ash to effective adsorbents for Cu removal from wastewater. *J. Hazard. Mater.* B 139, 254-259.

Wesche, K. (1991). *Fly ash in concrete: properties and performance*, 1st ed. E & FN Spoon, London.

Wu, C.Y., Yu H.F., Zhang, H.F. (2012). Extraction of aluminum by pressure acid-leaching method from coal fly ash. *Trans. Nonferrous Met. Soc.* China 22, 2282-2288.

Wu, C.Y., Yu, H.F., Zhang, H.F. (2012). Extracting of aluminum by pressure acid-leaching method from coal fly ash. *Nonferrous Met. Soc. China* 22, 2282-2288.

Xiong, Z.W., Yu, Y.L., You, M., Guo, C.l., Zhou, S.K., Yu, Z.X. (2007). Analysis of environment contamination from concomitant radioactivity of coal mine source. *Journal of China Coal Society* 32, 762-766.

Xu, H., Li, Q., Shen, L.F., Zhang, M.Q., Zhai, J.P. (2010). Low-reactive circulating fluidized bed combustion (CFBC) fly ashes as source material for geopolymer synthesis. *Waste Manage.* 30, 57-62.

Xu, H., van Deventer, J.S.J, Jannie, S.J. (2002). Geopolymerisation of multiple minerals. *Miner. Eng.* 15, 1131-1139.

Xuan, X.P., Yue, C.T., Li, S.Y., Yao, Q. (2003). Selective catalytic reduction of NO by ammonia with fly ash catalyst. *Fuel* 82, 575-579.

Yan, L., Wang, Y., Ma, H., Han, Z., Zhang, Q., Chen, Y. (2012). Feasibility of fly ash-based composite coagulant for coal washing wastewater treatment. *J. Hazard. Mater.* 203-204, 221-228.

Yang, C.H. *A study on the resources recovery of coal fly ash.* Master Dissertation, Cheng Kung University, Taiwan, 2000.

Yao, Z.T., Li, H.Y., Xia M.S. (2009). Hydrothermal synthesis of sodalite from coal fly ash and its property characterization. *The Chinese Journal of Nonferrous Metals* 19, 366-371.

Yao, Z.T., Tan, S.H., Xia, M.S., Ye, Y., Li, J.H. (2011). Synthesis, characterization and sintering behavior of indialite ceramic from fly ash. *Waste Manage. Res.* 29, 1090-1097.

Yao, Z.T., Xia, M.S., Ye, Y. (2011). Dilithium dialuminium trisilicate crystalline phase prepared from coal fly ash. *J. Mater. Eng. Perform.* 21, 877-881.

Yavuz, C.T., Prakash, A., Mayo, J.T., Colvin, V.L. (2009). Magnetic separations: from steel plants to biotechnology. *Chem. Eng. Sci.* 64, 2510-2521.

Ye, Y.P., Zeng, X.Q., Qian, W.L., Wang, M.W. (2008). Synthesis of pure zeolites from supersaturated silicon and aluminum alkali extracts from fused coal fly ash. *Fuel* 87, 1880-1886.

Yunusa, I.A.M., Eamus, D., Silva, D.L., Murray, B.R., Burchett, M.D., Skilbeck, G.C., Heidrich, C. (2006). Fly ash: an exploitable resource for management of Australian agricultural soils. *Fuel* 85, 2337-2344.

Zha F.S., Liu, S.Y., Du, Y.J., Cui, K.R. (2008). Behavior of expansive soils stabilized with fly ash. *Nat. Hazards* 47, 509-523.

Zhang, A.L., Wang, N.N., Zhou, J.T., Jiang, P., Liu, G.F. (2012). Heterogeneous Fenton-like catalytic removal of p-nitrophenol in water using acid-activated fly ash. *J. Hazard. Mater.* 201-202, 68-73.

Zhao, Y.L., Ye, J.W., Lu, X.B., Liu, M.G., Lin, Y., Gong, W.T., Ning, G.L. (2010). Preparation of sintered foam materials by alkali-activated coal fly ash. *J. Hazard. Mater.* 174, 108-112.

Zyryanov, V.V., Petrov, S.A., Matvienko, A.A. (2011). Characterization of spinel and magnetospheres of coal fly ashes collected in power plants in the former USSR. *Fuel* 90, 486-492.

In: Fly Ash
Editor: Prabir Kumar Sarker

ISBN: 978-1-62948-044-2

Chapter 2

EXPERIMENTAL CHARACTERIZATION AND HYDRODYNAMIC BEHAVIOR OF FLY ASH PARTICLES

Javier R. Sánchez,* Audelino Álvaro and Jesús Ma. Rodríguez
University of Salamanca, Department of Chemical and Textile Engineering,
Plaza de los Caídos, Salamanca, Spain

ABSTRACT

Particulate air pollution, especially that due to fine particles, is a matter of concern to the authorities and citizens of many countries. Among the many kinds of particle that can pollute the air here we focus on fly ash, which is produced in large quantities as waste. Such large quantities demand the reuse of the maximum amount possible.

This chapter presents the characterization of fly ash from coal combustion in a power plant, determining the mineralogical composition using X-ray techniques, and also particle shape by SEM microscopy and particle size distribution using a particle analyzer with laser diffraction technology. Also, determination of some important parameters to know its behavior, such as the mean density and the specific surface area of fly ash particles, as well as their terminal velocity are presented.

Following this, calculation of the minimum fluidization velocity for a fluidized bed containing fly ash and sand of larger particle size, considering that the former may be determined to be of the type Geldart-C and the last Geldart-B according to their characteristics and behavior in fluidization is presented. Finally, the elutriation constants of fly ash under the experimental conditions are determined.

These studies are useful to better understand the formation of a fly ash aerosol from a fluidized bed with a view to its use in subsequent studies concerning, for example, the filtration of particles carried in an aerosol. Regarding this issue we offer some comments at the end of the chapter.

Keywords: Fly ash; XRD; SEM; Fluidized bed; Elutriation; Fabric filters

* E-mail: jrsm@usal.es.

INTRODUCTION

Coal is a natural solid fossil fuel that comes from the decomposition of plant material and that can be used to generate electrical and thermal energy. However, while such phenomena are occurring, a large amount of combustion wastes is produced, such as furnace slag, bottom ash and fly ash, apart from air pollution. Thus, small fly ash particles are especially dangerous because of their low sedimentation rate, and they are able to remain in suspension in the air we breathe, submicron-sized particles being particularly dangerous. These small particles may also result in decreased visibility in some areas and cause damage to trees and plants owing to particle deposition.

Coal combustion products (CCP) include combustion residues such as fly ash and bottom ash from different types of boilers, boiler slag, and desulfuration products (Feuerborn, 2005). Fly ash is generally captured by electrostatic precipitators or other particle filtration devices before the flue gases reach the chimneys of coal-fired power plants (Wikipedia, n.d.).

The US production of CCPs was over 130 million short tons ($1.18 \cdot 10^{11}$ kg) in 2011, of which around 43% was profitably utilized. Fly ash contributed 60 million short tons ($5.44 \cdot 10^{10}$ kg), about 46% of the total CCP, of which a little more than 23 million tons ($2.09 \cdot 10^{10}$ kg), about 38%, were used in different applications, mainly (about half) as a cement or concrete additive (American Coal Ash Association, ACAA, 2011).

The treatment of smoke containing fly ash is not an easy task because of the high temperature at which the combustion and gasification processes take place. Accordingly, it is difficult to find a single procedure for gas cleaning with appropriate efficiency. The main use of electrostatic precipitators (ESP) is to reduce fly ash emissions from industrial boilers fired with lignite or bituminous coal. Sometimes an ESP can be coupled with the flue gas desulfuration and denitrification systems (Jedrusik, Swierczok, & Teisseyre, 2003, p. 295). Also, when the ESP is not capable of removing the finest particles, a baghouse can be used immediately after the ESP system, among other possibilities. We believe the above circumstances to be sufficient to justify a study of fly ash particles, most of which are spherical, smooth and small in size.

Currently, many applications for fly ash exist, among them the addition to cement and concrete products, addition to construction materials as a light-weight aggregate, structural filling and covering material, use on roads and pavements..., as mentioned by Ahmaruzzaman (2010, p. 331). Fly ash can also be used as a potential adsorbent and material for conversion into substances with a zeolite-type crystalline structure (Sarbak, Stanczyk, & Kramer, 2004, p. 82). It can also be added to the soil, improving the physical, chemical and biological properties of degraded soils, and may even increase soil fertility through increases in nutrients (Pandey and Sing, 2010). However, only a relatively small percentage of the material is used in such applications and hence a large part of the material generated each year is held in ash dams or similar dumps (Iyer and Scott, 2001, p. 217).

1. CHARACTERIZATION OF FLY ASH

The fly ash used in the experimental work came from coal combustion in a Spanish power plant and was collected with an electrostatic precipitator. When it is desired to characterize a type of solid particle it is necessary to determine several parameters, some of which are interrelated. Among the most useful, these parameters include the solid composition and the shape, size, density, specific surface area and terminal velocity of the particles.

The ASTM classifies fly ash in the C618 standard specification based on content of SiO_2+Al_2O_3+Fe_2O_3 (Standars, ASTM, 2003). When this content is greater than 70wt% the fly ash belongs to the F class, and is low in lime. If it is between 50 and 70% it is of class C. Low-calcium Class F fly ash is commonly produced from the burning of higher-ranking coals, i.e., bituminous coals or anthracites (Ahmaruzzaman, 2010, p. 330).

Prior to the determination of these properties, it is necessary to select a representative sample of the ash, using standard procedures. There are many possibilities, among which is the grid method (Hammerly, Marracino, & Piagenti, 1984, pp. 6-19) -which follows a zigzag pattern-, the quartering method, and others (Bermejo, 1981, pp. 31-49).

We chose the first one, which consisted in having enough solid on a flat smooth surface for this to be divided into smaller rectangles, forming a grid, after which the samples were chosen, taking care to collect them from the whole thickness of the fly ash at each sampling point. Once we had a representative sample of the ash, we proceeded to its characterization.

1.1. Qualitative and Quantitative Composition

Fly ash particles are abrasive, refractory in nature and highly contaminating, since they carry potentially toxic trace elements which condense from the flue gas, and their color may vary from tan to gray to black, depending on the amount of unburned carbon in the ash (Ahmaruzzaman, 2010, pp. 328-330). The color of the fly ash used by us was a medium grey, indicating the presence of a low amount of unburned carbon.

As mentioned by Sarkar, Rano, Udaybhanu and Basu (2006, p. 259), "the chemical composition of fly ash varies depending upon the kind of coal used in the combustion, the combustion conditions and the removal efficiency of air pollution control devices". The composition of the fly ash studied here was determined with an X-ray diffractometer (XRD), which is the tool most widely used to determine the mineralogy of fly ash. We performed the measurements with a Siemens diffractometer, which has a Difract-AT system and a DACO-MP microprocessor. A one-degree grid and a scanning velocity of one degree per minute were used during the measurements.

The results obtained using this technique are shown in Figure 1, in which the diffraction angle appears on the x-axis and the diffracted beam intensity on the y-axis, this latter measured in counts per second (cps). This figure shows the major components found in the samples analyzed. We found that the fly ash contained mainly SiO_2 as quartz and crystalline mullite forms, with a lower proportion of iron oxide, hematite and magnetite, as well as calcium oxide. The major peaks, which are characteristic of quartz and mullite, indicate the high proportion of these compounds in the fly ash in comparison with the other components.

The presence of other metals, whose signals are difficult to identify in this diffractogram, was confirmed by other techniques such as Energy-Dispersive X-Ray Spectroscopy (EDX). This method allowed us to obtain the quantitative composition of the fly ash characterized, expressed as a percentage of the oxide considered, as shown in Table 1.

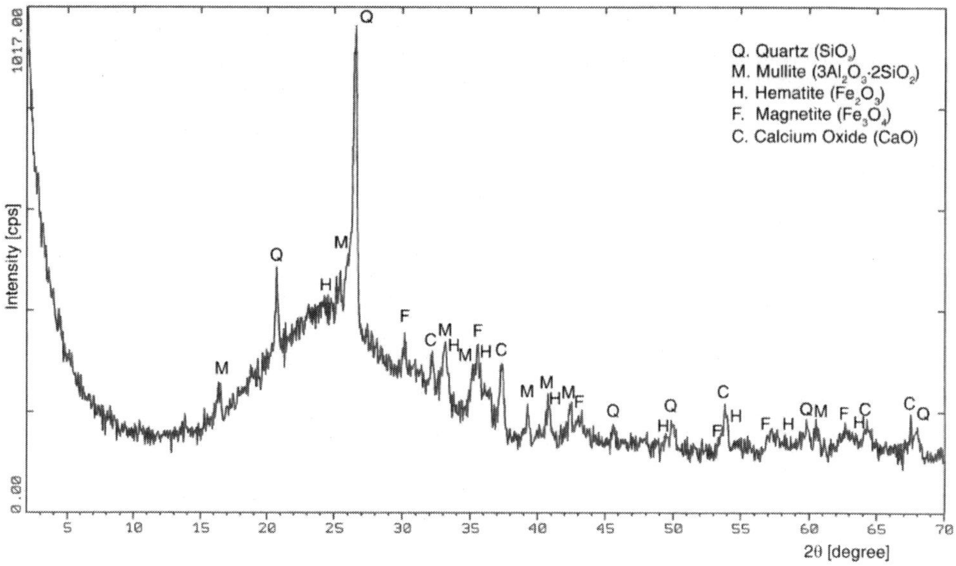

Figure 1. X-Ray Diffractogram (XRD) of a fly ash sample.

Table 1. Chemical composition of fly ash in wt% of major oxides

SiO_2	Al_2O_3	Fe_2O_3	CaO	K_2O	MgO	TiO_2	SO_3	unburned carbon
43-45	21-22	12-13	8-10	4-5	1-2	0-1	1-3	up to 5%

The values shown in Table 1 are consistent with the normal range of chemical composition for the low-calcium Class F fly ash, as discussed by Ahmaruzzaman (2010).

1.2. Determination of Particle Shape and Size Distribution of Fly Ash

The fly ash was previously sieved with a 63-μm mesh, mainly to remove any coarse particles and possible unburned material.

With SEM photography we noted that the fly ash particles were small, predominantly spherical in shape, and had a flat surface, as seen in Figure 2.

The predominantly spherical shape of fly ash is widely documented in the literature. Indeed, although they may display different morphologies, it is well known that, regardless of their origin, particles generated in the combustion of pulverized coal are mostly solid spheres as a result of the melting of mineral matter (R. Barea, 1998, pp. 3.26 and 3.31).

However, the fluidized beds used later were formed of two kinds of particles, fly ash particles and sand particles, the latter sieved to 149-177 μm. The particle size distribution of these two kinds of particle was measured using a Microtrac particle analyzer (Leeds & Northrup, 1993), model SRA 150, provided with a laser beam and a gas cell. The measurements are shown as the % cumulative mass fraction in Figure 3.

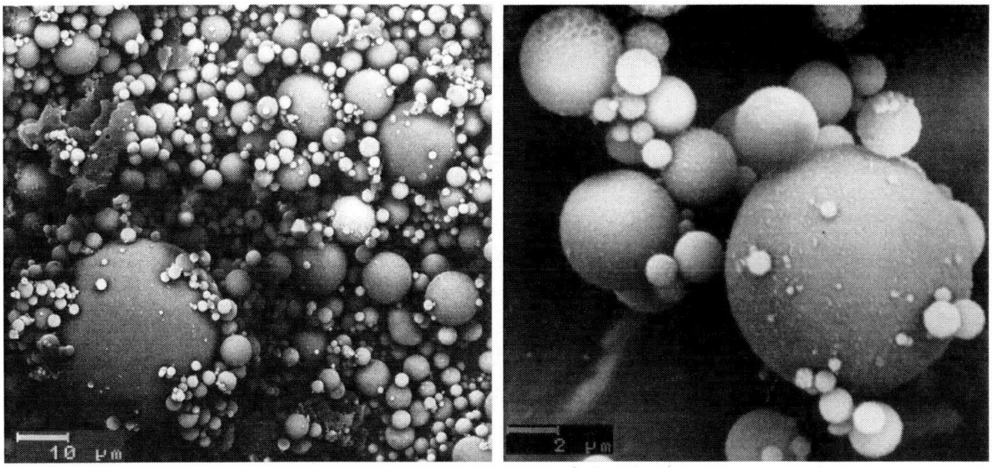

Figure 2. Scanning electron micrograph (SEM) of the fly ash used (left: x1000; right: x5000).

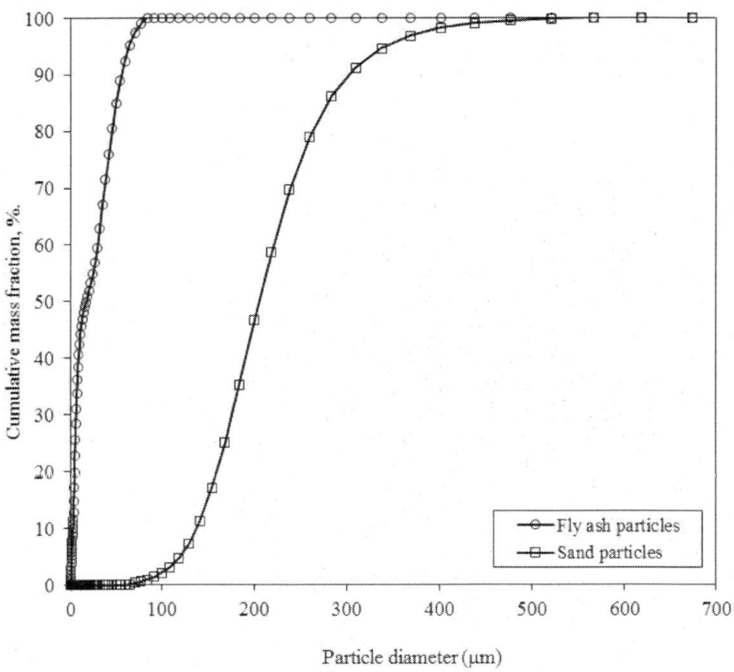

Figure 3. Initial fly ash and sand particle size distributions.

With the experimental data obtained from the graph, the mean diameters of fly ash were calculated, obtaining: $\overline{d}_{PS}$ = 6.8 μm and = 1.2 μm. The same was the case of the sand particles, yielding a mean diameter of $\overline{d}_{PS}$ = 202.5 μm and = 158.6 μm.

According to Arnaldos, Coll, Llop and Perales (1990), surface diameter, $\overline{d}_{PS}$, is the parameter most used since it is based on specific surface area, which is the parameter that best characterizes a system of particles with respect to its interaction with a fluid.

1.3. Determination of Density and Specific Surface Area of the Fly Ash Particles

Calculations of the density of solid particles involve no drawbacks when dealing with non-porous particles. Indeed, if the whole volume of the particle is filled by the solid both densities coincide, ρ_p.

The methodologies based on the displacement of a fluid using a pycnometer of mercury, water or another liquid are easily applicable techniques, satisfactory results being obtained for density, ρ_p (Arnaldos et al., 1990, p. 195). However, there are other methods for determining the density of porous particles, for example those based on the gas pycnometry technique or on the determination of the minimum fluidization velocity and porosity of a fluidized bed consisting of the particles under study and subsequent application of the Ergun equation (Margiatto & Siegell, 1983).

Since fly ash particles, at least the majority of them, are not porous, we chose the pycnometry procedure to determine the mean density. Accordingly, a small pycnometer (25 ml) and distilled water were used, applying ultrasound for better disaggregation of the fly ash particles in the water.

Thus, we obtained a mean density for fly ash particles, ρ_p, of 2400 kg/m^3, which is in agreement with the bibliography (Ahmaruzzaman, 2010, p. 330), since the values assigned to fly ash usually range from 2100 to 3000 kg/m^3. With this density and the mean particle diameter obtained previously it may be seen that this powder belongs to group C of Geldart's classification.

We attempted to obtain the specific surface of the fly ash particles using the nitrogen adsorption-desorption method at low temperature (BET method) (Sing et al., 1985), but the results were unsatisfactory. This could be explained in terms of the almost complete absence of pores on the sphere surface, which is practically flat, as shown in the SEM images (Figure 2).

In view of the above, laminar flow methods can be used. These only measure the surface area contributing to the pressure drop (Arnaldos et al., 1990, p. 194), which may be termed the external surface area and can be calculated as follows:

$$\text{Specific external surface} = \bar{a}_p = \frac{\text{Specific external surface area of the particle}}{\text{Particle volume}} \quad (1)$$

This is the specific surface area per unit volume of solid, in m^2/m^3.

The particle diameter of initial fly ash can vary between about 0.7 μm and slightly less than 90 μm, as can be seen in Figure 3. If we consider the particle's density as constant and equal to the mean density, calculated previously, and a mean superficial diameter of $\bar{d}_{PS} = 6.8$ μm, we would have that the mean specific surface area of the spherical fly ash particles could change as follows:

$$\bar{a}_p = \frac{4 \cdot \pi \cdot (\bar{d}_{PS}/2)^2}{(4/3) \cdot \pi \cdot (\bar{d}_{PS}/2)^3} = \frac{6}{\bar{d}_{PS}} = 8.82 \cdot 10^5 \, m^2/m^3 \quad (2)$$

The specific surface area, $\bar{a}'_p$, is also defined as surface area per gram of particles (Hinds, 1982, p. 103), therefore:

$$\bar{a}'_p = \frac{6}{\bar{d}_{PS}} \cdot \frac{1}{\rho_p} = \frac{6/(6.8 \cdot 10^{-6})\,\mathrm{m}^{-1}}{2.4 \cdot 10^3\ \mathrm{kg \cdot m^{-3}}} = 367.6\,\mathrm{m^2/kg} \tag{3}$$

And for nonspherical particles the sphericity factor (ϕ_S) must be added in the denominator.

This is only a mean value, but it would be necessary to consider that the specific surface area increases as the particle size decreases, and vice versa.

Ahmaruzzaman (2010) states that the surface area can change from 170 to 1000 $\mathrm{m^2/kg}$ for this kind of particle and the value obtained here lying within this range.

1.4. Terminal Velocity of Particles

The terminal (settling) velocity of an airborne particle is an important parameter for characterizing the settling behavior of the particle.

Baer (n.d.) suggests that when a particle is placed in a fluid and it begins to fall and settle by gravity, the drag force from the fluid originates some resistance. When these two forces (the drag force and gravity) become equal, the rate of settling is known as the terminal velocity. The terminal velocity of a particle is the point at which the velocity of a falling particle in a fluid no longer changes.

There are several procedures that can be used to calculate the mean terminal velocity of particles, u_{TS}. Kunii and Levenspiel (1991, pp. 80-81) describe a process in which the following equations are used:

$$u_{TS} = u_t^* \left[\frac{\mu(\rho_p - \rho) \cdot g}{\rho^2} \right]^{1/3} \tag{4}$$

$$u_t^* = \left[\frac{18}{\left(d_p^*\right)^2} + \frac{2.335 - 1.744 \cdot \phi_s}{\left(d_p^*\right)^{0,5}} \right]^{-1} \tag{5}$$

$$d_p^* = d_p \left[\frac{\rho(\rho_p - \rho) \cdot g}{\mu^2} \right]^{1/3} \tag{6}$$

u_t^* and d_p^* are dimensionless variables, being used in the calculation of u_{TS} and obtained according to equations (5) and (6); ϕ_s is the sphericity factor, dimensionless; ρ and μ are the density and viscosity of the gaseous fluid, respectively; ρ_p is particle density, and g is the acceleration of gravity.

We considered the density and viscosity of air at room temperature, a sphericity factor, ϕ_s, equal to one, due to the spherical shape of most fly ash particles, and a mean particle diameter $\overline{d}_{PS} = 6.8$ μm, as reported above.

The procedure was as follows: First we calculated d_p^* using equation (6) and then we obtained u_t^* using equation (5). Finally, we obtained the value of the terminal velocity for fly ash particles by solving equation (4), obtaining here a value of $u_{TS} = 0.336$ cm/s. This latter must be considered a mean value that decreases for smaller particles and increases for larger ones.

If the fluid velocity exceeds the terminal velocity, then the particles will remain entrained in the fluid and move upward with the fluid (Lupo and Tremblay, 2003, p. 103).

2. Hydrodynamic Behavior of Fly Ash

Study of the hydrodynamic behavior of particles in fluidized beds is important to understand the processes in which powders aimed at forming aerosols are used.

Fluidization is a process in which an upward fluid stream (in this case air) is used to suspend solid particles, which then also behave as a fluid. The set of fluidized particles is called fluidized bed.

Fluidization processes are successfully used in many industrial applications to enhance heat and mass transfer across beds, such as cement clinker production, uranium processing (Perry and Green, 1997), coal combustion and coal gasification in fluidized beds, fluidized bed bioreactors, and others.

The hydrodynamic behaviour of fluidized beds depends on the distributor chosen, on the properties of the gas and those of the solids and on the interparticle forces (Gupta et al., 2009, p.263).

Werther and Hartge (2003, p. 122) report that two terms are often used for the flow of solids out of fluidized beds: entrainment and elutriation. Entrainment refers to the flux of solids carried out of the fluidized bed by the gas. Elutriation refers to the classifying effect of the fluidized bed entrainment and it characterizes the selective removal of particles of individual size from the fluidized bed.

According to Wen and Hashinger (1960, p. 220), elutriation is a process involving the continuous removal of smaller particles coming from a bed with a broad range of particle sizes by means of a fluid current. The elutriation speed can be expressed as a first-order reaction that defines a constant, the so-called constant of elutriation, that has often been addressed in the literature. The behavior of this constant with different variables has also been reported by different authors using different methods to assess it.

In 1951 Leva found that the elutriation rate constant is only slightly dependent on composition and composite size of the large component in the bed. Osberg and Charlesworth (1951) found that only at 5% below the concentration of fines was a significant effect observed. On the other hand, Wen and Hashinger (1960, p. 226) stated that in a two-particle system the initial fines concentration seemed to have no effect on the elutriation constant up to a concentration of approximately 25%; above this, the constant tended to decrease as the concentration increased.

In a classical work, Geldart (1973) studied the fluidization of some kinds of solids based on their sizes and densities, classifying the particles according to their behavior in four groups, A, B, C and D. Geldart-C particles are the most difficult to fluidize, since they correspond to very fine cohesive powders, which tend to clump together owing to interparticle forces and hence form channels and agglomerates, leading to poor or even absent fluidization (Nguyen et al., as cited in Zhang et al., 2012, p. 235). The difficulty involved in the experimental treatment for the elutriation of small particles is why there are relatively few studies addressing this issue, although in the last two decades more have appeared.

According to Saleh, Cami, Thomas and Guigonet (2006, p. 135) there are several methods to improve fluidization processes in the case of Geldart-C particles: the use of fluidization additives, vibration, agitation, acoustic fields, magnetic fields, the addition of easy-to-fluidize particles (generally powders belonging to Geldart class B, having a size between 100 and about 1000 μm), among others. The addition of this latter type of product to class C powders improves fluidization due to the buffeting action of the larger particles which can be thought of as acting like a mechanical agitator.

Briens, Bergougnou, Inculet, Baron, and Hazlett, (1992, p. 57) have reported that the smallest particles are not the ones most easily elutriated from fluidized beds of mixed-sized particles. These investigators attributed this result to the agglomeration of fine particles with larger particles and showed that this phenomenon was not caused by electrostatic forces. In the same way Baeyens, Geldart, and Wu (1992, p. 76) worked with mixtures formed by adding Geldart type-C powders to group A particles and suggested that it is more likely that the finest particles adhere to each other to form agglomerates having, effectively, a larger size, or that they adhere to larger particles in the powder.

Liu and Kimura (1993, p. 190) found similar results. Kunii and Levenspiel (1991, p. 78) also reported that one way of processing these solids was to introduce them into a bed of the some material but with larger particle sizes, preferably Geldart B. This was the route selected here by us for experiments in fluidization, introducing sand into the bed.

It is also known that the addition of finer particles may improve the fluidization and handling of fine group C powders, although the mechanism underlying such improvement is not clear (Zhu, 2004, p. 286).

2.1. Minimum Fluidization Velocity

One of the most characteristic parameters of a fluidized bed is the minimum fluidization velocity, u_{mf}, whose calculation has been dealt with profusely in the literature (Kunii and Levenspiel, 1991; Lucas and Casal, 1984; Casal and Puigjaner, 1979; Wilkinson, 1995 ;...), and here we calculated it experimentally for the bedding used in the present work.

The aerosol was generated by the elutriation of fly ash in a fluidized bed, using air at ambient temperature as a fluidizing agent. All the particles were dried at 120°C before use.

The fluidization and elutriation tests were performed at room temperature in the set-up shown in Figure 4. The experimental procedure was as follows: fluidizing air from a compressor (1), previously dried by silica gel (2), was passed through a gas meter (3), allowing us to accurately determine the volumetric flow. Then, air was introduced in a 44-mm i.d. transparent cylindrical column, which was grounded, and its flow rate was adjusted via the corresponding valves until the desired superficial air velocity was obtained. Manometers,

M, were used to indicate the air pressure and thermometers, T, to determine temperature. The distributor (4) supported the bed (5), which was fluidized, and the gas leaving the top of the bed carried entrained particles with it, this being the beginning of elutriation. The elutriated particles were channeled through a tapered part (6), followed by an antistatic tube to a particle analyzer (7), in which the carry-over of fly ash particles was analyzed on-line, obtaining their particle size distribution. Then, they were collected in a filter of very small pore size (8), called the absolute filter, after which the air stream was passed through a rotameter (9) and a vacuum pump (10), this latter to avoid excessive increases in the line pressure.

The cumulative mass elutriated was determined by weighing the filter (8) before and after collecting the elutriated particles.

Silica gel drying was carried out to eliminate, or at least reduce, the influence of humidity changes on the elutriation because fly ash displays a significant degree of hygroscopicity, which would further increase the chances of interparticle adhesion.

Initially we prepared a bed of 100 g the fly ash itself, to ensure the impossibility of the fluidization only of Geldart-C type particles, in agreement with the literature. After several tests at different speeds it was observed that a kind of compact "stopper" was formed, floating up the freeboard. This "stopper" gave off irregular clusters of mixed-size particles, which then fell back to the bed. According to the criteria commonly used to describe this phenomenon, this cannot be considered fluidization with elutriation,

As a result, we decided to prepare a bed of 100 g, mixing the fly ash (9%) with sand (91%), this latter with a $\bar{d}_{PS}$ = 202.5 μm according to the particle analyzer. Its sphericity factor was ϕs = 0.40 and its terminal velocity u_{TS} = 0.91 m/s, calculated via equation (4). The characteristics of the sand allowed it to be classified within the Geldart-B type for fluidization. In this way, fluidization plus elutriation were possible, as we observed experimentally. The initial bed height was 54 mm.

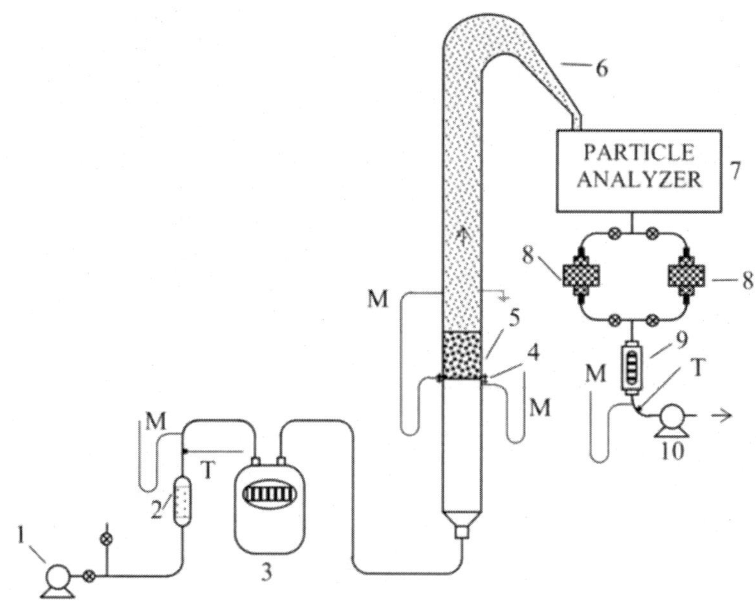

Figure 4. Experimental set-up for the fluidization and elutriation experiments.

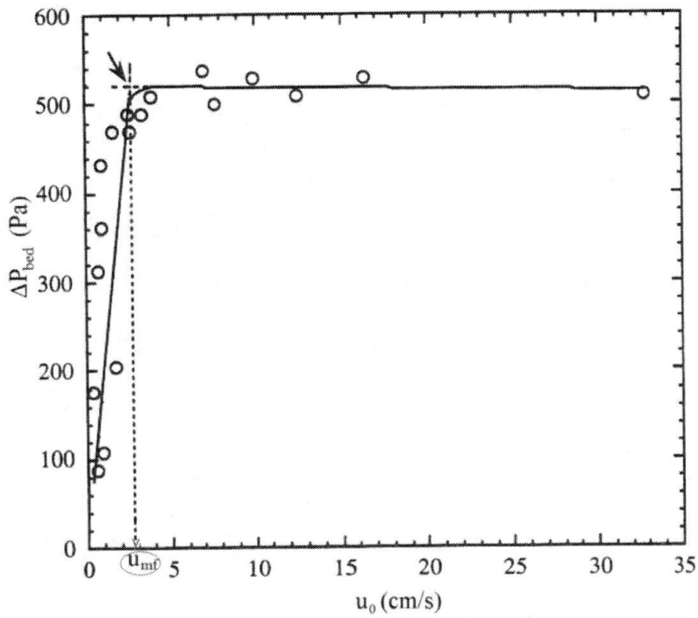

Figure 5. Graphical estimation of the minimum fluidization velocity of the sand and fly ash bed.

The significant difference between the terminal velocity of fly ash and sand indicated that the elutriated fly ash did not entrain the sand particles.

Several procedures exist for calculating the minimum fluidization velocity. In this sense, Gupta et al. (2009) listed seventy-nine correlations reported in the literature and they employed them to predict the minimum fluidization velocities for some selected materials.

However, the minimum fluidization velocity, u_{mf}, can be estimated approximately, as seen in Figure 5 graphically (see Kunii and Levenspiel, 1991, pp. 68-74). This figure shows that a progressive increase in the pressure drop can be observed with increasing superficial gas velocity. Saleh et al. (2006, p. 135) reported that regarding the superficial gas velocity, u_0, two regimes could be distinguished: "a fixed-bed region, where the pressure drop varied directly with the gas velocity, and a fluidized-bed zone, where the pressure drop remained constant. The transition velocity between these two regimes is known as the minimum fluidization velocity, u_{mf}". Here, a value of 2.74 cm/s was observed in Figure 5.

Ma and Kato (1998, p. 95) have indicated that upon increasing the concentration of fines in the bed, the u_{mf} of the mixture decreased.

2.2. Elutriation

When seeking a model of the physical behavior of a fluidized bed in the freeboard it is necessary to understand the entrainment dynamics of particles in this area. There are many studies addressing drag and elutriation, in which a large number of correlations have been proposed to calculate elutriation constants. These correlations have been listed by many different authors, such as Kunii and Levenspiel (1991), Baeyens et al. (1992), Wen and Chen (1982), Colakyan and Levenspiel (1984), Li and Kato (2001),

The problem is that these correlations were obtained under very different operating conditions, most of them using small-scale fluidized beds. Thus, extrapolation of the above empirical correlations to larger scales sometimes provides strange results. Li and Kato (2001, p. 209) reported that owing to "differences in the experimental conditions, investigators have obtained different results; the experimental conditions include column diameter, column height above the distributor, bed weight, particle sizes and distribution, type of particles, superficial gas velocity and gas humidity". The particle sizes of the fine fraction are especially important for the elutriation of particles from fluidized beds. However, according Leva (1951, p. 43), the large-sized fraction has comparatively little effect upon elutriation rates.

Geldart, Harnby, and Wong (1984, p. 33) reported that elutriation from fluidized beds is complex and that the mechanisms are poorly understood, even in the case of free-flowing solids. "They include the ejection of solids into the freeboard by bubbles, hydrodynamic interparticle effects, collision and momentum interchange, agglomeration and wall effects".

Colakyan and Levenspiel (1984, pp. 224-225) published a study on beds composed of coarse non-elutriable solids and a small amount of fines, never higher than 10%. Air at room temperature was used as the fluidizing gas. According to these authors, the elutriation of particles of size-i in a fluidized bed can be determined from the following expression:

$$-\frac{dX_i}{dt} = \frac{K_i^* S \cdot X_i}{W} = K_i X_i \tag{7}$$

K_i* and K_i are two forms of the elutriation rate constant for particles of size-i; X_i is the mass fraction of particles of size-*i* in the bed; *W* the total bed weight, and *S* is the cross-sectional area of the bed.

By integrating this equation 7, and assuming that the mass of solids in the bed remains approximately constant, we have:

$$W_{ie} = W_{i0}(1 - e^{-K_i t}) \tag{8}$$

W_{i0} is the initial mass of elutriable fine particles of size-i in the bed, and W_{ie} is the cumulative mass of size-i fly ash particles elutriated from the bed up to time t.

This would be so considering that no fragmentation occurs in the sand particles. To achieve this, we fluidized the sand bed previously for two hours and then sieved. In fact, we did not observe sand particles in SEM photos of the elutriated particles.

Experiments leading to the determination of the fly ash elutriation rate constants were carried out, attempting to overcome the difficulties of controlling the elutriation of the Geldart-C particles. Extensive studies have been performed on the elutriation of particles from groups A or B, but Li and Kato (2001, p. 209) state that the application of the empirical correlations used for these groups in the elutriation of group C particles usually affords unexpected results.

It was considered advisable to choose a small air superficial velocity so that elutriation would not occur too fast, because fly ash particles have a very low terminal velocity. Therefore, we gradually increase the superficial velocity up to 0.058 m/s, which afforded good results.

The constants K_i* obtained under the experimental conditions used by us (see equations 7 and 8) are depicted in Figure 6. Figure 6 shows two maxima for K_i*: one between 2 and 3 μm and another one between 13 and 16 μm, which could be interpreted as a greater ease of elutriation for these particle sizes at the superficial velocity at which the air tests were performed. This does not mean that smaller particles are not elutriated, but K_i* decreases when a size of 0.7 μm is approached, the lowest detection limit of our particle analyzer. Since K_i* is measured in $kg{\cdot}m^{-2}{\cdot}s^{-1}$ and these particles weigh very little, the K_i^* obtained will be small.

It was observed that under these experimental conditions the fly ash particles smaller than 28 μm were elutriated, which agrees with the diameter at which the particle terminal velocity is higher than the superficial velocity used here.

Geldart and Wong (1987) found that the entrainment flux of alumina, etc. first increases as the powder becomes finer and then decreases. They explained in qualitative terms the low entrainment rate of cohesive powders as being due to the presence of strong interparticle forces.

Finally, the experimental values of the K_i* elutriation rate constant versus particle size obtained here were compared with the values obtained via the different empirical correlations reported in the literature, in a similar way to Rodríguez, Sánchez, Álvaro, Florea, and Estévez (2000, p. 227) with iron oxide particles. The experimental results do not fit these correlations well, which is not surprising since they were proposed for particles in groups A or B. It is seen that the present experimental results fit the modified Yagi and Aochi correlation (equation 9) quite well when the two coefficients of this equation are amended.

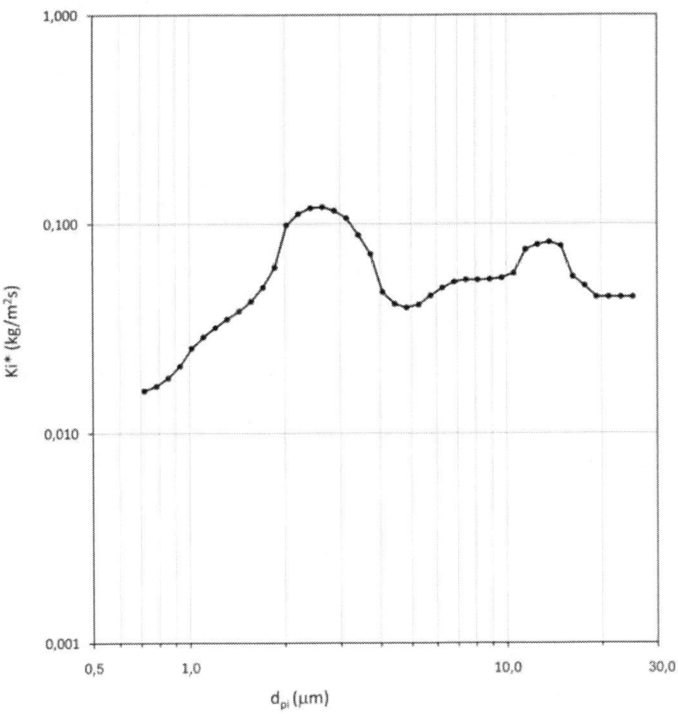

Figure 6. Variation in the elutriation rate constant of fly ash, K_i *, versus particle diameter ($u_0 = 0.058$ m/s).

$$\frac{K_i^* d_{pi}^2 g}{\mu(u_0 - u_{ti})^2} = 0{,}006\left(\frac{d_{pi} u_{ti} \rho}{\mu}\right)^{0{,}6} + \left(\frac{d_{pi} u_{ti} \rho}{\mu}\right)^{1{,}2} \tag{9}$$

Massimilla and Salatino (1987) state these coefficients depend on the nature of the particles.

These studies are useful to gain a better understanding of the generation of fly ash from a fluidized bed with a view to use in subsequent studies addressing, for example, the filtration of such particles.

3. Some Considerations on the Capture of Fly Ash Using Fabric Filters

Solid contaminant particles exhibit a large variety of sizes and a varied chemical constitution. According to Dibbs and Marier (1975), particles smaller than 1 μm in diameter are particularly harmful to human health because they can penetrate to the alveolar region of the lung where air-blood contact occurs. Moreover, these submicron-sized particles can stabilize as natural fogs and may aggravate their adverse effects on human health, crops and buildings, and they are very difficult to separate effectively with conventional commercially available equipment. However, there are different ways that can be employed to remove them more effectively (Stairmand, 1975).

In general, particle removal from gaseous streams can be performed by several operations of solid-gas separation, filtration being one of the methods most used, especially for fines, although there are other procedures. In all of them it is necessary to take into account that the most important feature in removing the particles is their size, and that the removal efficiency of different devices decreases with decreasing particle diameter (Nonhebel, 1972, p. 373).

Different techniques and devices are used to prevent the emission of particles into the atmosphere, and these have been reviewed by several authors, such as Sánchez, Estévez, Jaraíz, Rodríguez, and Macías (1994) and Sánchez, Álvaro, Rodríguez, and Estévez (2000). According to Phillips (2000), there are about 40 different types of gas-cleaning devices on the market and, based on shared characteristics, they can be grouped into five main types: mist eliminators, cyclones, wet dedusters, filters and electrostatic precipitators.

Fibrous filters are widely used for the high-efficiency cleaning of air containing suspended particles (Kirsh, 2007). Filter bags composed of filter fabrics are typically used for the collection of small particles and the type of filter fabric is critical for the effectiveness of the filter bag (Lin, Lou, Lei, & Lin, 2006).

In the industrial filtration of aerosols bearing solid particles with fabrics, three stages can be distinguished: first, filtration with a new fabric, which occurs only once in the useful life of the fabric and lasts a short time; second, the establishment of a residual layer of dust inside the fabric, which occurs after many cycles of filtration-cleaning; and third, permanent filtration (steady state), a stage in which the quantity of dust eliminated in each cleaning on average equals the quantity accumulated in each cycle.

Filter efficiency tends to be low until a cake has formed on the surface of the filter fabric. This cake is initially loose but strongly influences the separation of fine particles. However, the fabric serves not only as a base for the cake, but also plays a fundamental role in its formation and properties. Thus, it is known that the dust filter-cake formed is strongly dependent on the structure of the filter medium (Stöcklmayer, Krammer, & Höflinger, 1996) and the fabric structure has a significant influence on both collection efficiency and the pressure drop through the filters (Humphries, 1998, p. 247).

The textile materials most widely used for gas filtration are the so-called non-woven fabrics, in which the fibers are interlinked with one another to form a web, which is consolidated via different procedures. Among them, the needlefelt and melt-blown types are important. Sometimes, the needlefelt type is even reinforced by the introduction of a fabric inserted in intermediate layers (Sánchez, Rodríguez, Álvaro, & Estévez, 2010, p. 18).

Recently, interest in fibers with non-circular cross-sections for use in filters has increased. Thus, Sánchez, Rodríguez, Álvaro and Estévez (2007) have carried out experimental studies using non-woven fabric filters in order to assess their collection efficiency and pressure drop. They worked with fly ash particles carried by an aerosol and used different experimental variables, such as the superficial aerosol velocity, the particle size of the fly ash dust, the shape of the fiber cross-section and the thickness of the fibers making up the fabric filters. Figure 7 shows a SEM photo of a polyimide fabric filter used, showing the trilobal fibers. It is also possible to see the round shape and the small size of the fly ash particles. However, despite numerous studies carried out to calculate the efficiency and pressure drop of filters made with circular fibers, there are few experimental studies about trilobal fibers and still fewer theoretical studies. Regarding the latter, in a recent paper Fotovati, Tafreshi, and Pourdeyhimi (2011, pp. 1504-1505) established a definition for an equivalent circular fiber that can be used with the traditional filtration correlations (developed for circular fibers) for predicting the collection efficiency of filters made of trilobal fibers.

These trilobal fibers provide a larger specific surface area than traditional circular ones and therefore, under certain conditions, exhibit some advantages over the circular section in terms of collection efficiency and pressure drop when used for the filtration of gases. For these reasons, among others, they seem to have a good future in this field.

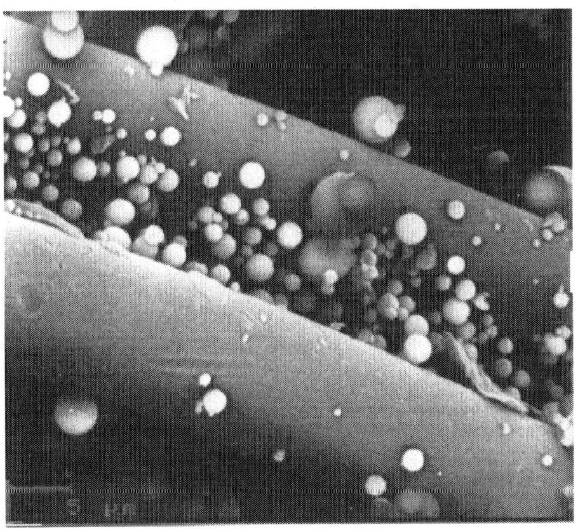

Figure 7. SEM photo of a commercial PI fabric filter (x2000) showing a channel of a trilobal fiber. The round shape of the fly ash particles is seen clearly.

CONCLUSION

A study addressing the characterization and elutriation of fly ash particles from a fluidized bed has been presented in the chapter.

First we characterized the fly ash, determining its mineralogical composition, shape and particle size distribution, density and specific surface area and terminal velocity.

In the characterization we observed that the fly ash consisted mainly of SiO_2 as quartz and crystalline mullite, also detecting other oxides but at much lower proportions. By SEM microscopy we noted that the fly ash particles were predominantly spherical and nonporous. Mean particle size was 6.8 μm, measured as surface diameter, but the majority of the particles were very small, as seen in their mean numerical diameter. We also determined the terminal settling velocity as referred to that mean particle diameter, this velocity proving to be 0.336 cm/s.

A fluidized bed was used to investigate the hydrodynamic behavior of the fly ash. This bed was formed of a mixture of fly ash powder (Geldart-C) and non-elutriable larger sand particles (Geldart-B) owing to the poor fluidization of the fly ash alone. From this bed we obtained the minimum fluidization velocity and elutriation constants. We observed that they afforded elutriated particles smaller than 28 μm under the experimental conditions employed and the particles that were elutriated in the greatest numbers were between 2-3 μm and 13-16 μm in size.

We have corrected the constants in the equation to fit the Yagi behavior of fly ash under our experimental conditions and, finally, we have offered considerations on the filtration of fly ash using non-woven fabric filters.

LIST OF SYMBOLS

$\bar{a}_P$: mean specific surface area per unit volume of solid, in m^2/m^3.

$\bar{a}'_p$: mean specific surface area per unit mass of solid, in m^2/kg.

$\bar{d}_P$: mean diameter for the size distribution actually present in the bed, m.

$\bar{d}_{PN}$: mean numerical diameter of particles, m, $\bar{d}_{PN} = \sum_{i=1}^{n}\left[V_i/d_{pi}^{\ 2}\right] \Big/ \sum_{i=1}^{n}\left[V_i/d_{pi}^{\ 3}\right]$.

$\bar{d}_{PS}$: mean superficial diameter of particles, m, $= \bar{d}_{PS} = 1 \Big/ \sum_{i=1}^{n}\left[X_i/d_{pi}\right]$

d_{pi}: diameter of fly ash particles of size-i, m.

d_p*: dimensionless particle size.

g: gravitational acceleration constant, $m \cdot s^{-2}$.

K_i: elutriation rate constant for particles of size-i, s^{-1}.

K_i^*: elutriation rate constant for particles of size-i, $kg \cdot m^{-2} \cdot s^{-1}$, $K_i^* = K_i \cdot W/A$.

S: cross-sectional area of the bed, m^2.

t: time, s.

u_{mf}: minimum fluidization velocity, $m \cdot s^{-1}$.

u_0: superficial air velocity, $m \cdot s^{-1}$.

u_{ti}: terminal velocity of particles of size-i, $m \cdot s^{-1}$.
u_{TS}: mean terminal velocity of particles, $m \cdot s^{-1}$.
u_t^*: dimensionless particle terminal velocity.
W: total bed weight, kg.
W_i: mass of the particles of size-i remaining in the bed at any time, kg.
W_{ie}: cumulative mass of size-i fly ash particles elutriated from the bed up to time t, kg.
W_{i0}: initial mass of elutriable fine particles of size-i in the bed, kg.
X_i: mass fraction of particles of size-i in the bed.

Greek letters
ρ_p: mean particle density, $kg \cdot m^{-3}$.
ρ: air density, $kg \cdot m^{-3}$.
μ: dynamic air viscosity, $kg \cdot m^{-1} \cdot s^{-1}$.
ϕ_s: sphericity

REFERENCES

Ahmaruzzaman, M. (2010). A review on the utilization of fly ash. *Progress in Energy and Combustion Science*, 36(3), 327-363.

American Coal Ash Association. (n.d.). (2011). *Coal Combustion Product (CCP)* Production & Use Survey Report. Retrieved February 25, 2013, from http://www.acaa-usa.org/associations/8003/files/Final2011CCPSurvey.pdf.

Arnaldos, J., Coll, T., Llop, M.F., & Perales, J.F. (1990). Caracterización de partículas sólidas. *Ingeniería Química*. 22, no 256, 191-198.

Baer, B. (n.d.). *How to Test for the Terminal Velocity of Particulates in Water*. Retrieved March 5, 2013, from http://www.ehow.com/how_12046929_test-terminal-velocity-particulates-water.html.

Baeyens, J. Geldart, D., & Wu, S.Y. (1992) Elutriation of fines from gas fluidized beds of Geldart A-type powders – Effect of adding superfines. *Powder Technology*, 71(1), 71-80.

Bermejo, F. (1981). *Tratado de Química Analítica general, cuantitativa e instrumental*. Madrid: Dossat.

Briens, C.L., Bergougnou, M.A., Inculet, I.I., Baron, T., & Hazlett, J.D. (1992). Size distribution of particles entrained from fluidized beds: Electrostatic effects. *Powder Technology*, 70(1), 57–62.

Casal, J., & Puigjaner, L. (1979) Cálculo de la velocidad mínima de fluidización. *Ingeniería Química*, 11(1), 53-59.

Colakyan, M., & Levenspiel, O. (1984). Elutriation from Fluidized Beds. *Powder Technology*, 38(3), 223-232.

Dibbs, H.P., & Marier, P. (1975). Some environmental effects, sources and control methods for fine particulates. *AIChE Symposium Series*, 71(147), 60-69.

Feuerborn, H.J. (2005, November 23-24). *Coal ash utilisation over the world and in Europe*. Paper presented at the International Workshop on Environmental and Health Aspects of Coal Ash Utilization. Tel Aviv. Retrieved January, 23, 2013, from http://www.coal-ash.co.il/sadna/Abstract_Feuerborn.pdf.

Fotovati, S., Tafreshi, H.V. and Pourdeyhimi, B. (2011). Analytical expressions for predicting performance of aerosol filtration media made up of trilobal fibers. *Journal of Hazardous Materials*, 186(2-3), 1503-1512.

Geldart, D. (1973). Types of gas fluidization. *Powder Technology* 7(5), 285-292.

Geldart, D., Harnby, N., & Wong, A.C. (1984). Fluidization of cohesive powders. *Powder Technology*, 37(1), 25-37.

Geldart, D., & Wong, A.C.Y. (1987) Fluidization of fine powders. *AIChE Symposium Series*, 83(255), 1.

Gupta, S.K., Agarwal, V.K., Singh, S.N., Seshadri, V., Mills, D., Singh, J., & Prakash, C. (2009). Prediction of minimum fluidization velocity for fine tailings materials. *Powder Technology*, 196(3), 263-271.

Hammerly, J.A., Marracino, J.M., & Piagenti, R.O. (1984). *Curso de Química Analítica*. Buenos Aires: Librería Ateneo.

Hinds, W.C. (1982) *Aerosol technology: Properties, behavior, and measurement of airborne particles*. New York: John Wiley and Sons.

Humphries, W. (1998, July 10-13). *Fabric filtration for flue gases*. Paper presented at the 1998 Annual World Conference of The Textile Institute, Sydney, Australia: Proceedings, 244-251.

Iyer, R.S., & Scott, J.A. (2001). Power station fly ash: a review of value-added utilization outside of the construction industry. *Resources, Conservation and Recycling* 31(3), 217–228.

Jedrusik, M., Swierczok, A., & Teisseyre, R. (2003). Experimental study of fly ash precipitation in a model electrostatic precipitator with discharge electrodes of different design. *Powder Technology* 135–136(1), 295–301.

Kirsh, V.A. (2007). Stokes flow in model fibrous filters. *Separation and Purification Technology*, 58(2), 288-294.

Kunii, D., & Levenspiel, O. (1991). *Fluidization Engineering*. Boston: Butterworth & Heinemann.

Leeds & Northrup. (1993). *Microtrac® X100 & SRA150: Operator´s Manual, 179551 Rev. B*; and *Microtrac® Dry Powder Sample: Operator´s Manual, 179528 rev A y 179529 rev. A*.

Leva, M. (1951). Elutriation of fines from fluidized systems. *Chemical Engineering Progress* 47(1), 39-45.

Li, J., & Kato, K. (2001). A correlation of the elutriation rate constant for adhesion particles (group C particles). *Powder Technology*, 118(3), 209–218.

Lin, J.H., Lou, C.W., Lei, C.H., Lin, C.Y. (2006). Processing conditions of abrasion and heat resistance for hybrid needle-punched nonwoven bag filters. *Composites Part A: Applied Science and Manufacturing*, 37(1), 31-37.

Liu, Y.D., & Kimura, S. (1993). Fluidization and entrainment of difficult-to-fluidize fine powder mixed with easy-to-fluidize large particles. *Powder Technology*, 75(2), 189-190.

Lucas, A., & Casal, J. (1984). Cálculo y diseño de lechos fluidizados: Parte I. *Ingeniería Química*, No 178, 135-143.

Lupo, J.F., & Tremblay, M.A.J. (2003). *Tailing injection deposition method theoretical model*. In A.A. Balkema (Pub.), Tailings and mine waste '03 (pp. 101-108). Lisse (NL): Balkema.

Ma, X., & Kato, K. (1998). Effect of interparticle adhesion forces on elutriation of fine powders from a fluidized bed of a binary particle mixture. *Powder Technology*, 95(2), 93-101.

Margiatto, C.A., & Siegell, J.H. (1983). Determination of porous particle density, *Powder Technology*, 34(1), 105-106.

Massimilla, L., & Salatino, P. (1987). A theoretical approach to the characterization of carbon attrition in a fluidized bed combustor. *Chemical Engineering Communications*, 62(1), 285-301.

Nonhebel, G. (1972). *Gas purification processes for air pollution control*. London: Butterworth Publishers.

Osberg, G.L., & Charlesworth D.H. (1951). Elutriation in a fluidized bed. *Chemical Engineering Progress* 47(11), 566-570.

Pandey, V.C., & Sing, N. (2010). Impact of fly ash incorporation in soil systems. *Agriculture, Ecosystems and Environmen*t 136(1-2), 16–27.

Perry, H.R., & Green, D.W. (1997), *Perry's Chemical Engineer's Handbook*, USA: McGraw-Hill Publishing Company.

Phillips, H.W. (2000). Select the proper gas cleaning equipment. *Chemical Engineering Progress*, 96(9), 19-40.

Rodríguez Barea, F. (1998). *Génesis de partículas en combustión de carbón y fueloil,* en II Curso de Tecnología de Partículas en Procesos Industriales y su Influencia Medioambiental. Madrid: CIEMAT.

Rodríguez, J.M., Sánchez, J.R., Álvaro, A., Florea, D.F., & Estévez, A.M. (2000). Fluidization and elutriation of iron oxide particles. A study of attrition and agglomeration processes in fluidized beds. *Powder Technology*, 111(3), 218-230.

Saleh, K., Cami X.B., Thomas A., & Guigonet P. (2006). An Experimental Study on the Fluidization Behaviour of Geldart C Glass Powders. *KONA-Powder and Particle Journal*, No.24 134-145.

Sánchez, J.R., Estévez, A.M., Jaraíz, E., Rodriguez, J.M., & Macias, A. (1994). Los filtros en la limpieza de gases: tipos y criterios de elección. *Ingeniería Química*, 26(303), 147-156.

Sánchez, J.R., Álvaro, A., Rodríguez, J.M., & Estévez, A.M. (2000). Limpieza de gases con partículas: Tecnologías actuales por vía seca. *Ingeniería Química*, 32(364), 127-134.

Sánchez, J.R., Rodríguez, J.M., Álvaro, A. and Estévez, A.M. (2007). The capture of fly ash particles using circular and noncircular cross-section fabric filters. *Environmental Progress*, 26(1), 50-58.

Sánchez, J.R., Rodríguez, J.M., Álvaro, A. and Estévez, A.M. (2010). Deduction and application of filtration and cleaning indices for aerosol filtration operations using nonwoven fabrics as filters. *Environmental Progress & Sustainable Energy*, 29(1), 17-24.

Sarbak, Z., Stanczyk, A., & Kramer-Wachowiak, M. (2004). Characterisation of surface properties of various fly ashes. *Powder Technology* 145(2), 82-87.

Sarkar, A., Rano, R., Udaybhanu, G., & Basu, A.K. (2006). A comprehensive characterisation of fly ash from a thermal power plant in Eastern India. *Fuel Processing Technology* 87(3), 259-277.

Sing, K.S.W., Everett, D.H., Haul, R.A.W., Moscou, L., Pierrotti, R., Rouquerol, J., & Siemienieswka, T. (1985). Presentación de datos de fisisorción en sistemas gas/sólido. *Pure Applied Chemical*, 85(57).

Stairmand, C.J. (1975). Gas/solids separation in pollution control - the wag ahead. *AIChE Symposium Series*, 71(147), 70-80.

Standars ASTM, USA, (2003). *Standard specification for coal fly ash and raw or calcined natural Pozzolan for use in concrete*; American Society for Testing Materials. Retrieved February 10, 2013.

Stöcklmayer, Ch., Krammer, A., & Höflinger, W. (1996, April 25). *Using a computer simulation method for calculating compression parameters from experimentally obtained pressure drop curves in dust cake filtratio*n. Paper presented at the International Symposium on Filtration and Separation of Fine Dust: Vienna: Proceedings, 104–113.

Wen, C., & Chen, L. (1982). Fluidized Bed Freeboard Phenomena: Entrainment and Elutriation. *AIChE Journal*, 28(1), 117-128.

Wen, Ch., & Hashinger, R. (1960). Elutriation of solid particles from a dense-phase fluidized bed. *AIChE Journal*, 6(2), 220-226.

Werther, J., & Hartge, E.U. (2003). Elutriation and Entrainment. In W.C. Yang (Ed.), *Handbook of Fluidization and Fluid-Particle Systems* (pp. 122-137). New York: Marcel Dekker, Inc.

Wikipedia (n.d.). Fly ash. Retrieved February 21, 2013, from: http://en.wikipedia.org /wiki/Fly_ash

Wilkinson, D. (1995). Determination of minimum fluidization velocity by pressure fluctuation measurement. *The Canadian Journal of Chemical Engineering*, 73(4), 562-585.

Zhang, L., Hou, J., Bi, X.T., Grace, J.R., Janke, T., & Arato, C. (2012). Fluidization characteristics and charging behavior of fly ash in a vibro-fluidized bed. *Powder Technology* 215-216(1), 235-241.

Zhu, J. (2004). Fluidization of fine powders. In J.S. Anthony, W. Hoyle, & Y. Ding (Eds.), *Granular materials* (pp. 270-295). Cambridge: The Royal Society of Chemistry.

In: Fly Ash
Editor: Prabir Kumar Sarker

ISBN: 978-1-62948-044-2

Chapter 3

PHYSICAL AND FLOW PROPERTIES OF FLY ASH FROM BIOMASS COMBUSTION

Christof Lanzerstorfer[*]
University of Applied Sciences, Upper Austria, Austria

ABSTRACT

This chapter presents a characterization of different fly ashes from grate-fired biomass combustion (cyclone fly ash, electrostatic precipitator fly ash and fabric filter fly ash) with respect to storage and flow relevant properties. Physical property measurements, including particle size distribution, bulk density, tapped bulk density and particle density, are presented. Powder flowability was measured using a ring shear tester. The flowability of the different fly ashes is compared and discussed with reference to their physical properties. The wall friction angles with different wall materials were measured using a ring shear tester for wall friction tests. Angles of repose were also determined. The application of powder flowability data is important in, for example, the design of hoppers of fly ash collection equipment or fly ash storage silos.

Keywords: Fly ash, biomass combustion, flow properties, physical properties

INTRODUCTION

The limited availability of fossil fuels and concerns about climate change caused by the carbon dioxide emissions from fossil fuel combustion have given rise to combustion of woody biomass for heat and power generation (European Biomass Association, 2012). Biomass combustion is considered to have almost no impact on carbon dioxide emissions because the emissions produced during combustion are compensated by the carbon dioxide fixed in the biomass during its growth.

---

[*] E-mail: c.lanzerstorfer@fh-wels.at.

However, one disadvantage of biomass combustion is the production of a considerable amount of ash, a bulk solid material with a broad range of particle size. Depending on the type of combustion process and the off-gas cleaning system, different ash fractions result from the combustion as residues. Bottom ash, the coarser ash fraction, is discharged from the firing grate (grate-fired combustion) or removed from the fluidized bed (in fluidized bed combustion). The finer ash fractions (fly ash) leave the combustion zone together with the flue gas. This fly ash usually accounts for one quarter of the total amount of ash (Van Loo and Koppejan, 2008). The fly ash has to be separated from the flue gas in a dust collector. In order to comply with low dust emission limits (< 20 mg/m³ (STP)), an electrostatic precipitator (EP) or a fabric filter has to be used as fly ash collector. The collection efficiency of cyclones is limited, especially for fine particles. Single stage dust collection by a cyclone is only sufficient for biomass combustion plants having a higher dust emission limit. However, cyclones are used as pre-collectors in two-stage dust collection systems.

The chemical composition of biomass fly ashes has been investigated; data are available, for example, in Steenari and Lindqviat (1997), Steenari, Karlsson and Lindqvist (1999), Van Loo and Koppejan (2008), Pöykiö et al. (2009) and Dahl, Nuresniemi, Pöykiö and Watkins (2009). Information about the physical properties of fly ash is surprisingly rare. Van Loo and Koppejan (2008) report values for the particle density and the bulk density of different biomass fly ashes. Results from particle size distribution measurements are available in Lind et al. (1999). However, data about the flow properties of fly ash from biomass combustion (yield locus, effective angle of internal friction, angle of wall friction, angle of repose) were not found in a review of the literature. Knowledge of the flow properties of the ash is necessary to make the design fly ash storage silos, hoppers of fly ash collection equipment and other ash handling equipment so that no flow problems occur (Schulze, 2006, p. 255). An indication of the flowability of bulk solids and powders can be derived from the Hausner ratio and from Carr´s compressibility, which are both calculated from the bulk density and the tapped bulk density (Stanley-Wood, 2008). However, in advanced bulk solids technology shear testers are used for the measurement of flow properties (Jenike, 1970). On the basis of such measurements a reliable design of bulk solids handling and storage equipment is possible.

1. METHODS

1.1. Sample Collection and Sample Preparation

Fly ash samples of approximately 10 dm³ were taken from the discharge system of the dust collectors of a 5.0 MW_{th} grate-fired district heating plant (plant A). This plant is equipped with a two-stage system for dust collection consisting of a cyclone and an EP installed in series. The biomass used for combustion in this plant is wood chips from the forest in a mixture of softwood and hardwood with about 80% softwood. The ratio of cyclone ash to EP ash is approximately 4 to 1. Additionally, an ash sample from the furnace (bottom ash) was collected. At plant B, a 35 MW_{th} grate-fired combined heat and power generation plant with a fabric filter for dust collection, a fly ash sample was taken from the dust discharge system of the filter. In this plant the biomass used for combustion is wood chips from the forest with about 70% softwood and 30% hardwood.

Results for the particle size distribution of fly ash are also available from previous measurements for plant A and from two other biomass combustion plants (plant C and plant D). Plant C is a 5.0 MW_{th} grate-fired district heating plant with a cyclone and an EP installed in series for dust collection. The biomass fuel in this plant is a mixture of about 90% softwood and 10% hardwood and consists of waste wood from the wood-processing industry (70%) and wood chips from the forest (30%). Plant D is a 1.2 MW_{th} grate-fired district heating plant with only a cyclone for dust collection. The fuel is mainly wood chips of hardwood from the forest.

The volume of the ash samples was reduced to a volume suitable for the various laboratory tests using sample dividers which were applied repeatedly (Haver RT 12.5, Retsch PT100, Quantachrome Micro Riffler). As the bottom ash containd some extremely coarse material (> 10 mm) that would block the sample deviders, the fraction > 10 mm was separated manually using a 10 mm sieve.

1.2. Measurement Procedures

The moisture content of the samples was determined gravimetrically. The samples were dried at 105°C for one hour.

The particle size distribution of the fly ash samples was measured using a laser diffraction instrument with dry sample dispersion from Sympatec, type HELOS/RODOS. Prior to the measurement, the ash samples were sieved at 500 μm using a laboratory sieve shaker (Fritsch ANALYSETTE 3 PRO) to remove the oversize fraction. The particle size distribution of the bottom ash sample was determined using the laboratory sieve shaker with sieves from 8 mm to 500 μm. The undersize material (< 500 μm) was analyzed using the laser diffraction instrument.

The density (particle density) of the ashes was determined according to ÖNORM EN ISO 8130-3 (Österreichisches Normungsinstitut, 2011). This method is based on a determination of the mass and the volume of a test portion using a liquid displacement pyknometer. The capacity of the pyknometers used was 50 cm³ and n-heptane (density: 0.684 g/cm³) was used for displacement of the air.

The bulk density of the fly ash samples ρ_{bulk} was determined according to ÖNORM EN ISO 60 (Österreichisches Normungsinstitut, 1999). The measurement is carried out in the following way: when the bottom cover of the funnel is removed, 120 cm³ of powder stored in the funnel flow by gravity into the coaxial 100 cm³ measuring cylinder. The excess material is removed by drawing a straightedge blade across the top of the vessel. The bulk density of the coarser bottom ash sample was determined according to ÖNORM ISO 697 (Österreichisches Normungsinstitut 1986). In this standard a bigger measuring cylinder (500 cm³) is used, thus reducing the wall effects, which show a greater influence for coarse material.

The tapped bulk density ρ_{tapped} of the samples was determined using a graduated measuring glass (250 ml), which was affixed on the bottom plate of a laboratory sieve shaker. The shaker was operated for one minute. The amplitude was 2 mm. The tapped bulk density is calculated from the mass and the final volume of the ash sample.

The angle of repose of the ash samples was determined according to DIN ISO 4324 (Deutsches Institut für Normung, 1983). The measurement is carried out in the following way: a cone of material is obtained by passing a given volume of the powder through a special funnel placed at a fixed height above a completely flat and level plate. The base angle of a cone is calculated from the diameter of the base plate and the height of the cone.

The yield locus for the fly ash samples was determined using a Schulze ring shear tester, type RST-XS with a 70 cm³ shear cell. For the bottom ash a Schulze ring shear tester type RST-01 with a 900 cm³ shear cell was used. In order to do a shear test with these testers, the bulk solid sample is loaded vertically at a normal stress and then a shear deformation is applied to the sample by moving the top platen at a constant velocity. This results in a horizontal shear stress in the sample. To measure a point of a yield locus two steps are necessary. First, the sample is consolidated in the pre-shear step. Here the point of steady-state flow with the pair of values for the normal stress σ and the shear stress τ is determined along with the bulk density. Subsequently, a point of the yield limit is measured at a reduced normal stress. The corresponding pair of values for the normal stress and the shear stress at a point of incipient flow is one point of the yield limit. By repetition of the procedure the whole yield locus can be determined. A Mohr stress circle which is tangential to the yield locus and runs through the point of steady-state flow can then be drawn. The slope of a tangent to this stress circle which runs through the origin of the σ-τ-diagram represents the effective angle of internal friction (Schulze, 2006, p. 84). The bulk density of the ash in dependence of the consolidation stress is also determined. For this, the volume of the ash sample is calculated from the measured hight of the sample in the shear cell.

The wall yield locus for the fly ash samples was determined using a Schulze ring shear tester type RST-XS with a wall friction shear cell, where the bottom ring of this cell is formed by a sample of the wall material. For the bottom ash a Schulze ring shear tester type RST-01 was used. The wall friction angles were determined for two wall materials, structural steel S235JR (1.0038) and wear-resistant steel Hardox® 450. For a shear test, the bulk solid sample is loaded vertically at a normal stress and then moved relative to the wall material surface with a constant velocity. To measure a point of a wall yield locus, corresponding pairs of values for the normal stress and the shear stress are determined. The slope of a straight line running through the origin of the σ-τ-diagram and a point of the wall yield locus is the kinematic angle of wall friction (Schulze, 2006, p. 95).

For the oversize material, the loss on ignition at a temperature of 550°C was determined. A sample of about 5 g was heated repeatedly in a furnace until no further loss of mass was measured.

1.3. Numerical Characterization of Flowability

The flowability of a bulk solid can be qualitatively rated by flowability indicators. In addition to the angle of repose, the Hausner ratio and Carr´s compressibility are two flowability indicators which are easy to determine. Both values are calculated from the bulk density ρ_{bulk} and the tapped bulk density ρ_{tapped}. The calculations are as follows (Carr, 1965; Hausner, 1967):

$$Carr's\ compressibility = \frac{\rho_{tapped} - \rho_{bulk}}{\rho_{tapped}}.100$$

$$Hausner\ ratio = \frac{\rho_{tapped}}{\rho_{bulk}}$$

The flowability indicators can be used to categorize the flowability of bulk solids and powders. The categories according to the United States Pharmacopeial Convention USP 29-NF24 (as cited in Stanley-Wood, 2008, p. 29) are shown in Table 1.

A quantitative characterization of the flowability of a bulk solid is possible with the ratio ffc of consolidation stress σ_1 to unconfined yield strength σ_c (Schulze, 1996):

$$ffc = \frac{\sigma_1}{\sigma_c}$$

The consolidation stress is equal to the major principal stress of the Mohr stress circle which runs through the point of steady-state flow and is tangential to the yield locus. The unconfined yield strength results from the stress circle which runs through the origin and is tangential to the yield locus (Jenike, 1970). The larger ffc is, the better a bulk solid flows. The usual classification used to define flow behaviour is as shown in Table 2.

Table 1. Categories of powder flow for various flowability indicators

Description of bulk solids flow	Angle of repose in °	Carr´s compressibility in %	Hausner Ratio
Very very poor; approx. non-flow	> 66	> 38	> 1.60
Very poor/very cohesive	56-65	32-37	1.46-1.59
Poor/cohesive	46-55	26-31	1.35-1.45
Passable	41-45	21-25	1.26-1.34
Fair	36-40	16-20	1.19-1.25
Good/free flow	31-35	11-15	1.12-1.18
Excellent/very free flow	25-30	< 10	1.00-1.11

Table 2. Numerical characterization of flowability (Schulze, 1996)

Description of bulk solids flow	ffc
Not flowing	ffc < 1
Very cohesive	1 < ffc < 2
Cohesive	2 < ffc < 4
Easy flowing	4 < ffc < 10
Free flowing	10 < ffc

The flowability of a bulk solid depends on the consolidation stress. For most bulk solids better flowability will be obtained at a greater consolidation stress. Because of this dependence it is not possible to describe the flowability of a bulk solid with only one numerical value. A good visualization of the flowability can be given in a diagram showing the unconfined yield strength dependent upon the consolidation stress when the diagram also includes lines of constant ffc ratio (Schulze, 1996).

2. RESULTS

2.1. Fly Ash Particle Size Distribution

The mass fraction of oversize material (> 500 μm) in the cyclone ash, the EP ash and the filter ash was 4.6%, 1.1% and 3.2%, respectively. The loss on ignition of the oversize material is in the range of 70%. The material consists of black pieces of charred wood as shown on the left side of Figure 1.

In the bottom ash the mass fraction of the oversize material (> 10 mm) which was excluded from the test sample was about 8.5%. This coarse material consists of molten slag and contains some foreign material like stones and nails. The colour of the fraction with particle sizes from 500 μm to 10 mm is red-brown to dark grey and the loss on ignition of this material is less than 0.01%. This material consists of slag granules as can be seen on the right side of Figure 1.

The particle size distribution of the different ash samples from plant A and plant B is shown in Figure 2. The mass median diameter for the EP ash, the cyclone ash and the bottom ash from plant A is 3.3 μm, 30 μm and 470 μm, respectively. The mass median diameter of the filter ash from plant B is about 20 μm.

In plant B the dust collection system consists of a single collection stage. Therefore, the filter ash contains also coarser particles compared to the EP ash from plant A, which contains only the particles which have passed through the cyclone. In the EP ash the mass fraction of particles larger than 10 μm was only about 5%. This low value indicates a high collection efficiency of the cyclone installed upstream from the EP.

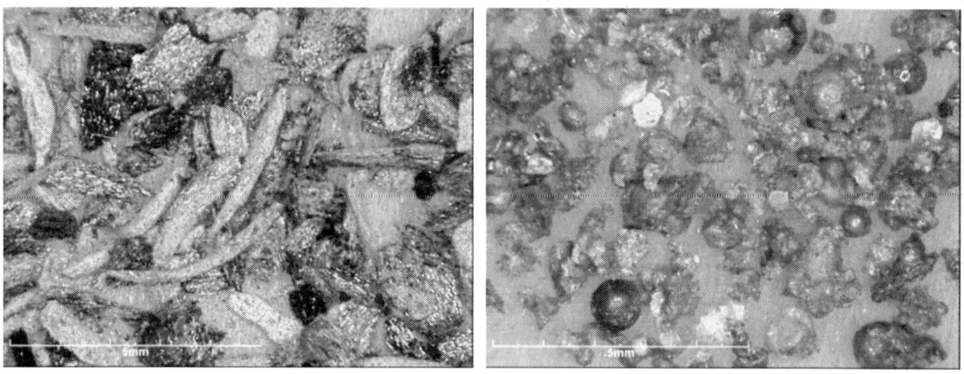

Figure 1. Microscopic image of the oversize fraction (> 500 μm) of the cyclone fly ash (left side) and the bottom ash (right side) from plant A.

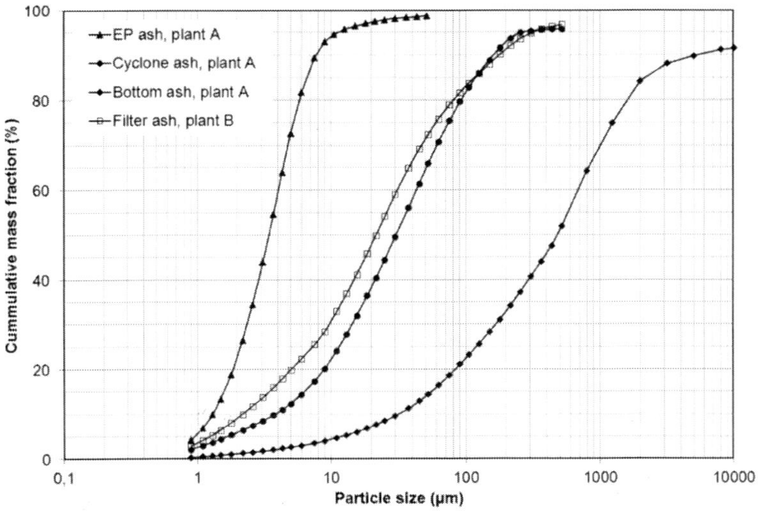

Figure 2. Particle size distribution of the different ash fractions from plant A and plant B.

In addition to the particle size distribution of the actual samples of cyclone ash and EP ash from plant A, results from particle size measurements from previous years for plant A, C and D are available. The results are shown in Figure 3. The variation in the particle size distribution is not very distinctive, neither for the EP ash nor for the cyclone ash. The average mass median diameter of the EP ash from plant A is 3.2 ± 0.4 μm. In plant C the average mass median diameter of the EP ash is in the same range (3.4 μm). Compared to the EP ash from plant A a higher mass fraction of coarse particles was found in the EP ash from plant C indicating the lower collecting efficiency of the cyclone in this plant. For the cyclone ash the variation in particle size distribution is less for the various samples from plant A (36 ± 6 μm) compared to the variation between the samples from the different plants for the year 2008 (35 ± 10 μm).

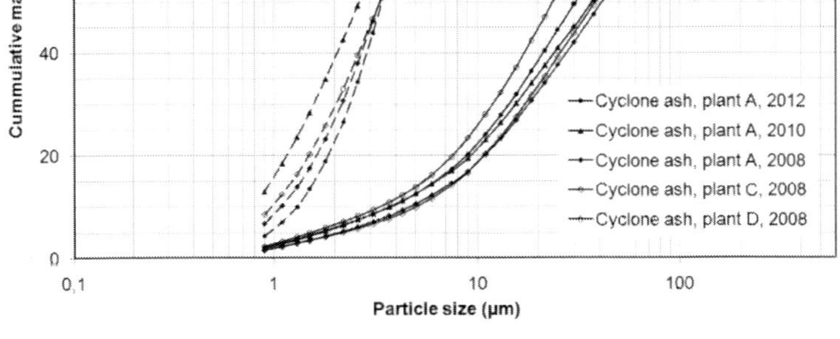

Figure 3. Particle size distribution of fly ash samples from plant A, C and D.

Table 3. Density of different ashes from plant A and B

	Humidity in % (w/w)	Density in kg/m³	Bulk density in kg/m³	Tapped bulk density in kg/m³
EP ash	0.7	2050	75	115
Cyclone ash	0.3	2330	660	970
Bottom ash°	0.1	2690	1130	1430
Filter ash	0.8	2460	345	500

° Oversize fraction (> 10 mm) excluded from the test sample

2.2. Fly Ash Density

The various densities of the fly ash samples are summarized in Table 3. The bulk density, the tapped bulk density and the density of the ashes from plant A increase with the particle size from the EP ash to the cyclone ash and the bottom ash. The values for the bulk density and the tapped bulk density of the filter ash from plant B are between the values for the cyclone ash and the EP ash whereas the density of the filter ash is higher than the density of both of these ashes.

In comparison with the data published by Van Loo and Koppejan (2008) the measured values for the density of the cyclone ash, the filter ash and the bottom ash are at the lower limits whereas the values for the bulk density are very similar. The measured values for the density and the bulk density of the EP ash are lower than the published values for filter ash. This is likely due to the fact that the EP ash was collected in a two stage dust collection system where the coarser part of the fly ash was already collected in the upstream cyclone.

2.3. Fly Ash Flow Properties

The measured parameters of the ash samples related to their flow characteristics are summarized in Table 4. For the ashes from plant A the angle of repose as well as the effective angle of internal friction decrease with increasing particle size from the EP ash to the cyclone ash and the bottom ash. For the filter ash from plant B the highest value for the angle of repose was measured, whereas the determined value of the effective angle of internal friction was between the values for EP ash and cyclone ash.

Table 4. Flow characteristics of different ashes from plant A and B

Ash	Angle of repose in °	Effective angle of internal friction in °	Wall friction angle	
			S235JR in °	Hardox® 450 in °
EP ash	48°	68°	38°	38°
Cyclone ash	47°	46°	28°	27°
Bottom ash°	38°	47°	21°	21°
Filter ash	50°	55°	30°	27°

° Oversize fraction (> 10 mm) excluded from the test sample.

The measured wall friction angles for each ash sample were almost identical for both wall materials. Only for the filter ash was a somewhat lower wall friction angle with Hardox® 450 was found. Similar to the effective angles of internal friction the wall friction angles also decrease with increasing particle size of the ashes.

The results for the ratio ffc are shown in Figure 4. The ffc of the filter ash increases from aprroximately 1.5 to 2.5 with increasing consolidation stress. The classification of the flowability improves therefore from Very cohesive to Cohesive. For the EP ash the ffc is between 1 and 2, the EP ash has to be classified as Very cohesive and is close to Not flowing at low consolidation stress. With increasing consolidation stress, the cyclone ash´s ffc increases from 3 to 4.5. Therefore, the cyclone ash is from Cohesive to Easy flowing. The ffc ratio for the bottom ash is even higher; this ash has to be classified as Easy flowing (close to Free flowing at higher consolidation stress).

The flowability categories according to the angle of repose would be Poor/cohesive for all fly ashes (filter ash, EP ash and cyclone ash) and Fair for the bottom ash.

The calculated values for Carr´s compressibility are 31, 35, 32 and 21 for filter ash, EP ash, cyclone ash and bottom ash, respectively. This would correspond to a characterisation of the flowability of the filter ash as Poor/cohesive, and for the EP ash as well as for the cyclone ash as Very poor/very cohesive. The resulting category for the bottom ash would be Passable.

The calculated values for the Hausner Ratio are 1.45, 1.53, 1.47 and 1.27 for filter ash, EP ash, cyclone ash and bottom ash, respectively. From these values the same characterisation of the flowability would result as with Carr´s compressibility.

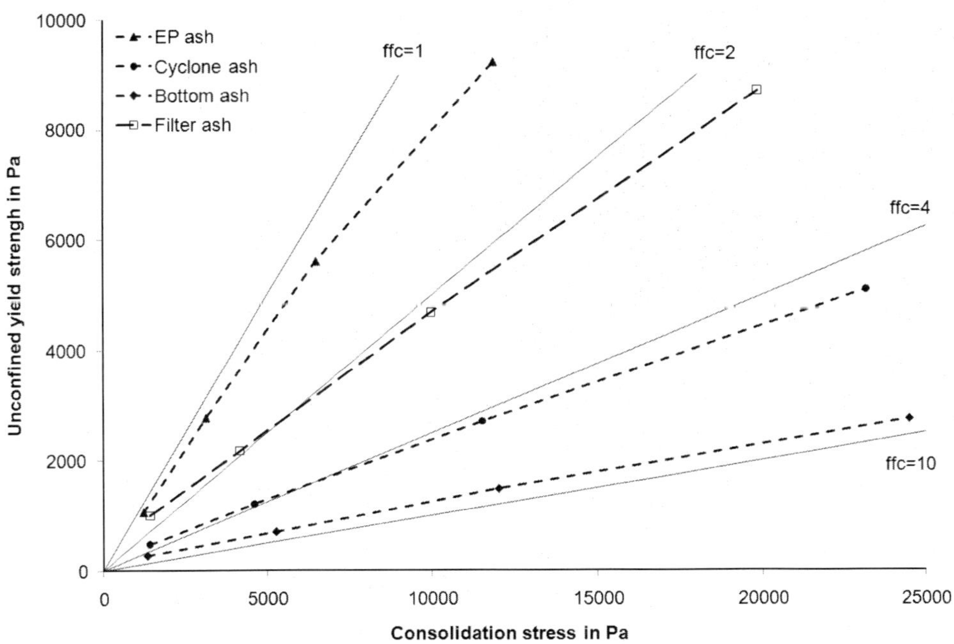

Figure 4. Flowability of ash samples from plant A and plant B.

A comparison of the resulting flowability categories for the different ashes according to the flowability indicators with the flow description resulting from the shear tests is shown in Table 5.

For the bottom ash the results suggested by the different flowability indicators and the shear tests are quite concurrent, although the wording is different. A comparison of the wording in Table 1 and 2 shows that the category Easy flowing in Table 2 corresponds to the categories Fair and Passable in Table 1.

Both fly ashes from plant A were assigned to the same flowability category by the different flowability indicators. However, according to the shear tests there is a great difference in flowability between the two fly ashes. For the cyclone ash all assignments of the flowability category by flowability indicators were inaccurate whereas for the EP ash only the angle of repose gave a inaccurate flow category. The flowability of the filter ash from plant B was slightly overestimated by all flowability indicators.

Table 5. Categories of the flowability of the ash samples of different ashes from plant A

	Angle of repose	Carr´s compressibility	Hausner ratio	ffc
EP ash	Poor/cohesive	Very poor/ very cohesive	Very poor/ very cohesive	Very cohesive
Cyclone ash	Poor/cohesive	Very poor/ very cohesive	Very poor/ very cohesive	Cohesive to Easy flowing
Bottom ash°	Fair	Passable	Passable	Easy flowing
Filter ash	Poor/cohesive	Poor/cohesive	Poor/cohesive	Very cohesive to Cohesive

° Oversize fraction (> 10 mm) excluded from the test sample

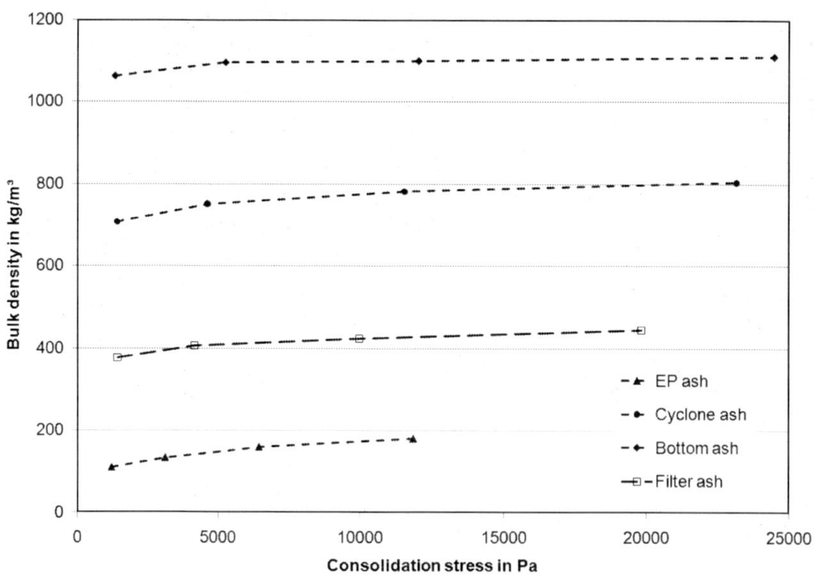

Figure 5. Bulk density of the ashes from plant A and plant B in dependence of the consolidation stress.

The bulk density dependent upon the consolidation stress is a function which is also required for silo discharge design. The measured results for the four ashes are shown in Figure 5. As is to be expected, for all ashes the bulk density increases with higher values of the consolidation stress.

CONCLUSION

Physical and flowability-related properties of different fly ash samples (filter ash, EP ash, cyclone ash) and a bottom ash sample from grate-fired biomass combustion plants were investigated. The three ashes from plant A differ considerably in particle size with an increase in the mass median diameter from one ash to the other by a factor of about ten. The particle size distribution of the filter ash from another grate-fired biomass combustion plant with a single stage dust collection system is located between the size distribution of the EP ash and the cyclone ash.

The values measured for density and bulk density are well within the range of reported values for the filter ash, the cyclone ash and the bottom ash while the measured values for the EP ash are considerably lower. Presumably, the published values do not include values of ash samples from the second stage of a two-stage dust collection system.

The results of the shear tests show a substantial difference in the flowability of the three fly ashes. Generally, the ashes with the finer particle size distribution display a worse flowability. The EP ash is Very cohesive (close to Not flowing). The filter ash is Very cohesive to Cohesive while the cyclone ash is Cohesive to Easy flowing, depending on the consolidation stress. As is to be expected, the relatively coarse bottom ash is Easy flowing.

The three flowability indicators produce results of inconsistent quality. While the flow categories obtained for the filter ash, the EP ash and the bottom ash correspond quite well with the results from the shear tests, the results obtained for cyclone ash differ greatly from the shear test result. Therefore, shear tests are strongly recommended for a reliable characterization of the flowability of fly ashes from biomass combustion.

REFERENCES

Carr, R.L. (1965). Evaluating flow properties of solids. *Chem. Eng.* 72(2), 163-169.

Dahl, O., Nuresniemi, H., Pöykiö, R., Watkins, G. (2009). Comparison of the characteristics of bottom ash and fly ash from a medium-size (32 MW) minicipal district heating plant incinerating forest residues and peat in a fluidized-bed boiler. *Fuel Proc. Tech.* 90, 871-878.

Deutsches Institut für Normung: DIN ISO 4324. Tenside; Pulver und Granulate; Bestimmung des Schüttwinkels. Berlin: Beuth Verlag, Dezember 1983.

European Biomass Association. (2012). *EuropeAn bioenergy outlook 2012*; Statistical Report. Brussels: Author.

Hausner, H.H. (1967). Friction conditions in a mass of metal powder. *Int. J. Powder Met.* 3(4), 7-13.

Jenike, A.W. (1970). Storage and flow of solids. Bulletin No. 123 Utah Engineering Experiment Station (4th ed.). Salt Lake City: University of Utah.

Lind, T., Valmari, T., Kauppinen, E.I., Sfiris, G., Nilsson, K., Maenhaut, W. (1999). Volatilization of the Heavy Metals during Circulating Fluidized Bed Combustion of Forest Residue. *Environ. Sci. Technol.* 33, 496-502.

Österreichisches Normungsinstitut: ÖNORM ISO 697. Tenside; Waschpulver; Bestimmung der Schüttdichte. Verfahren durch Messen der Masse eines gegebenen Volumens. Wien: Österreichisches Normungsinstitut, November 1986.

Österreichisches Normungsinstitut: ÖNORM EN ISO 60. Kunststoffe; Bestimmung der scheinbaren Dichte von Formmassen, die durch einen genormten Trichter abfließen können (Schüttdichte). Wien: Österreichisches Normungsinstitut, Dezember 1999.

Österreichisches Normungsinstitut: ÖNORM EN ISO 8130-3. Pulverlacke; Bestimmung der Dichte mit einem Pyknometer durch Flüssigkeitsverdrängung. Wien: Austrian Standards plus, Jänner 2011.

Pöykiö, R., Rönkkömäki, H., Nurmesniemi, H., Perämäki, P., Popov, K., Välimäki, I., Tuomi, T. (2009). Chemical and physical properties of cyclone fly ash from the grate-fired boiler incinerating forest residues at a small municipal district heating plant (6 MW). *J. Hazard. Mater.* 162, 1059-1064.

Steenari, B.-M., Lindqvist, O. (1997). Stabilisation of biofuel ashes for recacling to forrest soil. *Biomass and Bioenergy* 13, 39-50.

Steenari, B.-M., Karlsson, L.G., Lindqvist, O. (1999). Evaluation of the leaching characteristics of wood ash and the influence of ash agglomeration. *Biomass and Bioenergy* 16, 119-136.

Schulze, D. (1996). Measuring powder flowability: A comparison of test methods. Part I. *Powder and Bulk Eng.* 10(4), 45-61.

Schulze, D. (2006). Pulver und Schüttgüter. *Fließeigenschaften und Handhabung*. Berlin: Springer-Verlag.

Stanley-Wood, N. (2008). Bulk powder properties: instrumentation and techniques. In D. McGlinchey (Ed.), Bulk Solids Handling. *Equipment Selection and Operation* (pp. 1-67). Oxford : Blackwell Publishing.

Van Loo, S., Koppejan, J. (Eds.). (2008). *The Handbook of Biomass Combustion and Co-firing*. London: Earthscan.

In: Fly Ash
Editor: Prabir Kumar Sarker

ISBN: 978-1-62948-044-2

Chapter 4

FLY ASH RESISTIVITY: INFLUENCING FACTORS, PREDICTING MODELS AND ITS IMPACTS ON ELECTROSTATIC PRECIPITATOR PERFORMANCE

Shuran Li, Xiaoying Li, Yifan Huang, Zhen Liu and Keping Yan*
Key Laboratory of Biomass Chemical Engineering of Ministry of Education
Industrial Ecological and Environmental Research Institute
Zhejiang University, Hangzhou, P.R.China

ABSTRACT

This chapter discusses fly ash resistivity with a focus on its influencing factors and predicting models. It mainly depends on ash compositions, gaseous temperature and compositions. Li_2O plus Na_2O content greatly influences the maximum resistivity. Today, several models are available for predicting ash resistivity and ESP performance.

With regard to ash collection efficiency of electrostatic precipitator (ESP), high resistivity ash usually causes back corona inside the ESP, and low resistivity ash initiates leap collection. Both of them give lower collection efficiencies.

This chapter also presents typical ash micrographs collected from a number of Chinese coal-fired power plants, discusses the state of the art of ESP performance in terms of the so-called ESP index and particle grade collection efficiency. In order to improve the collection efficiency for high resistivity ash, this chapter also discusses ash conditioning and pulse power technologies.

Keywords: Fly ash; electrostatic precipitation; back corona; fine particle collection

* Corresponding Author address: 108 Main Building, Xixi Campus, Zhejiang University, Hangzhou, Zhejiang, China, 310028, Email: kyan@zju.edu.cn.

1. INTRODUCTION

Fly ash is a residue generated in combustion, and often comprises fine particles raised by flue gas. Fly ash exhausts from various sources, such as coal-fire boiler, metallurgical and cement production, fossil-fueled vehicles, living fuels burning and so on. For industrial applications, fly ash usually refers to ash produced during combustion of coal. The properties of fly ash particles from diverse sources are different as shown in Figure 1. Resistivity of common particles covers more than ten orders of magnitude, while the particle sizes spread from 0.1 μm to 1000 μm. Fly ash is referred to coal dust in this chapter for covering relatively large area in the figure, which represents characteristics of most fly ashes.

Fly ash doped with iron (Fe), magnesium (Mg), calcium (Ca) and sulfur (S) compounds over its aluminosilicate substrate demonstrates an adverse effect upon human health (Harrison & Yin, 2000). Exhausted fly ash is generally captured by electrostatic precipitator (ESP) or fabric filter (FF) before the flue gas reaches the desulfurization device (FGD), and then applied for building or refractory materials. ESPs have several advantages compared with other filtration equipment.

The form of the device can work at a high temperature (up to 850 °C) with low pressure loss, meanwhile requires lesser maintenance.

According to Figure 2, at the inlet of ESPs for numerous power plants, a bimodal number size distribution can be seen with a peak around 0.06 ~ 0.13 μm and 0.76 ~ 1.9 μm, respectively. The shift of peak values is most probably because of altered coal, mixed coal gas or fly ash recirculating combustion in power plant. The particle size distribution is a characteristic of the industrial process. Particles smaller than 2.5 μm (PM2.5) tend to emit from traditional cleaning devices, therefore reduce the removal efficiency (Ehrlich, Noll, Kalkoff, Baumbach, & Dreiseidler, 2007). Attempts, such as optimized real-time control of ESPs based on fly ash resistivity, bipolar pre-charger and hybrid ESP and FF, did give usable results.

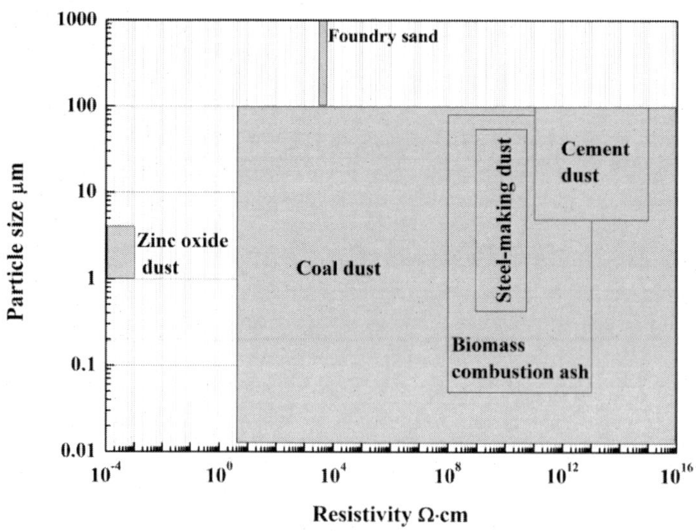

Figure 1. Characteristics of particle resistivity and sizes from various sources.

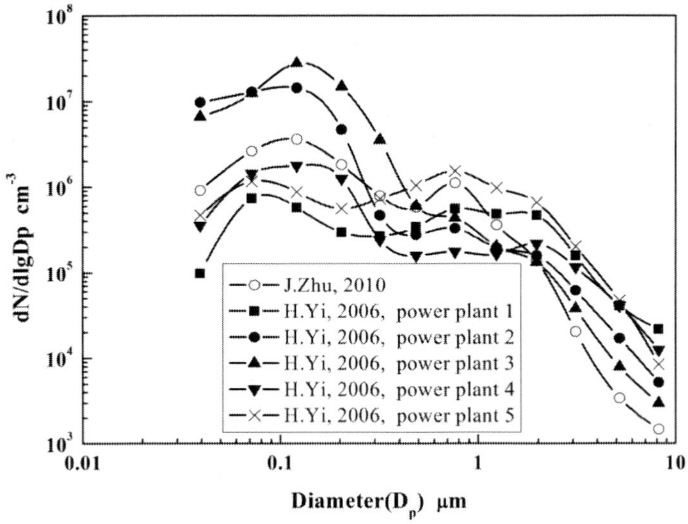

Figure 2. Size distribution of fly ash particle at the inlet of ESPs in various power plants (Yi, Guo, Hao, Duan, & Li, 2006; Zhu, Zhang, Chen, Shi, & Yan, 2010).

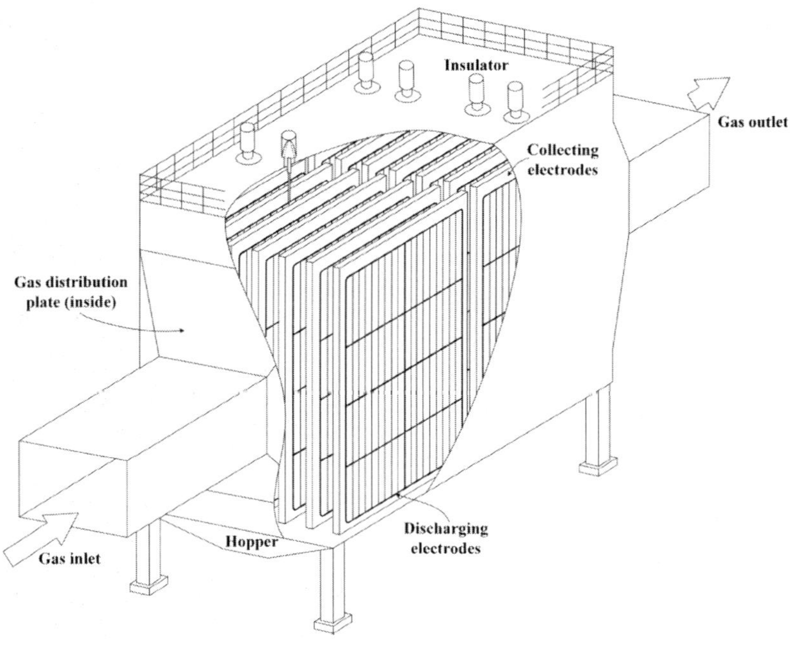

Figure 3. Schematic illustration of an ESP.

As shown in Figure 3, contaminated air stream flows into an ESP chamber from the gas inlet. The ESP ionizes the flue gas in which process produced ions charge particles in the gas stream. The discharging electrodes are usually small-diameter metal wires that create a strong electric field. The electric field forces charged particles towards collecting electrodes. Collecting electrodes are either a tube or a plate with opposite charge relative to that of discharging electrodes.

The electrodes are cleaned by intermittent impacts and then the fallen ash layers or agglomerates in hopper are removed. Industrial ESPs usually have several fields in one shell for increasing collecting area. Insulators, which separate earthed and discharging electrical conductors, are connected with the electrical system. Electrical system consists of high voltage components used to control the characteristics of voltage and current over the ESP. Hoppers are located at the bottom of the ESP. They collect and store the dust removed during the rapping temporarily. The rappers impact the electrodes for removing collected ash. Electrical resistivity, one of the critical factors influencing removal efficiency of ESPs, affects charging, collecting and rapping processes. The details will be discussed in following sections.

Electrical resistivity of fly ash quantifies how strongly fly ash opposes the flow of electric current. It often represents the resistance per square centimeter of a one-centimeter long/thick ash sample. Ash resistivity depends upon its chemical composition, temperature and constituents of flue gas. Ash of high resistivity causes back corona in ESPs, while that of low resistivity initiates leap collection.

To improve collection efficiency of ESPs, flue gas conditioning, pulse power generation and electrodes restructuring are all feasible solutions. The electrical resistivity of fly ash changes continuously during combustion. Therefore, prediction of ash resistivity is highly required for real-time control of ESPs.

2. Electrostatic Precipitator

2.1. Historical Development of ESPs

The study of fly ash resistivity accompanies the development of ESPs. History of ESP developments can be divided into three essential periods as shown in Figure 4. Commercial ESPs, especially those built at smelter, faced failure or bad performance because of the high ash resistivity in the first period. Investigators then began to observe the influence on voltage-current characteristics exerted by the ash resistivity. Many researches have been made for resistivity prediction and gas conditioning since 1937. The theory for ash conduction mechanisms is gradually better understood to guide industrial ESPs' design and operation.

Following the first successful commercial ESP at Pinhole in California, an ESP was installed at a lead smelter at Balaklala, California in 1910 (Cottrell, 1911). The precipitator, however, only achieved some 90% collection efficiency for the failure of controlling particulate electrical resistivity. This finding is considered to probably halt the investigation by O. J. Lodge in early 1880s (Lodge, 1884). In 1912, some work on this problem was done when cleaning the copper converter exhausts at a smelter at Garfild in Utah (Howard, 1914). It was found that conditioning by moisture injected into the gas stream or considerable sulfur trioxide presented improves the collecting efficiency significantly. Many hot gas dry ESPs failed to clean the blast furnace gases in the US and Europe for a lack of studies on high ash resistivity at a great temperature.

A successful hot dry ESP was installed at the UK Skinningrove Ironworks in 1917, while wet electrostatic precipitator was found to be more feasible for high-resistivity ash. (Parker, 1996).

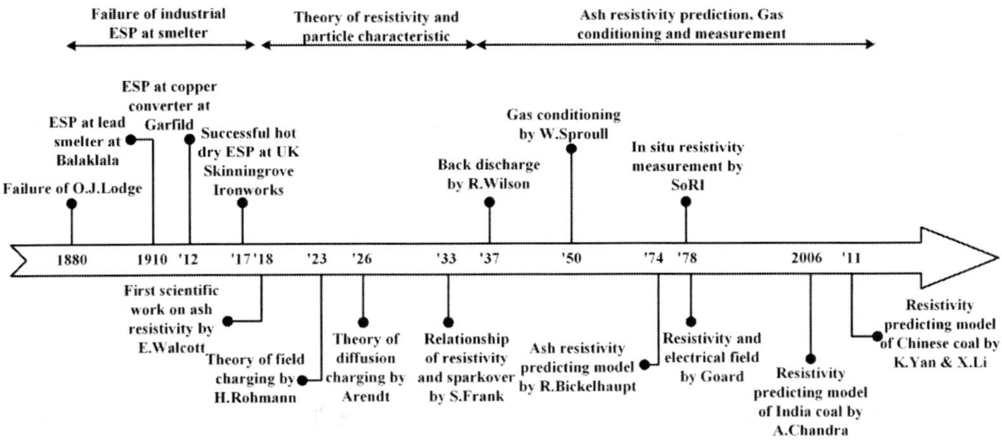

Figure 4. History of ESP and theory related to ash resistivity.

E. R. Walcott carried out the first scientific work on ash resistivity in 1918 (Wolcott, 1918). He found that a layer of dielectrics on the negative electrode reduces sparking voltage, suggesting high-resistance ash over the ground electrodes lowers the spark-over voltage in point-plate structure. Walcott's work also proved the sufficiently moist deposit or humidity gas flow would not lower the spark voltage, which is a feasible conditioning method. Later, in 1923, H. Rohmann developed the theory of field charging (Rohmann, 1923), and the diffusion charging process was described by P. Arendt and H. Kallmann in 1926 (Arendt & Kallmann, 1926). The magnitude of the charges was designated as a function of ion density, particle diameter, time and the electrical permittivity of the particle material. S. Frank investigated the relationship, proving Walcott's conclusion, between ash resistivity and reduction of spark-over voltage in 1933 (Frank, 1933).

During this period, most of the ESP installations were made on boilers burning high-sulfur coal. As low-sulfur coal being increasingly used, fly ashes produced by newly built boilers show higher resistivity, which hampers the collection. The problem was similar with ESPs for smelter exhausts. R. R. Wilson and H. J. White in 1937 reported back discharge from the high resistivity ash layer with a quantitative theory. High resistivity ash reduces collecting efficiency of ESPs, and back corona or spark-over might occur from the ground electrodes. Before White reported more detailed back corona effects in late '40s, he had already made in situ field resistivity measurements on dust in power plants and smelters in 1940 (White, 1974).

W. T. Sproull made studies in 1951 that indicates a high or a low temperature and moisture could improve ESP operation by reducing ash resistivity (Sproull & Nakada, 1951). Several researches about gas conditioning was carried out in 1940~60. Adding moisture, altering gas temperature and adding sulfur dioxide or ammonia were already practical procedures in the meantime. In 1974, R. E. Bickelhaupt observed that the electrical resistivity of ash is sensitive to its chemical composition (Bickelhaupt, 1979). He then developed his predicting method and computer program. P. R. C. Goard then showed resistivity is strongly field dependent (Goard & Potter, 1978). However, the laboratory measurements are difficult to compare with actual ash due to different packing density, and field strength from those of the actual condition. At nearly the same time, the U.S. Southern Research Institute (SoRI) developed a point-plane apparatus for in situ resistivity determinations (Nichols & Spencer, 1978). In 1991, H. Krigmont invented a patented measurement for resistivity by collecting particulate in the gas stream on the surface of a porous ceramic cylinder (Coe & Krigmont, 1995). In recent year, A. Chandra and K. Yan & X. Li have developed new predicting models for Indian and Chinese coal fly ash, respectively (Andrabi, Chandra, Aslam, & Hassan, 2013; Li, Zhang, Zhu, Feng, & Yan, 2012).

2.2. ESP Fundamentals

In ESPs, raw gas is treated passing through channels between discharge electrodes and earthed collecting electrodes, where a high voltage is applied for charging and capturing. Accumulated dust particles over collecting electrodes are then impacted down into the hopper as exhibited in Figure 5.

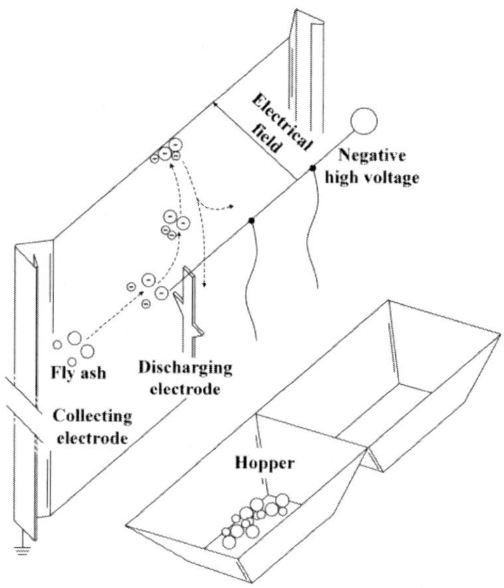

Figure 5. Schematic illustration of ESPs and the collection process of fly ash.

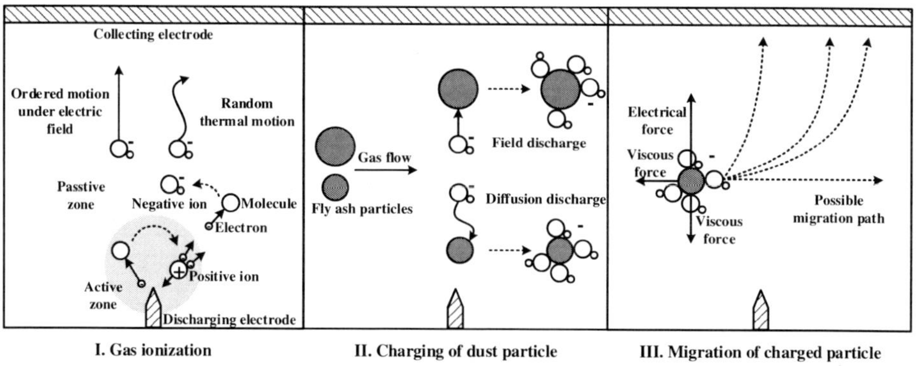

Figure 6. Schematic illustration of the first three collection steps in ESPs.

ESPs capture dust particles in five essential steps as follow, I. gas ionization, II. charging of dust particle, III. migration of charged particle, IV. particle deposition, V. removal of collected particle. The steps that have no separate existence can occur all in one certain discharge period. Figure 6 and Figure 10 illustrate these five steps.

Gas ionization appears in the neighborhood of tips on the thin electrode wires, where the electric field is enhanced greatly. Electrons and ions excite molecules, and the spatial distribution of them grows like an avalanche. A corona, a luminous discharge, is preferred for producing large amount of negative ions and electrons in the avalanche. To ignite a corona, a certain fairly high voltage is required. If the voltage is getting higher, the remaining gap between wires and plates, or wires and concentric cylinders, breaks down with a spark jumps across it. The spark, through which a large current flows, are harmful for industrial operation. As shown by Figure 6-I, The interspace between two polarities consists of two regions. The ionization process occurs in the active zone near the curved surface of discharging electrodes. Electrons accelerated by the high voltage potential collide with neutral gas molecules. One positive ion and two electrons drift in different directions during the collision. The electrons moving in the low field strength region are not able to ionize any further molecules. And numbers of electrons get entrained by the molecules forming additional negative ions which move between interelectrode spaces. This region is called the passive zone (Parker, 1996).Dust particles suspended in the gas streamer get charged by collision of ions. The two main mechanisms for charging are often referred to as field charging (particle parameter>1 μm) and diffusion charging (particle parameter<0.1 μm) (Parker, 1996). Field charging requires the ordered motion of ions under the presence of electric field, while diffusion charging based on random ionic collision driven by thermal motion as shown in Figure 6-II. Considerable work has been done on the theory of particle charging. Several models and equations suggest charging on the particle is related to the electrical permittivity of the particle material. The work, which are mainly equations for particle charging calculation, of B. Y. H. Liu and H. C. Yeh (Liu & Yeh, 1968), A. T. Murphy (Murphy, Adler, & Penney, 1959) and W. B. Smith (Smith & McDonald, 1976) are given in this section.

The saturation charge of field charging (Oglesby & Nichols, 1978) is

$$q_s = 3\pi\varepsilon_0 d_p^2 E_0 \frac{\varepsilon_r}{\varepsilon_r + 2} \tag{1}$$

where ε_0 is the permittivity of free space, d_p is the diameter of ash particle, E_0 is the uniform electric field strength, ε_r is the relative permittivity of ash particle, $\varepsilon_r = \frac{\varepsilon_p}{\varepsilon_0}$, ε_p is the permittivity of ash particle. The equation (1) is also called Pauthenier charging limit (Pauthenier & Moreau-Hanot, 1932).

The charge of particles $q_f(t)$ can be expressed as a function of time by the following equation:

$$q_f(t) = q_s \frac{1}{1+\frac{\tau}{t}} = 3\pi\varepsilon_0 d_p^2 E_0 \frac{\varepsilon_r}{\varepsilon_r + 2} \cdot \frac{1}{1+\frac{\tau}{t}} \tag{2}$$

where t is the charging time and τ is the time constant. τ can be designated as $\tau = 4\varepsilon_0 / N_0 eK_i$, N_0 is the unipolar ion concentration, e is electronic charge and K_i is mobility of ion.

White also expressed the diffusion charging as following function (White, 1963):

$$q_d(t) = \frac{d_p kT}{2e} \ln(1 + \frac{\pi d_p v N_0 t}{2kT}) \tag{3}$$

where k is Boltzmann constant, v is rms thermal velocity of the ions.

Combined effects of field and diffusion charging for particles with size region between 1 and 0.1 μm are often expressed by complex differential equation, which can be hardly solved. Thus, simplified models such as Cochet model and Robinson model are built for numerical solutions. In this chapter, we use Cochet model for expression of charges on these particles:

$$q_m = \left\{ \left(1 + \frac{2\lambda}{d_p}\right) + \left(\frac{2}{1 + 2\lambda / d_p}\right) \cdot \left(\frac{\varepsilon_r - 1}{\varepsilon_r + 2}\right) \right\} \cdot \pi \varepsilon_0 d_p^2 E_0 \tag{4}$$

Equation (4) gives the particle saturation charge, where λ is mean free path of ions (Parker, 1996).

Fly ash can be equivalent to a resistor and a capacitor connected in series. Both have the same value as that of ash. The value of them could be calculated by resistivity ρ, permittivity ε, thickness d and area A of the ash as follows, $R=\rho \cdot d/A$ and $C=\varepsilon \cdot A/d$. The product of R times C represents the time constant τ of a RC circuit, $RC=\rho \cdot \varepsilon=\tau$, which implies the permittivity can be written as

$$\varepsilon = \frac{\tau}{\rho} \tag{5}$$

Charged particles, with negative charge mostly, deposit to collecting electrodes under an applied electric field. The steady-state drift velocity of the particles is theoretical migration velocity (ω_{th}). Equation (6) is a valid expression for migration velocity of particles in ESPs (Parker, 1996; Riehle & Löffler, 1992):

$$\omega_{th} = \frac{qE_p}{3\pi d_p \mu} \tag{6}$$

In which q is the charge of the particle, E_p presents electric field adjacent to the collecting electrode and μ is the viscosity of the medium gas stream. Substituting equation (2), (3) and (4) for q in equation (6)1.6 gives:

$$\omega_{th} = \begin{cases} \dfrac{\varepsilon_0 d_p E_0 E_p}{\mu} \cdot \dfrac{2}{1+2\rho N_0 e k_i} \cdot \dfrac{t}{t+\tau} & \left(d_p > 1\,\mu m\right) \\ \left\{\left(1+\dfrac{2\lambda}{d_p}\right)+\left(\dfrac{2}{1+2\lambda / d_p}\right)\cdot\left(\dfrac{4-\rho N_0 e k_i}{4+2\rho N_0 e k_i}\right)\right\}\dfrac{\cdot\varepsilon_0 d_p E_0 E_p}{3\mu} & \left(0.1\,\mu m < d_p < 1\,\mu m\right) \\ \dfrac{kTE_p}{6\pi e \mu}\ln(1+\dfrac{\pi d_p v N_0 t}{2kT}) & \left(d_p < 0.1\,\mu m\right) \end{cases} \tag{7}$$

For extremely small particles, above equations should be multiplied by a Cunningham factor ($1+2A\lambda / d_p$) for modification. As equation 1.7 shows, migration velocity of particles larger than 0.1 μm is sensitive to the ash resistivity. Collection efficiency η relates to the specific collecting electrode area *A/Q* and migration velocity in the form of the Deutsch equation:

$$\eta = 1 - \exp\left(-\frac{A}{Q}\omega\right) \tag{8}$$

Therefore, migration velocity shows how easily the particles can be collected by ESPs. There are several forces acting on a charged particle as shown in Figure 6-III, electrical and viscous forces are most significant, while gravitational and inertial forces are in general negligible.

The expression of migration velocity is obtained by analyzing particle motions under these forces. The migration directions rarely follow the electric field strictly. Most particles move directly towards the collection plates to form a thin layer on the collecting electrodes in this figure. Meanwhile, a few particles flow out of the ESP in the random paths.

For engineering applications, the efficiency η is closely related to the useful corona power supplied to a precipitator, P_c (White, 1963):

$$\eta = 1 - \exp\left(-k_1 \frac{P_c}{Q}\right) \tag{9}$$

where k_1 is a parameter relevant with the condition of flue gas, ESP and particle.

The Deutsch equation is also modified for decreasing average diameter of entrained particles due to that more and more large particles have been collected continuously along the airflow direction. The efficiency η is given by equation (10):

$$\eta = 1 - \exp\left[-\left(\frac{A}{Q}\omega\right)^{k_2}\right] \tag{10}$$

where k_2 is also a parameter related to the operating conditions. In application, the value of k_2 is often set as 0.5, thus the equation can be rewritten as:

$$\eta = 1 - \exp\left(-\sqrt{\frac{A}{Q}\omega} \right) \tag{11}$$

The researches of industrial ESPs indicate the grade collection efficiency $\eta(r)$ is a function of applied electrical field strength:

$$\lg\left(\frac{1-\eta(r)}{\beta} \right) = -\alpha \cdot E_a \cdot E_p \cdot S \tag{12}$$

where E_a is the average electric field, E_p is the peak electric field, S is the specific collection area and α, β is the coefficient related to the running condition and particle size. The product of $S \cdot E_a \cdot E_p$ is called ESP index, which shows the collection capacity of ESP.

Equation (12) can be rewritten for mass of penetrated particles:

$$\lg m = \lg(\beta \cdot M_0) - \alpha \cdot E_a \cdot E_p \cdot S \tag{13}$$

where M_0 is the inlet mass concentration. Figure 7, Figure 8, Figure 9 prove equation (13). The grade penetrations show linear relationship with specific collection area, square electric field strength and their product. For collection of particles smaller than 0.5 μm, ESP index around 1300 $(kV/cm)^2 \cdot (m^2/m^3/s)$ is needed for controlling emission under 10 mg/m^3. That is the proposed value for specific collection area and average electric field strength is 100 $m^2/m^3/s$ and 3.6 kV/cm, respectively. The improvement requires matched high-voltage power supply for various fly ashes.

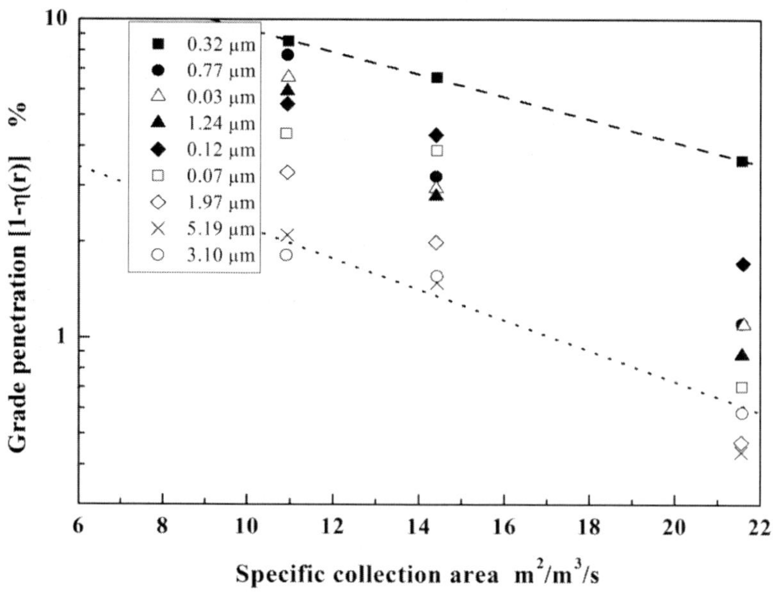

Figure 7. Grade penetration versus specific collection area (Zhu, 2010).

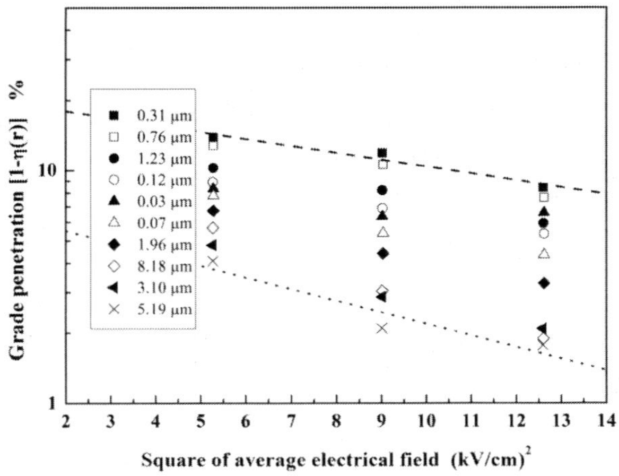

Figure 8. Grade penetration versus square of average electrical field (Zhu, 2010).

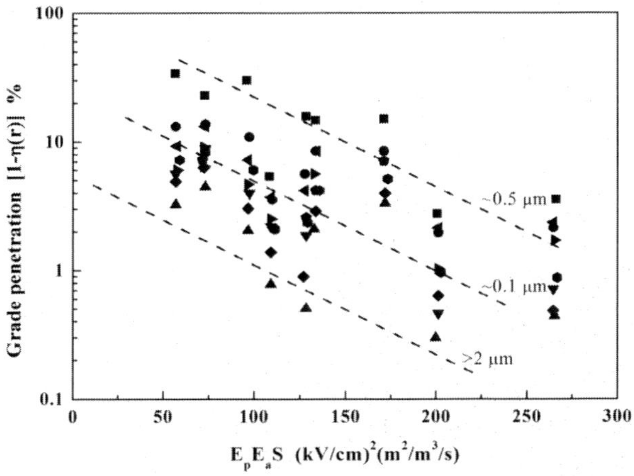

Figure 9. Grade penetration versus ESP index.

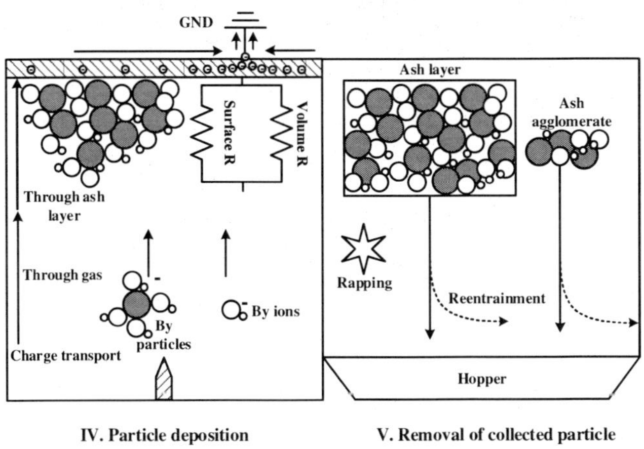

Figure 10. Schematic illustration of the last two collection steps in ESPs.

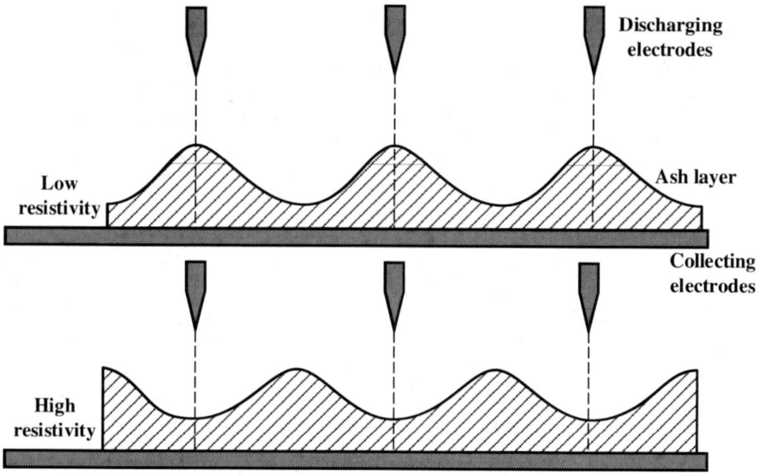

Figure 11. Various deposition sharps of ashes with different resistivity.

Charged particles agglomerate to form a thin dust layer after drifting to collecting electrodes. The electric charges pass through the dust layer, and flow from particles into the ground electrodes, as illustrated in Figure 10-IV. The transportation occurs by two mechanisms, surface and volume conductivity, which crucially determine the ash resistivity. Figure 10-V shows the removal of collected particles is achieved by impacts of hammers or sonic wave.

The common adhesive force, which holds the dust layer over ground electrodes, is enhanced by the electric field greatly. Both electric current and ash resistivity affect electric field strength, thus affect the cleaning of electrodes. Electrical force F_e holding the ash layer has been expressed by G. W. Penney and E. H. Klinger (Penney & Klingler, 1962) as:

$$F_e = \frac{1}{2}\varepsilon_0\left[\left(\frac{J\rho\varepsilon}{\varepsilon_0}\right)^2 - \left(\frac{v - J\rho l}{s - l}\right)^2\right] = \frac{1}{2}\varepsilon_0\left[\left(\frac{J\tau}{\varepsilon_0}\right)^2 - \left(\frac{v - J\rho l}{s - l}\right)^2\right] \tag{14}$$

where J is current density, v is voltage, l is thickness of ash layer and s is interelectrode spacing. For ashes with different resistivity, the deposition over collecting electrodes shows opposite sharp in Figure 11.

For high-current density flows through the area which discharging electrodes point at, low-resistivity ash deposition is thicker under these positions. While the deposition thickness on this area for high resistivity ash is the thinnest. The charges of them cannot easily release and the resulting electric field hampers the collection. The deposition shapes vary in different field of an ESP as shown in Figure 12. The closer the field placed with gas inlet, the thicker the ash deposition is. And the deposition circles overlap one another for field near the gas inlet.

This is because of the ever-changing fly ash concentration principally during collecting. Figure 13 pictures that collected ash can also deposit over the discharging electrodes. The collected ashes agglomerate as a layer around tens of millimeters thick like the photographs show in Figure 14.

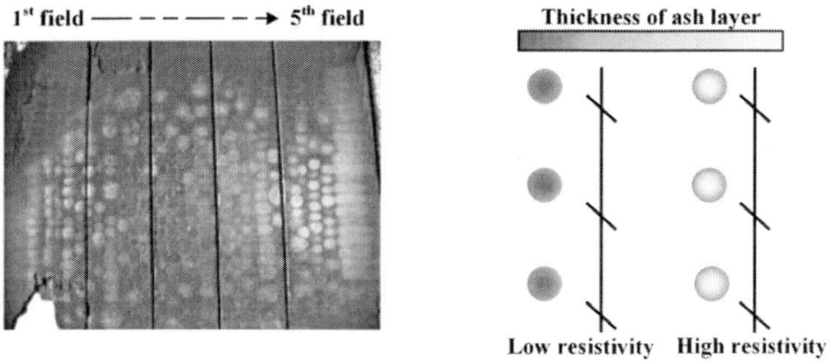

Figure 12. Photograph of collected fly ash layer on the collecting plate with 22 mm discharge electrode distance (left); various thickness of ash layer with different resistivity (right).

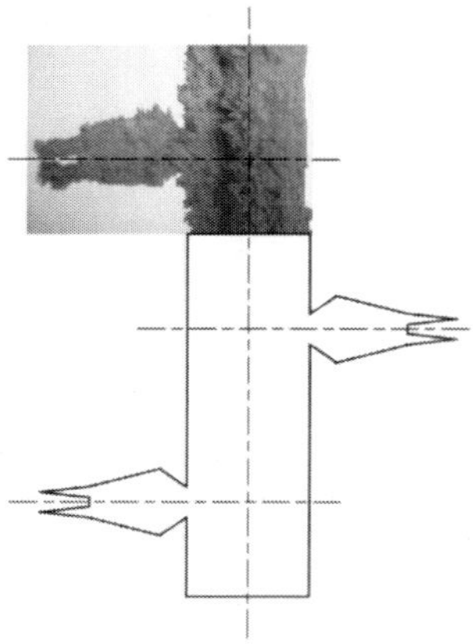

Figure 13. Ash particle deposition on the discharging (arista) electrodes.

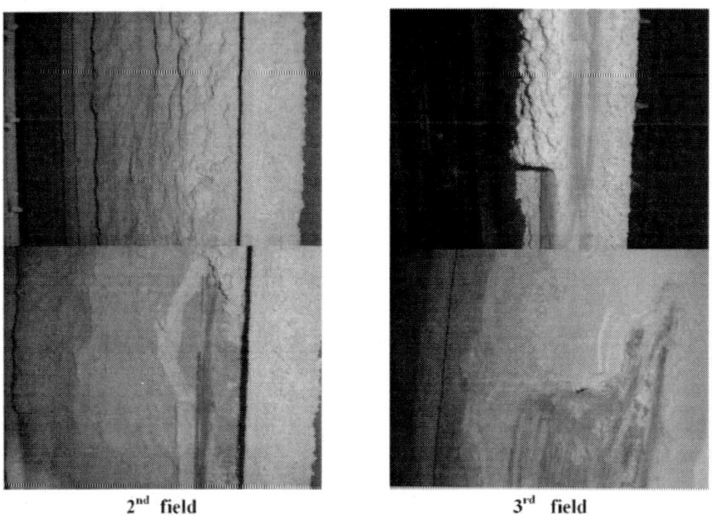

Figure 14. Photographs of collected ash layer within different field.

3. Morphology of Fly Ash

Morphological properties of fly ash have been studied by many investigators. G. Fisher demonstrated microcrystalline structures on the surface of ash samples by scanning electron micrographs, and that most particle spheres are packed with smaller ones (Fisher, Chang, & Brummer, 1976).

He then quantified eleven major morphological classes of coal fly ash by light microscopy (Fisher et al., 1978). A. Ramsden recognized seven categories of fly ash by using light and scanning electron microscopy (SEM) soon thereafter (Ramsden & Shibaoka, 1982). Numerous other classification methods are also made subsequently (Carpenter, Clark, & Su, 1980; sun, Han, Yao, & Xu, 2001).

In this chapter, light microscopy was used for defining six major classes of coal fly ash particles: A. cenospheres; B. dense, rounded particles; C. porous, vesicular particles; D. lamellar, angular particles; E. amorphous particles; F. microspheres.

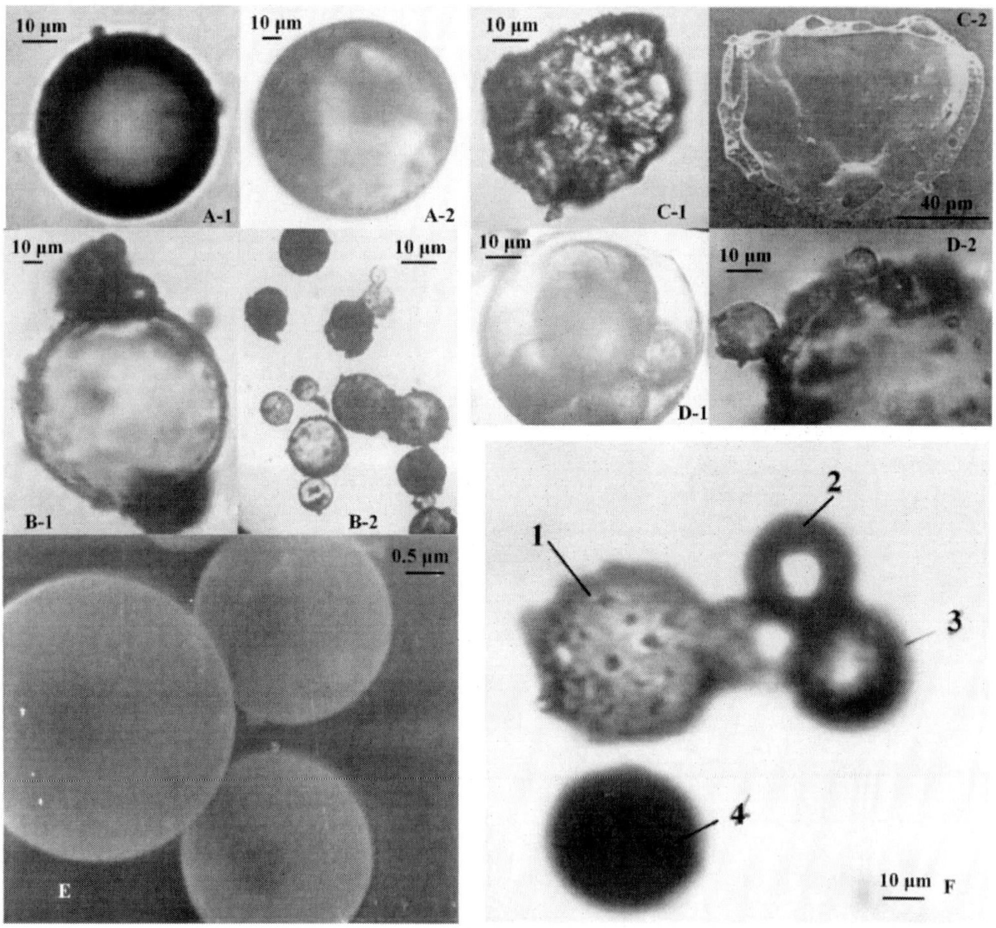

Figure 15. Optical micrograph of typical cenospheres of fly ash samples from Chinese power plants (Optical classification types: 1. nonopaque; 2. light-gathering; 3. semi- transparent; 4. opaque).

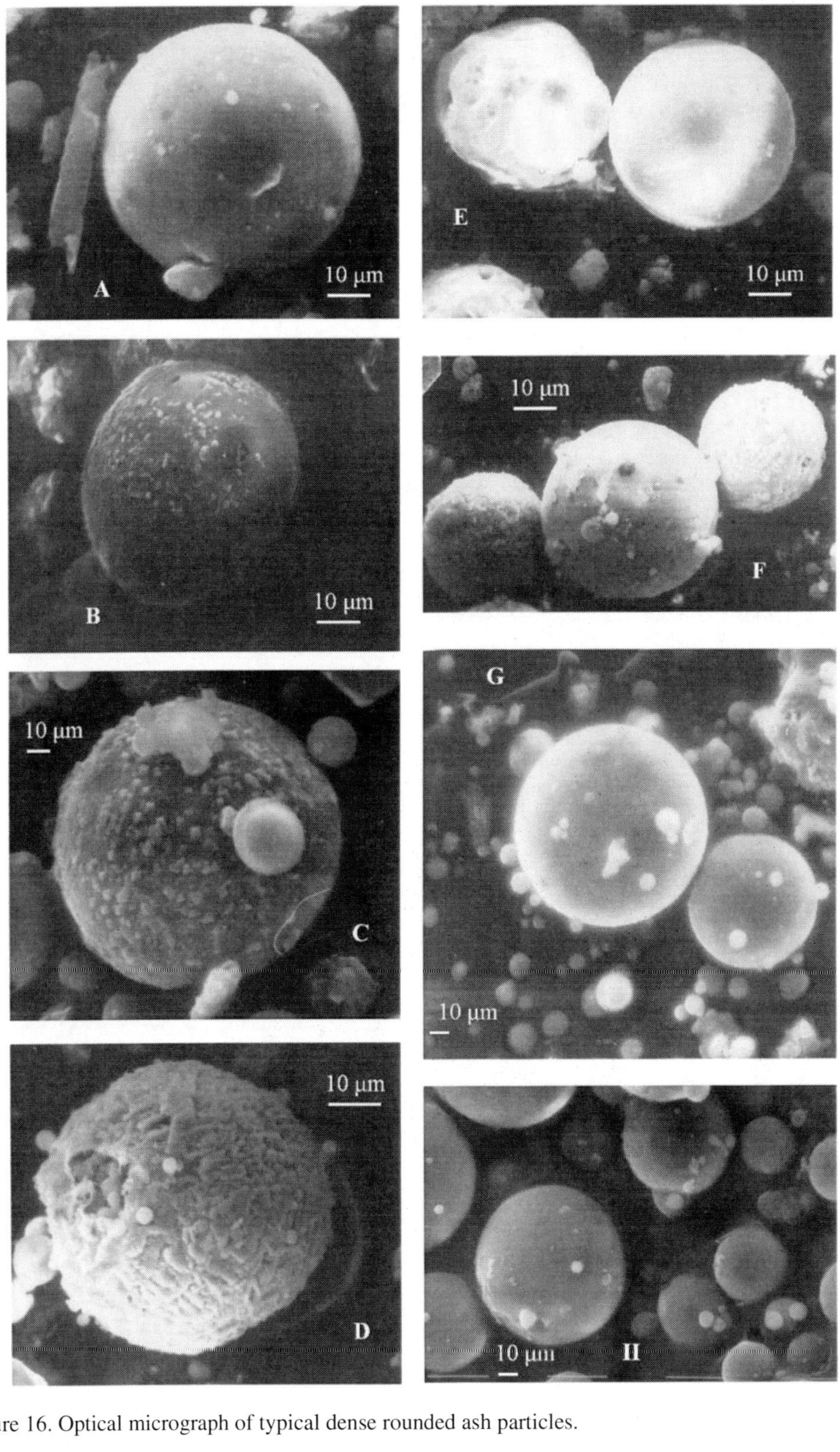

Figure 16. Optical micrograph of typical dense rounded ash particles.

As shown at the lower left in Figure 15, cenospheres can be grouped into four categories by optical transparency: 1. nonopaque; 2.light-gatering; 3.semi- transparent; 4.opaque. The rest micrographs in Figure 15 shows various shapes of ash particle. A-1/2 shows dense rounded cenospheres with smooth edge; B-1/2 shows dense rounded cenospheres with adhering microcrystals; C-1/2 shows angular cenospheres. D-1/2 shows plerospheres and E is the SEM micrograph of dense rounded cenospheres.

Figure 16 illustrates dense rounded particles. A, E shows dense rounded particles with a smooth edge; B, C, F, G, H shows dense rounded particles with adhering microcrystals; D shows dense rough rounded particles. Figure 17 is the micrographs of porous, vesicular ash particles. Figure 18 is the micrographs of lamellar ash particles. These particles can be rounded or angular. Figure 19 is amorphous particles often with flocculent structure. In pulverized coal boilers, amorphous particles are relatively rare, while they are the major trapped particles in circulating fluidized bed (CFB) boilers. Figure 20 shows microspheres. Most of the microspheres are rounded, while there are also angular ones. For a coal-fired power plant, ash particles are not limited to a specific shape, Figure 21 shows the mixed various types of fly ash particles. Fly ash particle shapes are controlled by varying the combustion conditions, vaporization and cooling rate of the combustible materials.

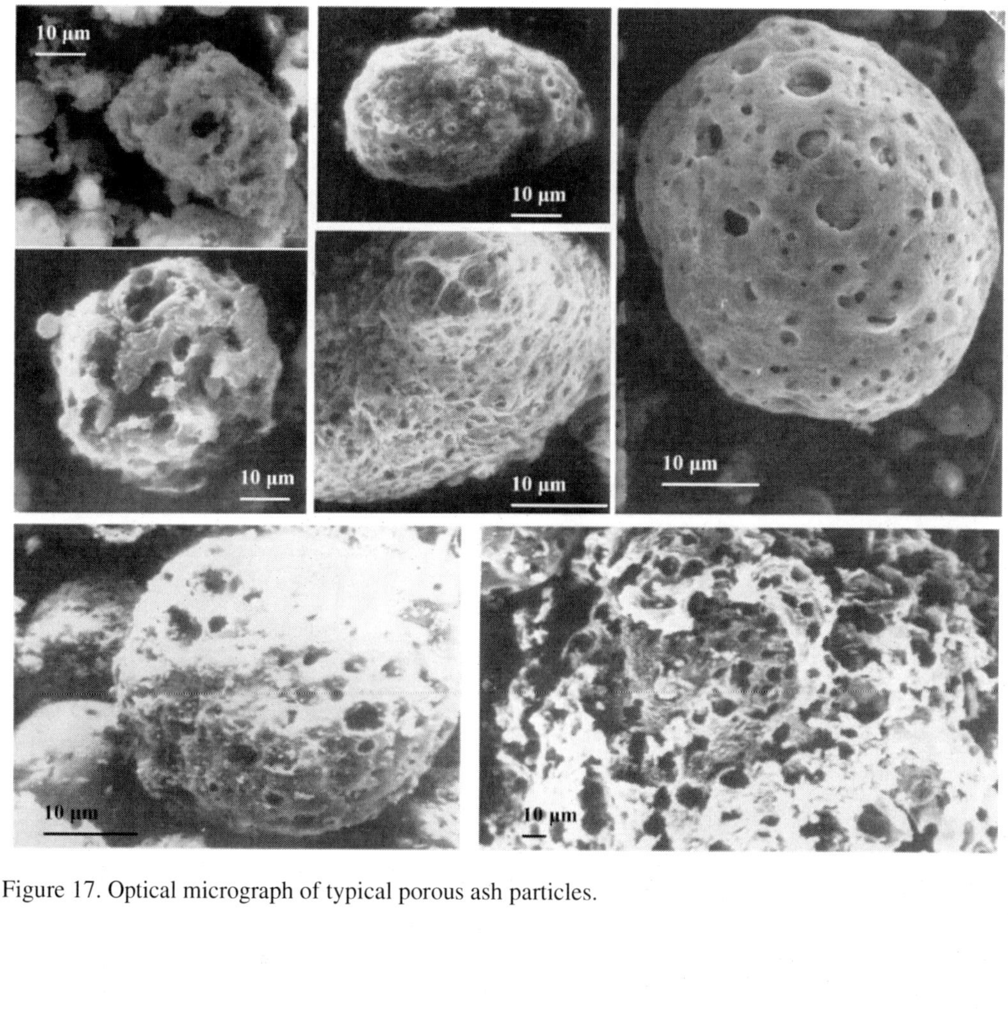

Figure 17. Optical micrograph of typical porous ash particles.

The morphological observation gives some preliminary information of the ashes that can be helpful for subsequent studies (Vassilev & Vassileva, 2005). The major compositions of coal can be deduced from the micrographs. For example, opaque rounded particles can be aluminosilicate or calcareous. Depending on its composition, the particle can end up as a finely crystalline material like a glass or a coarsely crystallized ceramic.

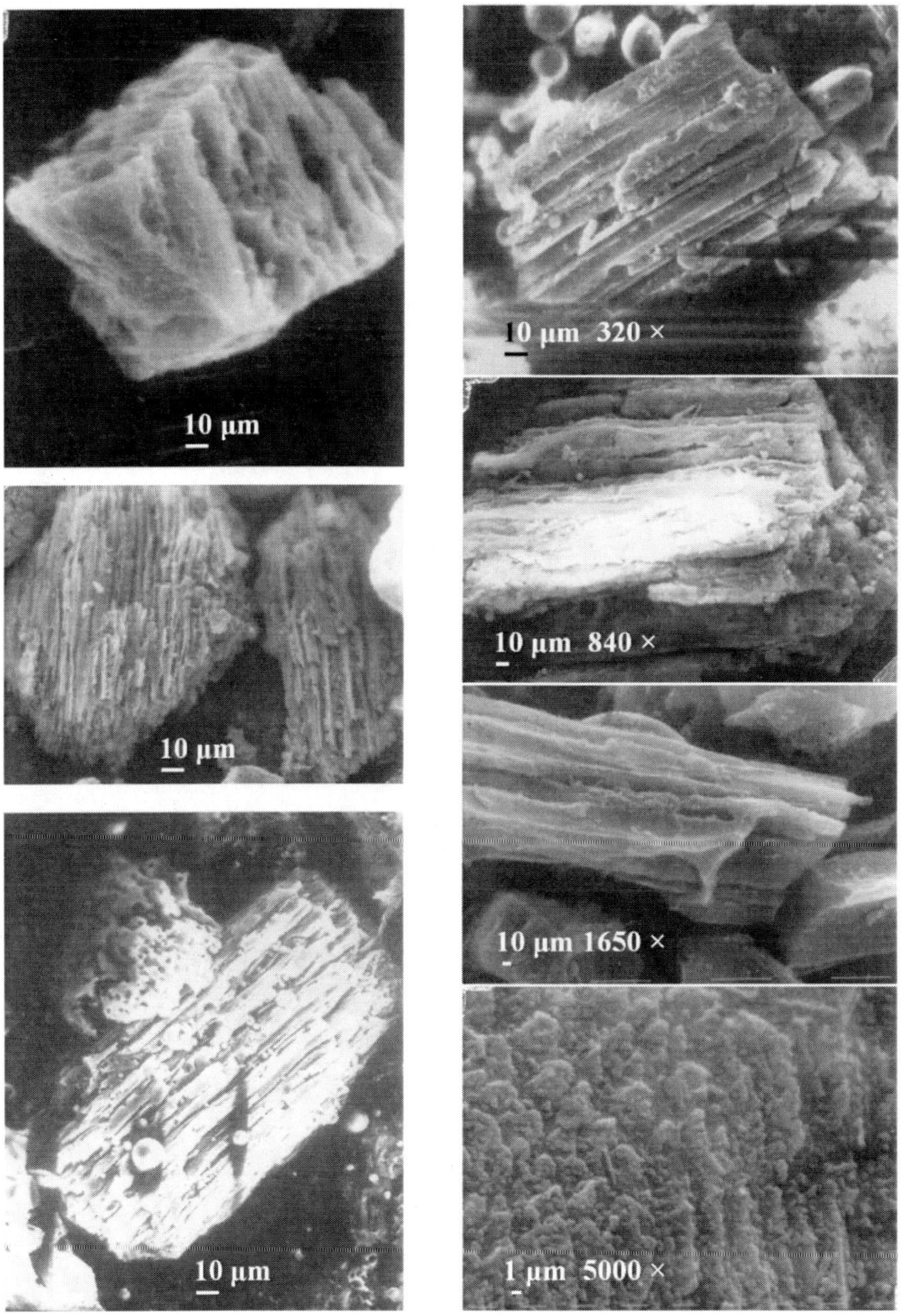

Figure 18. Optical micrograph of typical lamellar ash particles.

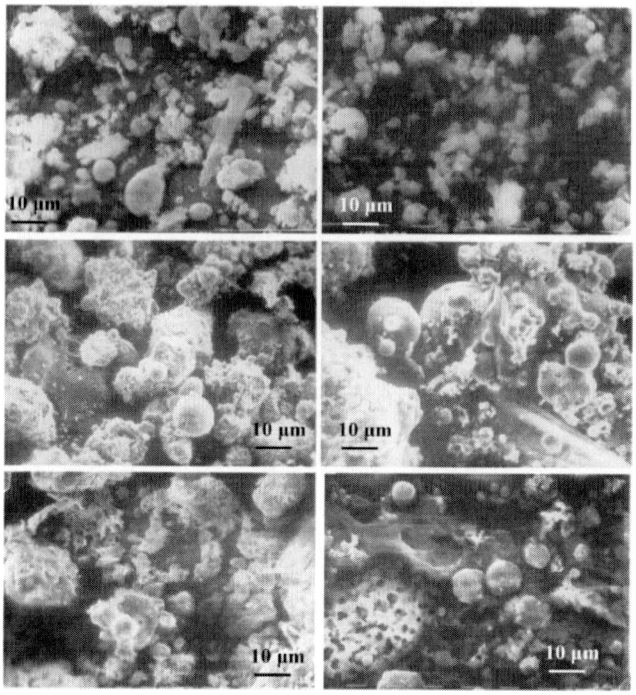

Figure 19. Optical micrograph of typical amorphous ash particles.

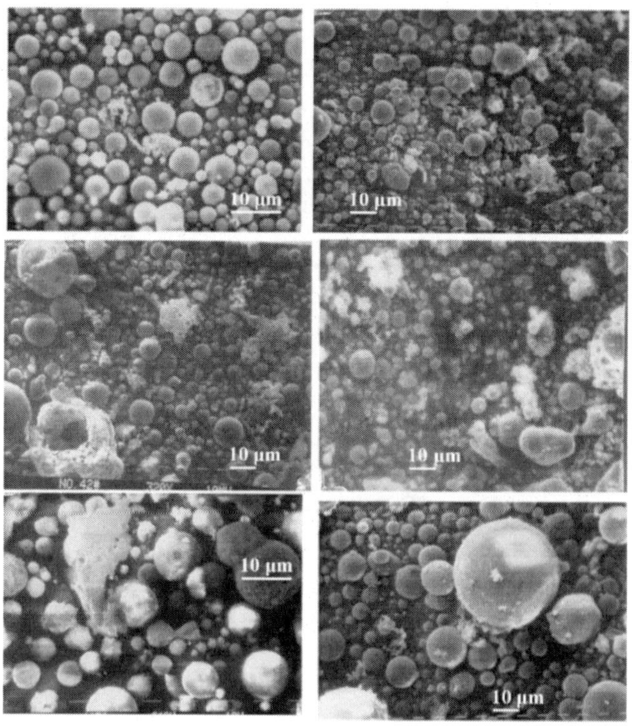

Figure 20. Optical micrograph of typical microsphere ash particles.

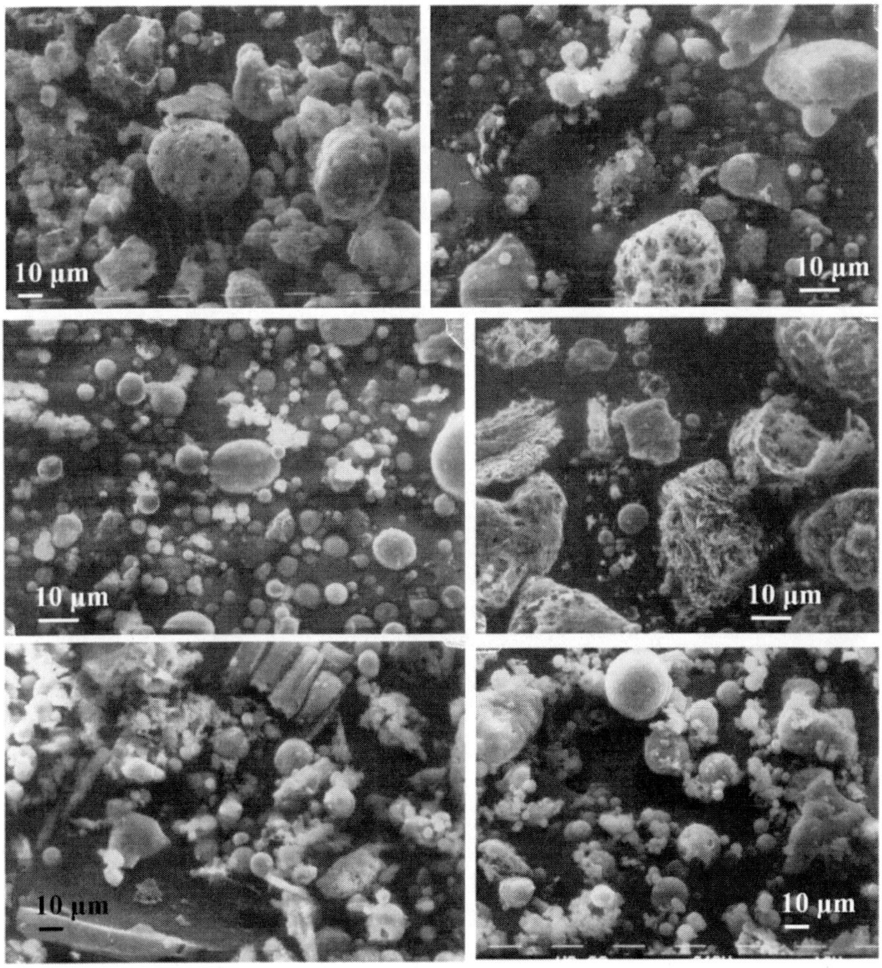

Figure 21. Optical micrograph of various types of ash particles.

4. Ash Resistivity and Its Effect on Particle Collection

4.1. Definition

Ash resistivity ρ, often expressed in the form of ohm·cm, is the degree of electrical conductivity of ash. It is the ratio of resistance and length to a cross-sectional area for a given ash layer. That is: $\rho = R \cdot A / l$, in which R is the resistance, l is the length and A is the cross-sectional area.

The resistivity of a collected layer of fly ash is sensitive to chemical, electrical factors and environmental conditions. It is known the electrical resistivity is determined by two mechanisms, the volume resistivity ρ_v and the surface resistivity ρ_s. These two processes achieve conduction paths through the ash layer separately. And the total resistivity can be calculated in the form of the products of their resistivity divided by their sum. The equivalent resistivity is:

$$\rho_{vs} = \frac{\rho_v \rho_s}{\rho_v + \rho_s} \tag{15}$$

Both the volume and surface conduction mechanisms depend on the chemical composition of fly ash, such as alkali ions, potassium and iron. The surface conduction resulting from adsorption of sulfuric acid vapor can be considered as a parallel resistor with surface conduction caused by ash surface and water vapor. That is $\rho_s = \dfrac{\rho_a \rho_{sw}}{\rho_a + \rho_{sw}}$, where ρ_a is acid resistivity and ρ_{sw} is the latter one. The maximum resistivity was defined by Bickelhuapt as the peak value of the total resistivity, like the curve vertexes in Figure 48. The maximum resistivity is commonly used for comparisons among data from different investigators for representing the equilibrium state of volume and surface conduction.

4.2. Migration Velocity

Migration velocity represents the precipitator efficiency. It drops with increasing resistivity for the pilot-scale tests (for cement dust) as the solid line shows in Figure 22. The ESP performance tends to be good for ash with resistivity ranging from 10^7 to 10^{10} Ω·cm. For higher resistivity, which depresses the electric field and lowers levels of particle charges, the velocity drops rapidly. And the performance for this resistivity range can be predicted to be poor. The data of White supported this hypothesis, ashes with resistivity higher than 10^{10} Ω·cm show relatively low migration velocity. However, collection of low-resistivity ash is also inefficient as shown in this figure. Its drift velocity reduces with decreasing resistivity lower than 10^7 Ω·cm. Obviously, a good collection efficiency of an ESP requires the dust resistivity falling in a certain optimum range. And it is apparent the collection efficiency can be improved by adjusting ash resistivity.

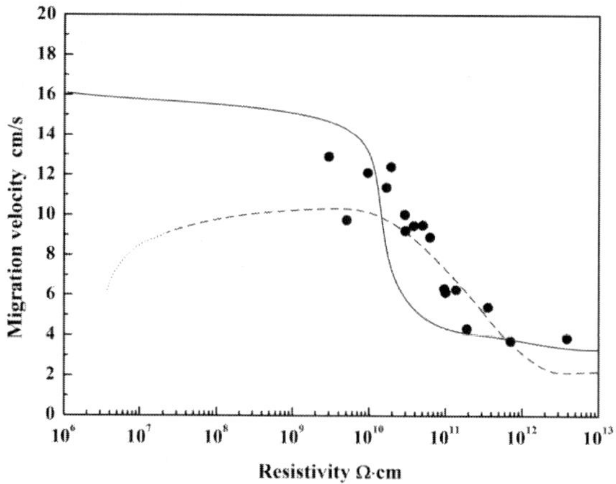

Figure 22. Migration velocity versus ash resistivity. (solid line-laboratory pilot ESP (Sproull, 1954); dash line-full-scale ESP (Coe & Lagarias, 1986); solid point-field performance testing (White, 1974)).

4.3. Spark-Over and Back Corona

The ionic current of ESPs must pass through the porous layer formed by the collected ash. The resistance of ash induces a voltage drop across the layer as shown in Figure 23. If this voltage drop is relatively low, the overall dust removal process is only affected slightly. However, if the electrical resistivity of ash layer is high enough, it can cause an electrical breakdown of the dust layer. The positive discharge, which is reverse to the initial discharge of ESP, occurs from the layer and injects into the inter-electrode space. The inter-electrode voltage below sparking voltage can limit the total particle charges and reduce the particle collection performance seriously. This bipolar equilibrium condition that results in an electrical breakdown of the ash layer at fairly low current density is called back corona or reverse ionization. If the interelectrode voltage drop is relatively high, the positive ions produced by the breakdown of the ash layer can initiate a spark in a condition called spark-over (Lawless & Sparks, 1980).

Back corona are usually observed on ash layers with resistivity higher than 5×10^{10} Ω·cm (Czapka, 2011). The total charge of collected fly ash layer can be expressed as a function of releasing time in an equivalent circuit: $Q(t)=Q_0\exp\left(-\frac{t}{\rho\varepsilon_0\varepsilon_r}\right)$. The charge of high resistivity ash layer is, apparently, more difficult to release. Back corona will effect a considerable reduction in collection efficiency of ESPs. The phenomenon affects submicron particles especially, for which the grade efficiency can be reduced over 60% (Chang & Bai, 1999). Optical emission spectroscopy and visual forms of back corona have been investigated by many researchers (Czech, Sobczyk, & Jaworek, 2011; Hoferer & Schwab, 2000; Masuda & Mizuno, 1977). Several mathematical models have also been built for describing the discharge characteristics. These models are either based upon the charges of collected particles and their contribution to the adjacent electric field over collecting plate or the simplified equivalent circuit of the electric-discharge plasma column, which often consists of numerous parallel RC two-terminals.

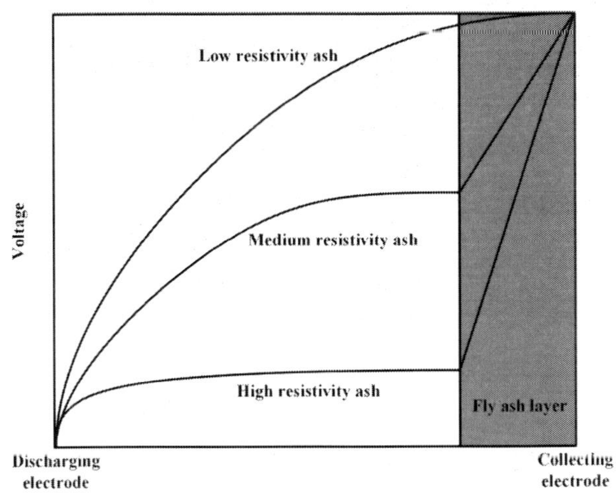

Figure 23. Schematic diagram of influence over voltage in inter-electrode space by resistivity.

I. L. Zou yielded the critical dust layer thickness for back corona occurrence by using mathematical models of charges and an additional electric field in the collected ash layer (Zou, 2005). The total charge of the ash layer Q can be stated as:

$$Q=2\pi\varepsilon_0\varepsilon_r J\rho l^2 \tag{16}$$

where l is the thickness of ash layer.

The ash layer can generate an electric field with the strength E_p near the plate:

$$E_p = \frac{Q}{2S\varepsilon_0\varepsilon_r} = \frac{\pi J\rho l^2}{S} \tag{17}$$

By substituting the layer area formula into equation (17), the critical thickness of ash layer l_{cr} for back corona occurring can be gained:

$$l_{cr} = \frac{c}{2}\sqrt{\frac{E_a}{J\rho}} \tag{18}$$

where $2c$ is the distance between arista on the electrodes, and E_a is the breakdown voltage of air.

II. T. Czech shows a simplified equivalent circuit for the ESPs as follows (Czech, Sobczyk, Jaworek, & Krupa, 2012). The voltage response of the RC circuit is an exponential decay of the initial voltage:

$$U_s(t) = U_0 \exp\left(\frac{t}{C_2(R_0+R_1)}\right) \tag{19}$$

The potential at the surface of the layer can increase up to a threshold magnitude resulting from equation (20). And the threshold breakdown electric field across the layer $E_{breakdown}$ is obtained.

$$E_{breakdown} = \frac{U_{threshold}}{l} \tag{20}$$

The back corona discharge takes place when the electric field across the ash layer is higher than the breakdown strength:

$$\rho J = \frac{U_s(t)}{l} = E_p \geq E_{breakdown} \tag{21}$$

III. Tamás Iváncsy calculates the formation of back corona using a simplified model (Ivancsy, Kiss, & Berta, 2009), which expresses the E_p as a time function of the electric field strength at the surface of the collecting plate interface:

$$E_m = E_s(ht) + \left[J\rho + E_s(ht)\right]\left(1 - e^{-\frac{t}{T}}\right) \tag{22}$$

where E_s is the field strength at the surface of the ash layer, $J = J_{dust} + J_{ion}$, and $T = \rho\varepsilon_r$. According to equation (22), it is obvious that after a given time, E_p will reach to the breakdown limit $E_{breakdown}$. The current densities are calculated according to the donor cell method.

IV. The collecting process can be analyzed by the equivalent R-C circuit as Figure 24 shows. For the formation of back corona, the simplified circuit determined U_m, the instantaneous voltage across the ash layer using the following assumptions: Average resistivity ρ, relative permittivity ε_r is considered constant. Growth rate of the dust layer thickness is considered constant for each analyzed section. Total applied voltage of the analyzed section is also considered constant.

For the first ash layer film, the instantaneous voltage $Us(t)$ of it can be expressed as:

$$U_s(t) = U_0 \exp\left[\frac{t}{C_1(R_D + R_1)}\right] \tag{23}$$

where U_0 is the original voltage applied across certain ash layer film, t is releasing time for an ash layer film, R_d is the total intrinsic resistance of ESP, high-voltage generation and the interelectrode space, R_l is resistance of a single ash layer film. $R_1 = R_2 = R_3 = \rho\frac{d_a}{A}$, d_a is the thickness of a single ash layer, A is the area of the ash layer.

The total instantaneous voltage $U_{ts}(t)$ is the sum of instantaneous voltage for each film:

$$\begin{aligned} U_{ts}(t) &= U_0 \exp\left[\frac{n\Delta t}{C_1(R_D + R_1)}\right] + U_0 \exp\left[\frac{(n-1)\Delta t}{C_2(R_D + R_1 + R_2)}\right] \\ &\quad + \ldots + U_0 \exp\left[\frac{\Delta t}{C_n(R_D + R_1 + R_2 + \ldots + R_n)}\right] \\ &= U_0 \exp\left[\frac{n\Delta t}{C_1(R_D + R_1)}\right] + U_0 \exp\left[\frac{(n-1)\Delta t}{C_1(R_D + 2R_1)}\right] \\ &\quad + \ldots + U_0 \exp\left[\frac{\Delta t}{C_1(R_D + nR_1)}\right] \end{aligned} \tag{24}$$

where Δt is the forming time interval. The field strength $E_{ts}(t)$ can be expressed as:

$$E_{ts}(t)=\frac{U_{ts}(t)}{nd_f}=\frac{U_{ts}(t)}{n(d_a+d_g)}=\frac{U_{ts}(t)}{l} \tag{25}$$

when $E_{ts}(t)$ is higher than the critical breakdown field strength, the breakdown of the ash layer occurs that induces back corona. By several simplifications, the equation (24) is approximately equal to:

$$\begin{aligned} U_{ts}(t) &\approx nU_0\times\frac{2\varepsilon AR_d+d_pl}{2\varepsilon AR_d} \\ &=U_t\times\frac{\rho l}{\rho l+AR_d}\times\frac{2\varepsilon AR_d+d_pl}{2\varepsilon AR_d} \end{aligned} \tag{26}$$

where U_t is the total applied voltage of the ESP.

The criterion determining whether back corona would occur is:

$$E_{ts}(t)=U_t\times\frac{\rho}{\rho l+AR_d}\times\frac{2\varepsilon AR_d+d_pl}{2\varepsilon AR_d}\geq E_{breakdown} \tag{27}$$

All the models illustrate that back corona occurrence is determined by the resistance of ash layer. Besides matched high-voltage generation, real-time control of ESPs based on resistivity prediction also improves its performance.

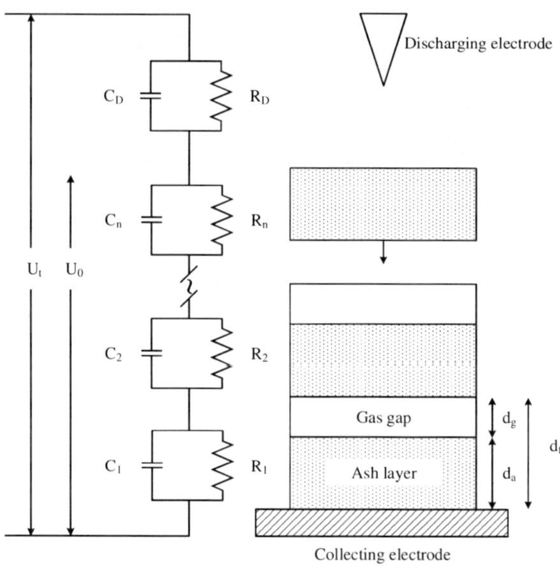

Figure 24. Schematic of equivalent R-C circuit for discharge to fly ash layer.

The ESP and its electrical system can be simplified as an equivalent circuit as shown in Figure 25. The discharging area can be regard as a variable resistor and a capacitor in parallel. Figure 26 illustrates the current flows through an ESP during the rising period of applied voltage. The current sustains during the decay period at a relatively low value. If back corona occurs, the sustained current grows higher. It is harmful for both power supply and collecting process as shown in Figure 27. To avoid the happening of back corona, pulsed power supply is probable in an industrial application.

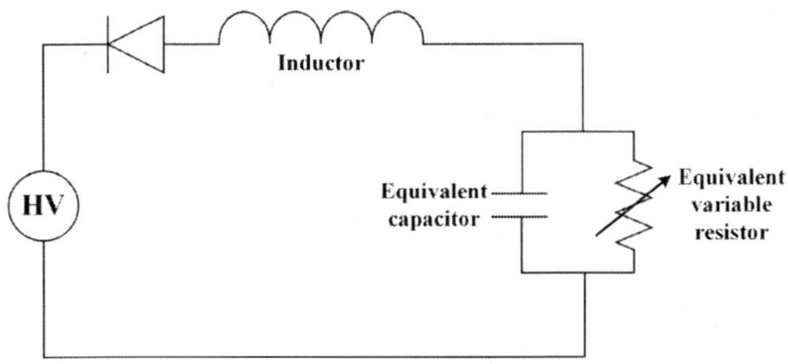

Figure 25. Schematic of equivalent circuit for ESP.

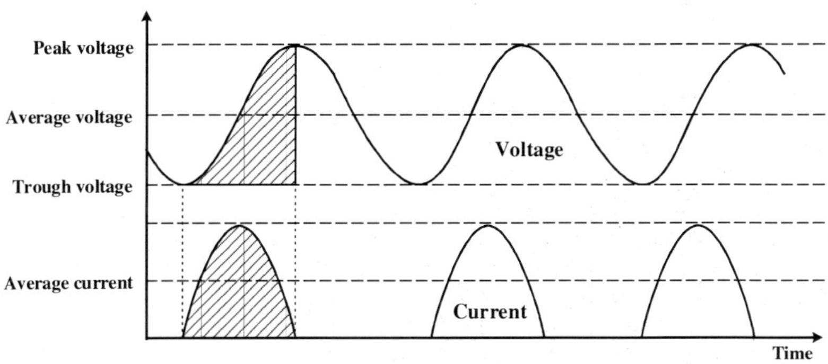

Figure 26. Typical voltage & current waveforms for several discharging periods.

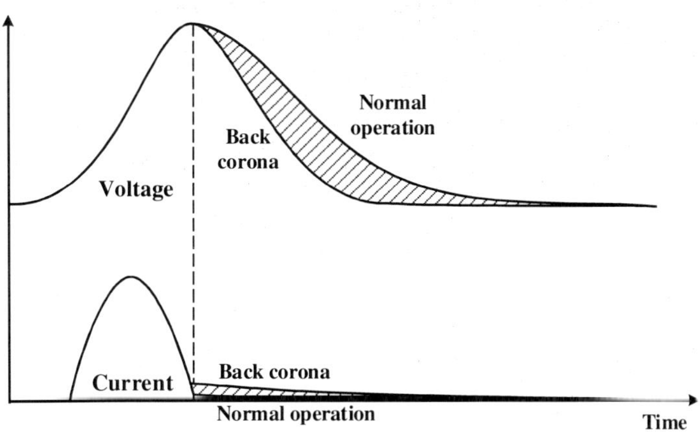

Figure 27. Typical voltage & current waveforms for one certain discharging period.

4.4. Leap Collection and Reentrainment

In the presence of low resistivity ash (often lower than 10^4 Ω·cm) and low current density through the plates, the collected particles may lose their charges shortly after reaching the collecting surface. They reentrain the gas stream due to lack of electric force holding them on the electrodes.

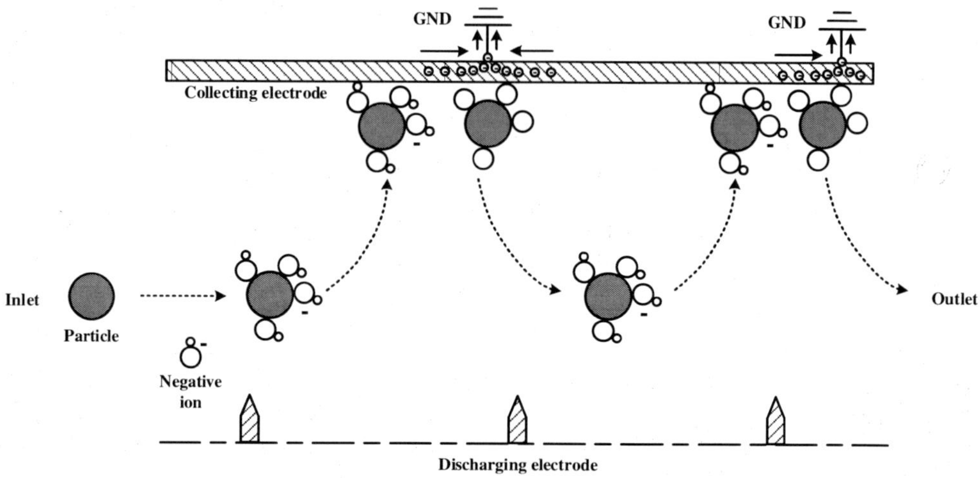

Figure 28. Schematic illustration of the leap collection process in ESPs.

The low resistivity particles could leave the collecting surface several times like jumping. This condition is commonly called leap collection or jumping phenomena. These particles have an interesting name: "dancing particles". As the particles approach the collecting electrodes, they are repelled back into the gas stream to be recharged and the process repeats. Most of them escape from the ESPs as they reach the outlet. During the rapping over the collecting plate, lumps of ash fall down towards the hopper under gravity. Some of them are caught again while most of them fall into the hoppers. The process in which collected ash gets back to the gas stream is called reentrainment.

4.5. Voltage & Current

Voltage-current characteristics of ESPs are commonly used for determining the operation conditions.

In general, the characteristics line of a precipitator collecting ash lies under that of clean condition. When back corona occurs, the characteristic curves up much more sharply. Typical voltage-current characteristics of ashes with different resistivity are shown in Figure 29.

It is known that, before the onset of back corona, the current density through collecting plate reduces with increasing resistivity at the same applied voltage. Once back corona starts, the net effect of a whole matrix of back corona is to produce an open loop characteristic with a sharp curve. Hysteresis will be exhibited when the applied voltage is raised and then lowered.

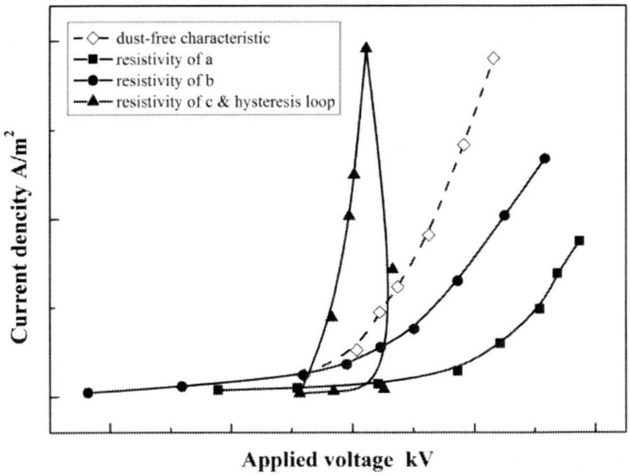

Figure 29. Typical voltage-current characteristic of an ESP for various ash resistivity (Resistivity: a<b<c) (McLean, 1988).

4.6. Electric Force

Figure 30 shows the magnitude of the electric forces for ash layer with various resistivity. The solid line by Durham indicates there are apparent electric repulsive forces acting on low resistivity ashes. The repulsion promote process called electrical reentrainment during which ashes reenter gas stream. For many full-scale ESPs, the reentrainment does not appear to be a problem as the exhaust temperature approaches to dew point. The slight difference makes the dust more adhesive in spite of its low resistivity (Parker, 1996). The dash line in this figure suggests the attractive force holding particles on the plates is predominant for higher resistivity. And ashes with relatively high resistivity can hardly be removed by rapping on collecting electrodes.

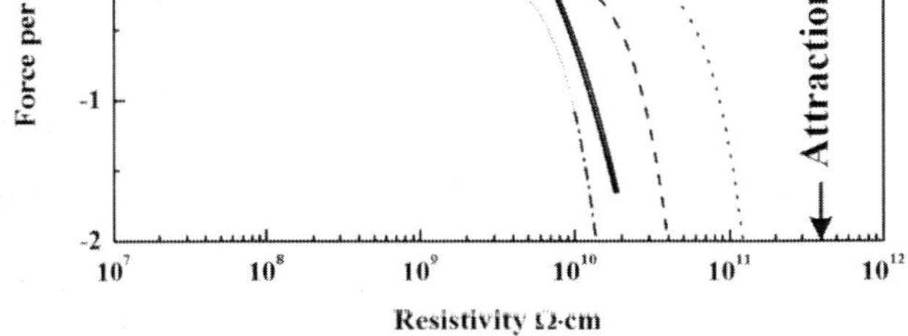

Figure 30. Force acting on particle surface for different electrical resistivities, current densities and field strengths (Durham, Rugg, Rhudy, & Puschaver, 1990; Oglesby & Nichols, 1978).

4.7. Collecting Efficiency

Figure 31 illustrates ESPs collect fewer particles with high resistivity for certain series of ash. The downward trend is different from one another. It has been discussed in the subsection of migration velocity. High resistivity depresses the electric field and lessens the particle charge. Therefore, collecting performance is lower with increasing ash resistivity. For ash particles with resistivity lower than 10^4 Ω·cm, leap collection phenomenon would hamper the collection.

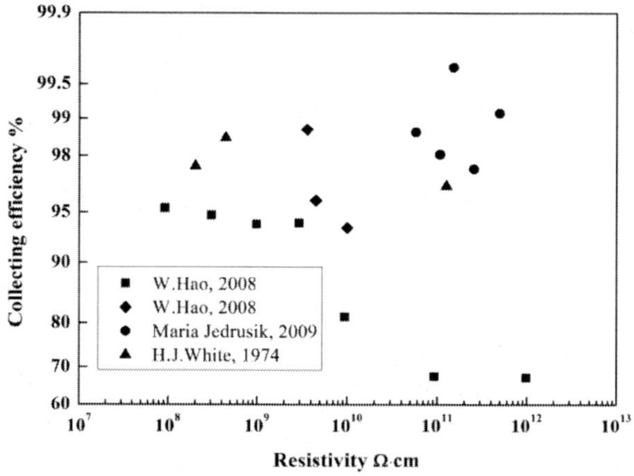

Figure 31. Characteristics of collecting efficiency for ashes of various resistivity. (Hao & Hou, 2008; Hao, Hou, & Xiong, 2008; Jędrusik & Świerczok, 2009; White, 1974).

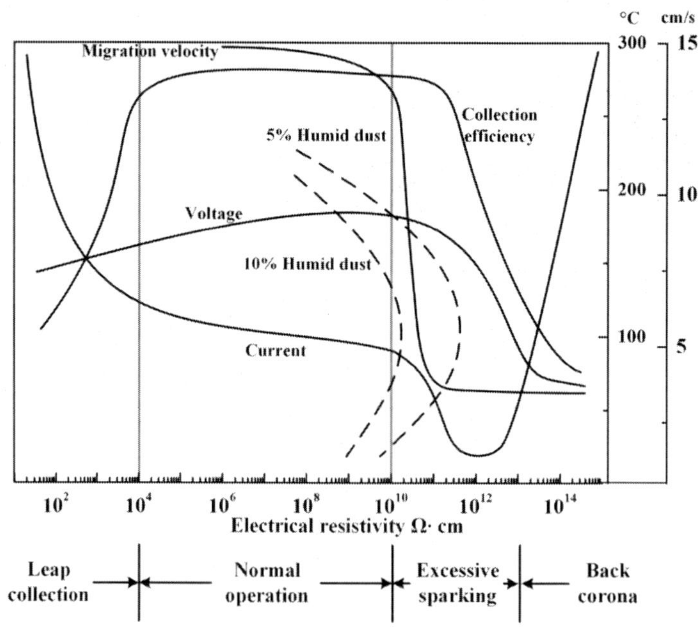

Figure 32. Relationship of the particle resistivity and operation parameters in the ESPs. (Walker, 1968).

In summary, characteristics of ESP operation and collection are dependent on resistivity of fly ashes entrained in the gas stream. The general relationship between electrical resistivity and operational variables such as collecting efficiency, migration velocity, voltage and current of ESPs are shown in Figure 32 (Walker, 1968). If the resistivity is too low, below 10^4 Ω·cm, collected particles will release their charges rapidly and be reentrained in the gas. It can surely increase the operating current for charging and losing charges more easily. If the resistivity is too high, for example above 10^{10} Ω·cm, the input power will be limited for that these particles are hard to charge. After depositing on the plates, particles lose their charges slowly due to the low conductivity of ash layer. These characteristics increase the voltage gradient across the layer, reduces the charging and collecting fields, and decrease migration velocity of particles. For fly ash with a resistivity higher than 10^{13} Ω·cm, the voltage gradient across the ash layer will lead to back corona. During this breakdown phenomenon, positive ions neutralize the unipolar space charge at interelectrode space, and reduce collecting efficiency severely.

5. Measurement of Fly Ash Resistivity

Measurements of fly ash resistivity are all based on its definition, which is $\rho = \frac{U \cdot A}{I \cdot l}$.

All methods involve at least two electrodes. A voltage superimposed between them results in a current to be measured. The ratio of applied voltage and current represents resistance of tested fly ash. The fly ash resistivity is earned when multiplied the geometric parameters, which indicates the cross-section area and length/thickness of the ash sample, by the calculated resistance. In general, ash resistivity can be measured either in laboratory or in situ. The major difference between them is the procedure of placing ash sample. The slight difference can make substantial influences on the stacking density and measured resistivity.

5.1. Laboratory Techniques

There are three standard techniques of laboratory measurement methods for ash resistivity: the American Society of Mechanical Engineers Power Test Code 28-1965, Determining the Properties of Fine Particulate Matter (ASME PTC28); International Electrotechnical Commission 1241-2-2, Test Methods-Section 2: Method For Determining The Electrical (IEC1241-2-2); resistivity of dust in layer standard 5958-1, Code Of Practice For Control Of Undesirable Static Electricity (BS5958-1).

In ASME PTC 28, ash samples are placed in a conductive cell like a shallow cup. By drawing a straight edge across the top of cell in the same direction, the upper surface of filled cell is smoothed. This procedure keeps the thickness of ash sample, and stacks that dust sample naturally. The movable circular electrode smaller than the cell is then lowered onto the sample surface freely under its own gravity. It presses the ash with a pressure about 10 g/cm^2. The environmental chamber maintains the temperature and humidity till equilibrium. After these processes, a record of applied voltage and measured current is obtained. Ash resistivity under different temperature, humidity and electric field can be calculated from the data.

In IEC 1241-2-2, the measuring cell consists of two cuboid stainless steel electrodes and two shorter cuboid isolated electrodes on an isolated plate. The four electrodes enclose a rectangular cell, in which ash samples are filled. The resistance of empty cell like a parallel resistor with ash sample should be measured first. The rest filling and measuring procedures are quite similar with ASME PTC 28. The calculation should exclude the contribution by parallel resistance of measuring cell.

Apparatus used by BS5958-1 fix two circular stainless steel electrodes inside an acrylic cell. Ash sample fills the whole cell naturally. The ash powder is then tamped down by gently knocking the cell on the bench. A voltage applied between two electrodes evokes current flowing through ash sample. Initial polarization transients should decay for sufficient time before measuring the steady current. This method assumes current density between electrodes are uniform and there is no distortion of the current flow at the edges of them. To reduce the error because of edge effects, a dielectric ring should be adopted for the electrodes.

The structure of measuring electrodes can also be an earthed cylinder electrode with a concentric column electrode inside it (Barranco et al., 2007). The general arrangement of above apparatuses is shown in Figure 33. These methods measure ash sample stacked naturally or artificially. The stack density different from dust layer over the collecting plate in ESP can affect the measuring results greatly.

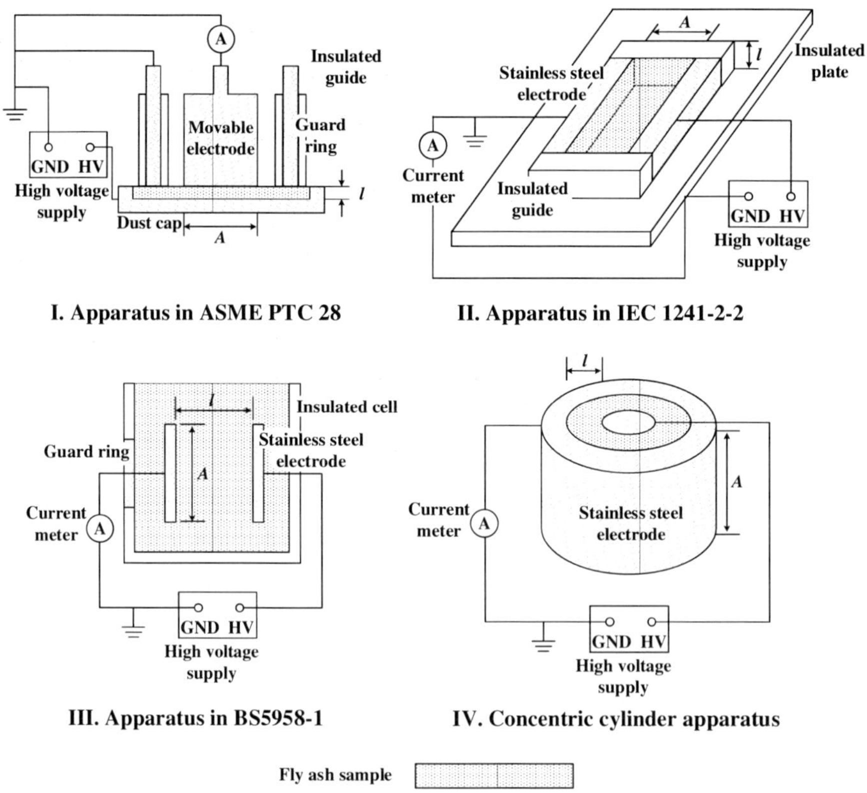

Figure 33. General arrangement of electrical resistivity apparatus.

Another type of apparatus simulating ash collection processes in ESPs is also feasible for measuring. The point-plane typed apparatus is the earliest device simulating the precipitator operation process by the collecting ash layer on its plane electrodes. Determination of ash resistivity can be made either by changes of current-voltage characteristic of the corona discharge or by the measuring leakage current through the ash layer. For the same purpose, comb-type electrodes superimposed with high voltage are used for collecting ash between the comb teeth. By applying another high voltage between the teeth after collection, resistivity can be calculated. Apparatus that is more complex is a bench scale ESP-like device, through which gas stream flows in a close circle. The measuring device is often point-plane structure.

5.2. In Situ Techniques

Several instruments are available for achieving in situ resistivity measurement. The above point-plane structure is frequently used for industrial applications. The measuring condition is most close to operating mode while the thickness of ash layer can be hardly determined. Comb-type instruments can also be inserted into the flue duct of ESPs for in situ measurement.

The concentric cylinder method was developed in Europe and U.S. by inserting directly into the duct like Figure 34. The dust particles in the gas stream flow into the cyclone cell. They are collected by gravity and slide downward to the bottom. A mechanical vibration compacts the particle layer. At the bottom of the cyclone, a column electrode (inner electrode) is fixed at the center of the cylinder electrode (outer electrode) vertically up. The resistivity of ash sample is calculated as $\rho = kR$, where R is the resistance of the dust layer and k is a constant for a specified measuring cell. The constant k is defined by $k = \dfrac{2\pi L}{\ln(r_2 / r_1)}$, where L is the length of cell, r_1 is radius of inner electrode and r_2 is radius of outer electrode.

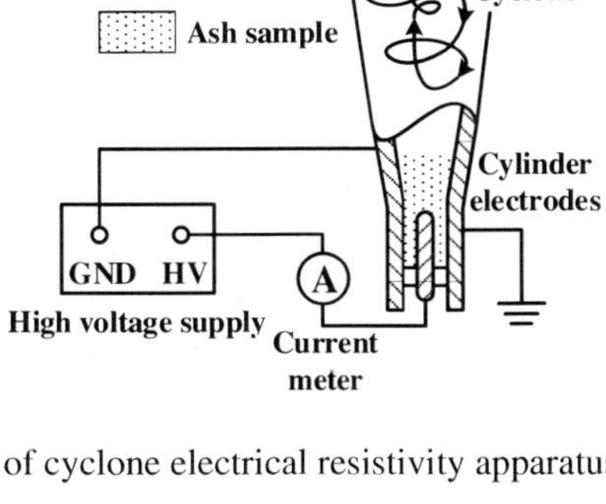

Figure 34. General arrangement of cyclone electrical resistivity apparatus.

5.3. Calibration

Because Li_2O plus Na_2O, Fe_2O_3 and CaO plus MgO concentrations significantly affect the resistivity, a normalization method has been proposed based on their average concentration to evaluate the influences of the other ash compounds on the resistivity. For example, if the average value of the atomic percentage lithium plus sodium is approximately 0.4%. The normalization consists of two following steps (Li et al., 2012):

1) Firstly, the β value for a given ash with the maximum resistivity ρ_{max} and lithium plus sodium atomic concentration of A_{ls} is calculated as:

$$\beta = \lg \rho_{\max} + 1.21 \lg A_{ls} \quad (28)$$

2) Then, its normalized resistivity ρ_N is defined as:

$$\lg \rho_N = -1.21 \lg(0.4) + \beta \quad (29)$$

Or substituting the equation (28) into the equation (29), we get the normalized resistivity as:

$$\lg \rho_N = \lg \rho_{\max} + 1.21 \lg A_{ls} + 0.481507 \quad (30)$$

The calibration procedures for Fe_2O_3 and CaO plus MgO are similar with the above steps.

6. Factors Influencing Resistivity

Ash particles encountered in ESPs vary in composition from metallic oxides to silica as the major ingredient. Meanwhile, the gas streams will also show different temperature, humidity and sulfur dioxide content due to various types of fuel and operating processes. The current conduction through the ash layer of course varies under different depositing conditions. Resistivity is principally governed by the following factors: the amount of diverse constituents, the temperature, the humidity, the electric field across the ash layer and so on.

6.1. Chemical Compositions

There are two mechanisms determining ash electrical resistivity, the surface conductivity, which is achieved by releasing alkali metal ions from the surface and forming absorbed surface layer to act as charge carriers, and bulk volume conduction. The latter is governed by transportation of electrical charges through the individual fly ash particles, while oxides of alkali metals, iron, aluminum, can act as charge carriers. Relationships between various constitute concentrations and measured ash resistivity are important for predicting resistivity of undefined ash and operating conditions of ESPs.

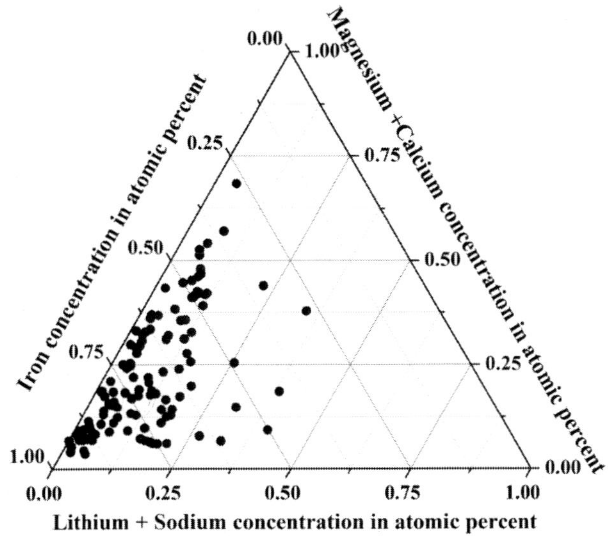

Figure 35. Chemical composition of the fly ashes from Chinese power plants (122 samples) (Li et al., 2012).

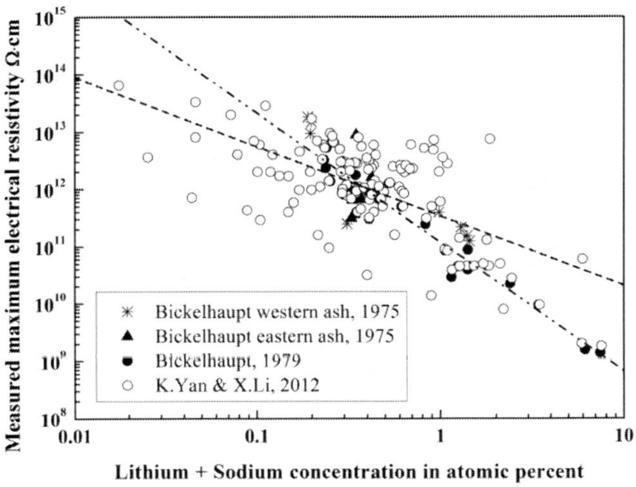

Figure 36. Maximum ash resistivity versus lithium plus sodium concentration. (dash line is fitting line for K. Yan & X. Li's data, dash dot line is fitting line for R. Bickelhaupt's data) (Bickelhaupt, 1975, 1979; Li et al., 2012).

The alkali metals (lithium plus sodium), iron and calcium plus magnesium concentrations of 122 fly ash samples from Chinese power plants are shown in Figure 35. Most of the lithium plus sodium, the potassium, the iron, the magnesium plus calcium, the aluminum and the sulfur conponents for these samples occur in the range of 0.3~0.5%, 0.25~0.8%, 0.6~2.5%, 2%~5%, 6.3~10% and 0.05~0.4%, respectively (Li et al., 2012). It is obvious the lithium plus sodium concentration of Chinese fly ash samples spreads in a narrow range, while iron concentration scatters over a large range.

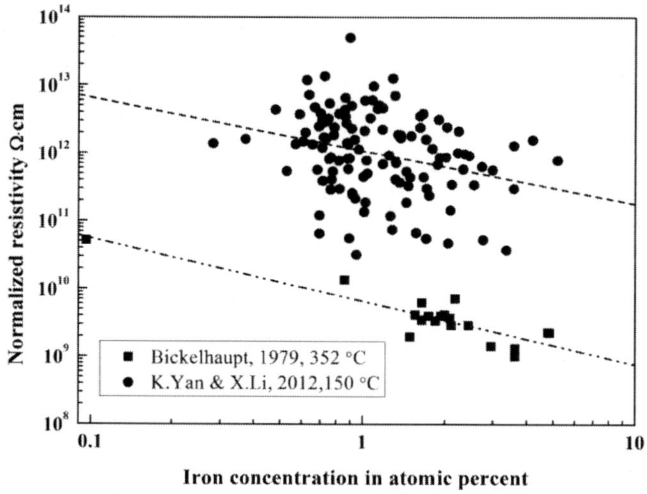

Figure 37. Normalized resistivity versus iron concentration (Bickelhaupt, 1979; Li et al., 2012).

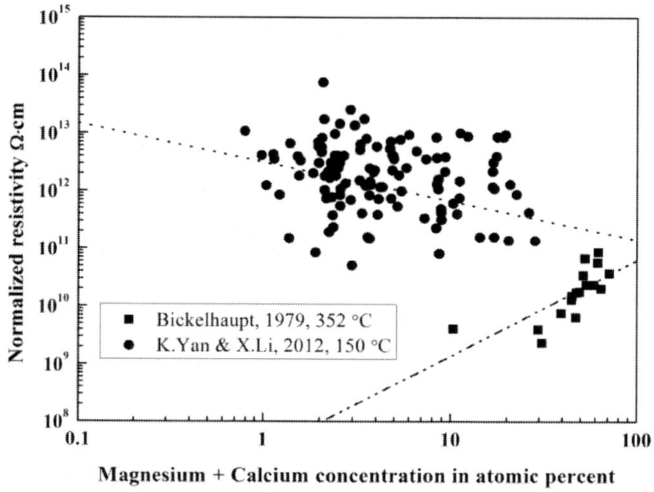

Figure 38. Normalized resistivity versus magnesium plus calcium concentration (Bickelhaupt, 1979; Li et al., 2012).

Alkali metals can easily lose their electrons for electric conduction. Resistivity data measured by different investigators all show inverse proportionality between resistivity and the sum of lithium and sodium concentration as illustrated in Figure 36. The number of charge carriers is proportional to the atomic concentration of ions, which is functioning as mobile ions, not oxide molecules. Data of R.Bickelhaupt were all tested at 60 °C in an atmosphere of air containing 9 vol% water while data of K.Yan & X.Li keep the testing temperature at around 150 °C with humidity of 3.39 vol%. The inverse correlations demostrate the alkali metal ions serve as charge carriers during the conduction process.

As Li_2O and Na_2O can significantly affect the ash resistivity, the measured values are normalized to 0.4 atomic percent lithium as shown in Figure 37.

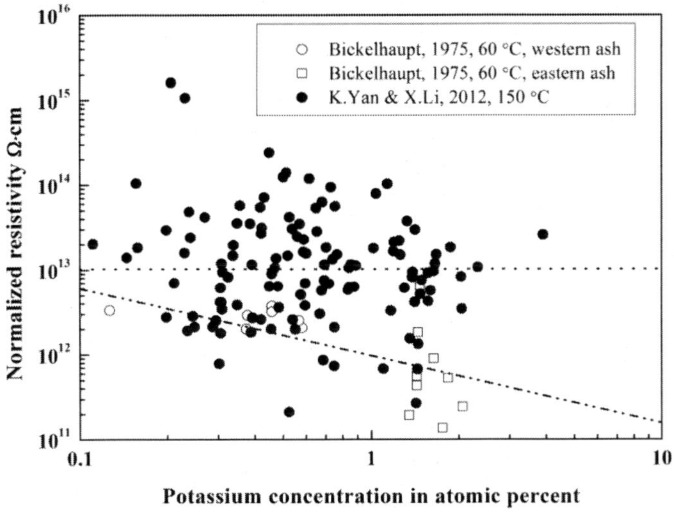

Figure 39. Normalized resistivity versus potassium concentraion (Bickelhaupt, 1975; Li et al., 2012).

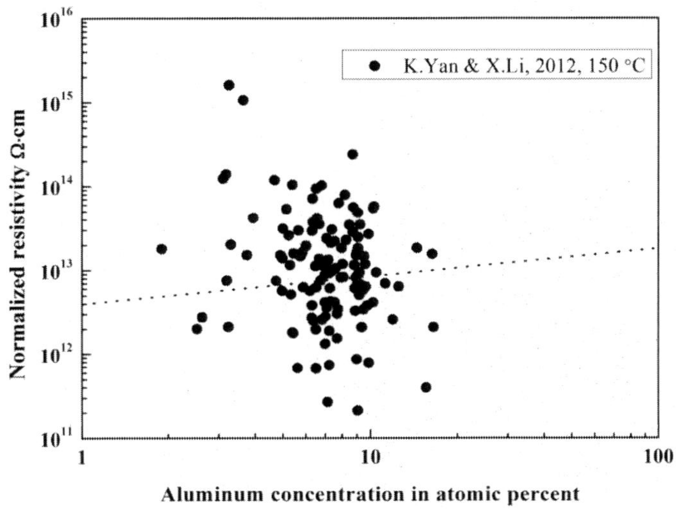

Figure 40. Normalized resistivity versus aluminum concentration (Li et al., 2012).

Several hypotheses have been made for explaining the decrease in resistivity with an increase in iron. First, higher iron concentration can reduce the fusion temperature of ash and enhance particle bonding for certain ash sample, especially for sintered samples. Second, the iron constituent, usually in the form of oxide, is a good electronic component itself. Third, iron may affect the mobility of alkali metals serving as charge carriers by changing the amorphous phase of the fly ash. Y. Yuan reported that fly ash mixed with ferric oxide or ferroferric oxide shows lower resistivity by an order of magnitude. And the breakdown voltage is also reduced (Yuan, Zhao, & Xue, 1996). Normalized resistivity of Bickelhaupt shows contradictory trend to that of K. Yan & X. Li in Figure 38. MgO, CaO shows similar resistivity higher than fly ash with similar electron activation energy. Therefore, the presence of MgO and CaO is ought to increase the ash resistivity. Reason for the opposite trends in this figure remaines unclear.Potassium is an alkali metal with lower electronegativity than sodium, therefore, may lose its electron easier. Resistivity data shown in Figure 39 presents an incongruous result for different ashes. Data of R. Bickelhaupt for the western ashes are clustered at one level of resistivity, while the data for the eastern ashes are spreaded over two orders of magnitude for a narrow band of potassium concentraion. The total trendency of discrete Bickelhaupt's data shows a similar inverse proportionality like that in Figure 36, which does not fit the data very well with a small determination coefficient. However, data of K. Yan & X. Li can be spread over five orders of magnitude for a wide band of potassium concentration, which indicates potassium may not affect the conduction process like other alkali metals. Bickelhaupt indicates the normalized resistivity have a good correlation with combined potassium and iron concentration. It should be noted the data of Bickelhaupt is normalized to 0.4 atomic% lithium plus sodium, while that of K. Yan & X. Li is normalized to 0.4 atomic% lithium plus sodium, 0.52 atomic% iron and 3.5% calcium plus magnesium.

Resistivity of high-alumina fly ashes is two orders of magnitude higher than that of common fly ash. Data of K. Yan & X. Li shows normalized resistivity of various aluminum concentration scatter over several orders of magnitude in Figure 40 without an obvious trend.

The linear regression analyses in Figure 36, Figure 37, Figure 38, Figure 39 and Figure 40 show chemical components have different influences to fly ash resistivity. They are reproduced in the Figure 41. The slopes of each line show the sensitivity of the chemical compositions influence on resistivity. It is apparent the lithium plus sodium concentration greatly influences the electrical resistivity of fly ash.

Sulfur dioxide or trioxide can enhance the conduction process over the ash surface. Bickelhaupt recorded typical resistivity versus time of exposure to an environment consisting of air, 9 vol% water and 9 ppm SO_3 at 147 °C as shown in Figure 42. At time equal to 0.5 hour the SO_3 was injected, and after around 2 hours, the resistivity begins to decrease.

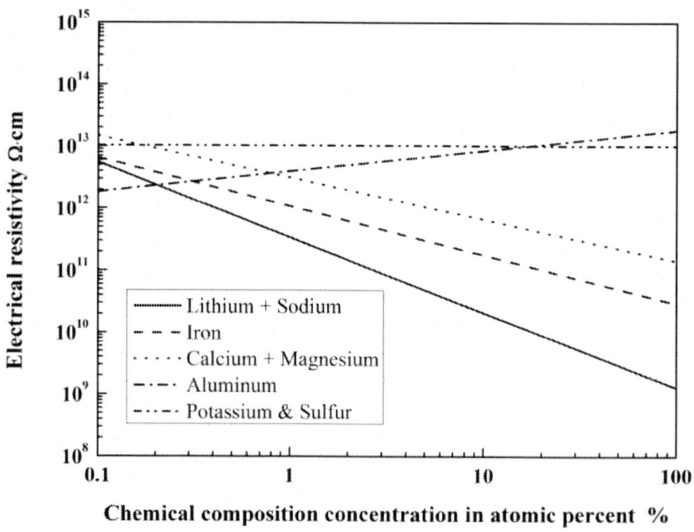

Figure 41. Resistivity as a function of the atomic percentage of lithium plus sodium, iron, calcium plus magnesium, aluminum, potassium and sulfur.

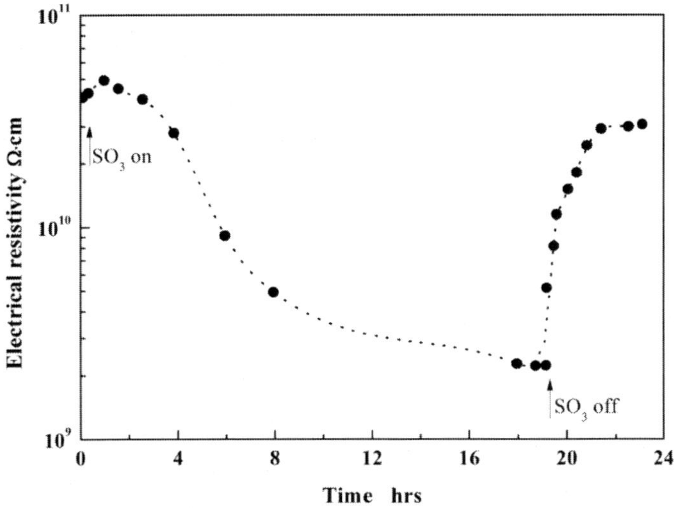

Figure 42. Resistivity versus time after injection of 9 ppm sulfur trioxide in an environment of air containing 9 vol% water (Bickelhaupt, 1979).

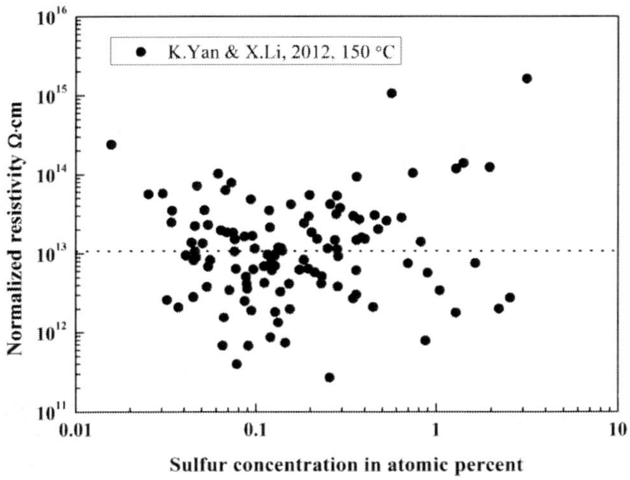

Figure 43. Normalized resistivity versus sulfur concentration (Li et al., 2012).

After 18 hours, the SO_3 injection was stopped. The effect of SO_3 dissappeared immediately and eliminated in 2 to 3 hrs. The eliminating time required is not related to the injecting time. The rapid disappearance of the effect of SO_3 after injection was stopped suggests that the absorbed sulfate or sulfuric acid participates throughout the surface conduction process directly.However, the sulfur component hardly shows any effect on ash resistivity, as illustrated in Figure 43. Sulfur component can enhance the conduction process in the form of salt or acid. Figure 45 suggests resistivity of fly ash in the presence of sulfuric acid vapor is lower than that with water vapor only at the same relatively low temperature. When the temperature is higher than the acid vapor dew point, the logarithm of resistivity is a linear function of absolute reciprocal temperature.

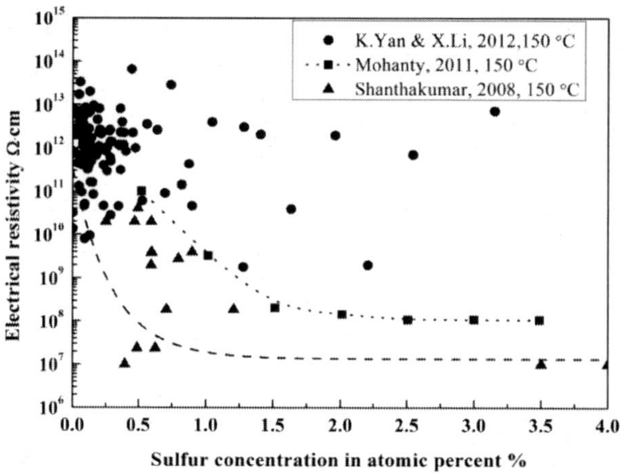

Figure 44. Resistivity versus sulfur concentration, (Li et al., 2012; Mohanty et al., 2011; Shanthakumar et al., 2008).

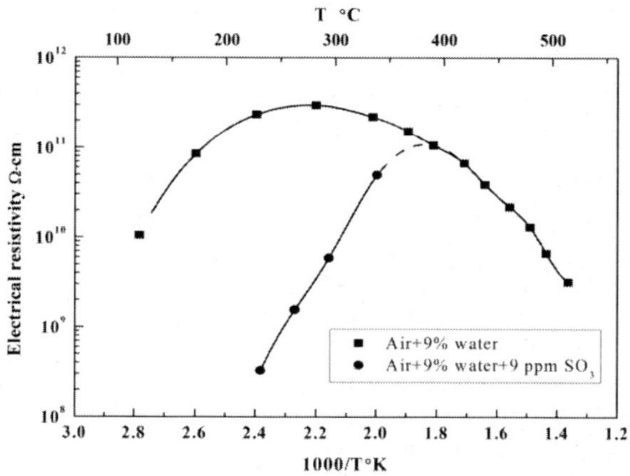

Figure 45. Resistivity versus temperature with and without sulfur trioxide injection (Bickelhaupt, 1979).

It should be noted several researchers reported the ash resistivity reduces with increasing sulfur content (Mohanty, Swai, Meikap, & Sahu, 2011; Shanthakumar, Singh, & Phadke, 2008). However, X. Li's data shows little changes with different sulfur contents as demonstrated in Figure 44. The normalized resistivity should be given for a more detailed comparison. The chloride content has been found to be an important parameter affecting ESPs performance. The formation of calcium chloride, a deliquescent salt, increases the affinity of the ash to hold moisture. And ash resistivity is reduced by increasing surface conductivity because of increased moisture.

The resistivity of low chloride content (0.1%) ash shows steeper relationship with temperature than that of high chloride content (0.3%). A series of tests was also made in which calcium chloride was added to keep the chloride level approaching to 0.3%. The plots in Figure 46 shows the effect of chloride on resistivity is independent of whether the chloride is inherent in the coal or added (Durham et al., 1990).

In addition, high chloride level also increases the particle cohesion. Resistivity of fly ash added with carbon decreases gradually as the round dots show in Figure 47 (Yuan, 1989). Carbon content higher than 10% usually creates excessive reentrainment and leap collection for low resistivity as the square points illustrate in Figure 47. The ash resistivity is lowered by carbon particles in layer.

The carbon acts by a physical admixture effect in which the current is carried by the carbonaceous particles in the mixture of fly ash and unburnt carbon as a whole (Parker, 1996, p 435).

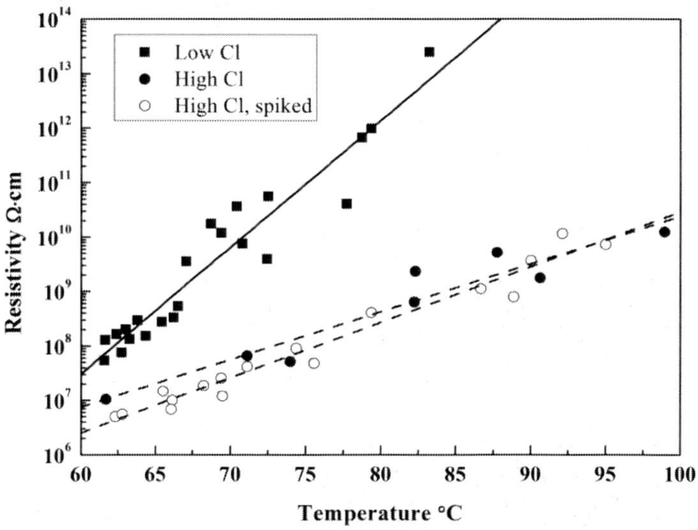

Figure 46. In situ resistivity measurements as a function of chloride level (Durham et al., 1990).

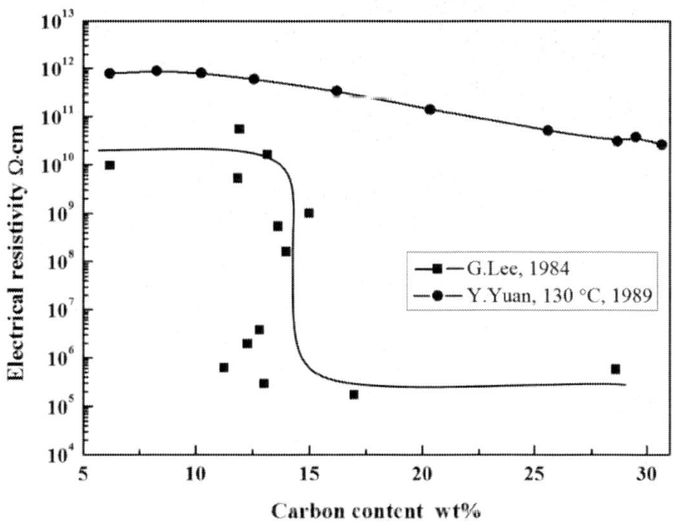

Figure 47. Electrical resistivity versus carbon content of ash (Lee & Whaley, 1983; Yuan, 1989).

6.2. Temperature, Humidity and Field Strength

The curve in Figure 48 draws the characteristic inverted V-shape of resistivity in a large range of temperature. The curve is linear in agreement with the Arrhenius equation for resistivity. The equation is :

$$\log \rho = \log \rho_0 + \frac{(\theta / k)\lg e}{T} \tag{31}$$

where, ρ is resistivity, ρ_0 is a complex material parameter including the number of mobile charge carriers. θ is experimental activation energy, k is Boltzmann's constant and T is the absolute temperature.

As lower temperatures are approached, the curve departs from the linear relationship because of increasing surface conduction. After passing through a maximum, resistivity decreases rapidly with lower temperatures and concomitantly greater humidity. There is a nadir for ash resistivity within the temperature range from 20 to 130 °C as shown in Figure 49. Water among the inner micro interspace of ash particles evaporates to form a water film from room temperature to around 80 °C. Therefore ash resistivity is reduced by the conductive film. As the temperature getting even higher, ash resistivity grows with less water content (Qi, Yan, & Yuan, 2006).

Fly ash resistivities are sensitive to environmental water concentration as shown in Figure 50. Higher environmental moisture usually induces lower ash resistivity at relative cold temperature. At a certain temperature and in an environment containing no sulfuric acid, the surface conductivity of fly ash depends directly on the water vapor concentration and its adsorption over the surface. The effect, however, is not as significant at high temperature as that at cold side temperatures. Curves in Figure 51 prove these deductions, which represent linear correlations with temperature after passing through a certain temperature.

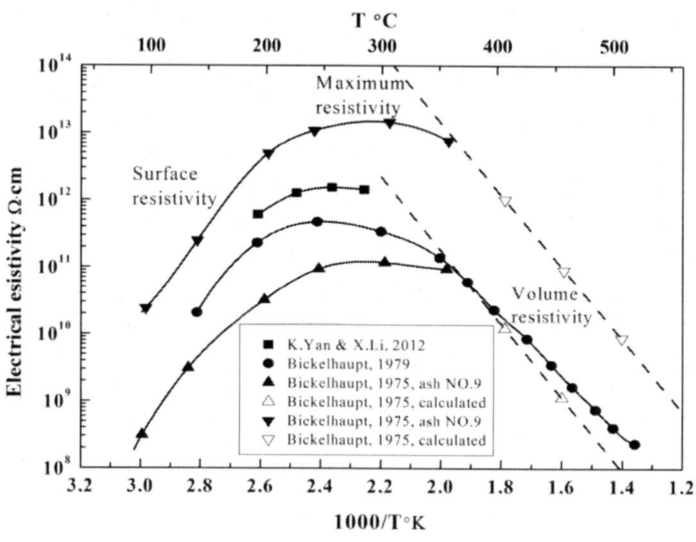

Figure 48. Resistivity versus temperature (Bickelhaupt, 1975, 1979; Li et al., 2012).

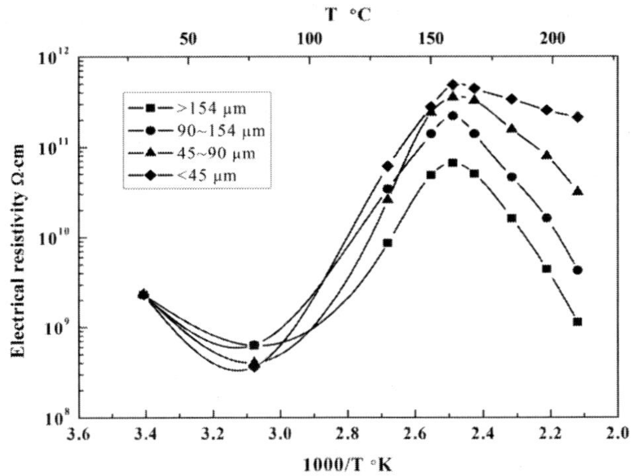

Figure 49. Resistivity of fly ash for lower temperature range (Qi et al., 2006).

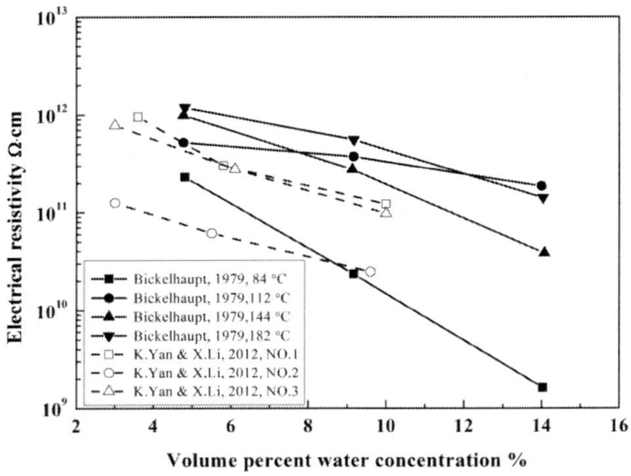

Figure 50. Resistivity versus environmental humidity (Bickelhaupt, 1979; Li et al., 2012).

The data in Figure 52 suggest resistivity is decreased with greater average electric field strength across a specific fly ash layer. It furthermore suggests the breakdown occurred at the same applied voltage, 4.65 kV for Bickelhaupt ash samples of two different thicknesses. The breakdown may relate to the fly ash itself or to the gas phase in the ash interstices. The maximum resistivity shows a negative relationship to the field strength. Neither the humidity nor the concentration of sodium, iron and calcium has any effect on above relationship (Li et al., 2012).

The particle size distribution and porosity also affect ash resistivity. When the humidity and temperature effects are not considered, the intrinsic resistivity of fly ash can be expressed as , where ρ_0 is the original resistivity and ε_p is porosity of a given ash layer.

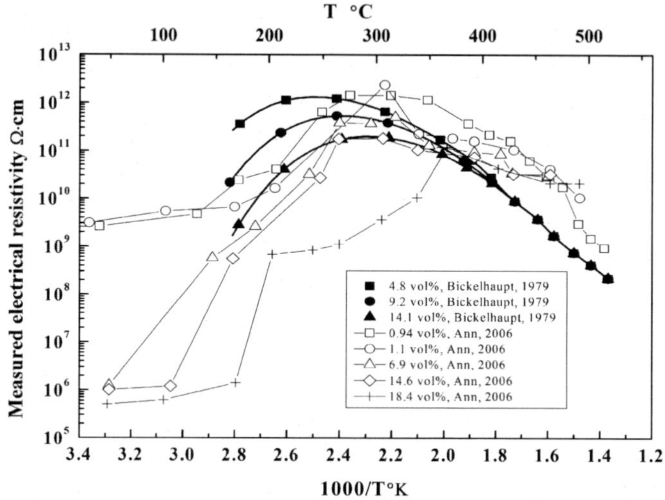

Figure 51. Resistivity versus temperature with different humidity. (Ahn & Lee, 2006; Bickelhaupt, 1979).

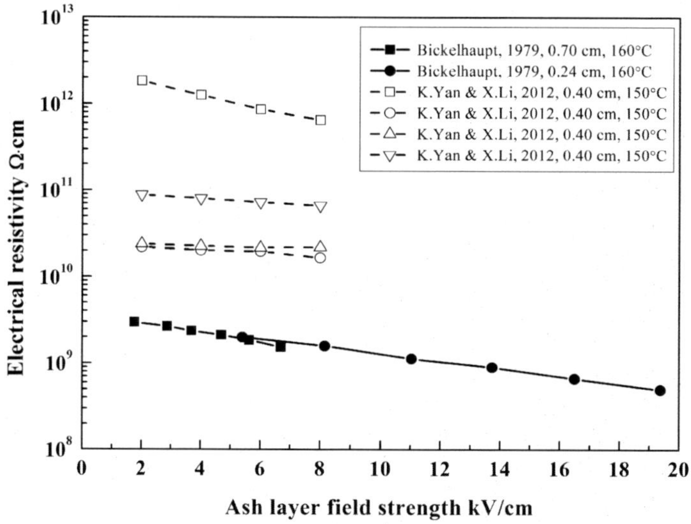

Figure 52. Resistivity versus field strength for ash layers of different thickness. (Bickelhaupt, 1979; Li et al., 2012).

The right fraction part of the equation is often referred as porosity coefficient, *C* or compacting factor, *H*. The ash layer is assumed containing plenty of particle chains. Current flows through the particle chains and contact points among them. The total current can be expressed as , where *K* is the ratio of chains to contact points, *N* is the average number of contact points, n is the number of chains in unit area, *I* is average current through each chains and *A* is the area of layer. The number of contact point depends on the layer porosity. For spherical particles with the same diameters, $N = 1.61\varepsilon_p^{-1.48}$.

Figure 53 shows that resistivity of relatively larger particles is higher than that of smaller particles for temperature lower than around 120 °C. It is likewise proved by 4 series of points in Figure 49.

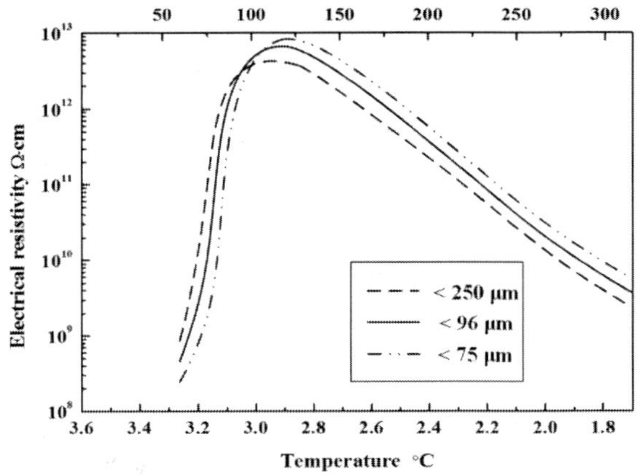

Figure 53. Electrical resistivity versus temperature for ash with different diameters (Y. Liu, Ma, & Zhang, 1994).

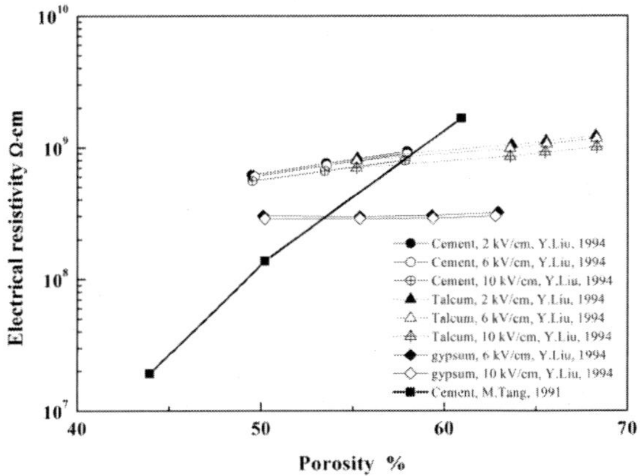

Figure 54. Electrical resistivity versus porosity of stacked particle sample (Y. Liu et al., 1994; Tang, 1991).

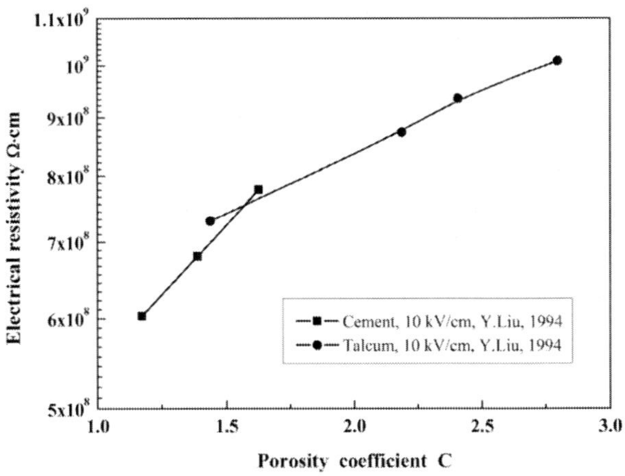

Figure 55. Electrical resistivity versus porosity coefficient of stacked particle sample (Y. Liu et al., 1994).

When temperature reaches 120 °C or higher, the resistivity curve of smaller particles lies above that of larger one. In the low temperature range, smaller particles absorb more water vapor increasing surface conduction for their larger specific area. When temperature is high enough that volume conduction is predominant, larger particles with lower porosity shows relatively low resistance. The resistivity is nearly linear with the porosity coefficient C as shown in Figure 54 and Figure 55.

7. Predicting Model of Ash Resistivity

7.1. R.E. Bickelhaupt Model

Bickelhaupt's model consists of three parts, i.e. volume resistivity ρ_v, surface resistivity induced by ash surface and water vapor ρ_{sw} and acid resistivity ρ_a as being expressed by equation (32), (33) and (34), respectively. The total resistivity ρ_{vs} is derived as equation (15). When there is no significant amount of sulfuric acid vapor, the ρ_s in equation (15) can be substitute as ρ_{sw}. If a significant amount of the acid vapor exists, the ρ_s should be calculated as section 4.1 shows.

$$\lg \rho_v = -1.8916 \lg A_{ls} - 0.9696 \lg A_i + 1.237 \lg A_{mc} - 0.03E + \frac{4334.515}{T} + 1.57595 \tag{32}$$

$$\lg \rho_{sw} = -2.233348 \lg A_{ls} - 0.000764 C_w - 0.03E - 0.000321 C_w \exp\left(\frac{2303}{T}\right) + 11.98555 \tag{33}$$

$$\lg \rho_a = 25.65278 - 0.371201 C_{SO_3} - \frac{5667.313}{T} - 0.03E \tag{34}$$

where, A_{ls}, A_i, and A_{mc} are the atomic concentrations of lithium plus sodium, iron, and magnesium plus calcium in unit of %, respectively. E is electric field intensity in unit of kV/cm. T is absolute temperature in K. C_w is water concentration in %. C_{SO3} is SO_3 concentration in dry volume in parts per million (ppm)

7.2. A. Chandra Model

Chandra's revised model is based on Bickelhaupt's one. While the coefficients in his model are revised according to the ash samples from Indian power plants. The sulfur concentration is usually low for Indian coals, so the acid resistivity is ignored in the Chandra's model. The revised volume ρ_v, surface ρ_s and total ρ_{vs} resistivity are expressed by the following equations.

$$\lg \rho_v = -3.6695 \lg A_{ls} - 2.1861 \lg A_i + 2.5514 \lg A_{mc} - 0.058847E + \frac{3394.117}{T} + 1.461271 \tag{35}$$

$$\lg \rho_s = -2.6756 \lg A_{ls} - 0.024929 C_w - 0.084383 E - 0.001269 C_w \exp\left(\frac{1870.1284}{T}\right) + 11.72543 \quad (36)$$

Chandra modified his model in 2012 with slight changes of the constants in the formula as follows (Andrabi et al., 2013):

$$\rho_v = \exp\left(-1.5193 \ln A_{ls} - 1.12633 \ln A_i + 1.89907 \ln A_{mc} - 0.06884 E + \frac{9635.16}{T} + 3.62368\right) \quad (37)$$

$$\rho_s = \exp\left(-2.7319 \ln A_{ls} - 0.00176 C_w - 0.06884 E - 0.000793 C_w \exp\left(\frac{2303.3}{T}\right) + 25.516\right) \quad (38)$$

7.3. K. Yan & X. Li Model

K. Yan and X. Li analyzed over 120 groups of Chinese ashes to determine the maximum resistivity of them. They found the maximum resistivity shows a simple relationship with Na_2O, Fe_2O_3 and water concentration. Other ash compositions such as aluminum, potassium, calcium plus magnesia and sulfur show ignorable effects on the maximum resistivity.

$$\lg \rho_{\max} = -1.2101 \lg A_{ls} - 0.782 \lg A_i + (0.132 A_{ls} - 0.141) C_w - 0.051 E + 11.40685 \quad (39)$$

7.4. Other Models

The conductivity of fly ash during combustion is related to characteristics of coal. It can be calculated as a function of the sulfur content, W_s, the ash content, W_a as $C = \frac{W_S}{W_a}\left(\frac{K_2O \times Fe_2O_3}{k} + Na_2O\right)$, where k is a specific coefficient (Yang, et al, 2002). EDF R&D and IRS srl developed a software named ORCHIDEE 2, which contains the module of resistivity calculation. The resistivity model used in ORCHIDEE 2 is based on that proposed by Bickelhaupt, while it has been improved to manage the discontinuities encountered in the initial model and to extend the interpolation ranges. It can also calculate the resistivity of ashes produced by combustion of coal blends (Arrondel et al., 2006).

7.5. Comparison of Predicting Models

The comparisons of the above predicting models are exhibited in Figure 56. All the models only indicate the changing trends of the resistivity with electric field strength. The calculated resistivity can deviate from the measured resistivity in a range of several orders of magnitude. According to these reports (Andrabi et al., 2013; Li et al., 2012), specific models should be used for coal from different regions. For example, Bickelhaupt model is better for

American coal, Chandra model is better for Indian coal and K. Yan & X. Li model fits the Chinese coal best.

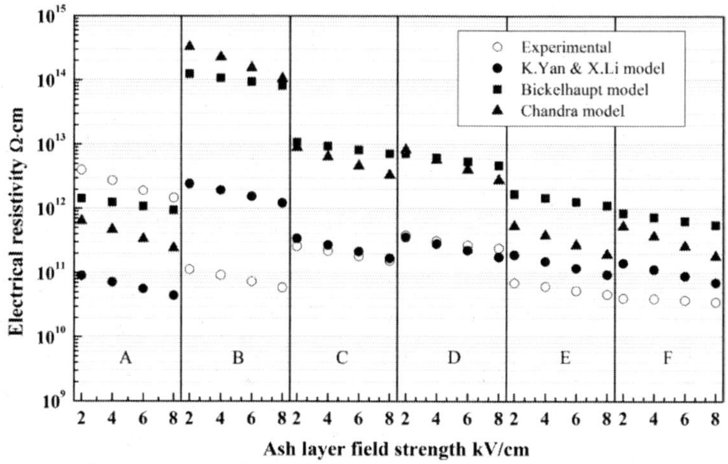

(Ash sample from A. Jiaxing power plant, Zhejiang; B. Junggar power plant, Inner Mongolia; C. 1st field of Zhanjiang power plant, Guangdong; D. 2nd field of Zhanjiang power plant, Guangdong; E. 4th field of Zhanjiang power plant, Guangdong; F. Huadian Zhuozi power plant, Inner Mongolia.)

Figure 56. Comparison of predicting models.

8. Precipitator Conditioning for Resistivity Changing

Relying on modifying the surface electrical conductivity of the dust, additives have been widely used to ESP for particle collection improvement. Sulfuric trioxide, ammonia and salts of sodium are all commonly used additive agents, while water can be sprinkled individually or combined with these agents. Sulfur trioxide conditioning reduces the electrical resistivity of fly ash particles, and ammonia conditioning increases the surface charge density and cohesive properties of the dust (Külaots, Gao, Hurt, & Suuberg, 2001). Costs of flue gas conditioning vary with conditioning type and power plant size. The capital cost for sulfur trioxide conditioning is \$6.24/kW, \$0.25/kW for ammonia conditioning and is \$2.47~4.78/kW for sodium conditioning (USEPA, 1998). Flue gas conditioning also leads to additional emissions from compounds formed by the conditioning agents.

8.1. Sulfur Trioxide Conditioning

Sulfur trioxide is the most widely used agent for fly ash conditioning. The sulfur, which accounts for roughly 0.3~4% of the coal by mass, are oxidized to SO_2 during the combustion process. SO_2 can be further oxidized to SO_3 by catalysts or heterogeneous oxidation. The amount of SO_3 is usually insufficient for reducing the ash surface resistivity in case of low-sulfur coal. Injection of additional SO_3 to the flue gas in the duct is feasible for this purpose. Molten sulfur is pumped from storage into the sulfur burner, which produces the additional

SO_3. And heated air enhances the conversion efficiency by atomizing the sulfur. The molten sulfur source can be replaced b y a device in which sulfuric acid is evaporated.

The major problem of SO_3 injection is that temperature and moisture control costs much, and the handling of SO_3. Acid vapor at a temperature below its dew temperature condenses to corrode the pipes and injection nozzles.

8.2. Ammonia Conditioning

Some fly ashes do not absorb sulfuric acid vapor from above process or combustion. For such ashes, surface of which are glassy due to the presence of silica, alumina and iron, ammonia conditioning is often used for modification. Ammonia reacts with natural sulfur trioxide over the fly ash, and produces ammonium bisulfate or sulfate. By addition of ammonia compounds, particles grow larger, cohesive force of ash layer and electric field is correspondingly enhanced. The improvement of additions of SO_3 and NH_3 is shown in Figure 57.

8.3. Sodium Conditioning

Addition of chemical compounds, like salts of sodium, increases charge carriers sites on the surface of particles. This conditioning process alters the composition or properties of coal. Normally, sodium carbonate and/or sodium sulfate are added directly to the coal for modification. The sodium compound mixes with the ash and serves as a conductive medium. It also reduces the resistivity of the ash layer on the plates. Another mean of sodium conditioning is to inject sodium compound into the boiler in which it combusts with coal. Sodium additives reduce ash resistivity as the presence of natural sodium composition would do. A 1.0 to 1.5% concentration of sodium carbonate has been demonstrated to reduce resistivity from 2.1×10^{10} to 3.7×10^{9} Ω·cm in a pilot scale ESP.

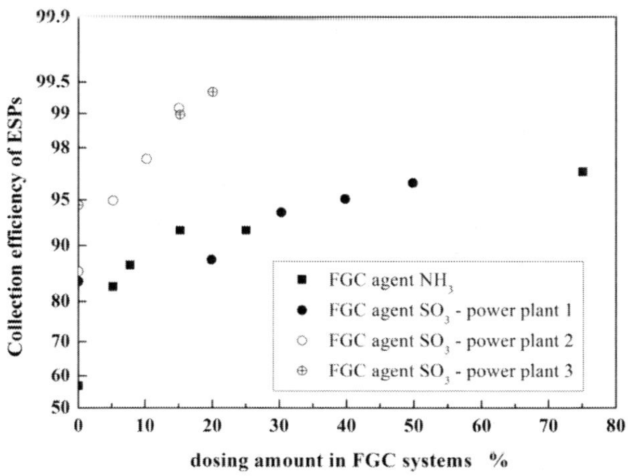

Figure 57. Effect of sulfur trioxide and ammonia conditioning on the efficiency of ESPs (Shanthakumar et al., 2008).

8.4. Water/Steam Conditioning

Injecting of water to the flue gas stream can reduce its temperature while increase its humidity. The water vapor absorbed on the particles forms a thin film, in which SO_3 dissolves. More charge carriers appear and surface resistivity decreases as more charge carriers appear.

However, this method lowers the flue gas temperature down to acid dew point and causes corrosion within the ESPs or downstream devices. And the ash lumps make it difficult to remove fallen ash in the hopper.

8.5. Improved High-Voltage Supplies

As subsection 4.3 illustrates, for high resistivity fly ash, pulse power source is more feasible for its steep decay edge.

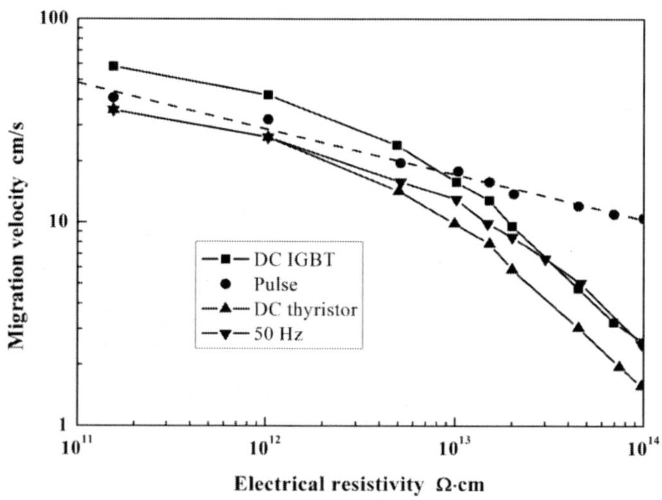

Figure 58. Migration velocity versus electrical resistivity for various power supply (Grass et al., 2004).

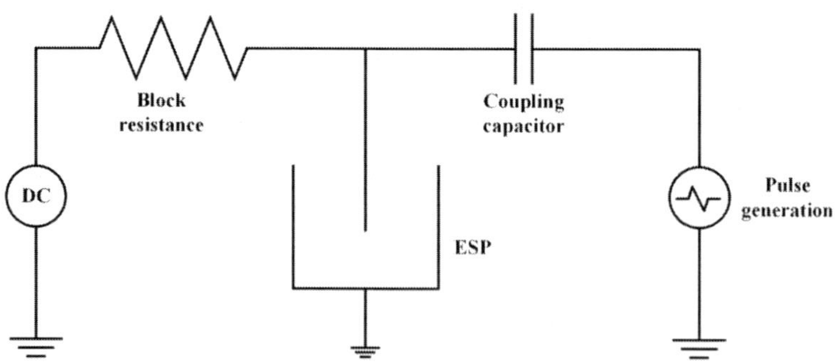

Figure 59. Simplified circuit of combined DC/pulse energization.

Figure 58 proves the migration velocity of ashes with electrical resistivity above 10^{12} $\Omega\cdot cm$ for pulse energization is significantly higher than that for conventional energization (Grass, Hartmann, & Klockner, 2004). For industrial application, pulse energization can be achieved by pulse voltage generation (Rea & Yan, 1995) or combined pulse and DC energization.

Figure 59 shows the simplified circuit of the latter. The pulse energization can generate high charge density that leads to increased charging of the fly ash particles. Flashover and back corona can be mainly avoided because of the short pulse duration and steep decay edge.

CONCLUSION

Electrostatic precipitators should work under real-time monitoring of ash layer resistivity and voltage-current characteristics for good performance. As online tests of ash resistivity are difficult and inaccurate, its prediction based on the combusted coal is now commonly used. Although each major boiler manager will have his own data bank of coal composition and ash electrical resistivity, on which he will base the design and operation of any new application, the prediction approach of ash resistivity will be similar to the outlined.

Precise predictions of ash resistivity depend on processing of huge amounts of data. So far, both external conditions, such as flue gas temperature, humidity, electric field and sulfur content, and chemical compositions show effects on ash resistivity clearly. However, the mechanisms of their effects on ash conduction remain unclear. Inconsistent or contradictory results from previous studies undoubtedly hinder further researches.

For a certain fly ash sample, resistivity shows inverted V-shape with the environmental temperature in a large range. Meanwhile, increasing lithium plus sodium, iron concentration or humidity lowers the ash resistivity. Effects of other chemical compositions, such as magnesium plus calcium, potassium, aluminum and sulfur, are still not completely understood. Therefore, subsequent investigations should provide detailed data. Data on chemical compositions, testing conditions (temperature, humidity, electrical field strength, porosity and size distribution of the ash sample), normalized resistivity can all improve the predicting model. In the meantime, more work of restraining back corona and high-voltage generation based on ash resistivity is preferred for industrial applications.

Methods for overcoming high resistivity problems in ESPs include conditioning with certain chemicals or water, improving high-voltage generation, restructuring the electrodes and changing the operating temperature and airflow velocity. Wet ESP is also a method of choice for the same purpose; in the meantime, it can trap ashes with low resistivity.

ACKNOWLEDGMENTS

This study is supported by the Chinese 863 fund via the project 2013AA065001, Hangzhou Sci. & Tech. fund via the project 20080213A15, and the R&D fund of Zhejiang Provincial Engineering Research Center of Industrial Boiler & Furnace Flue Gas Pollution Control.

REFERENCES

Ahn, Y. C., & Lee, J. K. (2006). Physical, chemical, and electrical analysis of aerosol particles generated from industrial plants. *Journal of Aerosol Science*, 37(2), 187–202.

Andrabi, S. J. A., Chandra, A., Aslam, M., & Hassan, I. A. (2013). Influence of chemical composition on the electrical resistivity of fly ash generated from Indian coal based thermal power plants. *Research Journal of Applied Sciences*, 5(6), 2284–2288.

Arendt, P., & Kallmann, H. (1926). The mechanism of charging mist particles. *Zeitschrift Für Physik*, 35, 421.

Arrondel, V., Bacchiega, G., Caraman, N., Gallimberti, I., Hamlil, M., Jacob, D., & RENARD, A. (2006, June). A friendly tool to assist plant operators and design engineers to control fly ash emissions: *ORCHIDEE*. Presented at the International Conference on Electrostatic Precipitation X, Australia.

Barranco, R., Gong, M., Thompson, A., Cloke, M., Hanson, S., Gibb, W., & Lester, E. (2007). The impact of fly ash resistivity and carbon content on electrostatic precipitator performance. *Fuel*, 86(16), 2521–2527.

Bickelhaupt, R. E. (1975). Surface resistivity and the chemical composition of fly ash. *Journal of the Air Pollution Control Association*, 25(2), 148–152.

Bickelhaupt, R. E. (1979). A technique for predicting fly ash resistivity. U.S. Environmental Protection Agency, Washington, D.C., EPA/600/7-79/204 (NTIS PB80102379).

Carpenter, R. L., Clark, R. D., & Su, Y. F. (1980). Fly ash from electrostatic precipitators: characterization of large spheres. *Journal of the Air Pollution Control Association*, 30(6), 679–681.

Chang, C. L., & Bai, H. (1999). An experimental study on the performance of a single discharge wire-plate electrostatic precipitator with back corona. *Journal of Aerosol Science*, 30(3), 325–340.

Coe, E. L. J., & Lagarias, J. S. (1986). Experience in conditioning electrostatic precipitators in the United States. Presented at the Joint Power Generation Conference, Oregon, Portland.

Coe, J. E. L., & Krigmont, H. V. (1995). Measurement of electrical resistivity of particulate entrained in a gas stream. U.S. Patent 5008628 A

Cottrell, F. G. (1911). The electrical precipitation of suspended particles. *Journal of Industrial & Engineering Chemistry*, 3(8), 542–550.

Czapka, T. (2011). Back-corona discharge phenomenon in the non-thermal plasma system. *IEEE Transactions on Plasma Science*, 39(11), 2242–2243.

Czech, T., Sobczyk, A. T., & Jaworek, A. (2011). Optical emission spectroscopy of point-plane corona and back-corona discharges in air. *European Physical Journal D*, 65(3), 459–474.

Czech, T., Sobczyk, A. T., Jaworek, A., & Krupa, A. (2012). Corona and back discharges in flue-gas simulating mixture. *Journal of Electrostatics*, 70(3), 269–284.

Durham, M. D., Rugg, D. E., Rhudy, R. G., & Puschaver, E. J. (1990). Low-resistivity related esp performance problems in dry scrubbing applications. *Journal of the Air & Waste Management Association*, 40(1), 112–124.

Ehrlich, C., Noll, G., Kalkoff, W.-D., Baumbach, G., & Dreiseidler, A. (2007). PM10, PM2.5 and PM1.0—Emissions from industrial plants—Results from measurement programmes in Germany. *Atmospheric Environment*, 41(29), 6236–6254.

Fisher, G. L., Chang, D. P. Y., & Brummer, M. (1976). Fly ash collected from electrostatic precipitators: microcrystalline structures and the mystery of the spheres. *Science*, 192, 553–555.

Fisher, G. L., Prentice, B. A., Silberman, D., Ondov, J. M., Biermann, A. H., Ragaini, R. C., & McFarland, A. R. (1978). Physical and morphological studies of size-classified coal fly ash. *Environmental Science & Technology*, 12(4), 447–451.

Frank, S. (1933). Spark discharges in air-dust mixtures. *Zeitschrift Für Physik*, 87, 323.

Goard, P. R. C., & Potter, E. C. (1978). Operational resistivity measurements on freshly generated fly ashes. Presented at the CSIRO Conference on Electrostatic Precipitation, Leura, N.S.W.

Grass, N., Hartmann, W., & Klockner, M. (2004). Application of different types of high-voltage supplies on industrial electrostatic precipitators. *IEEE Transactions on Industry Applications*, 40(6), 1513–1520.

Hao, Wen, & Hou, Y. (2008). Non-static collection process of the electrostatic precipitator. *Acta Scientiae Circumstantiae*, 28(10), 2059–2063.

Hao, Wenge, Hou, Y., & Xiong, H. (2008). Study on operation parameters of electrostatic precipitator based on non-steady electrostatic collection theory. *Chinese Journal of Environmental Engineering*, 2(3), 378–382.

Harrison, R. M., & Yin, J. (2000). Particulate matter in the atmosphere: which particle properties are important for its effects on health? *Science of the Total Environment*, 249(1–3), 85–101.

Hoferer, B., & Schwab, A. J. (2000). Local occurrence of back corona at the dust layer of electrostatic precipitators. Presented at the Electrical Insulation and Dielectric Phenomena, 2000 Annual Report Conference on (Vol. 1, pp. 93–96).

Howard, W. H. (1914). Electrical precipitation at garfield. *Transactions of the American Institute of Mining and Metallurgical Engineers*, (49), 540–560.

Ivancsy, T., Kiss, I., & Berta, I. (2009). Improved model for the analysis of back corona in pulse energized electrostatic precipitators. *Journal of Electrostatics*, 67(2-3), 146–149

Jędrusik, M., & Świerczok, A. (2009). The influence of fly ash physical and chemical properties on electrostatic precipitation process. *Journal of Electrostatics*, 67(2–3), 105–109.

Külaots, I., Gao, Y.-M., Hurt, R. H., & Suuberg, E. M. (2001). Adsorption of ammonia on coal fly ash. Presented at the 2001 International ash Utilization Symposium, Kentucky, U.S.

Lawless, P. A., & Sparks, L. E. (1980). A mathematical model for calculating effects of back corona in wire-duct electrostatic precipitators. *Journal of Applied Physics*, 51(1), 242–256.

Lee, G. K., & Whaley, H. (1983). Modification of combustion and fly-ash characteristics by coal blending. *Journal of the Institute of Energy*, 56(429), 190–197.

Li, X., Zhang, X., Zhu, J., Feng, W., & Yan, K. (2012). Sensitivity analysis on the maximum ash resistivity in terms of its compositions and gaseous water concentration. *Journal of Electrostatics*, 70(1), 83–90.

Liu, B. Y. H., & Yeh, H. (1968). On the theory of charging of aerosol particles in an electric field. *Journal of Applied Physics*, 39(3), 1396–1402.

Liu, Y., Ma, L., & Zhang, G. (1994). Effect of porosity on fly ash resistivity. *Labor protection of water resources & hydroelectric engineering*, (1), 22–25.

Lodge, O. J. (1884). Dust-free spaces. *Nature*, 30, 54.

Masuda, S., & Mizuno, A. (1977). Initiation condition and mode of back discharge. *Journal of Electrostatics*, 4(1), 35–52.

McLean, K. J. (1988). Electrostatic precipitators. *Physical Science, Measurement and Instrumentation, Management and Education - Reviews, IEEE Proceedings A*, 135(6), 347–361.

Mohanty, C. R., Swai, A. K., Meikap, B. C., & Sahu, J. N. (2011). Studies on factors influencing fly ash resistivity from electrostatic precipitator with reference to India. *Journal of Scientific & Industrial Research*, 70(9), 795–803.

Murphy, A. T., Adler, F. T., & Penney, G. W. (1959). A theoretical analysis of the effects of an electric field on the charging of fine particles. *American Institute of Electrical Engineers, Part I: Communication and Electronics, Transactions of the*, 78(4), 318–326.

Nichols, G., & Spencer, H. (1978). Test Methods and Apparatus for Conducting Resistivity Measurements. Southern Research Institute for the U.S. EPA.

Oglesby, S., & Nichols, G. B. (1978). Electrostatic precipitation. New York: Marcel Dekker.

Parker, K. R. (1996). Applied Electrostatic Precipitation. London: Chapman & Hall.

Pauthenier, M. M., & Moreau-Hanot, M. (1932). Charging of spherical particles in an ionizing field. *J. Phys. Radium*, 3(7), 590–613.

Penney, G. W., & Klingler, E. H. (1962). Contact potentials and the adhesion of dust. *American Institute of Electrical Engineers, Part I: Communication and Electronics, Transactions of the*, 81(3), 200–204.

Qi, L., Yan, W., & Yuan, Y. (2006). An experimental study of the mechanism governing the impact of the size of coal-fired fly ash particle on specific resistance. *Journal of engineering for thermal energy and power*, 21(4), 391–393.

Ramsden, A. R., & Shibaoka, M. (1982). Characterization and analysis of individual fly-ash particles from coal-fired power stations by a combination of optical microscopy, electron microscopy and quantitative electron microprobe analysis. *Atmospheric Environment (1967)*, 16(9), 2191–2206.

Rea, M., & Yan, K. (1995). Evaluation of pulse voltage generators. *IEEE Transactions on Industry Applications*, 31(3), 507–512.

Riehle, C., & Löffler, F. (1992). May an Increase of the Effective Migration Rate Observed in Electrostatic Precipitators with Wider Plate Spacings or Faster Gas Streams Really Be Termed “Non-Deutschian”? *Aerosol Science and Technology*, 16(1), 1–14.

Rohmann, H. (1923). A method for measuring the size of suspended particles. *Zeitschrift Für Physik*, 253–265.

Shanthakumar, S., Singh, D. N., & Phadke, R. C. (2008). Flue gas conditioning for reducing suspended particulate matter from thermal power stations. *Progress in Energy and Combustion Science*, 34(6), 685–695.

Smith, W. B., & McDonald, J. R. (1976). Development of a theory for the charging of particles by unipolar ions. *Journal of Aerosol Science*, 7(2), 151–166.

Sproull, W. (1954). *Laboratory Performance of a Special Two-Stage Precipitator for Collecting High Resistivity Dust and Fume*. Presented at American Chemical Society Meeting, New York, New York (September 1954).

Sproull, W. T., & Nakada, Y. (1951). Operation of Cottrell precipitators. *Industrial & Engineering Chemistry*, 43(6), 1350–1358.

Sun, J., Han, D., Yao, Q., & Xu, X. (2001). Classification and characteristics of micro-particles in fly ash from coal-fired power plants. *Journal of Chinese electron microscopy society*, 20(2), 141–146.

Tang, M. (1991). Fly ash deposition and its resistivity. *Jiangxi Nonferrous Metal*, 5(4), 235–238.

U.S.EPA. (1998). Stationary source control techniques document for fine particulate matter U.S. Environmental Protection Agency, Washington, D.C., No. EPA-452/R-97-001.

Vassilev, S. V., & Vassileva, C. G. (2005). Methods for Characterization of Composition of Fly Ashes from Coal-Fired Power Stations: A Critical Overview. *Energy & Fuels*, 19(3), 1084–1098.

Walker, A. B. (1968). Operating principles of air pollution control equipment and guidelines for their application. Presented at Proceedings MECAR Symposium, Design and Operation for Air Pollution Control, New York, New York.

White, H. J. (1963). Industrial electrostatic precipitation. New York: Addison-Wesley

White, H. J. (1974). Resistivity problems in electrostatic precipitation. *Journal of the Air Pollution Control Association*, 24(4), 313–338.

Wolcott, E. R. (1918). Effects of dielectrics on the sparking voltage. *Physical Review*, 12(4), 284.

Yi, H., Guo, X., Hao, J., Duan, L., & Li, X. (2006). Characteristics of Inhalable Particulate Matter Concentration and Size Distribution from Power Plants in China. *Journal of the Air & Waste Management Association*, 56(9), 1243–1251.

Yuan, Y. (1989). A study of the effect of dust fineness and carbon content on dust resistivity. *Journal of north china institute of electric power*, (3), 102–107.

Yuan, Y., Zhao, Y., & Xue, A. (1996). Effect of Fe_3O_4 and Fe_2O_3 on the specific resistivity and the field strength of electrical breakdown for coal fired fly ash. *China Environmental Science*, 16(4), 297–361.

Zhu, J., Zhang, X., Chen, W., Shi, Y., & Yan, K. (2010). Electrostatic precipitation of fine particles with a bipolar pre-charger. *Journal of Electrostatics*, 68(2), 174–178.

Zhu, J. (2010). Fine particle collection with electrostatic precipitation. PhD Thesis. Zhejiang University, Zhejiang.

Zou, L. (2005). Mechanism of back corona and anticoagulation in electrostatic precipitator. Master Thesis. Wuhan University of Science and Technology, Hubei.

In: Fly Ash
Editor: Prabir Kumar Sarker
ISBN: 978-1-62948-044-2

Chapter 5

COAL FLY ASH FROM BRAZILIAN POWER PLANTS: CHEMICAL AND PHYSICAL PROPERTIES AND LEACHING CHARACTERISTICS

Denise A. Fungaro[1,*], Juliana C. Izidoro[1], Fernando S. Santos[1] and Shaobin Wang[2]

[1]Nuclear and Energy Research Institute (IPEN – CNEN/SP),
Av. Prof. Lineu Prestes, Cidade Universitária, Rio de Janeiro, Brazil
[2]Department of Chemical Engineering, Curtin University, Perth, Australia

ABSTRACT

This chapter provides information about the Brazilian coal fly ash characteristics that are relevant to issues of ash utilization and safe disposal. Several coal fly ash samples were collected from baghouse filter, electrostatic precipitators and cyclone filter of five different Brazilian thermal power plants: Figueira, Jorge Lacerda, President Médici, São Jerônimo and Charqueadas. The materials were characterized in terms of pH, conductivity, real density, specific surface area (BET method), morphological analysis (SEM), chemical composition (XRF), mineralogical composition (XRD), cation exchange capacity, humidity, and loss-on-ignition. Leaching and solubilization tests were carried out following the Brazilian NBR 10004 code. Analyses of major elements indicate high SiO_2, Al_2O_3 and Fe_2O_3 contents. The mineralogical compositions of fly ashes include mainly mullite and quartz, some amounts of magnetite and hematite and an amorphous phase. Most elemental concentrations were of the same magnitude, independent of the power plant, except Figueira. The fly ash from São Jerônimo power plant presents a higher content of unburned carbon and specific surface area. According to their mineralogical and chemical compositions, Brazilian fly ashes can be used as raw materials for synthesis of zeolites using hydrothermal methods.

* Email address: dfungaro@ipen.br.

1. INTRODUCTION

Brazil and Colombia have the most proved reserves of coal in South America. Brazil has important reserves of lignite and sub-bituminous coal of 4559 Mt (BP Statistical, 2012). Coal mines in Brazil are mainly in the south and the power plants are located near these mines in the states of Rio Grande do Sul (RS), Santa Catarina (SC) and Paraná (PR). Data about Brazilian thermal power plants are provided by the Brazilian National Electricity Energy Agency (ANEEL, 2011).

The principal use of the coal mined in Brazil is the combustion for power generation. Approximate 1.5% of the electricity generated in the country is derived from coal-fired power plants, which accounts for 1.9 GW of total installed capacity. In the generation of electricity, these power plants produce approximately 4 M tonnes of ashes every year, which are composed of 65 to 85% of fly ash and 15 to 35% of bottom ash. Brazilian coals are richer in ash (content of 20–50 wt.%) and poorer in carbon when compared to worldwide coal and coal ash disposal is a serious environmental concern. Only 30% of fly ash is applied as a raw material for cement and concrete production. The remaining solid wastes are disposed in on-site ponds, nearby abandoned or active mine sites, or landfills. The properties of coal fly ash are strongly dependent on the geological origin and the combustion process of the coal. It is important to characterize regional coal ashes in detail to ascertain their potential uses as a raw material in the production of high value products and in evaluation of their environmental impacts (atmospheric emissions and ash leaching). Previously, some characteristics of Brazilian coal fly ashes have been investigated (Depoi et al., 2008; Levandowski and Kalkreuth, 2009; Pires and Querol, 2004; Rohde, Zwonok, Chies and Silva, 2006), but more studies are of fundamental importance for the development of various applications of fly ash utilization and evaluation of ash disposal.

In this work, coal fly ash samples were collected from baghouse filter, electrostatic precipitators or cyclone filter of five different Brazilian thermal power plants: Figueira (PR), Jorge Lacerda (SC), President Médici (RS), São Jerônimo (RS) and Charqueadas (RS). The materials were characterized and their properties were compared. Its potential as a raw material in the production of high value products was evaluated for increasing the productive chain of coal.

2. MATERIALS AND METHODS

2.1. Materials

The coal fly ash samples used in this study were collected from baghouse filter, electrostatic precipitators and cyclone filter of five different Brazilian thermal power plants: Figueira, Jorge Lacerda, São Jerônimo, Charqueadas and Presidente MédicI. The samples were labeled as: FG, JL, SJ, CH and PM, respectively.

The coal fly ash sample collected by cyclone from Figueira power plant was labeled FGC. The ashes were collected directly from retention systems from each power plant and were used without any pretreatment.

2.2. Characterization of Fly Ashes

The experimental details of characterization of Brazilian fly ashes samples (XRD, XRF SEM, BET surface area and real density, cation exchange capacity (CEC), pH, conductivity, humidity, loss-on-ignition) have been described in a previous paper (Izidoro et al., 2012a).

Instrumental neutron activation analysis (INAA) was used to determine concentrations of elements in the materials. About 100 mg of sample and reference material were weighted in polyethylene bags.

The accuracy of the results was determined using IAEA standard reference material SOIL-7. For irradiation, the samples, reference material and element standards were simultaneously submitted to a thermal neutron flux in the range of 3.5 to 5.0 x 10^{12} n cm^{-2} s^{-1} for 8 h at the nuclear research reactor IEA-R1 of the Nuclear and Energy Research Institute, (IPEN/CNEN-SP), São Paulo, Brazil.

After appropriate decay periods, γ-ray spectra of samples, reference material and element standards were measured using High-purity Ge (HPGe) detector Model POP-TOP (EG&G Ortec, Oak Ridge, TN, USA). This detector was coupled to an EG&G Ortec card and associated electronics. Spectrum analysis was carried out using VISPECT software, in TURBOBASIC language.

The particle size of the materials was measured using a laser based particle size analyzer, namely a Malvern MSS Mastersizer 2000 Ver. 5.54. Infrared (IR) spectroscopy analysis performed with a FTIR Nexus 670 Thermo Nicolet was carried out to study their structural features.

2.3. Leaching and Solubilization Tests

Brazilian fly ash samples were submitted to leaching and solubilization tests for waste classification according to Brazilian regulation (Standard Brazil ABNT NBR 10004, 2004). The procedures for leaching extraction and solubilized extraction of solid wastes are described at ABNT Norms (Standard Brazil ABNT NBR 10005, 2004; Standard Brazil ABNT NBR 10006, 2004).

Ten grams of material with diameter particles less than 9 mm were placed in a bottle and 193 mL of acetic acid solution (pH 4.93 ± 0.05) were added and agitated for 18 h at 30 rpm. After filtration, leachable compounds were extracted, and the elements were analyzed.

In the solubilization test, 100 mL of ultrapure water were added to 25 g of material and this suspension was agitated for 5 min and after it was allowed to stand for 7 d.

After filtration, the soluble compounds were extracted and the elements were analyzed by inductively coupled plasma optical emission spectrometry (ICP-OES) in a SpectroFlame M120 spectrometer from Spectro Analytical Instruments and by graphite furnace atomic absorption cold vapor spectrometry (GFAAS) in an Analyst 800 from Perkin Elmer.

3. Results and Discussion

3.1. Comparison of Fly Ash Properties

3.1.1. Chemical Composition

The chemical compositions of five Brazilian fly ash samples (by % weight) are given in Table 1. Usually, the characteristics of fly ashes varied depending on their sources and chemical composition of fly ash affects the properties of the end products.

Fly ash is mainly composed of metal oxides derived from inorganic compounds in the coal. It can be observed from Table 1 that the major constituents for all samples of the Brazilian fly ashes are: SiO_2, Al_2O_3, and Fe_2O_3. The contents of SiO_2 and Al_2O_3 are above 70% for all samples. Quantities below 5% of K_2O, CaO, TiO_2, SO_3, and MgO are also observed.

Table 1. Composition (wt%) of fly ash obtained from different Brazilian thermal power plants

Components	FG	JL	SJ	CH	PM
SiO_2	48.4	50.3	54.5	57.5	62.4
Al_2O_3	23.8	29.8	26.7	32.6	22.9
Fe_2O_3	10.2	6.70	5.00	3.60	4.90
K_2O	4.60	5.30	3.00	2.00	3.00
CaO	4.70	2.70	2.80	1.40	2.40
TiO_2	1.70	2.20	3.80	1.60	1.10
SO_3	2.10	1.40	2.90	0.40	0.500
MgO	1.70	1.10	1.00	0.70	1.10
Na_2O	1.30	-	-	-	1.30
P_2O_5	0.200	0.170	0.110	0.0900	0.0900
ZnO	0.300	0.100	0.0300	0.0200	0.0200
MnO	0.100	0.0700	0.0400	0.0200	0.0300
PbO	0.110	0.0300	-	0.0160	-
SiO_2/Al_2O_3	2.03	1.69	2.04	1.76	2.72

According to the classification of the American Society for Testing and Materials (Standard American, 1992), the fly ash samples investigated in this study could be classified as a class F fly ash due to their high (SiO_2 + Al_2O_3 + Fe_2O) > 70% and very low lime (CaO) contents (< 5%). Class F fly ash is normally produced from burning anthracite or bituminous coal. The SiO_2/Al_2O_3 ratios were calculated for the fly ashes and the values ranged from 1.69 to 2.72, indicating good source to synthesize zeolites (Shigemoto and Hayashi, 1993).

The chemical composition of major oxides in Brazilian coal fly ashes was similar to European, English, Spanish and Mexican fly ashes (Medina et al., 2010).

It is noted that FG sample presented the highest content of Pb, Fe, Ca and Zn. The highest concentration of iron oxide observed is related to the high content of pyrite in the coal (Flues et al., 2008). Elements such as As, Tl, and Pb are known to have high affinities to pyrite and ferric oxides.

The concentrations of minor and trace elements (mainly As, Cd, Mo, Pb, TI, U, Zn, and Hg) are, in general, the highest in the coal and combustion-by-products from Figueira, with statistical analysis showing that their coal and fly ash are significantly different from the other areas. Most element concentrations are of the same magnitude for other power plants (Depoi et al., 2008).

In another study, Levandowski and Kalkreuth (2008) showed that the highest concentrations of trace elements were Zn and As for fly ash samples from Figueira Power Plant. When compared with trace element concentrations in the feed coal, fly ashes showed a significant enrichment in most trace elements (As, B, Be, Cd, Co, Cr, Cu, Li, Mn, Mo, Ni, Pb, Sb, Tl, and Zn), suggesting a predominantly volatile nature for these elements.

Previous studies with analysis of coal, fly ash and bottom ash from Presidente Médice Power Plant were reported (Pires and Querol, 2004). The fly and bottom ashes were characterized by relatively high contents of SiO_2 (56–59%) and Al_2O_3 (36–38%) and low contents of alkaline oxide. Elements were classified into three groups based on the analysis of trace element concentrations in fly and bottom ashes, and enrichments or depletions of these concentrations in relation to the coal: Group I (volatile elements with subsequent condensation): As, B, Bi, Cd, Ga, Ge, Mo, Pb, S, Sb, Sn, Tl and Zn; Group II (no volatile elements enriched in bottom ash vs. fly ash): Ca, Fe, Mn, P, Ti and Zr; Group III (low volatile elements with no partitioning between fly and bottom ashes): Al, Ba, Be, Co, Cr, Cs, Hf, K, Li, Mg, Na, Ni, Rb, Sr, Th, U, W, Y and most of heavy rare earth elements.

The feed coals, fly ashes and bottom ashes collected from seven different units in Jorge Lacerda power plant have been subjected to mineralogical, geochemical, and petrographic studies, to investigate the relationship between feed coal and ash characteristics (Silva et al., 2010). The concentrations of major elements of fly ash samples, recalculated at a LOI-free basis, have quite similar elemental compositions.

Taking into account the major oxide components in the fly ashes, the samples were positioned in the ternary system, Fe_2O_3-SiO_2-Al_2O_3, as shown in Figure 1. The samples are grouped meaning that the constituents are weakly variable in their composition.

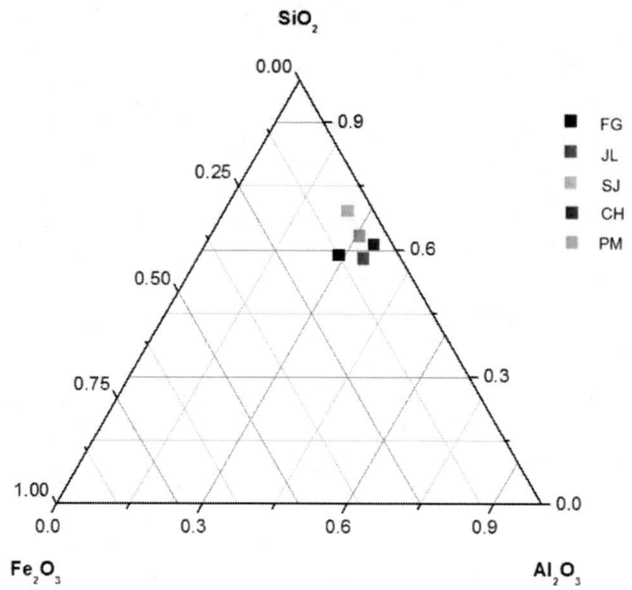

Figure 1. The phase diagram for the ternary system Fe_2O_3-SiO_2-Al_2O_3.

3.1.2. Physico-Chemical Properties

The physical properties of fly ashes vary depending on the type of coal burned, boiler type, and collector setup. The chemical composition of fly ash varies widely, depending on the type of coal burned, the particle size of the ash, and the efficiency of the collectors. The physicochemical properties of Brazilian coal fly ash samples are summarized in Table 2.

The pH of the coal ash is directly related to the availability of macro and micro nutrient and indicates whether the coal ash is acidic or alkaline in nature. The pH values of fly ash suspension were very close and ranged from 7.6 to 8.2 (Table 2). These values are justified by the presence of compounds formed by the cations, K^+, Na^+, Ca^{2+} and Mg^{2+} combined with carbonates, oxides or hydroxides which cause the alkalinity of the suspension (Paprocki, 2009). Theis and Wirth (1977) investigated factors affecting the pH of fly ash leachate and found that the mass concentration of Fe and Ca in the leachate roughly delineates the acidic or basic character of the fly ash leachate.

Based on the pH, coal ash has been classified into 3 categories, namely; slightly alkaline 6.5–7.5; moderately alkaline 7.5-8.5 and highly alkaline > 8.5 (Sijakova-Ivanova et al., 2011). The pH values of samples indicate that ashes were moderately alkaline in nature.

The conductivity values are related to some compounds present in coal fly ashes that can undergo dissolution.

The conductivity values of fly ash suspension ranged from 188.8 μS (SJ) to 538.0 μS (FG). The fly ash sample from Figueira power plant (FG) presented a higher value if compared to other samples. This result is in accordance with another study reporting the values of conductivity of different Brazilian fly ashes (Depoi et al., 2008). This is due to the low ash content present in the coal used in Figueira power plant, which enables the solubilization of a higher number of compounds and consequently higher amount of ions present in solution.

Table 2. Physicochemical properties of fly ashes samples

Sample	pH	C (μS)	Humidity (%)	Loss on ignition (%)	$R_{density}$ (g cm^{-3})	S_{BET} (m^2g^{-1})	CEC (meq g^{-1})
FG	8.2	538.0	0.800	9.13	2.40	8.7	0.035
JL	8.0	259.0	1.40	15.1	2.26	9.6	0.026
SJ	7.9	188.8	3.00	15.2	2.46	30.0	0.027
CH	7.8	201.0	0.600	2.60	2.54	3.3	0.026
PM	7.6	203.0	0.600	1.44	2.31	1.4	0.026

The loss on ignition (LOI) for fly ashes is usually due to the presence of unburned carbon (indicating the burning efficiency of a thermoelectric power plant) and mineral phases not stable at high temperatures (Umaña-Peña, 2002). Carbon has a low density and can adsorb significant amounts of water. This means that the maximum dry density and optimum moisture content of fly ash are influenced by the LOI. Higher LOI ash is lower in density, but has higher optimum moisture contents.

SJ and JL fly ashes presented the highest values (15.2% and 15.1%, respectively); FG fly ash presented an intermediate value (9.13%) and CH and PM fly ashes presented the lower values: 2.60% and 1.44%, respectively (Table 2). According to the values of loss of ignition, the decreasing burning efficiency of coal-fired power plants in Brazil is as follows: PM> CH > FG > JL > SJ.

As shown in Table 2, all fly ash samples presented less than 3% of humidity. This was because the samples were collected directly by the filters or electrostatic precipitators at high temperature. The values are in accordance with the results obtained from different Spanish and South African fly ashes (Umaña-Peña, 2002). The samples with higher loss of ignition showed higher humidity due to the presence of unburned carbon which will easily adsorb water.

The specific surface areas of the fly ash samples are given in Table 2. As also seen from Table 2, the density values for all samples were very close with average of 2.39 g^{-1} cm^{-3}. The low bulk density makes these fly ashes to be a good candidate for lightweight building material.

The specific surface area of fly ash samples ranged from 1.4 m^2 g^{-1} to 30.0 m^2 g^{-1}. The higher specific surface area of SJ is ascribed to a large amount of unburned carbon (Depoi et al., 2008). Higher amount of unburned carbon in the fly ash particles contributes to higher specific area due to the porous structure (Wang and Wu, 2006).

The CEC values of the ashes are much low, ranged from 0.026 and 0.035 meq g^{-1} and are similar to the values reported in the literature (Paprocki, 2009). Values of CEC between 0.023 and 0.154 meq g^{-1} have been recorded in fly ash from coal combustion (Pathan et al., 2003).

3.1.3. Mineralogical Composition

Figure 2 shows X-ray diffraction patterns of different Brazilian coal fly ashes. The identification and interpretation of PXRD patterns of the materials are prepared by comparing the diffraction database provided by "International Centre for Diffraction Data / Joint Committee on Power Diffraction Standards" (ICDD/JCPDS).

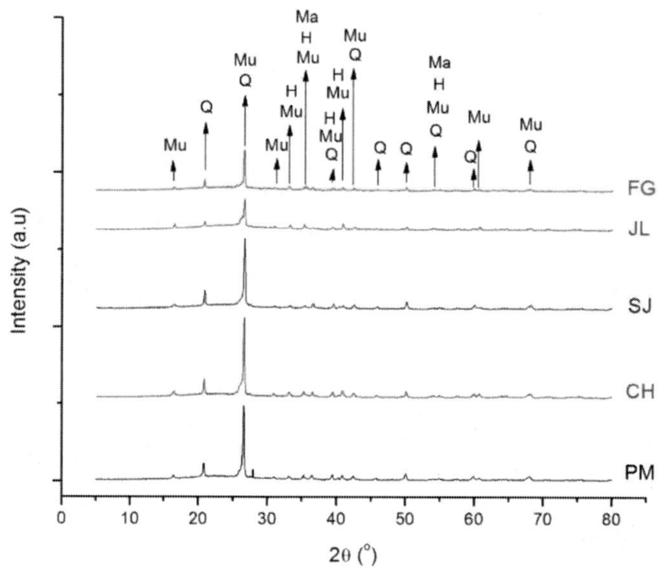

Figure 2. XRD Patterns of Brazilian fly ashes (Q = Quartz, M = Mullite, H = Hematite, Ma = Magnetite).

According to Figure 2, the Brazilian fly ashes used in this study are very similar and are composed mainly of mullite (JCPDS 74-4143) and quartz (JCPDS 85-0796) and some amounts of hematite (JCPDS 33-0664) and magnetite (JCPDS 71-6336), which are also confirmed from their chemical compositions (Table 1). All samples presented the amorphous halo between 10 and 30 in the 2θ angular zone which corresponds to the vitreous component of ashes.

3.1.4. Morphological Analysis

The scanning electron micrographs (SEM) of the original fly ashes are shown in Figure 3. Coal fly ash particles typically had the predominance of spherical shapes at different sizes and smooth surfaces, similar to previous observations in other studies (El-Naggar et al., 2008; Ojha et al., 2004; Pires and Querol, 2004; Singh and Kolay, 2002) Different physical states of silica are responsible for the particles of irregular size (Sarbak et al., 2004).

The particles of SJ fly ash presented a different aspect (Figure 3C). The spongy appearance of this particle is the characteristic of fly ash samples with high content of unburned coal, which contributes to the high values of loss of ignition and surface area (Umaña-Peña, 2002). This morphology is due to low efficiency of the burning system used in São Jerônimo power plant.

3.2. Characterization of Cyclone Fly Ash

The fly ash from Figueira Power Plant is collected from the flue gas by means of baghouse using fabric filters and cyclone collector. Cyclones are low-cost, low-maintenance centrifugal collectors that are typically used to remove particulates in the size range of 10–100 microns. The cyclone fly ash is the ash fraction that has been carried through the flue gas channels into the cyclone and precipitated there. Technically speaking, cyclone ash is the ash fraction that has been trapped in the cyclone.

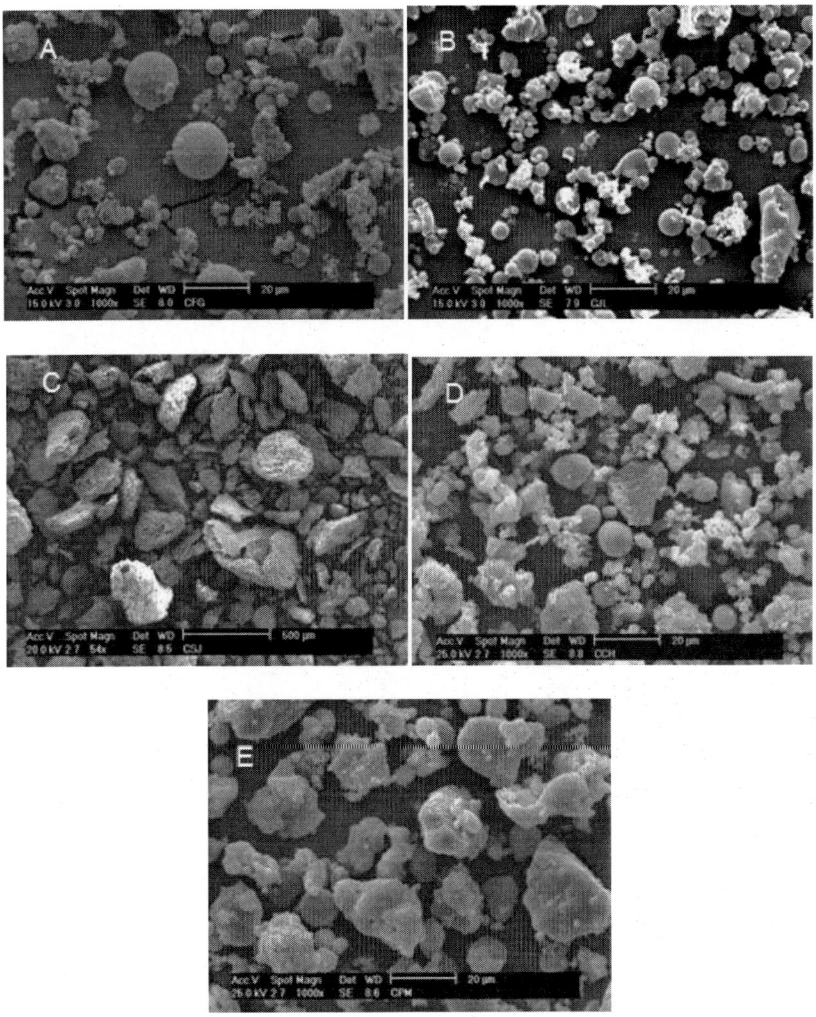

Figure 3. SEM micrographs of fly ashes from A) FG; B) JL; C) SJ; D) CH; E) PM.

During combustion most of the heavy metals in the fuel are retained in the ash. The various ash fractions such as bottom ash, cyclone ash and baghouse filter ash may contain relatively high concentrations of heavy metals (Silva and Boit, 2011). Combustion acts like a thermodynamic separation process for the different inorganic materials in the fuel. Elements with a low volatility will concentrate in the bottom ash, while more volatile elements will concentrate in the fly ash (Font et al., 2012).

The consequence of this separation process in combustion plants is that the different ash fractions have a different chemical composition. Before ash residue can be utilized or disposed of in a landfill, its chemical as well as extraction (leaching) characteristics must be known.

Fly ash from baghouse filter produced in the Brazilian power plant is mostly recycled as cement raw material to be used in civil engineering materials. However, cyclone ash does not find a commercial application at present time; rather, it is usually stored in abandoned surface mine or dumped in landfills in the vicinity of the power plant.

Very few studies have been carried out to assess the environmental impact of Brazilian cyclone ashes. Cyclone ash samples contained larger particles than fabric filter and electrostatic precipitator ash samples and showed lower cytotoxicity, very little mutagenicity and lower concentrations of trace elements (Depoi et al., 2008; Mumford and Lewtas, 1984).

3.2.1. Physico-Chemical Properties

The sample of coal fly ash collected by cyclone from Figueira Power Plant (FGC) was characterized. Some important physicochemical characteristics of this material are shown in Table 3.

Table 3. Physicochemical properties of cyclone ash

Characteristics	Obtained Values
SiO_2 (wt. %)	40.4
Al_2O_3 (wt. %)	15.2
Fe_2O_3 (wt. %)	10.6
K_2O	2.23
CaO (wt. %)	1.60
Others (wt. %)	$\leq$ 0.98
pH in water	8.18
Conductivity ($\mu S\ cm^{-1}$)	201
Humidity (%)	1.59
Loss og ignition (%)	8.18
Specific surface area ($m^2\ g^{-1}$)	3.31
Real density ($g\ cm^{-3}$)	2.33
CEC ($meq\ g^{-1}$)	0.0791

Fly ash samples were collected in baghouse filter and cyclone in the Figueira power plant and it was observed that the concentrations of most elements were higher in the fly ash taken from baghouse filter (Table 1). This fact means that these elements are mostly associated with smaller particles of fly ash. The position of the filter and temperature also influence composition, where the decrease in temperature increases the potential for trace element capture (Meij, 1994).

The trace element concentrations observed in cyclone ash include As, B, Be, Cd, Co, Cr, Cu, Li, Mn, Mo, Ni, Pb, Sb, Tl, and Zn (Levandowski and Kalkreuth, 2009).

3.2.2. Laser Granulometric Analysis

Laser granulometric analysis results for FGC are reported both as absolute and cumulative volumetric distributions (Figure 4). From the cumulative distributions the following values for d_{10}, d_{50} and d_{90} can be observed: 13.6, 43.6 and 174.8 μm, respectively. The Sauter mean diameter (D[3,2]) was 24.3 μm and is indicative of the active surface area of the material.

Cyclone fly ash has a broad size distribution with diameters spanning more than three orders of magnitude. The median diameters of ash samples from six representative coals collected in cyclones and baghouses of American coal fired power plants were between 1.2-3.4 μm and 0.4-1.2 μm, respectively (Ghosal et al., 1995).

The grain size distribution of fly ash sample from a Brazilian power plant (Presidente Medici Power Plant) collected from electrostatic precipitator was determined. The continuous interval of particle sizes was between 0.5 and 600 μm, and the average diameter was 49.30 μm. Ninety percent of the particles have diameters smaller than 232 μm, while the finest particles (< 5 μm) correspond to ~ 1% of the total mass of the sample (Pires and Querol, 2004).

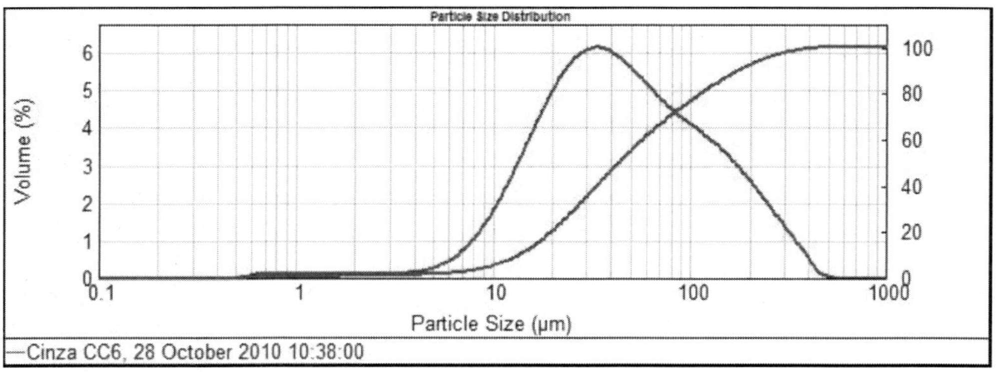

Figure 4. Particle size distribution of FGC.

3.2.3. Instrumental Neutron Activation Analysis

A total of 14 elements, i.e. As, Br, Co, Cr, Eu, Fe, La, Na, Sb, Sc, Sm, Th, U and Zn, were determined in Figueira coal-fired power plants by instrumental neutron activation analysis. The results of analysis for coal cyclone fly ash (FGC) are presented in Table 4, accompanied by standard uncertainties.

Radionuclides frequently undergo concentration in Brazilian coal ashes. The coal from Figueira coal-fired power plant is extremely rich in uranium (the average concentration is 71.6 mg kg^{-1}). Uranium occurrences are known in Permian rocks of Paraná Basin from Brazil to Uruguay. The area near Figueira is rich in uranium, with an average of 0.148% U_3O_8 in calcitic sandstone above the coal seam (Depoi et al., 2008; Flues et al., 2007).

It is suggested that the high concentrations of As and Zn are due to the elements association with the high content of pyrite in the coal from Figueira (Flues et al., 2008).

Table 4. Chemical elements in FGC samples analyzed by INAA

Element	Concentration ($\mu g\ g^{-1}$)[1]	Element	Concentration ($\mu g\ g^{-1}$)[1]
As	632 ± 95	Na	8293 ± 50
Br	< dl	Sb	8.3 ± 0.4
Co	24.0 ± 0.1	Sc	25.8 ± 0.6
Cr	93 ± 6	Sm	19 ± 1
Eu	4.1 ± 0.1	Th	23.9 ± 1.1
Fe	95100 ± 4076	U	227 ± 11
La*	121 ± 4	Zn	822 ± 82

(1) Mean and standard deviation of two determinations; (*) element that can be fission products; dl = detection limit.

3.2.4. Mineralogical Composition

The XRD analysis of FGD powder revealed the presence of quartz crystallite, mullite, iron oxides (magnetite and hematite). The presence of amorphous phases are identified as a broad diffraction 'hump' in the region between 18 to 30 degrees of 2θ (Figure 5). This result is similar to that for other Brazilian fly ash samples (Figure 2). The intensity of quartz is very strong, with mullite forming a chemically stable and dense glassy surface layer.

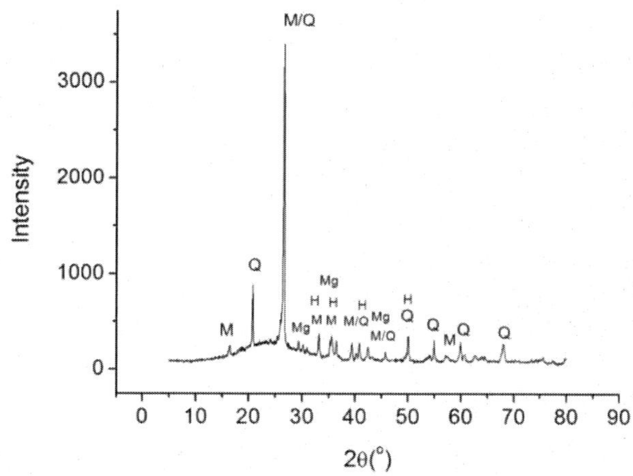

Figure 5. X-ray diffraction pattern of cyclone ash (Q = Quartz, M = Mullite, H = Hematite, Mg = Magnetite).

3.2.5. Morphological Analysis

FGC was a mixture of unburned organic materials and different inorganic phases. The morphology of a fly ash particle is controlled by combustion temperature and cooling rate. The majority of the particles ranged in size from approximately 5 to 58 μm and consisted of solid spheres (Figure 6A). Hollow cenospheres, pleurospheres (Figure 6B) and irregularly shaped unburned carbon particles tended to be in the upper end of the size distribution.

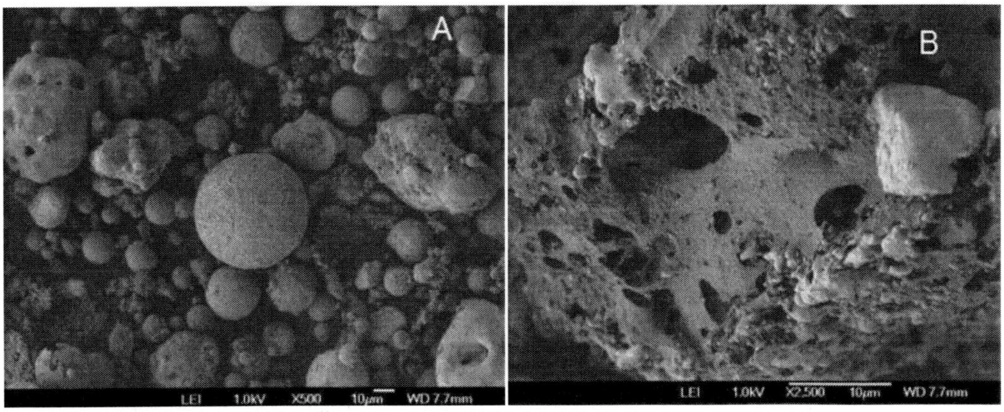

Figure 6. SEM micrographs of cyclone Fly Ash.

3.2.6. FTIR Spectroscopy

The FT-IR technique is an important tool to identify the characteristic functional groups on the material surface. The FTIR data validated the presence of Si-O-Si groups of quartz in cyclone fly ash. The IR spectrum for FGC shows three wide bands, characteristic of aluminosilicates: at 1078, 779 and 459 cm ^{1} (Figure 7). The bands at 1078 and 779 cm^{-1} are attributed to the asymmetrical stretching vibration and symmetrical stretching vibration of the Si-O-Si groups (Thongsang and Sombatsompop, 2006). The band at 459 cm^{-1} was associated with T–O bending vibrations.

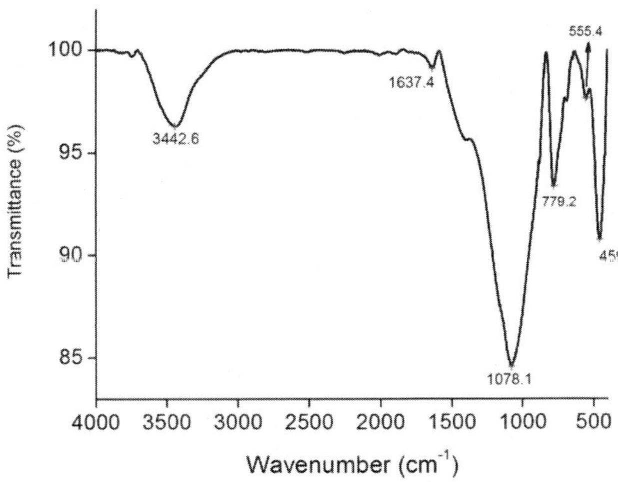

Figure 7. FTIR spectrum of FGC.

The FT-IR absorption bands in the range 3800–1600 cm^{-1} are generally assigned to vibration modes of hydrous or hydroxyl components on the fly ash (Mouhtaris et al., 2003). The band at 3442.6 cm^{-1} is attributed to the asymmetric and symmetric stretching vibrations ν(O–H), suggesting the presence of amorphous silicate material (glass) or possibly hydrated aluminum silicates. The band at 1637.4 cm^{-1} is attributed to the bending mode of H_2O

molecules. The bands at 779.18 and 555.4 cm^{-1} could be assigned to quartz as crystal phase in the coal fly ash.

3.3. Leaching and Solubilization Tests

The fly ash composition varies as a function of the coal source, but is generally enriched in a number of trace elements which may potentially become bio-available in environment applications (Levandowski and Kalkreuth, 2009; Silva et al., 2010).

Fly ash samples were subject to leaching and solubilization tests for solid waste classification according to Brazilian regulation (Standard Brazil ABNT NBR 10005, 2004; Standard Brazil ABNT NBR 10006, 2004). The chemical elements chosen for these tests are those that represent the highest risk to fauna and flora. They are used to evaluate the environmental behavior of waste materials. According to these tests the waste materials can be classified as hazardous (Class I), and Class II- Non-hazardous, being that this last one is divided into: Class II A- not inert and Class II B- inert (Standard Brazil ABNT NBR 10004, 2004).

Tables 5 and 6 show the results of the acetic acid leaching and water solubilization tests carried out with Brazilian fly ash samples. In these tables the chemical element values obtained from the samples and the limit values determined by the Brazilian Standard NBR 10004 can be observed.

According to the results obtained, the fly ashes from Figueira Power Plant can be classified as Class I (hazardous) and the other samples can be classified as Class II A (non-dangerous/non-inert) materials. Thus, the fly ash from Figueira Power Plant should be carefully disposed of in industrial landfills designed to handle hazardous waste.

Table 5. Concentration of elements leachated from fly ashes and the allowed limit values

Elements	FG	JL	SJ	CH	PM	FGC	Limit[(a)] (mg/L)
Ag	< 0.02	< 0.02	< 0.02	< 0.02	< 0.02	< 0.02	5.0
As	36.8	0.35	0.066	0.052	0.042	12	1.0
Ba	0.111	0.202	0.808	0.2403	0.1658	0.1168	70
Cd	0.0506	0.0053	0.0027	0.0014	0.0011	0.0119	0.5
Cr	0.041	0.050	0.038	0.028	0.053	< 0.02	5.0
Hg	< 0.001	< 0.001	< 0.001	< 0.001	< 0.001	< 0.001	0.1
Pb	0.015	0.0086	0.089	0.089	0.016	0.0085	1.0
Se	0.072	0.064	0.0025	0.0025	0.0025	0.023	1.0

(a) Established by Annex F of Norm ABNT NBR 10004 (2004).

Table 6. Concentration of elements solubilized from fly ashes and the allowed limit values

Elements	FG	JL	SJ	CH	PM	FGC	Limit[(a)] (mg/L)
Ag	< 0.02	< 0.02	< 0.02	< 0.02	< 0.02	< 0.02	0.05
Al	0.57	1.92	0.30	0.308	8.15	20.8	0.2
As	4.11	0.216	0.097	0.066	0.037	10.1	0.01
Ba	0.0691	0.0711	0.0285	0.0444	0.204	0.129	0.7
Cd	0.00033	0.0006	0.0005	0.0004	0.00077	0.0058	0.005
Cr	0.60	0.535	< 0.02	0.0654	0.188	0.0678	0.05
Cu	< 0.02	< 0.02	< 0.02	< 0.02	< 0.02	< 0.02	2
Fe	< 0.05	0.056	< 0.05	< 0.05	3.44	0.3807	0.3
Hg	< 0.001	< 0.001	< 0.001	< 0.001	< 0.001	< 0.001	0.001
Mn	< 0.02	< 0.02	< 0.02	< 0.02	0.101	0.0592	0.1
Na	87	32	2.366	8.19	15.53	55.7	200
Pb	0.0103	0.064	0.008	0.046	0.092	0.016	0.01
Se	0.065	0.126	0.012	0.035	0.033	0.009	0.01
Zn	0.014	0.029	0.0252	0.0140	0.138	0.0864	5

(a) Established by Annex G of Norm ABNT NBR 10004 (2004).

Conclusion

The fly ash properties determined in this work could be regarded as representative of all coal fly ash produced in Brazilian thermal power plants because they were taken from the region south of the country where mineral coal is concentrated.

The fly ash samples from the five Brazilian power stations have similar mineralogy, with quartz, mullite, hematite and magnetite as the principal crystalline components, and amorphous material. The samples also have similar concentration of major and trace elements, except the one from Figueira. The concentrations of most elements are higher in the fly ash from Figueira.

The fly ash from São Jerônimo Power Plant presented higher unburned carbon content and specific surface area.

Fly ash samples from Figueira Power Plant were classified as hazardous because they had chemical elements with concentrations above maximum limits established by Brazilian NBR 10004 code in extract leachates. Thus, these fly ashes should be carefully disposed. On the other hand, the solubilization tests indicated that all others samples can be classified as non-inert waste, based on NBR 10004/2004 (Class II A, non-dangerous/non-inert).

The SiO_2/Al_2O_3 ratios indicated that Brazilian fly ash samples were suitable for the synthesis of zeolites, geopolymers and silica extraction.

5. Fly Ash Utilization and Recommendations

Fly ash, although posing environmental pollution, is an important raw material for various applications. A review study showed the various uses of fly ash, including low-cost adsorbents for flue gas cleaning, removal of pollutants from wastewater, synthesis of zeolites, mine backfill, light weight aggregate, road sub-base, construction/cement applications, etc (Ahmaruzzaman, 2010).

Reuse of coal fly ash may provide a new source of revenue for coal-fired power plants and may offset expenses associated with its disposal. In the environmental aspect, the increased use of coal fly ash can reduce energy consumption, landfill space, and conserve natural resources. Utilization of coal fly ash can be in the form of an alternative to another industrial resource, process, or application.

The chemical and mineralogical compositions of Brazilian fly ashes led us to conclude that they were suitable for the synthesis of zeolites and for obtaining other silica and alumina compounds.

Zeolites are hydrated aluminosilicate minerals with a three-dimensional open structure making them very useful for solving the mobility of toxic elements in a number of environmental applications. This is strictly connected with their ability to exchange cations, large surface area, and typical structural characteristics (such as porosity), which facilitate pollutant adsorption and encapsulation. Zeolites can also act as molecular sieves and catalysts. In addition, zeolites from fly ash have potential to tie in well with coal fired systems.

In our group we have been developing zeolites from Brazilian coal fly ashes by conventionally hydrothermal treatment and fusion method. Pure zeolites X and A (Izidoro et al., 2013), zeolite from fly ash-iron oxide nanocomposites (Fungaro et al., 2012; Fungaro, Yamaura and Carvalho, 2012), surfactant-modified zeolites (Fungaro et al., 2013; Fungaro and Magdalena, 2012) and zeolitic materials (Izidoro et al., 2012a) were synthesized and characterized. The crystalline phases after classic hydrothermal treatment were identified to be hydroxy-sodalite, NaP1 and X zeolites with quartz and mullite remained after the treatment.

In addition, the zeolites from Brazilian coal fly ashes were used as low-cost adsorbents and have proven to be promising materials for the removal of dyes (Fungaro et al., 2009) and toxic metals (Izidoro et al., 2012b; Izidoro et al., 2013) from aqueous solution. Moreover, a considerable reduction in the elements contents and in the toxicity to living organisms (*D. similis and V. Ficheri*) was attained in the fly ash zeolitization (Carvalho et al., 2011, Fungaro et al., 2010).

Brazilian Fly ashes have been also applied as a support for catalysts (Muhammad et al., 2012) and as a by-product aggregate in the manufacture of ecological bricks (Silva, 2011).

Demand for electricity is expected to increase steadily in the future, and coal will continue to play a significant role in meeting this demand. However, the contribution of coal to the energy mix, and the overall cost of electricity, will increasingly depend on ability to find economical ways to reduce or eliminate any potential adverse environmental impacts associated with the disposal and utilization of coal combustion by-products.

The effective utilization of coal ashes is strongly desired from environmental, ecological and economical points of view. Therefore, it is highly recommended to carry out detailed research to evaluate the potential utilization of Brazilian fly ashes at an industrial scale. The development of new technology for efficient utilization of fly ash is also needed.

ACKNOWLEDGMENTS

The authors gratefully acknowledge the assistance of power plants in collecting fly ash samples and technical visits. The support of Conselho Nacional de Desenvolvimento Científico e Tecnológico (CNPq) and Capes Foundation (Ministry of Education of Brazil) is gratefully acknowledged.

REFERENCES

Ahmaruzzaman, M. A. (2010) review on the utilization of fly ash. *Progress in Energy and Combustion Science,* 36(3), 327–363.

ANEEL Agência Nacional de Energia Elétrica. (2011). Retrieved April 05, 2013, from http://www.aneel.gov.br/aplicacoes/audiencia/arquivo/2011/043/documento/nota_tecnica_034_2011_srg.pdf.

BP Statistical Review of World Energy. (2012). Retrieved April 05, 2013, from http://www.bp.com/assets/bp_internet/globalbp/globalbp_uk_english/reports_and_publications/statistical_energy_review_2011/STAGING/local_assets/pdf/statistical_review_of_world_energy_full_report_2012.pdf.

Carvalho, T. E. M., Fungaro, D. A., Magdalena, C. P., & Cunico, P. (2011). Adsorption of indigo carmine from aqueous solution using coal fly ash and zeolite from fly ash. *Journal of Radioanalytical & Nuclear Chemistry*, 289(2), 617-626.

Depoi, F. S., Pozebon, D., Kalkreuth, W. D. (2008). Chemical characterization of feed coals and combustion-by-products from Brazilian power plants. *International Journal of Coal Geology*, 76 (3), 227-236.

El-Naggar, M. R., El-Kamash, A. M., El-Dessouky, M. I., & Ghonaim, A. K. (2008). Two-step method for preparation of Na A-X zeolite blend from fly ash for removal of cesium ions. *Journal of Hazardous Materials,* 154(1-3), 963-972

Flues, M., Camargo, I. M. C., Figueiredo Filho, P .M., Silva, P. S. C, & Mazzilli, B. P. (2007). Evaluation of radionucletides concentration in Brazilian Coals. *Fuel*, 86(5-6), 807–812.

Flues, M., Sato, I. M., Cotrim, M. B., Figueiredo Filho, P. M., & Camargo, I. M. C. (2008). Evaluation of the influence of a coal plant operation on metal and as concentrations in the soil of Figueira, PR-Brazil. *Química Nova*, 31(1), 25–30.

Font, O., Córdoba, P., Leiva, C., Romeo, L. M., Bolea, I., Guedea, I., Moreno, N., Querol, X., Fernandez, C., & Díez, L. I. (2012) Fate and abatement of mercury and other trace elements in a coal fluidised bed oxy combustion pilot plant. *Fuel,* 95, 272-281.

Fungaro, D.A., Bruno, M., & Grosche, L. C. (2009). Adsorption and kinetic studies of methylene blue on zeolite synthesized from fly ash. *Desalination and Water Treatment*, 2(1-3), 231-239.

Fungaro, D. A., Grosche, L.C., Pinheiro, A. S., Izidoro, J.C., & Borrely, S. I. (2010). Adsorption of methylene blue from aqueous solution on zeolitic material and the improvement as toxicity removal to living organisms. *Orbital*, 2(3), 235-247.

Fungaro, D. A., Yamaura, M., & Craesmeyer, G. R. (2012). Uranium Removal from Aqueous Solution by Zeolite from Fly Ash-Iron Oxide Magnetic Nanocomposite. *International Review of Chemical Engineering (I.RE.CH.E.),* 4 (3), 353-358.

Fungaro, D. A., & Magdalena, C. P. (2012). Adsorption of Reactive Red 198 from aqueous solution by organozeolite from fly ash: Kinetic and Equilibrium studies. *International Journal of Chemical and Environmental Engineering Systems,* 3(3), 74-83.

Fungaro, D. A., Yamaura, M., Carvalho, T. E. M. (2012). Zeolite from Fly Ash-Iron Oxide Magnetic Nanocomposite: Synthesis and Application as an Adsorbent for Removal of Contaminants from Aqueous Solution. In M.K. Andreyev & O.L. Zubkov (Eds), Zeolites: Synthesis, Chemistry and Applications (pp.1-34). New York: Nova Science Publishers.

Fungaro, D. A., Borrely, S. I., & Carvalho, T. E. M. (2013). Surfactant Modified Zeolite from Cyclone Ash as Adsorbent for Removal of Reactive Orange 16 from Aqueous Solution. *American Journal of Environmental Protection,* 1(1), 1-9.

Ghosal, S., Ebert, J. L., & Self. S. A. (1995). Chemical composition, size distributions for fly ashes. *Fuel Processing Technology,* 44(1-3), 81-94.

Izidoro, J. C., Fungaro, D. A., Santos, F. S., & Wang, S. (2012a). Characteristics of Brazilian coal fly ashes and their synthesized zeolites. *Fuel Processing Technology*, 97, 38–44.

Izidoro, J. C., Fungaro, D. A., & Wang, S. (2012b). Zeolite synthesis from Brazilian coal fly ash for removal of Zn^{2+} and Cd^{2+} from water. *Advanced Materials Research*, 356-360, 1900-1908.

Izidoro, J. C., Fungaro, D. A., Abott, J. E., & Wang, S. (2013). Synthesis of zeolites X and A from fly ashes for cadmium and zinc removal from aqueous solutions in single and binary ion systems. *Fuel,* 103, 827–834.

Levandowski, J., & Kalkreuth, W. D. (2009). Chemical and petrographical characterization of feed coal, fly ash and bottom ash from the Figueira Power Plant. Paraná, Brazil. *International Journal of Coal Geology*, 77(3-4), 269-281.

Medina, A., Gamero, P., Querol, X., Moreno, N., De León, B., Almanza, M., Vargas, G., & Izquierdo, M. (2010). Fly ash from a Mexican mineral coal I: Mineralogical and chemical characterization. *Journal of Hazardous Materials*, 181(1-3), 82-90.

Meij, R. (1994). Trace Element Behavior in Coal-Fired Power Plants. *Fuel Processing Technology*, 39(1-3), 199-217.

Mouhtaris, Th., Charistos, D., Kantiranis, N., Filippidis, A., Kassoli-Fournaraki, A., & Tsirambidis, A. (2003). GIS-type zeolite synthesis from Greek lignite sulphocalcic.fly ashes promoted by NaOH solutions. *Microporous and Mesoporous Materials,* 61(1-3), 57–67.

Muhammad, S., Saputra, E., Sun, H. , Izidoro, J. C., Fungaro, D. A., Ang, M. H., Tade, M. O., & S. Wang. (2012). Coal fly ash supported Co3O4 catalysts for phenol degradation using peroxymonosulfate. *RSC Advances,* 2(13), 5645-5650.

Mumford, J. L., & Lewtas, J. (1984). Evaluation of fly ash collection methods for short-term bioassay studies of fluidized-bed coal combustion. *Environmental Science & Technology*, 18(10), 765–768.

Ojha, K., Pradhan, N. C., & Samanta, A. N. (2004). Zeolite from fly ash synthesis and characterization. *Bulletin of Materials Science*, 27(6), 555-564.

Paprocki, A. (2009). Síntese de zeólitas a partir de cinzas de carvão visando sua utilização na descontaminação de drenagem ácida de mina. Master Thesis. Pontifícia Universidade Católica do Rio Grande do Sul, Porto Alegre, Brazil.

Pathan, S. M., Aylmore, L. A.G., & Colmer, T. D. 2003. Properties of several fly ash materials in relation to use as soil amendments. *Journal of Environmental Quality*, 32(2), 687-693.

Pires, M. & Querol, X. (2004). Characterization of Candiota (South Brazil) coal and combustion by-product. *International Journal of Coal Geology*, 60(1), 57-72.

Rohde, G. M., Zwonok, O., Chies, F. & Silva, N. I. W. (2006). *Cinzas de carvão fóssil no Brasil*. Porto Alegre: CIENTEC.

Sarbak, Z., Stanczyk, A., & Kramer-Wachowiack, M. (2004). Characterization of surface properties of various fly ashes. *Powder Technology*, 145(2), 82–87.

Shigemoto, N., & Hayashi, H. (1993). Selective formation of Na-X zeolite from coal fly ash by fusion with sodium hydroxide prior to hydrothermal reaction. *Journal of Materials Science*, 28(17), 4781–4786.

Sijakova-Ivanova, T., Panov, Z., Blažev, K., & Zajkova-Paneva, V. (2011). Investigation of fly ash heavy metals content and physico chemical properties from thermal power plant, Republic of Macedonia. *International Journal of Engineering Science and Technology*, 3(12-13), 8219-8225.

Silva, L. F. O., Ward, C. R., Hower, J .C., Izquierdo, M., Waanders, F., Oliveira, M. L. S., Li, Z., Hatch, R. S., & Querol, X. (2010). Mineralogy and Leaching Characteristics of Coal Ash from a Major Brazilian Power Plant. *Coal Combustion and Gasification Products*, 2, 51-65.

Silva, L. F. O., & Boit, K. M. (2011). Nanominerals and nanoparticles in feed coal and bottom ash: implications for human health effects. *Environmental Monitoring and Assessment*, 174(1), 187-197.

Silva, M. V. (2011). Development of bricks with incorporation of coal ash and sludge from water treatment plant. Master Thesis. Instituto de Pesquisas Energéticas e Nucleares. Retrieved April 05, 2013, from São Paulo University Digital Thesis.

Singh, D. N., & Kolay, P. K. (2002). Simulation of ash water interaction and its influence on ash characteristics. *Progress in Energy and Combustion Science*, 28(3), 267- 299.

Standard American. (1992). ASTM C618-92a, Standard specification for fly ash and raw or calcinated natural pozzoland for use as mineral admixture in Portland cement concrete. American Society for Testing and Materials, annual book of ASTM Standards, Vol 04.02. Pennsylvania.

Standard Brazil ABNT NBR 10004. (2004). Solid Waste – Classification. Retrieved April 05, 2013, from Standard Brazil Online database.

Standard Brazil ABNT NBR 10005. (2004). Procedure *for obtaining leaching extract of solid wastes*. Retrieved April 05, 2013, from Standard Brazil Online database.

Standard Brazil ABNT NBR 10006 (2004b). Procedure for obtaining soluble extract of solid wastes. Retrieved April 05, 2013, from Standard Brazil Online database.

Theis, T. L., & Wirth, J. L. (1977). Sorptive Behavior of Trace Metals on Fly Ash in Aqueous Systems. *Environmental Science & Technology*, 11(12), 1096-1100.

Thongsang, S., &Sombatsompop, N. (2006). Effect of NaOH and Si69 treatments on the properties of fly ash filled NR composites. *Polymer Composites*, 27(1), 30-40.

Umaña-Peña, J. C. (2002) Síntesis de zeolitas a partir de cenizas volantes de centrales termoeléctricas de carbón. PhD Thesis. Universitat Politécnica de Catalunya, Barcelona, Espanha.

Wang, S., & Wu, H. (2006). Environmental-benign utilisation of fly ash as low-cost adsorbents. *Journal of Hazardous Materials*, B136(3), 482–501.

In: Fly Ash
Editor: Prabir Kumar Sarker ISBN: 978-1-62948-044-2

Chapter 6

CHARACTERIZATION AND UTILIZATION OF COAL COMBUSTION BY-PRODUCTS IN MONGOLIA

Jadambaa Temuujin*
Institute of Chemistry and Chemical Technology, Mongolian Academy of Sciences, Ulaanbaatar, Mongolia

ABSTRACT

Mongolia is rich in coal with up to 160 billion tonnes of reserve. Coal is also the cheapest raw material for producing electricity and more than 90% of electricity in Mongolia is produced by coal fired thermal power stations. At present, there are five operating power plants in the centralised energy system which generates both heat and electricity. These plants burn 5-6 million tons of coal resulting in more than 600,000 tons of coal combustion by-products per year. Three thermal power plants are located in Ulaanbaatar city and produce more than 80% of the total electricity in Mongolia. Within these plants, one generates around 300,000 tons of fly and bottom ash and another around 100,000 tons of bottom ash.

The physical and chemical characteristics of fly ash and boiler slags determine how to make of use them. The physical properties of these residues are determined by the technology, design and operational conditions of the power stations. The chemical composition of the fly ash and slag depends completely on the inorganic matter within the source coal deposits and cannot be controlled. Comprehensive characterisation of coal combustion by-products were performed using XRD, XRF, SEM, BET and gamma-spectroscopy methods. The thermal power plants in Ulaanbaatar city uses coal from the Baganuur and Shivee ovoo deposits. The chemical composition of these fly ashes suggests they belong to class C fly ashes.

The content of radionuclides in the fly ash also depends on the original coal. Radiation measurement of these ashes also revealed significant variation with time. For instance, ash from burning Shivee-ovoo coal collected in 2011 was measured to have 342 Bq/kg, the same ash collected in 2012 had 275 Bq/kg. Ash from the pond of one power plant has a radiation equivalent of 568 Bq/kg, which is too high for use as building

* E-mail: jtemuujin@yahoo.com.

materials. Other samples from the same source had 179-351 Bq/kg, low enough for use in building materials.

Fly ashes from the Mongolian thermal power plants were used for zeolite synthesis and geopolymer type concrete and paste preparation. The properties of the prepared products are discussed in this chapter.

Keywords: Ash utilization; bottom ash; dust; fly ash; geopolymer type binder; radioactivity; zeolites

1. INTRODUCTION

General Trend of Fly Ash and Slag Utilization in the World

The most common use of coal is for burning as a solid fuel to produce electricity. It is estimated that around 40% of global electricity is generated by coal burning (Wikipedia, 2013) and this is unlikely to decrease in the near future. Approximately 10-30% of the original coal is noncombustible and remains as solid by-products. Generally, the chemical, mineralogical composition and volume of the coal combustion by-products are determined by the content and composition of the inorganic constituent of the original coal and design and operation of the power station (Jha et al., 2008). Coal combustion by-products (CCP) also can be divided into different types; namely bottom ash or boiler slag, fly ash and flue gas desulphurisation materials. In a review on fly ash utilisation Ahmaruzzaman (2010) states that the world's current annual production of coal ash is around 600 million tonnes of which fly ash is about 500 million tonnes. Ash products are usually disposed of in landfill or in ash ponds, but with such enormous volumes it is creating environmental problems and becoming a costly overhead for coal fired power generation. Thus, there is a strong incentive to find useful and sustainable ways to utilise ash products (Wang and Wu, 2006). Currently worldwide utilisation is about 16% of total the total ash produced, though it varies from country to country. For instance, Japan utilises over 97% of coal ash (Japan Coal Energy Center, 2012) while in Australia over 40% of the ash is utilised (Australian Ash Development Association, 2012). Based on statistics issued by European Coal Combustion Products Association (ECOBA) in 2009, the total CCP production for 2007 in the European Union was about 61 Mt of which 89.3% was utilized. About 52.7% of the CCPs produced was used in construction industries (concrete, cement, gypsum panel, etc.) and civil engineering (road base, embankment, flowable fill, etc.) and 36.5% was used in restoration of open cast mines with 2.5% were stockpiled for future use and 8.3% were disposed of (Fu, 2010).

Historical Overview of Fly Ash and Slag Utilization in Mongolia

After 1921, Mongolian electricity production increased greatly. Mongolia had adopted the policy for electrification of industrial and cultural centers, provincial centers and factories and mines at local level. In 1934, Mongolia produced 1130 thousand kW/h electricity, by 1939 it has increased tenfold to 12193 thousand kW/h. By 1960, electricity production of Mongolia reached to 106,4 million kW/h and total installed capacity to 60 MW. At that time the biggest power plant had a capacity of 24 MW. In 1960, 80% of the total electrical power was produced by locomobile and diesel electrogenerators (Mongolian Ministry of Energy, 2012).

After 1960 Mongolia began to use Russian designed combined heat and power plants which also generated steam for industrial use. All these power plants were coal fired and since then there has been a considerable amount of fly ash produced. The concept of the heat and power networks is technically efficient, providing electricity, industrial steam, space heating and domestic hot water from a single energy source.

In Mongolia over 90% of electricity is produced by coal fired power stations. The maximum energy production of all the power stations is 4,256 million kWh, however this is still not sufficient to supply all of Mongolia's needs necessitating importation of 214 million kWh electricity from Russia. According to studies conducted on large industrialized and settlement regions and cities it can be estimated that between 2015 and 2030, the country's energy consumption rate will be doubled (Yeren-Ulzii, 2012). As the coal is the main source for power generation the amount of coal combustion by products is expected to increase, substantially. Recent reports have indicated that Mongolia has total coal reserves of over 160 billion tonnes located in more than 200 coal deposits within 12 coal basins (Mongolian Ministry of Energy, 2012).

Table 1 shows the main thermal power plants in Mongolia and their capacity and location.

Table 1. The main thermal power plants in Mongolia

№	Coal-fired Power Plants	Capacity (MW)	Available (MW)	Share in CES (%)	Location	Year installed	Efficiency (in 2009)
1	TPS2	21.5	18	2.7	Ulaanbaatar	1961	21
2	TPS3	136	105	17.5	Ulaanbaatar	1968	38.6
3	TPS4	580	452	70.2	Ulaanbaatar	1983	40.1
4	Erdenet plant	28.8	21	3.6	Erdenet city	1987	40.8
5	Darkhan plant	48	39	6	Darkhan city	1965	28.5
	CES Subtotal	814.3	615	100			
6	Dornod plant	36	-	-	Dornod aimag	1969	19.4
7	Umnogovi plant	6	-	-	Umnogovi aimag	2001	-
	Total	856.3					

Since the numbers of thermal power plants are expecting to rise the amount of residue will also increase. The majority of coal firing residue from Mongolian power plants cannot be considered as fly ash. Fly ash is composed of fine particles which are transported within the flue gas and recovered using an electrostatic precipitator. At this time, only the TPS4 has an electrostatic separator to collect fly ash. Other thermal power stations are producing bottom ash or boiler slags. The two largest thermal power stations in Ulaanbaatar city are using a huge volume ash pond to keep the removed coal combustion by-products wet. However, due to a land shortage in Ulaanbaatar city, there is problem to build a new ash pond. There have been several attempts to use power station residues for building materials. From the 1980's residues were mixed with cement for building block production. Current Mongolian and international standards for cement production allow fly ash to be applied. Also one national standard (Mongolian standard, 1986) approved and controlled the use of power station residues in building materials production. However, the use of such residues for production of building materials wasn't popular in Mongolia because of the presence of radionuclides within the residues. The poor public perception of the health hazards involved as meant that this type of material hasn't been fully utilized.

Since the Thermal Power Station 4(TPS4) uses an electrostatic filter, fly ash from it is the most suitable for the cement industry in the current Mongolian situation.

Fly ash from power plants used in cement industry has two important properties.

- Due to ash's spherical shapes, cement slurry's rheological property is improved and the water content is decreased.
- Due to Amorphous Aluminum Silicate content of the ash, it reacts with the calcium oxide and then forms a calcium aluminosilicate connecting phase which improves strength of the cement.

However, fly ash use is not efficient without knowledge of the slag and fly ash structure, the combustion temperature and the technology procedures. Generally, the fly ash generated at relatively high combustion temperature is suitable for the cement industry.

Domestically Mongolia uses about 6 million tonnes of coal per annum generating approximately 500-600 thousand tonnes of ash. However, the utilisation rate of the ash is very low and is possibly less than 10% based on survey from the thermal power stations. Planning is underway to build several new thermal power stations to meet electricity demand within Mongolia and to provide spare capacity for export of electricity to neighboring countries. Once the new power stations come to operation the amount of coal combustion by-products will dramatically increase. It is thus imperative for Mongolia to identify beneficial ways to utilise coal ash. Therefore, rational utilization of coal combustion by-products is one of the major research themes for our researchers.

Problems Associated with Coal Combustion By-products in Ulaanbaatar

According to research data, air and soil quality within the Mongolian capital city has reached a disastrous level. The main reasons for the air pollution in Ulaanbaatar are reported to be coal fired stoves of the Gher residential district, thermal power stations, small and medium sized low pressure stoves and motor vehicles (Mongolian President Office website, 2013).

An air pollution study indicated that the thermal power stations and Gher district of Ulaanbaatar city consumed annually approximately 6 million tonnes of coal which released in to air 71.9 thousand tonnes of SO_2, 57.3 thousand tonnes of NO_2, 174.1 thousand tonnes of fine dust and 13.9 million tonnes of CO_2. Research on air dust quality study of Ulaanbaatar city revealed that 50% of total dust was from the soil/earth and 35% from the smoke and ash pollution resulted from coal burning (World Bank Report, 2010).

Research by the Ulaanbaatar Pediatric Hospital has shown that in sources of overall diseases 38.3% was accounted for air dust source, and overall polluting substance based diseases accounted as 43.14% from heavily polluted soil. Not only human beings but also plants, animals, as well as environmental degradation and constructions and buildings may receive certain damages due to air pollution.

Therefore, to find a way to reduce the coal ash quantity which is one of the main reasons of air and soil pollution, especially ashes from thermal power station, is a necessary work for future development of the Ulaanbaatar city. Adding electrostatic precipitators to the existing power stations would help a lot.

This chapter summarises the latest results on characterization of coal combustion by-products in Mongolia and their potential applicability for value added products such as zeolites and geopolymer type binders.

2. METHODS

2.1. Raw Materials

Fly ash from the 4th thermal power plant of Ulaanbaatar city was used as the raw materials. Since the two boilers of the 4th Thermal power station used Baganuur coal and another 2 boilers Shivee ovoo coal, two different fly ashes were collected from this power station. Ashes were also taken from the filled ash pond of 3rd thermal power station (Ulaanbaatar city) and Darkhan thermal power station (Darkhan city). The 3rd thermal power station is using Baganuur coal while Darkhan station is using Shariin gol hard coal as fuels. For the experiments ash samples of about 100-500 kg were taken and statistically sampled for chemical and mineralogical characterisations.

For the zeolite synthesis Baganuur fly ash was used because of lower Ca and higher Al and Si content than Shivee ovoo fly ash. Fly ash and aluminium oxide (gamma alumina) mixtures were mixed with sodium hydroxide solutions (2 or 4 M) in a Teflon lined hydrothermal bomb and heated at 100, 150 or 200°C for 4, 8, 16 or 24h. The volume of the hydrothermal bomb was 20 ml and was used 3 g of fly ash + γ-Al_2O_3 mixtures and 20 ml activating solutions. After the desired time, the hydrothermal bomb was removed from the oven and quenched in cold water to stop the reaction. The contents were then filtered, washed several times with distilled water and dried at room temperature (Temuujin, 2011, 2012a).

Geopolymer pastes were synthesised with a targeted Si:Al ratio between 2.78 to 3.9. Alkali activation of the ash was undertaken with 2, 4, 6, 8 and 10 mol/L NaOH solutions. The water to solids mass ratio varied between 0.28-0.35. Geopolymer pastes were poured in 2cm cubic moulds made from iron. After curing at 70°C for 1 day, samples were removed from the mould.

2.2. Characterisation

The chemical composition of the ashes was determined by X-ray fluorescence (XRF) (Rigaku, Primini), granulometry by particle size analyzer (Shimadzu, SALD-2100) and specific surface area by BET analysis (Belsorp, mini). Prior to making the surface area measurement, samples were outgassed at 200°C for 4h and high purity helium gas was used as adsorbate. The mineralogical composition of the fly ashes and synthesised samples was determined by X-ray diffraction (XRD). XRD patterns of the powders were obtained with a Bruker "D8 Advance" diffractometer and Shimadzu "MAXima_X XRD-7000" diffractometer using Cu-Kα radiation. Mineralogical characterization of the some ash products were performed by Rietveld method with corundum was used as internal standard. Morphology of the fly ash was observed by SEM and FESEM (Evo 40 XVP and JEOL JSM-6701F). Ash surfaces were studied with a Zeiss EVO 40XVP scanning electron microscope with EDS X-ray detector (Energy Dispersive Spectroscopy, Oxford Instruments). Radiation measurements were performed by γ-spectroscopy according to Mongolian standard MNS 5072:2001. Water adsorption was measured by weight ratio of both dry samples and after immersion in boiling water for 30 min. Compressive strength of prepared samples was determined with a Universal Testing Machine (Temuujin, et al, 2012b). For the compressive strength testing were used at least 3 samples.

3. Results and Discussion

3.1. Chemical and Mineralogical Characterization of Fly Ashes

Table 1 shows chemical composition of the used coal combustion by products.

Table 1. Chemical composition of ash products in wt%

Samples	SiO_2	Al_2O_3	Fe_2O_3	CaO	MgO	MnO	SO_3	K_2O	Cl	SrO	TiO_2	P_2O_5	V_2O_5	LOI
Shivee ovoo fly ash	33.85	12.15	9.89	30.8	6.41	1.185	3.65	0.73	-	0.44	0.35		0.025	0.3
Baganuur fly ash	55.2	14.15	10.55	15.0	1.56	0.34	1.23	1.31	0.135	0.25	0.25	-	-	1.05
3rd TPS	52.3	15.45	10.1	16.55	1.94	0.112	1.36	1.52	0.018	0.24	0.299	-	-	11.01
Darkhan TPS	51.2	28.55	11	3.4	1.655	-	0.315	1.72	0.055	0.09	1.53	0.385	0.03	7.83

The dash in the table shows below detection limit.

The chemical composition, size distribution, classification of fly ash and bottom ash or slag depends on the source coal deposit used for the thermal power plants. At present there is no official use of either fly ash or bottom ash from the TPS4 power plant. This plant uses coal from two different deposits, Baganuur and Shivee ovoo. The variability of these deposits results in the residues having highly variable chemical and mineralogical compositions making utilization difficult. The mineralogical composition of the residues varies even for a single coal deposit depending when the ash was collected i.e night or day, winter or summer. The chemical composition also varies within one coal deposit and usually chemical composition of ashes collected at one time differs with that collected another time though difference might be within 5%. According to the international classification (ASTM, 2012) Baganuur and Shivee ovoo fly ashes from TPS4 are Class C fly ash. Except Darkhan TPS's bottom ash all ashes contain relatively low content of aluminum oxide and high content of calcium oxide. Chemical composition of the ash taken from ash pond of 3rd TPS resembles the chemical composition of Baganuur fly ash from 4rd TPS. According to chemical composition, pond ash from the Darkhan TPS can be considered as class F ash, though, this classification applied for fly ash (ASTM, 2012).

XRD revealed (Figure 1) that all the ash samples are complex and contain at least 6 phases, while some of them have more than 10 phases which makes it very difficult to fully characterise the ashes. The most intense peaks belong to quartz. The second largest chemical element in the ash is Ca and most of the calcium in Baganuur and Shivee ovoo fly ash is present as lime CaO, merwinite $Ca_3Mg(SiO_4)_2$, gehlenite, $Ca_2Al[AlSiO_7]$ and akermanite $Ca_2Mg[Si_2O_7]$. By XRD it was determined that about 65% of Baganuur and Shivee ovoo fly ashes were amorphous.

Figure 2 shows scanning electron micrographs and EDS spectra of the Baganuur and Shivee ovoo fly ashes.

The EDS spectra suggest that Ca bonds with Si, Al and Mg and strongly support the XRD results. There are some morphological differences between Baganuur and Shivee ovoo ashes. Baganuur fly ash consists of mainly rods and spherical particles while Shivee ovoo fly ash consists of mostly spheres. Some rod in Baganuur fly ash contains Ca and P atoms and most probably represents apatite. Spherical particles of the both fly ashes represent aluminosilicate and calcium aluminosilicate.

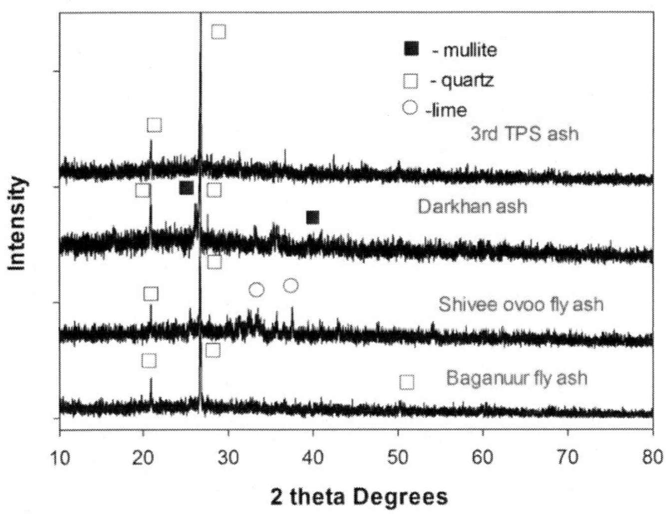

Figure 1. XRD patterns of the coal combustion by products.

3.2. Granulometrical Characterization

Particle size distribution of ash samples is shown in Figure 3. The average d_{50} particle size of the Shivee ovoo fly ash is 14.5 μm while for Baganuur fly ash it is 20 μm. The average sizes of the ash pond samples are larger than the fly ash samples, namely 90.6 μm for Darkhan ash and 42.7 μm for the 3rd thermal power station (TPS).

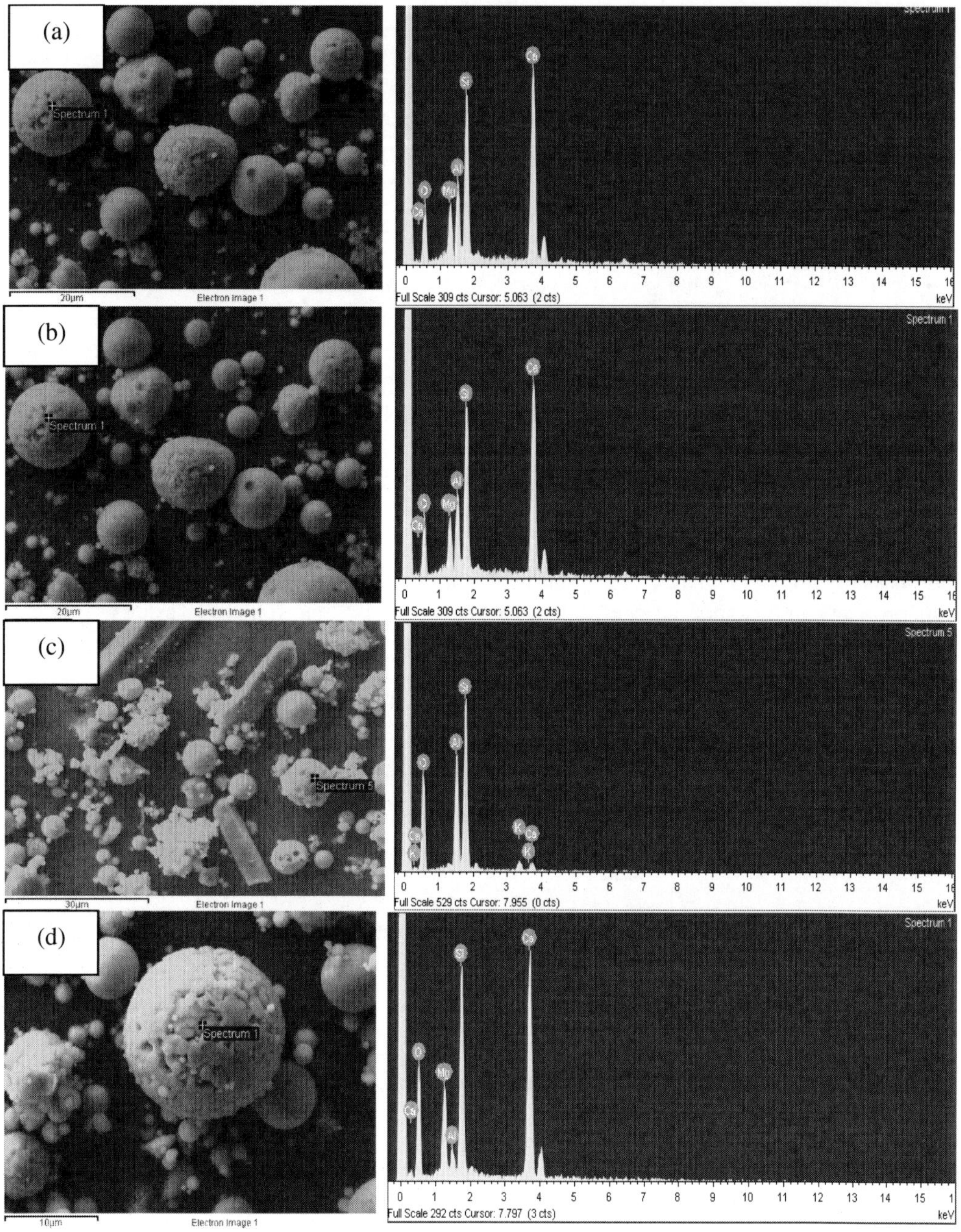

Figure 2. SEM images and EDS spectra of the Baganuur (a, b) and Shivee ovoo (c, d) fly ashes.

SEM observation of the samples showed in Figure 4 also supports particle size distribution data.

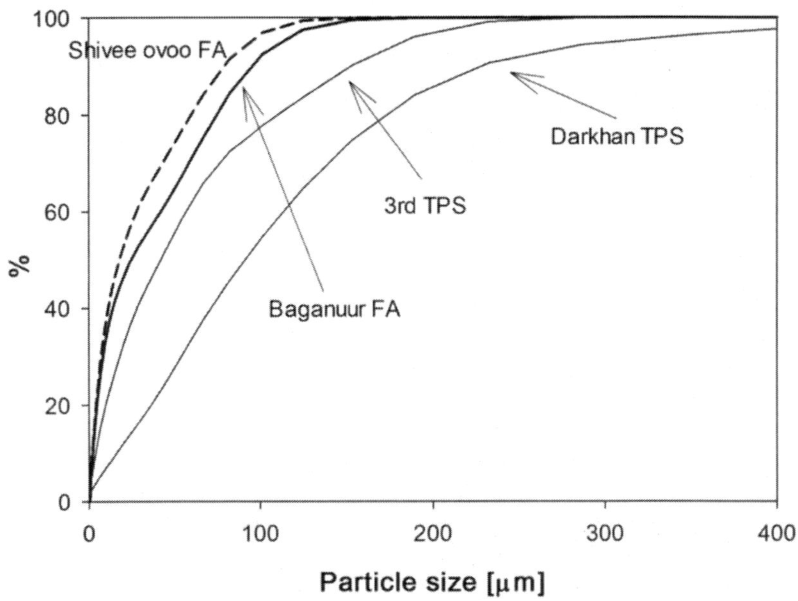

Figure 3. Particle size distribution of the experimental ashes.

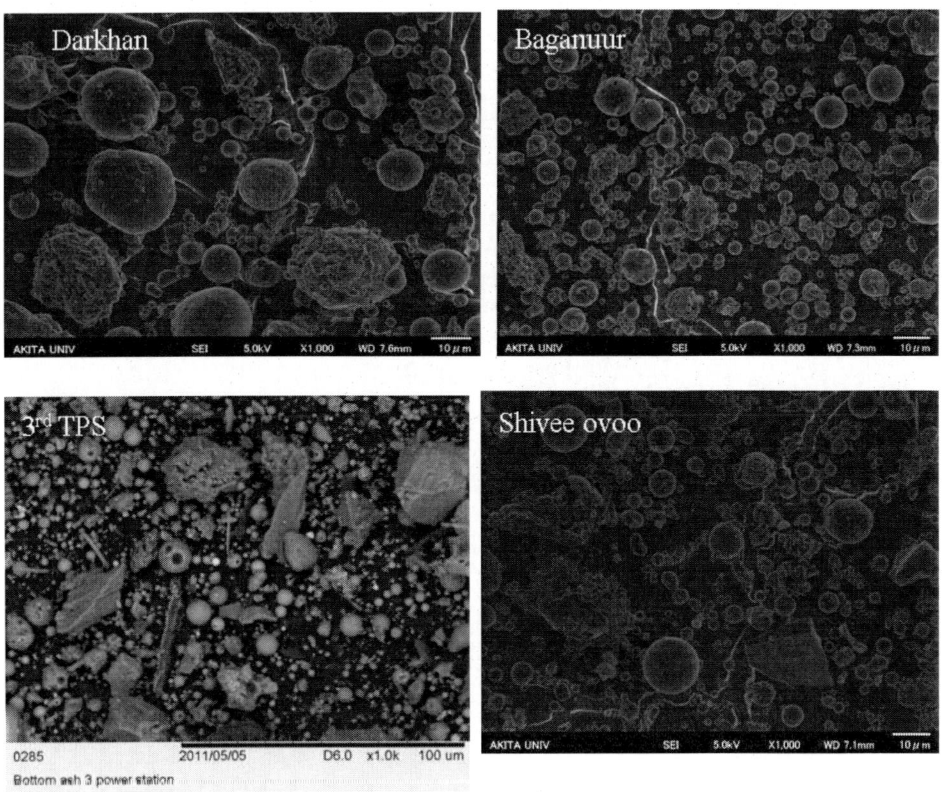

Figure 4. SEM micrographs of the ash samples.

The micrographs above showed fly ash contains more spherical particles than bottom ash or slag. Ash pond samples have a significantly larger specific surface area and pore volume than the fly ashes obtained directly from the power station. Table 2 shows porous properties of the ash products. Specific surface area of the ash pond samples are 10 to 30 times higher and pore volumes are 25 to 60 times greater than fly ashes obtained directly from the power station. These differences also suggest that ashes submerged in ponds have undergone a significant chemical or physical change by leaching out soluble phases. The loss on ignition of the ashes differs largely. Loss on ignition of Baganuur fly ash was 1.05%, Shivee ovoo 0.3%. The loss of ignition of the ash pond ashes of 3rd TPS was 11.01 wt.% and of Darkhan 7.83 wt.%, respectively. Keeping of ash in an ash pond increases its moisture content possibly causing some hydration or hydroxylation. It also possibly caused increasing of the surface areas of ash pond ashes.

Table 2. Porous properties of the coal combustion by products

Sample name	Specific surface area, BET m^2/g	Pore volume, ml/g $(x10^{-4})$
Fly ash, Shivee ovoo	1.04	1.49
Fly ash, Baganuur	2.75	2.12
Ash pond, 3rd TPS, Baganuur	30.6	87.30
Darkhan TPS, ash pond	9.76	57.70

3.3. Radiation Characteristics

According to the Mongolian radiation safety standards, the radionuclides present in coal combustion by-products is a very important factor when considering utilization. Since coal combustion by-products are mainly used for preparation of building materials, their radiation characteristics are necessary factors to be determined in order to use them. It is indicated that chronic exposure of human beings to low doses of ionizing radiation can cause health effects which may only appear after 5-30 years (Kovler, 2009).

Radiation characteristics of building and construction materials permitted according to Mongolian standard (2001) are shown in Table 3.

Table 3. Radiation characteristics of the building and construction materials

№	Radium equivalent (Bq/kg)	Type of building/construction
I	<370	Living house, all kind of public buildings
II	<740	Industrial building and road construction in populated area
III	<1500	Road and building far from populated area
IV	1500<...<4000	It will be decided by radiation and health safety organizations
V	>4000	Shall not be used

As it shown in the Table, for the living house and public buildings radiation activity must be below 370 Bq/kg. Important is to remember, that this requirement is for the building materials. But coal combustion by-products are by themselves not building materials, they are raw materials for building materials preparation and are typically diluted by the addition of other components.

Table 4 shows radiation data of the used fly ashes.

Table 4. Radiation characteristics of the coal combustion by products

№	Sample	Isotope activity concentration, Bq/kg			Radium equivalent, Bq/kg
		A_{Ra-226},	A_{Th-232},	A_{K-40},	
1	Baganuur	242.4	31.1	381.6	314.4
2	Shivee ovoo	262.8	48.7	215.6	342.7
3	3rd TPS	502.4	27.7	355.9	567.9
4	Darkhan TPS	152.1	47.85	422.6	248.7

The composition of radionuclides in fly ash depends on the radioactivity of the original coal and is concentrated many times in the fly ash during combustion. Radiation measurement of these ashes also revealed variation of the radiation data with the same ashes collected at different time frame. For instance, Shivee ovoo and Baganuur ashes collected in 2011 was measured to have 342 Bq/kg and 314 Bq/kg respectively. Table 5 shows the radiation values of Mongolian coal combustion by products taken from open sources (Caritas Czech, 2012).

Table 5. Radioactivity of some coal combustion by products taken from open sources

No.	Sample ID	Isotope Activity (Bq/kg)			Radium Equivalent (Bq/kg)
		^{226}Ra	^{232}Th	^{40}K	
1	Shivee ovoo fly ash, K-6	183.7	50.1	240.4	267.5
2	Shivee Ovoo fly ash, K-7	210.2	41.4	295.6	287.8
3	Baganuur fly ash, K-1	228.5	33.7	267.7	294
4	Baganuur fly ash, K-2	208.3	33.5	239.9	271.1
5	Baganuur fly ash, K-4	181.3	27.4	239.4	236.4
6	3rd TPS, used ash pond-1	284.7	20.6	472.2	351.3
7	3rd TPS, used ash pond-3	109.7	29.7	379.1	179.7
8	3rd TPS, used ash pond-5	111.2	40.2	359.5	192.8
9	Darkhan TPS, used ash pond 2009	101.7	20	340.4	156.2

These data indicate that there is significant variability in the radioactivity in the ash from a single power station. Consequently, it is impossible to know the chemical composition or associated radioactivity of each ash without measuring every batch. The radioactivity of Baganuur and Shivee ovoo fly ashes is higher than Darkhan ash pond ashes. Therefore, it seems clear that the original radioactivity of Shariin gol coal deposit is lower than Baganuur and Shivee ovoo coal deposits.

All coals around the world have radioactive materials. Generally, relatively younger coal has higher radioactive content. The radioactive materials in the coal concentrate into the ash after combustion. Therefore, radioactive content in the fly ash is different depending on the radioactive content in the coal. The radioactive materials in the inorganic compounds concentrate and then remain in the ash during the coal combustion.

Otherwise, the higher radioactive content the coal comprises, higher is the radioactive content concentrates in the ash. However, there is a reference which described reducing of the radioactivity of the fly ash utilized in the construction industry by using snow as an additive to compacting water of fly ash (Baykal and Saygili, 2011). They explained it by increased void ratio after melting of ice, increased densification of matrix around the pores due to higher level of cementitious mineral formation. Such a novel technique can help reduction of ash radioactivity and finding a way of using coal combustion by-product that is considered as not useful in its original form by the construction industry.

3.4. Fly Ash as Source of Dust

The air pollution in Ulaanbaatar, the capital city of Mongolia, is very serious, especially during winter time when Ulaanbaatar is the most highly polluted city in the world. Pollutants such as particulate materials (PM) and SO_2 are much higher than international standards. The measured PM_{10} levels in Ulaanbaatar are 2-5 times higher than Mongolia's Air Quality Standard of 50 $\mu g/m^3$, 5-10 times higher than World Health Organization (WHO) Guideline Value of 20 $\mu g/m^3$, and 3-7 times higher than European limit value of 40 $\mu g/m^3$ (CDM-Mongolia, 2013). It is well documented that particles (primary PM_{10}, $PM_{2.5}$, and secondary PM due to SO_2 and NO_x emissions) cause negative health effects when inhaled by people. Inhaling PM can severely affect the lungs and heart.

As described in the introduction only TPS4 is equipped with electrostatic filters for fine particle capture, the other power plants are equipped with wet scrubbers. The boilers for thermal stations are equipped with cyclones. No emission control equipment is used on heat only boilers. Uncontrolled pollutants emit from these boilers and power plants pollute the air in surrounding areas.

Typical emission factors are 19.5 kg/tonne for PM10 and 7.8 kg/tonne for PM2.5. The efficiency of the flue gas cleaning for particles is set to 80% for TPS 2 and 3, and 95% for TPS 4. The TPS emissions take place at a high level. (World Bank Report, 2010)

3.5. Utilisation of Fly Ash for Zeolite Synthesis

It was found that under the experimental conditions trialled it was difficult to obtain pure zeolitic compounds. Although zeolitic compounds were formed at temperatures as low as 100°C the fly ash and Al_2O_3 were not completely reacted at this temperature. Therefore, hydrothermal temperatures were increased to 150 and 200°C. XRD patterns indicated (Temuujin, 2011) that there were several different zeolitic compounds formed, although it was difficult to assign phases to every reflection. Faujasite and sodium aluminium silicate hydrate ($Na_3Al_3Si_3O_{12}\cdot 2H_2O$) were most readily identified. Higher crystallinity and greater conversion to zeolites was achieved by using more concentrated NaOH, higher temperatures and longer heating times. The porous properties of the starting materials and reaction products are shown in Table 6.

Table 6. Porous properties of the raw materials and reaction products

Sample name	Temperature, °C	NaOH conc.,mol/L	Time, h	BET surface area, m^2/g	Pore volume, ml/g
Al_2O_3				97.2	0.1324
Fly ash, Baganuur				2.75	0.000433
2M NaOH-T150-4	150	2	4	56.2	0.1972
2M NaOH-T150-24	150	2	24	17.7	0.0439
2M NaOH-T200-4	200	2	4	24.8	0.0928
2M NaOH-T200-24	200	2	24	12.9	0.0407
4M NaOH-T150-4	150	4	4	31.5	0.1366
4M NaOH-T150-24	150	4	24	10.8	0.0243
4M NaOH-T200-4	20	4	4	13.6	0.0471
4M NaOH-T200-24	200	4	24	11.3	0.0442

BET surface area of short dwell time samples are larger than long dwell time samples due to the slow dissolution of starting Al_2O_3 which had a very large surface area. In contrast to previous research by Querol et al. (2002), the presented fly ash contains higher amounts of calcium. However, there were no Ca containing zeolitic compounds identified.

Adsorption of Cr^{6+} from simulated waste water indicates that synthesised zeolitic compounds have at least 50% higher adsorption ability than pure fly ash samples.

3.6. Utilisation of Fly Ash for Geopolymer Type Binder Synthesis

Preliminary study of both ash pond ashes (Darkhan and 3rd TPS) for geopolymer synthesis revealed weak mechanical properties i.e. compressive strengths were less than 5 MPa and a low density. Therefore, ash pond ashes were not used for the further study. Tables 7 and 8 show compressive strength and water adsorption of Baganuur and Shivee ovoo based geopolymer samples. Since the starting material contains higher amounts of Ca than is usually accepted for geopolymer synthesis, the material produced is more appropriately referred to as an alkali activated material rather than a geopolymer. For the geopolymers low calcium class F fly ash is preferable source material, because high calcium in class C fly ash may interfere with geopolymerisation reaction (Temuujin et al., 2009b).

Shivee ovoo fly ash based alkali activated material showed flash setting by increasing alkali concentration, thus the workability reduced in contrast to Baganuur fly ash based material.

As expected, alkali activated Baganuur and Shivee ovoo fly ash samples both show a clear trend of increasing strength with increase in mol/L NaOH used to activate the ash. With increasing alkali concentration prepared sample becoming with more compact structure as it can be suggested by water adsorption data. In other words, improvement of geopolymerisation or/and self-cementing reaction occurred. Previously by Temuujin et al (2009b) was described possible formation of C-S-H (calcium silicate hydrate) or calcium aluminosilicate either in amorphous or poorly ordered crystalline form which difficult to detect by XRD in the calcium added geopolymer system. Such calcium containing compounds may also form in the present system. XRD patterns also indicated that (Temuujin et al., 2012b) with increasing alkali concentration begin to form alumosilicate zeolitic compound such as hydroxysodalite in the alkali activated material.

Based on the experimental results, geopolymer type concretes of acceptable quality were also produced.

Table 7. Compressive strength and water adsorption of Baganuur fly ash based alkali activated material, adapted from reference (Temuujin et al., 2012b)

NaOH concentration, mol/L	Compressive strength, MPa	Water adsorption, %
2	2.1	28.4
4	9.4	25.4
6	11.6	19.9
8	31.2	18.3
10	32.8	15.0

Table 8. Compressive strength and water adsorption of alkali activated Shivee ovoo fly ash based paste, adapted from reference (Temuujin et al, 2012b)

NaOH concentration, mol/L	Compressive strength, MPa	Water adsorption, %
2	10.4	23.5
4	12.6	18.7
6	17.3	16.8
8	18.6	16.7
10	23.8	15.7

CONCLUSION

The main thermal power stations of Mongolia are using lignite type brown coal for combined thermal energy and electricity generation. Only the 4th Thermal power station in Ulaanbaatar city is equipped with electrostatic separator and produces fly ash. Both Baganuur and Shivee ovoo fly ashes obtained from the 4th Ulaanbataar thermal power station are class C ash. Content of radionuclides within these fly ashes doesn't exceed minimal requirements for building materials production. However, there is a particulate matter problem with coal combustion in Ulaanbaatar city.

Alkaline activation of these fly ashes causes either geopolymerisation or/and formation of CSH phases. Increasing of alkali concentration results improved compressive strength of cured samples. Increased concentration of curing solution caused appearance of hydroxysodalite after the curing.

Formation of various types of zeolitic compounds from class C fly ash based samples was found to occur at temperatures as low as 100°C. Higher level of crystallinity products and greater conversion to zeolites was achieved using more concentrated NaOH, higher temperatures and longer dwell times. The phases positively identified were faujasite-Na and a sodium aluminium silicate hydrate.

ACKNOWLEDGMENT

This research was supported by Ministry of Education and Sciences and Science and Technological Foundation for the research grant, science and technological and innovation projects. The author wishes to acknowledge contribution of all his co-workers from Laboratory of Materials Science and Technology, Institute of Chemistry and Chemical Technology, Mongolian Academy of Sciences and Curtin University (Australia) to this chapter.

REFERENCES

Ahmaruzzaman, M. (2010). A review on the utilization of fly ash. *Progress in Energy and Combustion Science*, 38, 327-363.

ASTM (2012). Standard specification for coal fly ash and raw or calcined natural pozzolan for use in concrete. (ASTM C618-08a), Retrieved 2012 from ASTM database.

Australian ash development association, Accessed in 2012. From http://www.adaa.asn.au/documents/ADAA_Mship_Report_2010.pdf.

Baykal, G.,.Saygili, A. (2011) A new technique to reduce the radioactivity of fly ash utilized in the construction industry. *Fuel* , 90 1612–1617.

Caritas Czech tender on Study of fly ash waste from Mongolian power plants and the practical uses of that fly ash as a component of construction materials. Accessed on February, 2013 from caritasczech.org/res/data/007/000992.pdf

CDM-Mongolia organized workshop. Khorolmaa, G., National standards of air pollutants in coal fired power plants and thermal plants, Report of Ministry of Environment and Green development, Clean technology and science division, 24 January 2013.

Fu, J. (2010). *Challenges to increased use of coal combustion products in China*, Master's thesis, Linköping University, Sweden, 2010, Digital thesis.

Japan coal energy center, (2012). Accessed on January, 2013 from http://www.jcoal.or.jp/coaltech_en/coalash/ash01e.html.

Jha, V, Matsuda, M, Miyake, M,(2008). Resource recovery from coal fly ash waste: an overview study, *Journal of the ceramic society of Japan*, 116, 167-175.

Mongolian ministry of energy, accessed on May 2012. From http://www.mmre.energy.mn/branch/history/104/135/).

Mongolian ministry of energy, Accessed on May 2012. http://www.mmre.energy.mn/branch/now/103/detail/1004090003/ .

Mongolian president office. Air pollution of Ulaanbaatar city. Accessed on 5th January, 2013. From http://air.president.mn/index.php?option=com_content&view=article&id=85:2011-11-04-05-17-35&catid=2:news&Itemid=3.

MNS 5072:2001, Mongolian standard on Determination of radioactive elements content in building materials, soil and earth crust by Gamma spectrometer's method.

MNS 3927:1986, Mongolian standard on utilisation of thermal power stations ash for building materials production. Technical requirement.

K. Kovler, Radiological constraints of using building materials and industrial by-products in construction, *Construction and Building Materials* 23 (2009) 246–253.

Temuujin, J, van Riessen, A,(2009a). Effect of fly ash preliminary calcination on the properties of geopolymer, *Journal of Hazard Materials*, 164, 634-639.

Temuujin, J., van Riessen, A, and Williams, R, (2009b). Influence of calcium compounds on the mechanical properties of fly ash geopolymer pastes, *Journal of Hazardous Materials*,167, 82-88.

Temuujin, J, (2011). Preparation of zeolitic compounds from high calcium fly ash, International forum on strategic technology IFOST-2011, 22-24 August, 2011, Harbin, China, *IEEE Conference Proceedings,* vol.1, pp159-161.

Temuujin, J. (2012a). Preparation of zeolitic compounds from class C fly ash and its characterisation, *5th Asian Particle Technology Symposium APT-2012,* July 2-5, 2012, Singapore, Conference Proceedings pp.125-130.

Temuujin, J., Rickard, W., Van Riessen, A. (2012b). Characterisation of various coal combustion products from Mongolian thermal power stations and their application for preparation of geopolymers and zeolitic compounds, 26th International Mineral Processing Congress, September 24-28, 2012, New Delhi, India, *Conference Proceedings*, Paper 1136, pp.5455-5461.

Wang, S, Wu, H, 2006. Environmental-benign utilisation of fly ash as low-cost adsorbents, *Journal of hazardous materials B*, 316, 482-501.

World bank. Air Pollution in Ulaanbaatar Initial Assessment of Current Situation and Effects of Abatement Measures, Discussion Paper. Accessed on 4 the March, 2013, from http://www-wds.worldbank.org/external/default/WDSContentServer/WDSP/IB/2010/02/18/000333037_20100218233600/Rendered/PDF/529700REPLACEM1paper0FINAL002110110.pdf).

Wikipedia.org. Accessed on 2013 from http://en.wikipedia.org/wiki/Coal.

Querol, X, Moreno, N, Umaña, J, Alastuey, A, Hernández, E, López-Soler, A and Plana, F, 2002. Synthesis of zeolites from coal fly ash: an overview, *International Journal of Coal Geology*, 50, 413-423.

Yeren-Ulzii.B, Current and future prospects of the electricity sector of Mongolia, Tenth Meeting of the Working Group on Energy Planning and Policy 5-6 April 2012. Accessed on 10th January 2013 from http://www.unescap.org/esd/Energy-Security-and-Water-Resources/energy/security/workingpmeeting/documents/Ulzii.pdf.

PART B: UTILIZATION OF FLY ASH

In: Fly Ash
Editor: Prabir Kumar Sarker

ISBN: 978-1-62948-044-2

Chapter 7

UTILIZATION OF FLY ASH IN THE CONSTRUCTION INDUSTRY

Salim Barbhuiya*

Curtin University, Perth, Western Australia, Australia

ABSTRACT

The use of fly ash improves the performance of concrete in its fresh and hardened states. This enables the concrete industry to use hundreds of millions of tonnes of this by-product material, which would otherwise be landfilled, as a waste. This also allows the concrete producer to design and modify the concrete mix to suit the desired application. The fly ash concrete system offers a holistic approach that can help us to achieve the goals of meeting the rising demands for concrete, enhancement of concrete durability with little or no increase in cost (in some instances reduced cost), and ecological disposal of large quantities of the solid waste products from coal-fired power plants. Several laboratory and field investigations involving concrete containing fly ash had reported to exhibit excellent mechanical and durability properties. However, due its conservative nature the construction industry is still not fully utilizing the potential benefits of using fly ash in concrete production. In this chapter an attempt is made to review first the origin, properties and current knowledge of fly ash as reported in the literature. This is followed by a discussion on the utilisation of fly ash in construction and the potential issues related to their use. Finally, areas of further research have been identified.

Keywords: Fly ash, workability, compressive strength, structural concrete, pumped concrete

1. INTRODUCTION

In many eastern European countries, the states of the former USSR, and other countries, which show dramatic industrialization trends like China, and India, there is a rising demand

* Corresponding Author address: Department of Civil Engineering, Curtin University of Technology, GPO Box, U1987, Perth, WA 6845, Australia. Email: salim.barbhuiya@curtin.edu.au.

for coal. A by-product of combustion of coal, fly ash is collected at a rate of about 20 t/h for a 660 MW unit and its production reaches 1 Mt/y in Italy and about 40 Mt/y in Europe. This implies that the total amount of coal waste in the form of fly ash dispensed by factories and thermal power plants has been on the rise. This in turn has led to fly ash posing as a grave environmental issue. Concerns such as low storage capacities of power plants, usually corresponding to about one week of fly ash production, and undesirable economical after-effects of fly ash landfill bring forth the need for a complete fly ash reutilization.

From the aspect of power generation, fly ash is treated as a waste component, but from the perspective of coal utilization, fly ash is an asset that remains to be completely utilised. In the field of thermal electricity, fly ash is yet to be exploited while in the cement industry, it is used as a raw material in the manufacturing of concrete. Studies are being conducted on its application in recycling materials for agriculture and its conversion into zeolite as well as engineering has been taken into account. Another alluring option using fly ash is its application as a low-cost adsorbent for the treatment of gas and water. Various investigations are reported in this literature on the usage of fly ash for adsorption of single pollutants in an aqueous mixture or flue gas. This chapter will examine the several uses of fly ash posing as low-cost adsorbents for cleaning of flue gas, treatment of wastewater, zeolite synthesis, light weight aggregate, mine backfill, cement applications as well as road sub-base construction.

2. Origin, Properties and Current Knowledge of Fly Ash

Fly ash is defined in ACI Committee 116 document on Cement and Concrete Terminology as "the finely divided residue resulting from the combustion of ground or powdered coal, which is transported from the firebox through the boiler by flue gases". Millions of tonness fly ash is generated worldwide from coal firing power since 1920. The current annual production of fly ash worldwide is estimated around 500 million tonneses. Although the present day utilisation of fly ash on worldwide basis varies from 3 to 57%, the world average only amounts to 16% of the total ash. Fly ash consists of fine, powdery particles predominantly spherical in shape, either solid or hollow, and mostly glassy (amorphous) in nature. The particle size distribution of fly ash is generally similar to that of silt. The specific gravity usually ranges from 2.1 to 3.0, while the specific surface area may vary from 170 to 1000 m^2/kg. The colour of fly ash can vary from tan to grey to black depending on the amount of unburnt carbon in the ash. Figure 1 is a **SEM** image **showing** the characteristic morphology of the original **fly ash**. This ash consists of a series of spherical vitreous particles of different sizes.

Table 1 shows typical compound analyses for two fly ashes and a natural pozzolan (Class F fly ash, Class C fly ash and Class N - Natural) and ordinary Portland cement (OPC). This table reveals that the same compounds exist in fly ash and OPC. Those of fly ash are amorphous (glassy) due to rapid cooling while those of cement are crystalline, formed by slower cooling. OPC is rich in lime (CaO) while the lime content in fly ash is low.

Figure 1. SEM **image** of fly ash.

Table 1. Typical chemical compounds in pozzolans and OPC

Chemical compound (%)	Pozzolan type			OPC
	Class F Fly ash	Class C Fly ash	Class N Natural	
SiO_2	54.90	39.90	58.20	22.60
Al_2O_3	25.80	16.70	18.40	4.30
Fe_2O_3	6.90	5.80	9.30	2.40
CaO	8.70	24.30	3.30	64.40
MgO	1.80	4.60	3.90	2.10
SO_3	0.60	3.30	1.10	2.30
Na_2O+K_2O	0.60	1.30	1.10	0.60

Fly ash contains high amount of reactive silicates while OPC has smaller amounts. A review of the chemistry of both materials makes it apparent that a blend of the two will enhance various properties of concrete. Through pozzolanic activity, fly ash combines with free lime to produce the same cementious compounds formed by the hydration of OPC (Figure 1).

Fly ash is reported to both accelerate and retard the setting time of cement. The general consensus seems to indicate that low-calcium fly ash tends to retard whereas high-calcium fly ash is less predictable and has been reported to either retard or accelerate the setting time. The hydration behavior of C_3A and C_2S in fly ash is the same as that in OPC, but the rate of formation of C-S-H from the glass phase is comparatively slow. Low-calcium fly ash, which has very little or no self-cementing properties, hydrates when alkalis and $Ca(OH)_2$ are added. The hydration products such as C-S-H, C_2ASH_8 and C_4AH_{13} are formed, and hydrogarnet is produced at a later stage. As more $Ca(OH)_2$ is supplied, more of it is fixed by silica and alumina in fly ash. The degree of hydration in fly ash is increased in the presence of gypsum. According to Liu *et al.*, (2005) this is because the surface is activated by the destruction of the structure of the glass and crystalline phases caused by the dissociation of Al_2O_3 reacting with SO_4^{-2}.

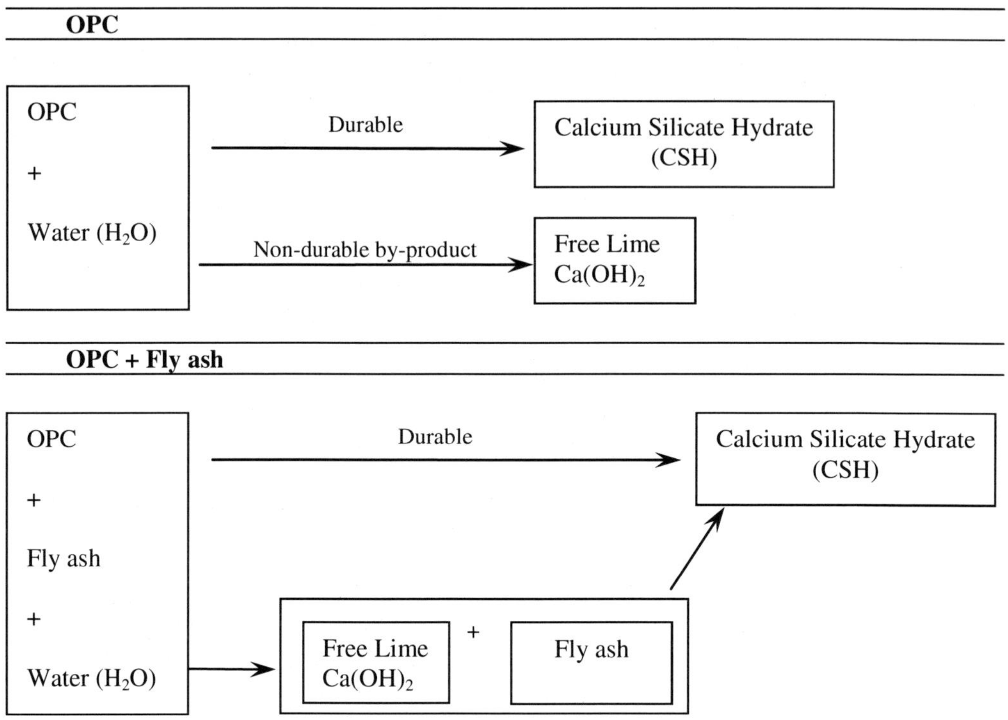

Figure 1. Hydration products of cementing binders.

3. Fly Ash for Structural Concrete

A major use of fly ash in the construction industry is in the production of high quality structural concrete. Fly ash contributes beneficial properties to the concrete while helping to maintain the economy. The properties include compressive strength, workability and durability. A fly ash concrete mix, designed for equivalent performance to conventional concrete at normal ages will generally gain strength more slowly at early ages, after about 7 days, the rate of strength gain of fly ash concrete exceeds that of conventional concrete, enabling equivalence at the desired age. This higher rate of strength gain continues over time enabling fly ash concrete to produce significantly higher ultimate strength that can be achieved with conventional concrete. Fly ash concrete designed for equivalent performance at 7 days or earlier will yield practically the same strength gain, prior to the design age. At all ages thereafter, fly ash concrete will exhibits much higher strength gain than the conventional concrete. High compressive strengths in the range of 50-100 MPa are often required in structural concrete. High quality fly ash complying with ASTM C 618 is most advantageous in achieving these strength levels.

Davis *et al.*, (1937) found that the addition of fly ash to concrete generally improves its workability. Since then several investigations have shown that fly ash produces a complex interaction within the fresh concrete system affecting both the solid and paste phases. A summary of these effects is given in Table 2. Considering the first of these phenomena in more detail, OPC grains are prone to exhibit some coagulation or flocculation in fresh concrete, which tends to produce an inhomogeneous and non-uniform hydrate structure. The addition of fly ash can physically disperse these cement flocs, thus freeing more paste to lubricate the aggregates and thus improves the workability. This is analogous with the result of dosing a concrete mix with a chemical plasticising admixture. However, unlike a plasticiser, this dispersion is stable and can, therefore, create a denser, uniform matrix as hydration proceeds.

By virtue of the spherical nature of the particles, a ball-bearing effect is widely quoted as the means by which fly ash improves concrete workability. While this phenomenon was first reported in 1954 (Anon, 1954), the exact mechanism still remains ill-defined. In addition to the above, each spherical glassy particle of fly ash in concrete carries an associated surface film of water as a consequence of its geometry and hydrophilic nature (Dhir *et al.*, 1986). The addition of fly ash to concrete will, therefore, achieve a form of water-entrainment, ensuring wide dissemination, and retention, of a significant proportion of the mix water throughout the plastic concrete. This physico-chemical effect also helps to inhibit bleeding and ensures an even distribution of water at fine loci, where it is most efficiently utilised.

Table 2. Influence of fly ash on freshly-mixed concrete (Swamy, 1986)

Influence	Effects
Physical Dispersion of cement flocs Increase in particle packing Ball bearing	Reduce mix water demand, bleeding and segregation
Physico-chemical surface Water entrainment	Improves workability, placeability, pumpability, homogeneity, stability and finishability, produces a denser matrix structure

Fly ash is commonly considered as a material that can be used to minimise or eliminate the risk of alkali-aggregate reaction in concrete (Building Research Establishment, 1982; Hobbs, 1982). Several reasons have been put forward for this beneficial effect. Firstly, straightforward dilution of the water-soluble alkali concentration in the mix occurs. Secondly, the reaction of fly ash with $Ca(OH)_2$ liberated by the hydration of OPC lowers the pH of the pore solution, thereby reducing the reactivity between the alkalis and silica in the aggregates. Third, the highly reactive amorphous phase (glass) of fly ash itself may rapidly consume the alkali in the cement in producing a non-expansive calcium-alkali-silica gel and, thus leaving no alkalis to produce hydroxyl ions to react with the silica in the aggregates. Finally, any reduction in the permeability of concrete due to the inclusion of the fly ash may beneficially reduce the moisture movement into the concrete. However, the extent to which the improvement can be effective for a given concrete appears to be related to the type and amount of fly ash used.

As far as the quality of fly ash on alkali-aggregate reactivity is concerned, its alkali content has some bearing on its effectiveness. Although it is generally assumed that the alkalis present in low-lime fly ash are combined in the amorphous phase (glass) and are not available for the reaction, it has been suggested (Hobbs, 1982) that in calculating the total alkali content (acid-soluble) of concrete, low-lime fly ash should be assumed to have an available alkali content of 0.2% by mass. For the high-lime fly ash, the situation appears to be different, because it contains large amounts of soluble alkali sulphates, and these have been reported to affect the alkali-silica reactivity adversely (Mehta, 1993).

In general, with low-lime fly ash, a cement replacement of at least 25% is required to be effective, but depending upon the alkali content of the OPC and the type of aggregate, a replacement level of 50% becomes necessary, in which case the resulting lower early age strengths may become a problem and the design of the mix needs to be carefully considered. Whilst the role of fly ash in minimising or eliminating the risk of alkali-aggregate reactions in concrete has long been recognised, and is now increasingly being accepted, this is not so in the case of the alkali-carbonate reaction, which can give rise to similar expansion and damage to the concrete. In fact fly ash has been shown to be of little help in this respect.

Atis (2003) reported that when 70% of PC was replaced with fly ash, the concrete carbonated more than an OPC mix, but with a 50% replacement level the carbonation was similar or less than that found in a 100% PC mix. When PFA was added at levels more than 50%, it is likely that there was insufficient OPC to form $Ca(OH)_2$ to facilitate the hydration of the fly ash component, thus a much more permeable hydrated cement paste (hcp) matrix is formed. However, the author concluded that a longer curing period was required to achieve a satisfactory concrete.

Sanjuan (2003) reported that the effect of natural carbonation could only allow absolute comparison between cement families, but within cement families, it was found that concretes made without fly ash performed better in resisting the carbonation. Whilst developing a model for resisting carbonation, Papadakis *et al.*, (1992) found that the rate of carbonation increased with the use of fly ash. Papadakis (2000) reported again that carbonation depth increases with fly ash replacement of OPC. Therefore, it appears that fly ash concrete is comparable to OPC concrete in its resistance to carbonation. However, the pozzolanic reaction of the fly ash occurs only after the hydration of the OPC and, therefore, a longer curing time can show improvement in the carbonation resistance of fly ash concrete.

4. Fly Ash for Pumped Concrete

Pumped concrete must be designed so that it can be easily conveyed by pressure through a rigid pipe or flexurable hose for discharge directly into the desired area. The use of fly ash can greatly improve pumpability while enhancing the quality of concrete and controlling costs. Pumped concrete must be more fluid with enough fine materials below 300 microns control the liquid under pressure, there must be more of these sizes than in a normal mix. Generally speaking, the finer the material, the greater the control is. Moreover, the coarse aggregate grading should be continuous and often the sand content must be increased by up to 5% at the expense of the coarser aggregate so as to balance the 500 micron- 5 mm fraction against the finer solids. It is possible to achieve a pumpable concrete using extra water and fine aggregate. However, adding extra water and fine aggregate leads to a weaker concrete. The usual remedial for this are either to increase the cement content, which is costly, or to use chemical admixtures, which also can be costly and may lead to segregation in marginal mixes. The use of fly ash in pumped or conventionally placed mixes can reduce the water requirement by 2 to10 % for any given slump. Moreover, if fly ash is utilized, its unique workability and pumpability properties permit a better balance of sand to coarse aggregate resulting in a more economical, pumpable concrete.

5. Fly Ash for Controlled Density Fill

The material and methodology used for void filling has remained virtually unchanged for centuries. The filling process has involved compacting granular materials into voids to provide stabilized fill. Modern technology has only provided for mechanized compaction versus the use of manual labour. Currently, there is an engineered product available designed to eliminate failure inherent in the traditional method. This product is called controlled density fill (CDF). The first reason to use CDF was likely one of being able to do a difficult job well. Since that time, the list of reason for its use has grown greatly and continuous to grow. The materials used in the production of CFD are the vary same utilized in the production of OPC concrete. Class F is used as a workability agent to provide mixes that can flow great distances without segregation. It also provides a slight cementing action.

6. Fly Ash for Pavement Concrete

Fly ash has been used in road paving for more than 2,000 years. The Romans used naturally occurring volcanic ash from Mount Vesuvius to cement paving stones in their roadways. Many miles of these ancient roadways-although rough by our standards still exists as useable highways. Today, in Europe, paving stones have been replaced by modern day pavement, but a product almost identical to volcanic ash is still in used. In fact, most European highways have been constructed with fly ash in all levels, including the wear course. Some of the reasons for which fly ash is used in concrete paving have more to do with the physical characteristics of fly ash than the chemical and strength gain characteristics. With modern construction technologies such as paving trains using slipform equipment, the fly ash facilitates placement of the concrete at lower slumps while maintaining excellent workability. This means less handwork for the paving contractor and better surface texture and edge characteristics for the design engineers. The use of fly ash also results in a denser concrete-one that will have much greater ultimate strength and durability.

7. Fly Ash for Block and Concrete Pipe Manufacturing

The manufacturing of concrete masonry units uses a dry, harsh concrete mixture compacted into moulds with great mechanical energy. When demoulded, these units maintain their shape during handling and transportation into a curing environment. Curing methods consists of the high temperature autoclave and the atmospheric pressure high temperature kiln. The use of high quality fly ash has become accepted practice in the industry.

Fly ash improves block manufacturing in two basic ways. It gives producers the strength required and at the same the added plasticity that fly ash contributes to the relatively harsh block mixtures-assuring improved finish and texture, better mould life and better sharper corners. Additional benefits of fly ash in block include and reduced permeability and shrinkage, increased durability and virtual elimination of effervescence.

Class F fly ash has been used successfully in the manufacturing of concrete pipe for more than 30 years. It has become an almost indispensable ingredient to the dry, harsh mixes typically used in modern pipe manufacturing. Concrete pipe is made essentially by two different processes, one using extremely dry mixtures and the other using plastic concrete mixtures. Dry cast concrete pipe is produced utilizing mechanical and//or vibratory compaction to consolidate dry concrete into a form. The form is removed from the pipe as soon as the casting is finished. With removal of the form, the green pipe is carefully transported to its place of curing. Atmospheric pressure curing at elevated temperature is typically used to obtain early age performance. Wet cast pipe uses plastic concrete placed and compacted in a form, which remains around the pipe until certain levels of performance are achieved. Although decline in popularity wet cast pipe may be manufactured by the spinning process (centrifugal) to remove excess water and air to produce great density and low permeability. Fly ash has found widespread use in the manufacturing of concrete pipe as a cementitious material and as an aggregate mineral filler to enhance quality and economy.

8. Potential Issues Related to the Use of Fly Ash

The pozzolanic reaction of fly ash being a slow process, its contribution towards the strength development occurs only at later ages and, hence, the early strength of fly ash concrete is significantly reduced. This is a barrier for the application of fly ash concretes in precast industries, where formwork turnover time is significantly reduced with slow strength development. Therefore, it is of prime importance to improve the early strength of fly ash concretes for realization of full benefits of fly ash to precast industries. Different approaches have been used to accelerate the pozzolanic reaction of fly ash and, thereby, to increase the early strength of concrete containing fly ash. These include: mechanical treatment (grinding), accelerated curing and autoclaving and chemical activation (alkali and sulphate activation). However, alkali activation used in concrete may lead to alkali–silica reaction, while the sulphate activation may impair the durability of concrete due to the formation of large quantities of ettringite. The main reason for the low early strength development of fly ash concrete is the unavailability of sufficient quantities of $Ca(OH)_2$ to react with the fly ash particles. Hence, there is a possibility of improving the pozzolanic reaction of fly ash by adding hydrated lime. Another reason for the poor early strength development of fly ash concrete is that the reactive silica in fly ash accumulates in the interior part of its particles, which is very slowly corroded in the hydrated cement paste environment. Silica fume, having a greater surface area and higher SiO_2 content, has been found to be more reactive than the fly ash and, hence, it has been used in combination with fly ash in low levels of replacement of Portland cement with fly ash. However, the benefit of this combination for fly ash is not well understood.

9. Areas for Further Research

In the aspect of environment and as a challenging alternative to substituted zeolites or activated carbon for adsorption, the research on the utilization of fly ash has enormous scope. The behavior of fly ash on the property of adsorption relies intensely on the chemical treatment applied for activation and on its origin. No application has been recognized yet on the industrial calibration and in the aspects of high volume and high value utilization of fly ash for industrial applications, a lot of economic obstacles are to be surpassed. Changes have to be brought in the properties of fly ash to enrich its adsorption capacity (Lin, 1971) since raw fly ash has the property of poor adsorption. Aluminosilicates are a component of fly ash, which are essential for the synthesis of zeolites, which has number of uses as ion exchangers and adsorbents, thus exhibiting higher adsorbent capacity relative to the raw fly ash. The hydrothermal conversion of fly ash into different types of zeolites has been studied by making use of the extricated supernatant solution. The conversion of the mass fly ash into pure zeolite has been successfully studied by some researchers (Magnuson and Malenka, 1970). Fly ash contains phases of silico-aluminate and therefore it is a suitable substance for producing geopolymers (Kumar and Patil, 2006; TIFAC, 2005), quartz–cristobalite–tridymite and mullite. Another feature of fly ash is its amorphous aluminosiliate property making it suitable for various industrial applications like synthesizing ultramarine blue. The repeated accumulation of foodcrops deprives the soil of its trace elements and nutrients. Adding fly ash to soil improves its infiltration characteristics, moisture-holding capabilities, and plant nutrients and though this has been stressed upon, yet the full-scale application of using fly ash as a soil amendment has not been substantiated. Fly ash also has uses in the control of water pollution and this consists of neutralization of acidic wastewaters, phosphorus removal from wastewater, physical conditioner for sludge dewatering, sorbent for various organics, sealing of contaminated sediments, etc. There is plenty of room for more research to be carried out in these areas.

Conclusion

To present a better understanding of fly ash utilization, this chapter has aimed to cover a broad domain of information. To have a better knowledge of alternate use of fly ash, a study of coal formation as well as a combustion procedure is essential. In most cases till date, it has been noted that recorded work in these areas has been primarily of laboratory scale, and advanced development is of utmost need. Fly ash has found uses in construction, zeolite synthesis, dyes and heavy metals. These usages can go a long way in averting the effects of environmental pollution. To maximize the benefits and for efficiency of fly ash, there is a need for latest technologies to be put into practice. The chemical changes in fly ash are necessary to enhance the removal efficiencies and the capacity of adsorption. The unburnt carbon in fly ash exhibits a significant nature for adsorption and when converted to activated carbon, will improve the adsorption capacity. Separation of unburnt carbon from the minerals can be advantageous for the synthesis of zeolite as well as for the production of cement. For covering different aspects at various levels, the studies on fly ash utilization must be intensely taken up so as to cut down the effects of environmental pollution. There is an indication from the review that that the prime factor of crucial comparison between fly ash related products, and the ones obtained from earlier usual methods has not been properly attended to. The study of environmental effects of fly ash as a component in the manufacture of materials is yet to receive proper incorporation of knowledge.

REFERENCES

Atis, C.D., Accelerated carbonation and testing of concrete made with fly ash, *Construction and Building Materials,* Vol. 17, 2003, pp. 147-152.

Anon, Fly ash use as a concrete additive questioned by building inspectors, *Engineers News-Record,* 152, 12, 1954, pages. 22.

Building Research Establishment, Alkali-aggregate reactions in concrete, *BRE Digest* 258, HMSO, London, 1982.

Davis, R.E., Jones, P.W. and Shreve, R.N., Properties of cements and concrete containing fly ash, *Proceedings of ACI,* Vol. 33, 1937, pp. 577-612.

Dhir, R.K., Munday, J.G.L., Ho, N.Y. and Tham, W.K., PFA in structural precast concrete: measurement of permeability, *Concrete,* 1986, pp. 4-8.

Hobbs, D.W., Influence of pulverised-fuel ash and granulated blast furnace slag upon expansion caused by the alkali-silica reaction, *Magazine of Concrete Research,* Vol. 34, 1982, pp. 83-93.

Kumar, S., Patil, C.B., Estimation of resource savings due to fly ash utilization in road construction. *Resour Conserv Recycl* 2006; 48: 125–40.

Lin Y, K., Compressibility, strength, and frost susceptibility of compacted fly ash. PhD thesis, *University of Michigan,* Michigan; 1971.

Liu, B., Xie, Y. and Li, J., Influence of steam curing on the compressive strength of concrete containing supplementary cementing materials, *Cement,* Vol. 35(5), 2005, pp. 994-998.

Magnuson, M.O. and Malenka, W.T., *Utilization of fly ash for remote filling of mine voids.* In: United States Department of the Interior, Bureau of Mines Information Circular 8488. Washington, DC: US Department of the Interior; 1970.

Mathur, A.K., Ash utilisation in NTPC. In: *Proceedings of the workshop on fly ash utilisation: issues and strategies;* 2000. p. 41–45.

Mehta, P.K. and Monteiro, P., *Concrete: Microstructure, Properties and Material, Second Edition*, The McGraw-Hill Companies Inc, 1993, 548 pp.

Pan, P.N., Prospect of utilisation of fly ash in coal mines – a prospective. In: *Proceedings on the workshop on fly ash utilisation: issues and strategies;* 2000. p. 57–62.

Papadakis, V.G., Fardis, M.N. and. Vayenas, C.G., Effect of composition, environmental factors and cement line mortar coating carbonation, *Materials and Structures,* Vol. 149(25), 1992, pp. 293-304.

Papadakis, V.G., Effect of supplementary cementing materials on concrete resistance against carbonation and chloride ingress, *Cement and Concrete Research,* Vol. 30(2), 2000, pp. 291-299.

Sanjuan, M.A., Concrete carbonation tests in natural and accelerated conditions, *Advances in Cement Research,* Vol. 15(4), 2003, pp. 171-180.

Swamy, R.N., *Cement replacement material,* Blackie and Sons Ltd., 1986, 259 pp.

TIFAC (Technology Information, Forecasting and Assessment Council), http://www.tifac.org.in/ news/flyindia.htm; 2005.

In: Fly Ash
Editor: Prabir Kumar Sarker

ISBN: 978-1-62948-044-2

Chapter 8

PROPERTIES OF HIGH STRENGTH CONCRETE CONTAINING CLASS F FLY ASH

Pradip Nath* and Prabir K. Sarker
Department of Civil Engineering, Curtin University, Perth, Western Australia

ABSTRACT

Fly ash, being a by-product, is a readily available material for potential utilization in various Civil Engineering applications. Utilization of fly ash as a supplementary cementitious material adds sustainability to concrete by reducing the greenhouse gas emission associated with cement production. However, different fly ash affects the properties of concrete differently because of the variation of its source, burning process and resulting chemical composition. This chapter discusses the properties of fly ashes produced in Australia and investigates the mechanical and durability properties of high strength concrete using a Class F fly ash. Fly ash concretes using 30% and 40% fly ash as cement replacement were designed with and without adjusting water to binder ratio (w/b) and total binder content. Concrete samples having high compressive strength varying in the range of 65 to 85 MPa were tested for shrinkage, water permeability and corrosion resistance properties. Concretes, in which mix proportions were adjusted for fly ash inclusion achieved equivalent control concrete strength and showed less drying shrinkage, sorptivity, water permeability, and chloride penetration than those in which adjustments were not made. Fly ash concretes showed similar carbonation as compared to the control concrete of similar grade. The durability properties improved with the increase of fly ash content from 30% to 40% of the binder and with the increase of age. The results show the suitability of using class F fly ash as substantial parts of the binder for making high strength concrete of good durability.

Keywords: Carbonation, chloride penetration, durability, fly ash, permeability, shrinkage

* Corresponding Author address: Dept. of Civil Engineering, Curtin University, Perth, WA 6845, Australia. Email: pradip.nath@curtin.edu.au.

1. INTRODUCTION

Concrete is the most widely used construction material in the world. In foreseeable future, the demand for concrete is anticipated to increase rapidly in the developing nations causing massive rise in the demand of cementitious materials. Due to wide availability, Portland cement is yet the prime choice as the binder for concrete. As a result, cement production is increasing as much as 3% annually (McCaffrey, 2002). Cement production involves an energy intensive kilning process and contributes significant amount of greenhouse gases, causing cement to be the main source of CO_2 emission in concrete industry. In 2011, total cement production in the world reached about 3.6 billion tons, which resulted in over 2 billion tons of CO_2 (Puyvelde, 2013). To ensure sustainability in concrete production, it is indispensable to replace cement with other environmentally friendly supplementary cementitious materials and find efficient alternative binder for concrete.

Fly ash is one of the supplementary cementitious materials being used as a replacement of cement over the past few decades. Fly ash is a pozzolanic material obtained as a by-product from coal-fired power plants. When mixed with Portland cement, the products of fly ash-cement interaction are very similar to those formed by cement hydration.

In the context of sustainability, partial replacement of cement in concrete with fly ash is an effective way of reducing overall cement production, which in turn reduces greenhouse gas emissions. There are 80% to 90% less greenhouse gas emissions in the production of fly ash than cement (Duxson et al., 2007). Inclusion of fly ash in concrete also ensures effective utilization of currently under-utilized fly ash worldwide. In 2011, total surplus coal combustion products in Australia were 9.4 million tons which were stored into onsite ponds awaiting some future opportunity for economic reuse (Ash Development Association of Australia, 2012).

Incorporation of fly ash in partial replacement of cement is believed to enhance durability characteristics of concrete; however, it depends on a wide range of factors. The properties of fly ash vary with the source and type of coal and its combustion process. Due to the variation in composition, different fly ash affects the properties of concrete differently. For this reason, the results published in the literature vary with the variation of fly ash and local conditions, which may not be applicable for the other fly ash available elsewhere. Hence, it is essential to investigate the properties of concrete utilizing local fly ash prior to real application to build confidence of the local construction industry. This chapter represents a brief analysis on the nature and types of Australian fly ash and a study of durability properties of high strength concrete using fly ash from a Western Australian power plant.

2. PROPERTIES OF FLY ASH

2.1. Production

Fly ash is a by-product of the combustion of pulverized coal in thermal power plants. Coal is fed to a series of mills within a power station that pulverize the coal to a very fine powder. The pulverized coal is fed into a boiler to combust and to generate heat, which is then used to produce steam required for power generation. During the coal combustion process, minerals in the coal fuse to form glassy alumina-silicate spheres. These spheres remain in suspension within the flue gas produced in the boiler and are ultimately collected downstream by either electrostatic precipitators or other scrubbing systems such as a mechanical dust collector. When coal is burnt in a modern pulverized fuel furnace, two types of ash are produced. The fine ash, which is recovered from the flue gas, is called fly ash (FA). This material comprises up to 90% of the total ash produced. The remainder consists of similar particles that have fused together into aggregate sized lumps. These fall to the bottom of the furnace and are known as furnace bottom ash (FBA). Other by-products of coal combustion include cenospheres, molten inorganic boiler slag and flue gas desulfurization materials.

2.2. Physical and Chemical Properties

Fly ash appears as fine gray powders consisting mostly of spherical glassy particles that are produced at very high temperature of about 1400°C in the furnace. The color of fly ash can vary from tan to gray to black, depending on the amount of unburned carbon in the ash. Particles of fly ash vary in a wide-range of shapes, the majority being spherical particularly in the smaller size range. The specific gravity of fly ash usually ranges from 2.1 to 3.0, while its specific surface area may range from 170 to 1000 m^2/kg as measured by the Blaine air permeability method. Fly ash has pozzolanic properties and it reacts chemically with calcium hydroxide at ordinary temperatures to form compounds possessing cementitious properties. Fly ash can significantly improve concrete performance in various ways.

The chemical and mineralogical compositions of fly ash are governed by the interaction of a variety of factors, principally the type and source of coal, minerals and primary elements present in the coal, furnace type, plant age and the specific combustion conditions exposed to the coal. The mineral phases those found in fly ash include hematite, maghemite, magnetite, mullite and quartz. It mainly consists of relatively large percentage of amorphous aluminosilicate glass and crystalline phases (Bucea, Cao & Sirivivatnanon, 1996).

Three types of fly ash were defined in ASTM Standard C618-08a (2008): Class N, Class F and Class C.

- Class N fly ash includes raw or natural pozzolans such as volcanic ash, calcined clays and shales etc.
- Class F fly ash is typically produced from burning anthracite or bituminous coal, but may also be produced from subbituminous coal and from lignite. It is also known as low-calcium fly ash (< 10% CaO) and it has pozzolanic property.
- Class C fly ash is typically produced from burning lignite or subbituminous coal, and may also be produced from anthracite or bituminous coal. Class C fly ashes contain total calcium contents, expressed as calcium oxide (CaO), higher than 10%; hence, it is known as high-calcium fly ash. Class C fly ash has self-cementing property.

The principal components of bituminous coal fly ash (Class F) are silica, alumina, iron oxide, and calcium, with varying amounts of carbon, as measured by the loss on ignition. Lignite and sub bituminous coal fly ashes (Class C) are characterized by higher concentrations of calcium oxide and magnesium oxide and reduced percentages of silica and iron oxide, as well as lower carbon content, compared with bituminous coal fly ash. Table 1 shows a comparison of typical chemical composition of fly ashes and Portland cement. Generally fly ash is rich in SiO_2 and Al_2O_3 content and contains a very less amount of CaO as compared to cement.

Table 1. Typical chemical compounds in pozzolans (Adopted from Technical Bulletin 2, 2005) and Portland cement

Cementitious material		SiO_2	Al_2O_3	Fe_2O_3	CaO	MgO	SO_3	Na_2O & K_2O
Fly ash type	Class N	58.2	18.4	9.3	3.3	3.9	1.1	1.1
	Class F	54.9	25.8	6.9	8.7	1.8	0.6	0.6
	Class C	39.9	16.7	5.8	24.3	4.6	3.3	1.3
Cement		22.6	4.3	2.4	64.4	2.1	2.3	0.6

2.3. Usage

There are several applications where fly ash is being utilized for engineering and non-engineering purposes. In some applications it is used for its physical characteristics while its chemical characteristics proved most effective usage (Bucea, Cao & Sirivivatnanon, 1996). Fly ash can produce a compacted mass as dense sand; therefore it is ideal for backfilling retaining walls and dams over soft soil because of its high internal angle of friction, low unit mass and reduced settlement (Heidrich, 2003). As by-products of a very efficient industry, for a given power station, the types of ash have consistent properties and many potential uses. Many new potential uses have been identified by research conducted in Australia and overseas. The following are some possible applications of fly ash: flowable fill, waste stabilization, mine and land fill, road base stabilization, cement and concrete products, structural fill and embankments, shotcrete, geopolymer concrete and in agriculture and horticulture as soil amendment, fertilizer and zeolites.

The use of fly ash in the manufacture of cement and concrete products represents a large sector for value added utilization. New products such as bricks and aggregates made of fly ash are gradually emerging in the concrete industry. It is anticipated to take a great share of natural aggregate resources. Further opportunities exist for increasing the use of fly ash in high volume fly ash concrete as well as geopolymer concrete with a better understanding of the material and its effect in the concrete. Geopolymer binder is produced by the alkaline activation of fly ash and has a great potential to eliminate the necessity of cement in concrete. Asphalt concrete is another potential application where fly ash can be used as a mineral filler to refine the voids and improve the stiffness of the asphalt matrix.

The coarse furnace bottom ash (FBA) is not highly pozzolanic in nature and is less suitable for cement or concrete products. It can be used as a sand replacement, aggregate for lightweight blocks, a road-base component, and as engineered bulk fill. In recent years agricultural applications have increased in volume because of its suitability as drainage medium.

3. FLY ASH IN AUSTRALIA

Australian energy sector is mainly dependent on natural resource such as coal and gas. There are about 40 coal fired power stations throughout Australia. In 2010-11, coal fired power stations generated around 78% of total electricity consumed by National Electricity Market (NEM) in eastern Australia and 51% of total electricity consumed by South West Interconnected System (SWIS) in Western Australia (Energy White Paper 2012). These power stations produce about 14 million tons of coal combustion products on an average every year. Based on the installed capacity in 1999, national electricity demand growth estimates and proposed new capacity, ash material availability is expected to increase at around 3-5% annually (Heidrich, 2003).

According to Annual membership survey report 2011 of Ash Development Association of Australia (2012), approximately 13.1 Mt (million tons) of Coal Combustion Products (CCP) were produced within Australia in 2011. Only 6.01 Mt (or 46%) of CCPs have been effectively utilized in various value added products or to some beneficial end over the period. This indicates a notable rise in the beneficial use of CCPs as compared to the previous periods of 2010, 2009 and 2008 with effective utilization being 41%, 34% and 31% respectively. The growth in utilization is mainly associated with the civil construction, cement and concrete production and mining remediation application. Approximately 2.1 Mt (or 12%) was used in high value-added applications such as cementitious applications or concrete manufacture which is 0.6 Mt higher than that used in 2007.

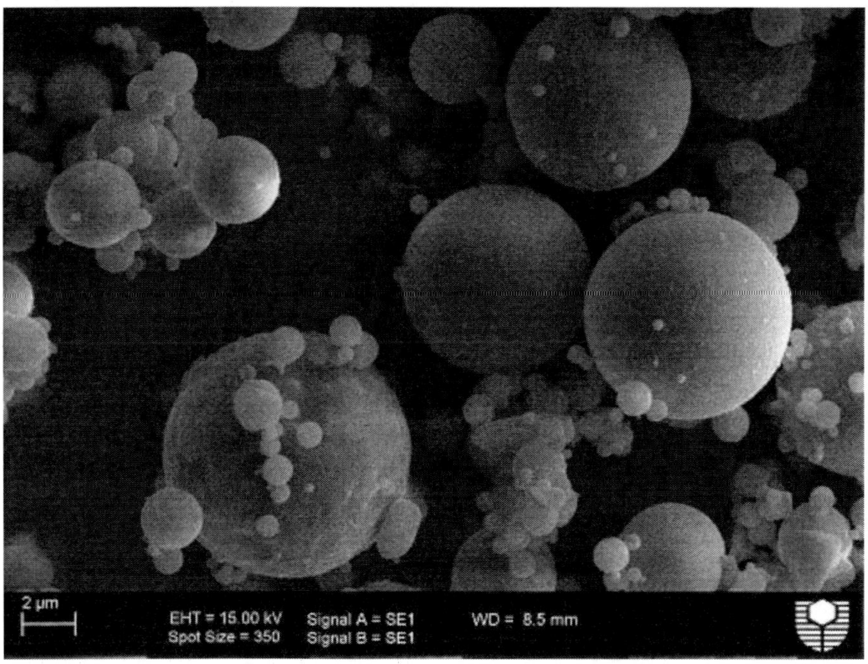

Figure 1. Scanning Electron Microscope image of fly ash from Western Australia.

3.1. Chemical Properties

The variation of mineralogy and inorganic geochemistry of the original feed coals and the different furnace operating conditions affect the chemical composition of fly ash. Australian power plants utilize mainly black coal (anthracite, bituminous and sub-bituminous coals) and brown coal (lignite, which is a low rank, high moisture content coal) for 54.9% and 21.8% of total electricity production respectively (Australian Coal Association, 2013). Fly ashes produced in Australian power stations are light to mid-gray in color and have the appearance of cement powder. Particle sizes range from less than 1 micron to 200 micron and are irregular to spherical in shape. Figure 1 is a typical micrograph of fly ash particles, taken using a Scanning Electron Microscope (SEM). The X-ray Fluorescence results show that fly ashes produced in Australia contain mainly silica and alumina (80-85%) and CaO (< 10%) with varying amounts of iron, potassium and titanium oxides. South Australian ash contains elevated concentrations of Na, Mg and Ca. The samples from central Queensland and West Australian coast ashes are almost similar and contain higher concentrations of iron oxide. Ash samples from Stanwell (Rockhampton) had elevated concentrations of iron and calcium oxides. In contrast, brown coal from the Gippsland region in Victoria is sulfidic and varies considerably with respect to the other elements (Ash Development Association of Australia, 2009a).

The majority of fly ash produced in Australia is classified as F type as per ASTM Standard C618-08a (Heidrich, 2003). Mean compositions of Australian fly ashes are shown in Table 2. Ward and French (2006) studied chemical characteristics of nine fly ash samples of different Australian power plants. Table 3 summarizes the chemical compositions and physical characteristics of these fly ash samples. The results indicate significant difference between the glass compositions of the ashes derived from lower-rank coals and those derived from higher-rank (bituminous) materials. The percentage of glass in the Australian fly ashes studied ranges from 45% to 80%. According to the glass composition data the fly ash samples was classified into two groups: group A has glass with a relatively low percentage of SiO_2 (45–60%) and a high percentage of Fe_2O_3 (> 10%) and group B has glass with apparently high SiO_2 (60–70%) and low Fe_2O_3 contents (< 5%). The ashes of group A were generally derived from sub-bituminous coals and those of group B were derived from combustion of bituminous coal. Glass composition of fly ashes plays a role in determining particle density and surface area characteristics among these ash samples. Fly ashes of Group A generally have higher particle densities (>2.0 g/cc) compared to Group B fly ashes. Most of the Group A ashes have significantly higher surface areas than the Group B ashes, whether or not micropores are included in the BET measurement (Ward and French, 2006).

Table 2. Principal chemical constituents of Australian coal fly ashes as percentages (Heidrich, 2003)

	SiO_2	Al_2O_3	Fe_2O_3	CaO	MgO	Na_2O	K_2O	SO_3	LOI*
Mean	61.1	26.2	4.2	1.7	0.9	0.8	1.1	0.2	2.6
Std. Deviation	6.62	2.48	3.31	1.74	0.54	1.05	0.62	0.19	2.78

* Loss on ignition.

Table 3. Chemical composition (normalized to a carbon-free basis) and physical properties of Australian fly ash samples (Ward & French, 2006; Killingley et al., 2000)

	Group A				Group B				
Power station no.	A1	A2	A3	A4	B1	B2	B3	B4	B5
Minerals (%)	52.7	55.0	53.9	20.0	39.0	20.9	33.7	34.0	26.3
Glass (%)	47.3	45.0	46.1	80.0	61.0	79.1	66.3	66.0	73.7
Physical Properties									
Particle density (g/cc)	2.1	2.1	2.2	2.2	1.7	2.0	1.9	1.8	2.0
BET surface area (m^2/g)	9.7	8.6	11.2	3.2	3.1	1.7	1.7	2.5	0.9
BET(non-micropore) (m^2/g)	4.2	3.9	3.4	2.0	2.0	1.1	1.0	1.5	0.6
Unburnt carbon (%)	2.17	3.29	2.25	4.71	3.67	1.45	2.65	3.55	0.6

3.2. Environmental Aspects

The use of fly ash into recognized and beneficial applications has substantial environmental benefits. Some of the most established environmental benefits gained from the use of fly ash include reduction of waste and associated space for landfill, conservation of natural resources such as gypsum, limestone and natural gas when fly ash is used as a replacement in cement production and reduction of greenhouse gas emissions when used as a cement replacement in concrete. Replacing one ton of cement with fly ash reduce up to one ton of CO_2 emission. In Australia, more than 32.9 million tons of fly ash has been used in cementitious applications or concrete manufacture from 1975 to 2011 (Ash Development Association of Australia, 2012). The use of fly ash may significantly contribute to further reduction of the carbon footprint of concrete. The introduction of carbon tax by Australian government since July 2012 puts additional cost for cement and concrete production. Certainly, this will act as a driving force to utilize more fly ash or other by-product materials in replacement of cement.

Fly ash is classified as waste in Australian legislation and the standards set by the states vary widely (French & Smitham, 2007). Although the beneficial use of fly ash is increasing every year, yet majority of the coal combustion products remain unused which is considered as a potential risk for current disposal and management practices. The surplus fly ash is destined for disposal in engineered surface impoundments and landfills, or in abandoned mines and quarries. The primary environmental concern associated with these disposal sites is the potential for contamination in soil, groundwater and atmosphere. The Ash Development Association of Australia (ADAA) regularly investigates the element concentration and leachability of selected metal species from coal combustion products (CCPs) under Environmental Monitoring Plan (EMP). The report presented in 2008-09 (Ash Development Association of Australia, 2009b) confirms the consistent benign (inert) nature of Australian CCPs as compared to the previous results. All total metal results for M17 elements (Arsenic, Cadmium, Manganese, Tin, Barium, Chromium, Molybdenum, Zinc, Beryllium, Cobalt, Nickel, Mercury, Boron, Copper, Selenium, Lead and Silver) are below limits specified by New South Wales & Queensland regulatory legislations. All leachate results are either below or just above the laboratory detection limits for each analyte, and accordingly were well within the acceptance criteria adopted for each jurisdiction. Hence use of these fly ashes in civil engineering applications and in agricultural sectors may not cause any adverse contamination and leaching of toxic heavy metals.

Fly ash may also contain radioactive elements originated from coal which is a potential hazard when they are dispersed in air and water or are included in commercial products. The various radionuclides in the original coal partition between the different forms of ash, with the more volatile radionuclides (^{210}Po and ^{210}Pb) accumulating in the fly ash and stack emissions, and the more refractory elements (U and Th) in the bottom ash and slag. Gaseous ^{222}Rn is released to the atmosphere during combustion. An assessment to determine the concentration of naturally occurring radionuclides present in 55 samples from different power plants of Australia was conducted by Ash Development Association of Australia (2009a). In general, higher Th and U assays were obtained in fly ashes from South Australia and the West Australian coast. Fly ash samples from the Gippsland region contain significantly lower concentrations of these elements. The results of this study have shown that all radionuclide is below the prescribed limit of 1 Bq/g for radionuclides of natural origin or 10 Bq/g for ^{40}K (IAEA Safety Guide No. RS-G-1.7, 2004). Thus, none of the fly ash samples is considered radioactive from a regulatory perspective. The radioactivity indices in fly ash are generally comparable to cement, sand, aggregate and bricks. While fly ash is mixed with other ingredients to make concrete, its low concentration of radioactivity is diluted significantly (Ash Development Association of Australia, 2009a).

4. Fly Ash Utilization in Concrete

Although the earliest noteworthy study of the use of fly ash was in 1937, its use as a pozzolanic ingredient was recognized as early as in 1914 (Fly ash, 2013). Volcanic ash, which possesses similar properties to fly ash, was used as pozzolan in Roman structures such as aqueducts or the Pantheon. Its use became more common since 1980s. At first, it was used as a cost-saving measure. But over the years, the concrete industry discovered its qualitative advantages.

In Australia, fly ash imported from USA was used first in 1949 in the grout of shaft lining at the power station of the Snowy Mountains Hydro-Electric Scheme in New South Wales (Ash Development Association of Australia, 2009c). The study on the properties of fly ash concrete was initiated since 1950 (Munt & Munroe, 1950). Fly ash concrete was first used in in Keepit Dam of New South Wales, a structure constructed by the Water Conservation and Irrigation Commission in 1959 (Roper, Baweja & Kirkby, 1984). The fly ash was supplied from the Pyrmont Power Station in Sydney. Since the early 1960's, fly ash has been used in many concrete structures successfully in Australia.

4.1. High Volume Fly Ash (HVFA) Concrete

High Volume Fly Ash (HVFA) concrete was developed at Canada Centre for Mineral and Energy Technology (CANMET) in the late 1980s (Mehta, 1999). Typically, HVFA concrete has a very low water to cementitious materials ratio (w/cm) and incorporates up to 55% Class F fly ash. Because of low water content, high-range water-reducing admixtures or superplasticizers are used to achieve workable slumps ranging from 150 mm to 200 mm (Malhotra, 2002). Considerable research has been conducted on the properties of concrete using higher proportions of fly ash, such as 30–70% of the total binder (Carette et al., 1993; Siddique, 2004). However, the recommended fly ash replacement level for high strength concrete is usually 15–25% (ACI committee 211, 2008).

The use of fly ash in concrete improves overall concrete performance. Firstly, fly ash reduces the amount of water needed to produce a given slump (Jiang and Malhotra, 2000). The spherical shape of fly ash particles creates a "ball-bearing" effect that fills voids and increases the workability of the mix. Its dispersive ability provides water-reducing characteristics similar to a water-reducing admixture. In addition, fly ash enables the mixture to flow better in pump hoses and makes it more workable under hand finishing. Fly ash produces relatively more cementitious paste due to its lower unit weight. By its shape and dispersive action, fly ash particles create better lubrication of the aggregates which increases the flow ability of concrete (ACI Committee 211, 2008).

Having reduced quantity of cement in the mix and slow hydration, fly ash concrete generates low heat of hydration (Neville and Brooks, 1987) as compared to Portland cement concrete. Hence it reduces the amount of thermal cracks and drying shrinkage of concrete. With the development of hydration the fly ash concrete becomes denser as compared to Portland cement concrete. Thus it achieves high long term strength (Sivasundaram, Carette & Malhotra, 1990) and higher resistance to water permeation (Naik, Singh & Hossain, 1994). This makes the fly ash concrete highly resistant to aggressive materials, because most of the aggressive agents finds its way to concrete in dissolved form with water. Fly ash concrete is efficient in increasing the initiation period and reducing the corrosion rate of steel reinforcement due to chloride ingress (Bucea, Cao & Sirivivatnanon, 1996). Hence, the use of HVFA concrete results in extended maintenance free service life of reinforced concrete structure in aggressive environments.

There are some limitations of using fly ash in concrete. If concrete is made with poor quality fly ash, inadequately proportioned and cured, it becomes highly permeable, particularly in the early ages (Ramezanianpour & Malhotra, 1995). Under such conditions, chlorides as well as moisture and oxygen penetrate the concrete and the protective layer at the reinforcing steel surface can be destroyed. Fly ash concrete shows slower setting time due to slow hydration (Bentz and Ferraris, 2010). Fly ash concretes also have low early strength due to the high alumina and iron oxide content (Mehta, 1985). This difficulty can be eliminated by using accelerating admixture in the mixture.

4.2. Mechanism of Fly Ash Reactivity in Concrete

The chemical compounds of fly ash and Portland cement are similar. However, fly ash compounds are generally amorphous (glassy) due to rapid cooling, while those of cement are crystalline, formed by slower cooling (Technical Bulletin 2, 2005). Fly ash is categorized as a normal pozzolan, a material consisting of silicate glass, modified with aluminum and iron (Papadakis, 1999). Portland cement is richer in lime (CaO) in comparison to fly ash. When cement starts hydration in the presence of water, about 20% of CaO is released in a free state

(Technical Bulletin 2, 2005). This free lime is available to react with fly ash in the form of water soluble $Ca(OH)_2$ (calcium hydroxide). When pozzolanic materials such as fly ash are added in concrete, the cement hydration product (calcium hydroxide) is transformed into secondary calcium silicate hydrate (CSH) gel, but with a lower Ca/Si ratio (Escalante-Garcia & Sharp, 2004). This additional CSH gel refines the larger pores into finer pores. Montgomery, Hughes and Williams (1981) suggest that, as the hydration proceeds around individual spheres of fly ash, the outward growth leads to mechanical interlocking of the hydration products. This enhanced interlinking aspect is thought to be fundamental to the strength imparting mechanism. Thus, the use of mineral admixture such as fly ash improves the quality of concrete by transforming relatively weak hydration product $Ca(OH)_2$ into stronger CSH gel (Papadakis & Tsimas, 2002).

4.3. Durability of Concrete Containing Fly Ash

Numerous published literatures are available which demonstrate the effect of fly ash on the strength and durability properties of concrete. Concrete containing fly ash exhibits excellent mechanical and durability properties, such as low permeability to chloride ions and other aggressive agents (Langley, Carette & Malhotra, 1989; Malhotra, 1990); however, poor performances of fly ash concrete have also been reported in literature (Berndt, 2009). The properties of fly ash concrete vary widely due to variable composition of fly ash and varying mixture proportions (Shehata, Thomas & Bleszynski, 1999; Burden, 2006).

The performance of concrete greatly depends on the role of fly ash in the mixture, either as aggregate or as cement replacement in concrete. Fly ash tends to have a low reaction rate when used as an admixture in concrete mixes. At 28 days, the degree of fly ash reaction rate is slightly more than 10% (Lam, Wong & Poon, 2000). Furthermore, only 20% of the fly ash reacts after 90 days. One explanation for the low activity of fly ash at room temperature is that the pH of the solution 13 does not meet the requirements of fly ash, 13.3 (Gopalan, 1993). Papadakis and Tsimas (2002) studied the efficiency factor and design of supplementary cementitious materials (fly ash) in concrete and reported that when fly ash substitute aggregates, the strengths are higher than that of the control concrete. When fly ash replaces cement, the strength is reduced at first, but this gap is gradually reduced as time proceeds. The strength exceeds the control strength for fly ash with higher active silica content than cement. An efficiency factor equal to Portland cement (k=1) was reported for pulverized fly ashes in terms of strength. In addition, Ganesh Babu and Rao (1993) reported that the overall cementing efficiency depends on the age and replacement percentage. The strength decreases continuously from 100 to 35% at replacement levels varying from 20 to 75% respectively.

The composition of fly ash influences the properties of concrete. Class C fly ash typically generates more heat of hydration than class F, not as effective as class F fly ash in mitigation of alkali-silica reaction and not resistant to sulfate attack. However class C fly ash generates more strength at early ages than class F fly ash. Papadakis (1999) and Papadakis (2000a) reported that when aggregates are replaced by low-calcium fly ash (class F), higher strengths are observed after 14 days; whereas in cement replacement, higher strengths are observed after 91 days (Papadakis, 1999). When aggregates are replaced by high-calcium fly ash (class C), significantly higher strengths are observed from the beginning of the hydration, as well as higher water binding and significantly lower porosity (Papadakis, 2000a). In the case of high calcium fly ash, due to replacement of cement the strength remained constant.

Fly ash in concrete reduces drying shrinkage (Atis 2003b), thus generates fewer cracks which ensure greater resistance to deterioration. Chindaprasirt, Homwuttiwong, & Sirivivatnanon (2004) found reduced drying shrinkage of mortars using fly ashes of different fineness. The inclusion of high volumes of fly ash in concrete with a low water-cementitious material ratio results in a reduction in the shrinkage values of up to 30% when compared to OPC concrete (Atis 2003b). Superplasticized concretes show about 50% higher shrinkage in comparison to non-superplasticized one. Kumar, Tike and Nanda (2007) reported, for superplasticized fly ash concrete with 20-60% replacement, drying shrinkage of concrete decreases with decreasing w/b ratio and increasing fly-ash content. In contrast, an increase of drying shrinkage when fly ash is included in concrete has also been reported (Erdog˘an, 1997 and Munday et al., 1982, as cited in Atis 2003b).

Fly ash influences the pore structure of concrete in different way. By replacing up to 45% class F fly ash, reduced pore diameter and porosity of concrete were observed at 28 days, whereas fly ash-cement paste revealed increased porosity (Poon, Lam & Wong, 2000). Papadakis (1999) observed increased porosity when Class F fly ash replaced cement and decreased porosity when fly ash replaced aggregate in mortar.

Concrete's durability, is governed to a great extent by its permeation properties. Naik, Singh & Hossain (1994) tested concretes with up to 70% Class C fly ash and obtained reduced air and water permeability of fly ash concretes at 91 days. They reported best performance for 50% fly ash replacement. Tasdemir (2003) investigated the effect of Class C fly ash as one of the mineral admixtures, replacing 10% of cement in concrete with fly ash and using w/b ratio of 0.60. Higher sorptivity coefficient was reported for fly ash concrete as compared to normal concrete at early age. However, Camoes, Aguiar, and Jalali (2003) obtained reduced sorption coefficient by using w/b ratio in the range of 0.25-0.40 and Class F fly ash content of as high as 60% of the total binder.

In a study conducted by Ueda et al. (as cited in Jerath & Hanson, 2007) the results showed that when fly ash was mixed as the substitute for fine aggregate, restrained permeation of CO_2 or Cl^- into concrete resulted in small corrosion loss for steel. However, when fly ash was mixed as a substitute for cement, corrosion loss of reinforcing steel was not restrained under the severe corrosion condition in this study. In other study, Papadakis and Tsimas (2002) concluded that, the concrete incorporating supplementary cementitious materials, whether used as partial replacement for aggregate or cement, exhibited significantly lower total chloride content. Higher efficiency factor values against chlorides as compared with the corresponding values for strength were presented ($k = 2–2.5$) for fly ashes.

Naik et al. (2003) used Classes C and F fly ashes for up to 70% and 67% cement replacement respectively to study the long term performance of concrete pavements. Long term performance tests conducted using core specimens from in-place pavements showed that the concrete mixtures containing Class F fly ash exhibited higher resistance to chloride-ion penetration than those containing Class C fly ash, revealing greater pozzolanic strength contribution of Class F fly ash relative to Class C fly ash.

Papadakis (2000b) reported that, the specimens incorporating fly ash, whether it substitutes aggregate or cement, exhibit significantly lower total chloride content. The use of supplementary cementitious material as an addition to a concrete mix, replacing either aggregates or cement, significantly lengthens the chloride-induced corrosion initiation stage.

In the study of Kouloumbi and Batis (1992) it is reported that the addition of the fly ash causes an increase of the total chloride content of the corresponding specimens over the OPC mortar specimens. Nevertheless, they also noted improved corrosion behavior of the steel reinforcing bars because fly ash reduces the free chloride content of the corresponding specimens.

Other factors that influence the behavior and efficiency of fly ash concrete include fly ash/cement ratio, optimum content of fly ash (Onera, Akyuzb & Yildiza 2005), water to cementitious material ratio (Srinivasan, Tiwar & Banchhor 2004), use of a superplasticizer (Kumar, Tike & Nanda 2007), aggregate grading (Jerath & Hanson 2007), and curing period and exposure (Toutanjia et al. 2004).

Fly ash concrete generally exhibits higher carbonation than OPC concrete (Jiang, Lin & Cai, 2000; Khan & Lynsdale, 2002; Sisomphon & Franke, 2007). However, Cabrera and Woolley (1985) reported reduced carbonation of old fly ash concrete. Atis (2003a) found higher carbonation for 70% fly ash and less carbonation for 50% fly ash incorporation as compared to control cement concrete. The study of Thomas, Matthews and Haynes (2000) shows that the carbonation of concrete with 15% to 30% fly ash is similar to or slightly higher than the concrete of equivalent strength without fly ash. Concrete containing 50% fly ash carbonated at a significantly faster rate than control specimens of the same grade and the differences were particularly marked for poorly cured concretes. Papadakis (2000b) observed that, the use of fly ash as a cement replacement may result in a shortened carbonation-induced corrosion initiation stage in contrast to that as aggregate replacement.

Thus it can be seen from the above studies that the application of fly ash in concrete can enhance durability features, but the extent of improvement is dependent on the mix proportioning and the properties of fly ash. The properties of high strength concrete containing class F fly ash as substantial parts of the binder designed following two different approaches of mixture proportioning are presented in the following sections.

5. Experimental Study

5.1. Materials

The materials used in this study were those commercially available in Australia. A General Purpose (GP) Portland cement conforming to Standards Australia (AS 3972) and Class F fly ash (ASTM C618-08a) sourced from Western Australia were used. The chemical compositions of cement and fly ash are shown in Table 4. The fine aggregate used was natural sand with a nominal maximum size of 1.18 mm. Coarse aggregates used were crushed granite with nominal maximum size of 7, 10 and 20 mm that met the Australian standard (AS 2758.1). Physical properties and size distribution of aggregates are presented in Table 5. Normal tap water was used and a superplasticizer (high range water reducer) was used to enhance workability.

5.2. Mixture Proportions and Casting

Two series (Series A and Series B) of concrete mixtures were designed in accordance with the recommendations of ACI 211.4R-08 (ACI Committee 211, 2008). Each series of mixture had a control mix (OPC concrete) and two mixes with 30% and 40% fly ash of the total binder content (cement + fly ash). Past researches indicate that the optimum cement replacement level for fly ash concretes is about 40% when high compressive strength is required (Kumar, Tike & Nanda, 2007; Onera et al. 2005). Thus, fly ash content was kept limited to 40% of total binder to achieve required strength grade at 28 days. The mixtures of series A were designed with varying total binder content and varying water-binder (w/b) ratio to adjust the proportion for fly ash inclusion. All three mixtures of series B were designed with a constant w/b ratio and constant amount of total binder. Superplasticizer was added to the mixture and its amount was adjusted to achieve a minimum slump of 150 mm. The mixture proportions and the measured slumps of the different batches of concrete are shown in Table 6. Thus, the properties of high strength concretes of similar workability proportioned in two different ways were determined and compared.

Table 4. Chemical composition of cement and fly ash

Parameter	SiO_2	Al_2O_3	Fe_2O_3	$SiO_2+Al_2O_3+Fe_2O_3$	CaO	MgO	SO_3	K_2O	Na_2O	P_2O_5	Chloride	Loss on ignition
Cement (%)	21.10	4.70	2.70	28.50	63.60	2.60	2.50	-	0.50	-	0.01	2.00
Fly ash (%)	50.50	26.57	13.77	90.84	2.13	1.54	0.41	0.77	0.45	1.00	-	0.60
Class F fly ash* (%)	-	-	-	70.00 min	10.00 max	-	5.00 max	-	-	-	-	6.00 max

* Specification as per ASTM C618-08a.

Table 5. Properties of aggregates

Properties	Coarse aggregate		Fine aggregate	
	Sieve (mm)	Percent passing	Sieve (mm)	Percent passing
Sieve analysis	19.00	100.0	4.75	100.0
	9.50	50.0	2.36	100.0
	4.75	8.9	1.18	99.7
			0.60	77.2
			0.30	24.2
			0.15	2.7
Fineness Modulus	6.40		1.96	
Specific Gravity (SSD)	2.67		2.62	
Absorption (%)	0.40		0.99	
Unit Wt. (kg/m^3)	1645		1687	

Table 6. Concrete mixture proportions (kg/m³)

Series	Mix ID	Fly ash (%)	Cement	Fly ash	Granite	Sand	Water	Super-plasticizer	w/b	Slump (mm)
A	A00	0	355	0	1185	740	145.5	5.11	0.41	140
	A30	30	308	132	1185	661	141.0	4.77	0.32	170
	A40	40	264	176	1185	665	136.5	4.75	0.31	185
B	B00	0	517	0	1185	594	150	6.77	0.29	150
	B30	30	362	155	1185	570	150	4.80	0.29	175
	B40	40	311	207	1185	561	150	4.24	0.29	160

Aggregates were prepared to saturated surface dry (SSD) condition. The ingredients were mixed in a laboratory pan mixer. Cylinders of 100 mm diameter and 200 mm height were cast for compressive strength, sorptivity, water permeability and chloride permeability tests. For dying shrinkage test, 75 × 75 × 280 mm prisms were cast. Cylinders of 150 mm diameter and 300 mm height for split tensile strength and prisms of 100 × 100 × 400 mm size were cast for flexural strength and carbonation test. Molds were filled in two equal layers and compacted using a vibrating table. The samples were de-molded after 24 hours of casting and then cured under water at 23°C for up to 28 days. At the age of testing, specific specimens were prepared in accordance with the specifications of the particular test.

6. Testing Methods

The durability of concrete principally depends on its resistance to permeation of aggressive materials. It was investigated by conducting water sorptivity and water permeability test. Resistance to corrosion was investigated by carbonation, chloride ion permeability and chloride diffusion tests. Mechanical strength and drying shrinkage were determined at different ages of concrete samples. Mechanical strength tests include compressive strength, indirect split tensile strength and flexural strength. The selected durability properties of high strength fly ash concretes were tested and compared with those of OPC concrete.

The drying shrinkage was measured in accordance with the standard AS 1012.13 (1992) to check the effect of fly ash on the volumetric stability of the concrete samples. Average length of three specimens was recorded for each mix at specified intervals up to 180 days. Drying shrinkage was calculated using the 7th-day readings as the reference length.

The sorptivity test was performed in accordance with ASTM C1585-04 (2004). The 50 mm thick specimen was cut from the top of a 100 mm diameter cylinder. Two identical specimens were conditioned by drying in the oven until a constant weight was achieved. The cut surface was then put in contact with water after sealing all other surfaces to prevent the evaporation of absorbed water. Change in mass due to sorption was measured at definite intervals for the first six hours. The initial rate of sorption was reported.

Water permeability of concrete samples was tested using a method developed by Taywood Engineering Limited (TEL) (GHD/TEL Method, n.d.; Nath & Sarker, 2010). In this method, water is pressurized through a 50 mm thick section of a 100 mm diameter concrete specimen at high pressure (950 KPa, equivalent to 96.9 m head of water). The test was continued for 7 days to have a steady rate of flow, and the mean value of 7th-day results of two identical specimens was reported. The permeability coefficient was measured using Darcy's equation in the following form (Eq. 1).

$$k = (Q \times L) / (A \times H) \tag{1}$$

where k = permeability coefficient (m/s), Q = flow rate (m^3/s), A = area of specimen (m^2), L = depth of specimen (m), H = head of water (m).

The accelerated reference test method recommended in the *fib* model code (fib CEB-FIP, 2006) was followed for carbonation test. After seven days of water curing, the samples were stored in the laboratory climate (reference temperature, T_{ref} 23°C and relative humidity, RH_{ref} 60 ± 10%). At the age of 28 days the specimens were placed in the carbonation chamber maintained with the standardized climate (T_{ref} 23°C, RH_{ref} 65%) and exposed to CO_2 with a concentration of 2.0% by volume. After 28 days of exposure, the specimen was split and depth of carbonation was measured spraying an indicator (1% Phenolphthalein solution) on the split surface. The carbonated area remains colorless while unaffected area turns purple.

The rapid chloride penetration test (RCPT) of the concrete samples was conducted in accordance with ASTM C1202-07 (2007). The likelihood of chloride ion penetration through concrete was measured in terms of total charge passed within a six-hour test.

Chloride diffusion test was carried out in accordance with the Nordtest Method (NT Build 443, 1995). The samples were about 50 mm thick and 100 mm in diameter. The test specimens were epoxy-coated on every face leaving only one surface open to be exposed to the sodium chloride solution (165 ± 1 g/L). After an extended period of exposure, eight layers were grinded from each sample at an interval of 2 mm from the exposed surface and dried in the oven. The acid soluble chloride content of oven-dried ground samples was measured by a potentiometric titration method. The chloride content in each layer was calculated as the average of test results of three samples. The values of surface chloride concentration (C_s) and apparent chloride diffusion coefficient (D_a) were determined by fitting Eq. 2 to the measured chloride contents and by means of a non-linear regression analysis with the method of least squares fit. The chloride penetration parameter (k_{cr}) was calculated using the Eq. 3 with the measured D_a (mm^2/yr), C_s, initial chloride concentration (C_i) and the reference chloride concentration (C_r) which is taken as 0.05 (% mass of concrete).

$$C(x,t) = C_s - (C_s - C_i).erf\left(\frac{x}{\sqrt{4.D_a.t}}\right) \tag{2}$$

$$k_{cr} = 2\sqrt{D_a}.erf^{-1}\left(\frac{C_s - C_r}{C_s - C_i}\right) \tag{3}$$

Where, $C(x, t)$ = chloride concentration measured at depth x and exposure time t (mass %), C_s = projected chloride concentration at the interface between the exposure liquid and test surface (mass %), C_i = initial chloride concentration in concrete prior to submersion in the exposure solution (mass %), x = depth below the exposed surface (m), D_a = apparent chloride diffusion coefficient (m^2/s), t = the exposure time (seconds) and erf = error function.

7. Results and Discussions

7.1. Mechanical Strength Properties

The mixtures prepared in this study were designed to achieve 28-day compressive strength more than 50 MPa as recommended by Standards Australia (AS 3600-2009) for the most severe environment. The compressive strength results up to 335 days are shown in Figure 2. The percentage strength developments of the fly ash concretes with respect to corresponding control concrete are plotted in Figure 3. The compressive strength, indirect tensile strength and flexural strength at 28 days are presented in Table 7.

In series A, the fly ash concretes (A30 and A40) gained similar strength of control concrete (A00) at 28 days. At 56 days, both the fly ash concretes gained more than 110% of the control concrete's strength (Figure 3). After 56 days, the fly ash concretes continued to gain strength, while the control concrete's strength remained steady until the age of 335 days. At 335 days, both 30% and 40% fly ash concretes achieved more than 120% strength of the control concrete.

Strength of concretes in series B decreased for the inclusion of fly ash in place of cement while keeping w/b ratio and the amount of total binder constant (Figure 2). However, from Figure 3 it is clear that, the fly ash concretes achieved more than 80% of the control concrete's strength at 28 days which reached 92% and 96% of control concrete's strength at 56 days for 40% and 30% fly ash content respectively. At 335 days, both 30% and 40% fly ash concretes gained similar strength which is 97% of the control concrete. Thus, compressive strength has not been affected significantly in the later age due to incorporation of fly ash. The results indicate that incorporation of fly ash in concrete has decreased strength at the earlier age as compared to that of control concrete, but at later stage they have either gained more strength (series A) or reached very close (series B) to the control concrete strength depending on the mixture proportions. Generally, concretes with 30% fly ash have shown higher strength gain than that of 40% fly ash. The strength of fly ash concretes in both series developed at a higher rate than that of control concrete until the age of 56 days. Significant part of strength of fly ash concretes was developed during 28 to 56 days. It implies the notable strength development capability of fly ash concrete due to pozzolanic reaction after 28 days in contrast to normal cement concrete. The trend of strength development of the fly ash concretes is in line with those reported by other researchers (Poon, Lam & Wong, 2000; Siddique, 2004).

Table 7. Compressive, tensile and flexural strengths at 28 days

Mixture		Compressive strength (MPa)	Indirect tensile strength (MPa)		Flexural strength (MPa)	
Series	Mix ID	Test	Test	Eq. 4	Test	Eq. 5
A	A00	69.0	3.8	3.3	7.2	5.0
	A30	73.0	4.4	3.4	8.0	5.1
	A40	65.5	3.8	3.2	6.1	4.9
B	B00	85.5	4.4	3.7	9.9	5.6
	B30	73.5	4.2	3.4	7.3	5.2
	B40	69.0	4.1	3.3	6.7	5.0

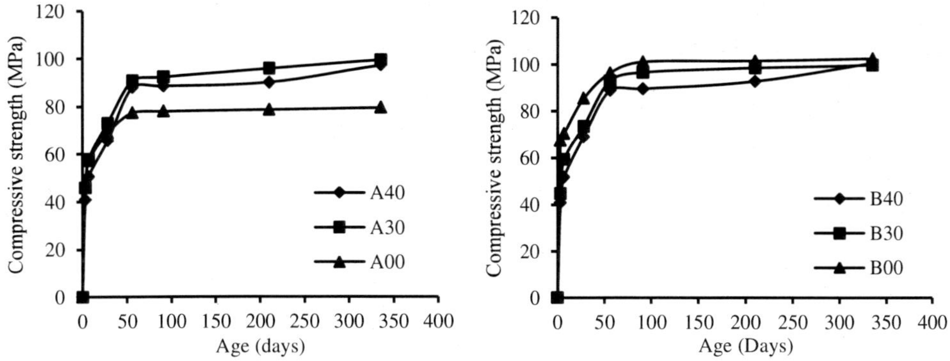

Figure 2. Development of compressive strength of concretes of series A (left) and series B (right).

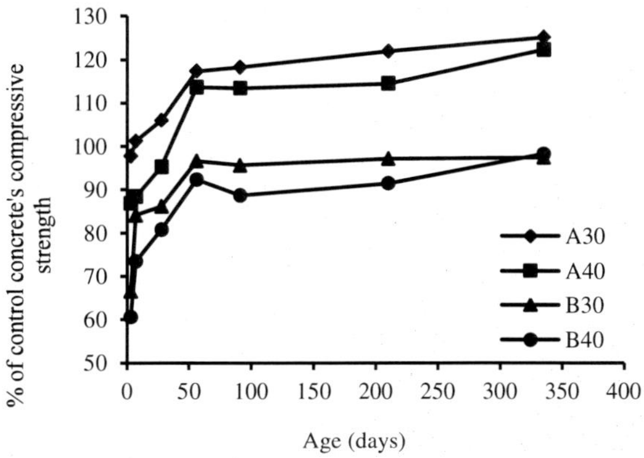

Figure 3. Strength development as compared to control concrete.

Tensile strength of concrete is related to the shrinkage, bond and shear behavior. The results of indirect tensile strength conducted in accordance with AS 1012.10 (2000) at 7, 28 and 335 days are plotted in Figure 4. The theoretical characteristic principal tensile strength (f'_{ct}) was calculated using Eq. 4 (AS 3600-2009).

$$f'_{ct} = 0.4\sqrt{f'_c} \tag{4}$$

It can be seen from the results (Table 7) that, for every mix, experimental value of 28 days tensile strength was more than that calculated theoretically. The tensile strength results of both series A and series B concretes followed similar trend of compressive strength. The tensile strength increased with the increase of age of the specimens (Figure 4).

At 7 days, mixture A30 and A40 reached 95% and 92% of 28-day strength. After 28 days, mixture A30 exceeded the control concrete strength and mixture A40 achieved similar strength of control concrete. Fly ash concretes surpassed control concrete's tensile strength by more than 10% after 335 days. Concrete with 40% fly ash had the most increment. The rate of strength gain was higher for the fly ash concretes than that of the control concrete.

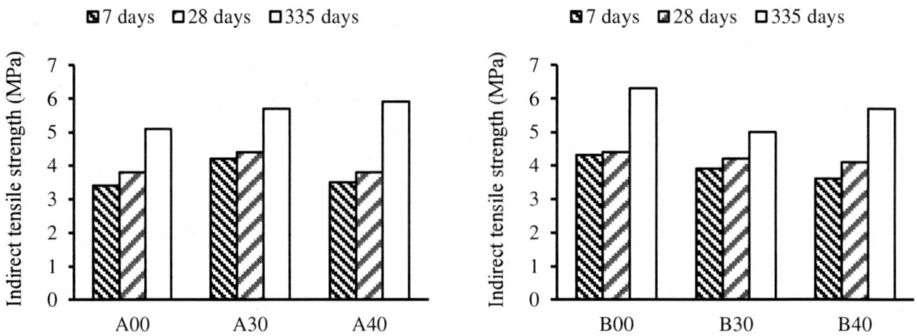

Figure 4. Indirect tensile strength of concretes of series A (left) and series B (right).

In case of series B, the tensile strength decreased with respect to control concrete due to incorporation of fly ash in place of cement without adjusting w/b ratio and binder content (Figure 4). Fly ash concretes developed strength at a rate slightly slower than that of control concrete. At 7 days, mixture B00, B30 and B40 gained 98%, 93% and 88% of 28-day strength respectively. These rates are faster than corresponding rates of compressive strength. At 28 day, concretes with 30% (B30) and 40% fly ash (B40) achieved 95% and 93% of control concrete's tensile strength respectively.

The flexural strength was measured at 28 days in accordance with the AS 1012.11 (2000). In addition, the theoretical flexural strength (f'_{cf}) was calculated using Eq. 5 as provided in AS 3600-2009 to determine flexural strength based on 28 days compressive strength. Results are shown in Table 7.

$$f'_{cf} = 0.6\sqrt{f'_c} \tag{5}$$

The effect of fly ash on the 28-day flexural strength of concretes in both series also followed similar trend of compressive strength. All mixtures showed higher strength in test than that predicted by Eq. 5. In both series, concrete with 30% fly ash has shown higher flexural strength than that of concrete with 40% fly ash. It is due to slower pozzolanic reaction of fly ash in concrete at 28 days.

7.2. Drying Shrinkage

The drying shrinkage of the samples was recorded up to 180 days. The effect of incorporation of fly ash in concrete on drying shrinkage of series A and series B are shown in Figure 5 (Nath & Sarker, 2011a). It can be seen from the figures that, most of the shrinkage in all of the concrete samples took place within 56 days after casting. The values of shrinkage in all of the concrete specimens at 56 days were well below 1000×10^{-6} as specified by the standard AS 1379 (2007) for normal and special class of concrete.

Concrete incorporating fly ash have shown less drying shrinkage than the control concrete when they were designed with variable w/b ratio and total binder content (series A). This is because of the lower w/b ratio in the fly ash concretes. Shrinkage was similar for fly ash concretes and control concrete of series A up to 21 days (Figure 5). After that, the rate of shrinkage decreased for the fly ash incorporated concretes to reach a 10% lower value than that of the control concrete at 56 days and afterwards. The concrete containing 40% fly ash (A40) achieved similar shrinkage as that of the concrete containing 30% fly ash (A30).

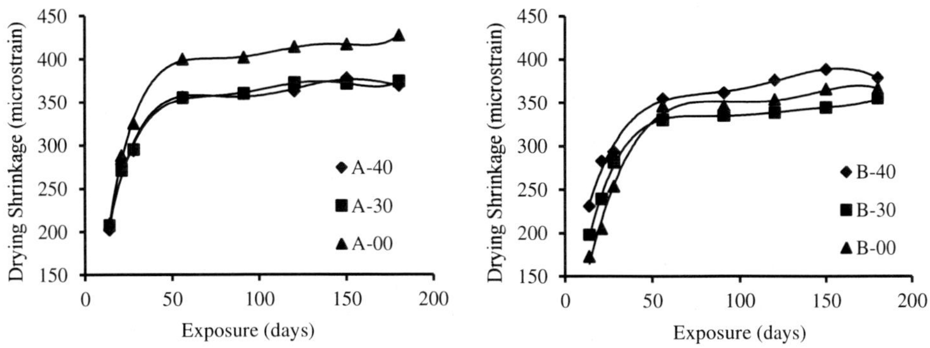

Figure 5. Drying shrinkage of concretes of series A (left) and series B (right).

In Series B, shrinkage values of the fly ash concretes (B30 and B40) were higher than that of the control concrete (B00) until 28 days (Figure 5). After that, the rate of shrinkage decreased for fly ash concretes and reached a value almost similar to that of the control concrete at 56 days. The shrinkage values of both of the fly ash concretes were very close (within ±4%) to that of control concrete up to 180 days. Concrete with 30% fly ash showed less shrinkage than that with 40% fly ash. The fly ash concrete requires less water to produce similar effect of cement concrete (Jiang & Malhotra, 2000). In series B, all the mixtures had same water content. This resulted in relatively higher amount of free water in the fly ash concretes. Hence drying shrinkage increased with the increase of fly ash. In their study, Munday et al. (as cited in Atis, 2003b) also found increase of drying shrinkage for fly ash incorporation in some concrete. This result signifies the need of adjustment in mix proportions when it incorporates fly ash as partial replacement of cement.

7.3. Permeation Properties

Concrete is a porous material, hence water permeation is considered as an important mechanism for the transport of harmful materials (chloride, sulfate, acids etc.) into concrete. In this study permeation properties were determined by sorptivity and water permeability test. The sorptivity test gives an indication of concrete's resistance against capillary permeation, whereas water permeability test indicates the resistance against the uniaxial flow of water passing under high pressure through the concrete sample. Both tests were conducted at 28 days and 180 days. Papworth and Grace (1985) recommended the typical values of water permeability and sorptivity for various performance classes of concrete (Table 8). These recommendations have been used in this study to rate the quality of the concretes.

Figure 6 shows the results of sorptivity test at 28 days and 180 days. Concretes in series A, those achieved similar strength at 28 days, have shown significant reduction of sorption due to inclusion of fly ash. At 28 days mixture A40 showed slightly higher absorption than mixture A30. However, after 180 days, the sorptivity of mixture A40 reached 17% lower than that of mixture A30. Sorptivity dropped by 25% and 37% of the control concrete's value for 30% and 40% fly ash concrete respectively.

Table 8. Concrete performance classification (Adopted from Papworth and Grace, 1985)

Test method	Performance limits		
	Poor	Acceptable	Very good
Water permeability, m/s	$> 10^{-10}$	10^{-10} to 10^{-12}	$< 10^{-12}$
Sorptivity, $mm/min^{1/2}$ ($\times 10^{-4}$ $mm/s^{1/2}$)	> 0.2 (> 258.2)	0.1 to 0.2 (129.1 to 258.2)	< 0.1 (< 129.1)

In series B, inclusion of fly ash has decreased absorption slightly (Figure 6). Although w/b and total binder content in the mixtures were constant, the sorptivity decreased with the increase of fly ash content in the concrete. For concretes with 30% and 40% fly ash sorptivity at 28 days reduced by 6% and 21% of that of the control concrete (B00) respectively. The rate of sorptivity of fly ash concrete tends to be similar to that of control concrete over the age up to 180 days.

The values of sorptivity coefficients of fly ash concretes are less than 0.1 $mm/min^{1/2}$ (129.1×10^{-4} $mm/s^{1/2}$), which can be rated as a concrete of 'very good' performance (Table 8) according to Papworth and Grace (1985).

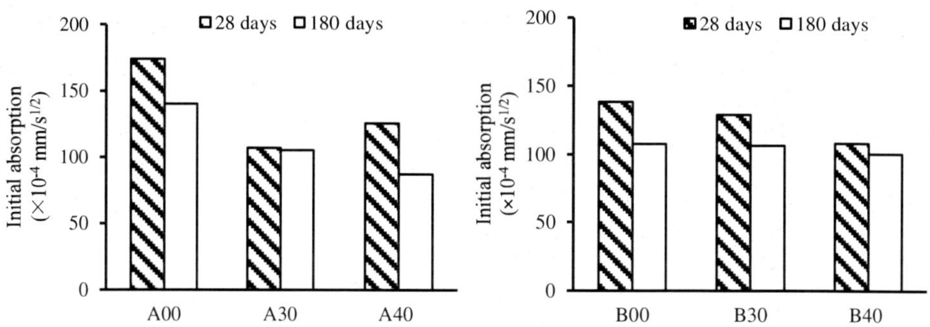

Figure 6. Initial sorptivity coefficients of concretes of series A (left) and series B (right).

Water permeability coefficients for different mixtures at the two ages are compared in the graphs of Figure 7. In series A, the 28-day water permeability values of the fly ash concrete samples (A30 and A40) were less than that of the control concrete (A00). The values of permeability coefficient of both the fly ash concretes were below 10^{-12} m/s which indicate 'very good' performance of concrete (Table 8) according to Papworth and Grace (1985). Permeability coefficients of the fly ash concretes decreased by about 65% of that of the control concrete (A00). Concrete with 40% fly ash (A40) had the similar permeability to that of 30% fly ash concrete (A30) at 28 days. But at 180 days, the permeability of mixture A40 was 15% less than that of mixture A30 and 68% less than that of mixture A00. The low water permeability of the fly ash concretes is mainly because of their lower w/b ratios as compared to that of control concrete mix. Thus, inclusion of fly ash up to 40 % of the total binder and reduced w/b ratio has resulted in high strength concrete of lower permeability.

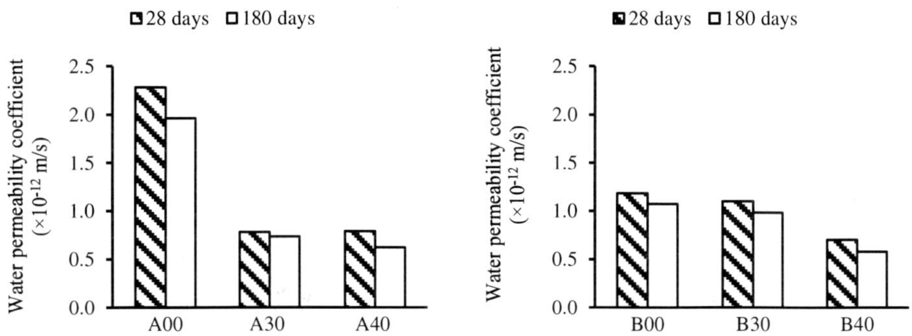

Figure 7. Water permeability coefficients of concretes of series A (left) and series B (right).

In series B, since the w/b ratio of the concretes of this series was constant, the effect of fly ash on the reduction of water permeability was less than that in series A. While mixture B40 (40% fly ash) achieved 'very good' performance limit according to Table 8, mixture B30 was in the margin of that limit at 28 days. At this age, water permeability coefficient decreased by 7% and 40% of control concrete's permeability for 30% and 40% fly ash contents respectively. At 180 days, the permeability decreased for every mix. The permeability coefficients reduced by 17%, 11% and 9% for B40, B30 and B00 respectively.

Sorptivity and water permeability test results at 28 days show similar trends for both series of concretes (Nath & Sarker, 2010). This implies the similitude of two different forms of water penetration in concrete; capillary suction and the penetration of water under pressure.

In general, the sorptivity and water permeability results generally indicated the positive effects of fly ash incorporation in concrete which varied with the mixture proportions. The reduction in sorptivity and water permeation was higher when inclusion of fly ash in the mix was combined with reduced w/b ratio and accordingly adjusted total binder content (series A). However, the presence of fly ash in series B with constant w/b and total binder content also restricted the flow of water. The sorptivity results are in agreement with that reported by Camoes, Aguiar and Jalali (2003). Fly ash contributed in reducing absorption by filling up the micro-pore through pozzolanic reaction. It also acts as a micro-aggregate (Poon et al. 2000). With this dual action, fly ash blocked the pore systems and reduced the volume of large pores in the concrete which eventually improved the resistance to permeation (Marsh et al. 1985).

7.4. Corrosion Resistance

7.4.1. Carbonation

The results after 28 days of exposure in carbonation chamber (2% CO_2) are shown in Figure 8. Mean carbonation depth was measured from the edge of split surface of the specimen. The readings from the troweled top surface and unexposed bottom surface edge were not included in the calculation.

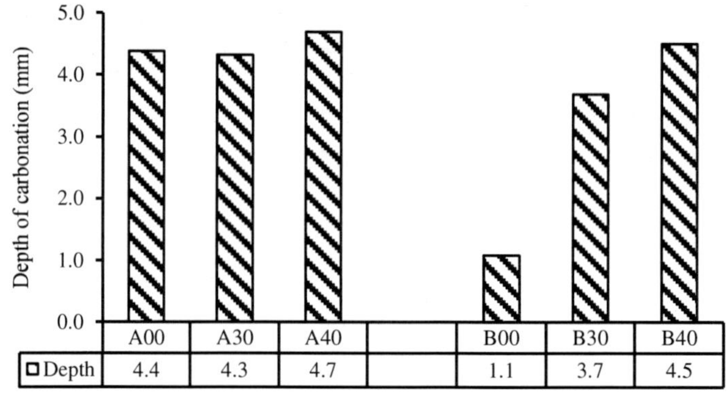

Figure 8. Depth of carbonation of concretes of series A and B.

In case of the concretes of series A, the depth of carbonation for three mixes resulted in very close average values. Fly ash concretes of series A, which were designed by varying w/b ratio and total binder content to achieve similar compressive strength at 28 days, also showed depth of carbonation similar to the control concrete. Thomas et al. (2000) also found similar carbonation trend with concretes of similar grade incorporating up to 30% fly ash.

Fly ash concretes of series B suffered higher carbonation as compared to control concrete (B00). The depth of carbonation increased with the increase of fly ash in the mix. The average carbonated depth of mix B30 and B40 were more than 3 and 4 times of that of mix B00 respectively for similar condition. Carbonation of mix B00 is less than that of mix A00 due to its (B00) significantly less w/b ratio and higher cement content.

Hence Fly ash incorporation in concrete influenced the depth of carbonation differently in the two series of concretes. Fly ash replacement without adjusting other mix variables (w/b and total binder content) does not improve the carbonation resistance at 28 days of age. This is because of relatively lower degree of hydration of the fly ash concretes as compared to the control concrete at the test age.

7.4.2. Chloride Penetration

Chloride-induced corrosion of reinforcement is one of the major causes of deterioration leading to a reduced service life of reinforced concrete structures. In the presence of chloride ions, the breakdown of the passive oxide layer, even at high pH levels, at the concrete-steel interface is the most common cause that initiates pitting corrosion of steel reinforcement. Among the several methods of chloride resistance tests, the rapid chloride permeability test (RCPT) has been an acceptable method for its quickness, repeatability and precision (Castellote & Andrade, 2006). In this method the likelihood of chloride ion (Cl^-) penetration through concrete is measured in terms of the electrical indication within a six-hour test. Total charge passed through the concrete sample is measured and the chloride ion penetrability level is assessed by following the recommendation of the ASTM standards C1202-07 (2007).

The RCPT test was performed at 28 days and 180 days. The total charge passed through different mixtures of concrete at different ages is plotted in Figure 9.

The fly ash concretes achieved better resistance than the corresponding control concretes in both series A and B. Chloride penetrability decreased with the increase of fly ash in the mixtures. At 28 days of age, fly ash concretes achieved 'Low' (>1000-2000 coulomb) level of chloride penetration as compared to the 'Moderate' (>2000-4000 coulomb) level for corresponding control concretes. At 180 days, the level of chloride penetration has decreased to 'Very Low' (>100-1000 coulomb) for fly ash concretes through attaining lower value of charge passed than that of corresponding control concretes.

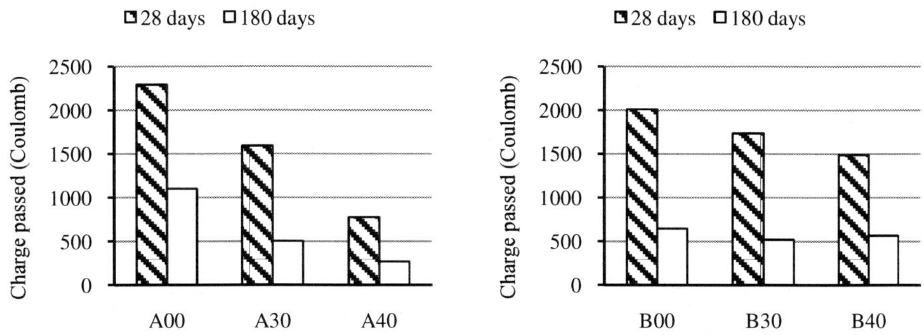

Figure 9. Chloride permeability levels of concretes of series A (left) and series B (right).

In series A, the fly ash concrete samples showed significant resistance to chloride ion penetration at early age of 28 days. Concrete with 30% and 40% fly ash achieved 35% and 45% less charge passed than that of the control concrete (A00) respectively. At 180 days of age, the resistance to chloride ion penetration improved in all the mixes, but fly ash concretes improved at a higher rate as compared to the control concrete. At this age, the total charge passed reduced by about 68% and 40% of the corresponding 28-day values in the fly ash and the control concrete specimens respectively.

In case of series B, replacing cement partially with fly ash using constant w/b ratio also improved the resistance to chloride ion penetration, but to a lesser extent as compared to series A. At 28 days, the charge passed was 10% and 24% less for 30% and 40% fly ash respectively than that of the control concrete (B00). At 180 days, while mixture B00 achieved 56% reduction of charge passed, mix B30 and B40 achieved over 64% reduction. Concrete with 40% fly ash showed slightly higher chloride ion penetrability than the 30% fly ash concrete at 180 days. However, concrete specimens of both 30% and 40% fly ash were in the range of 'very low' chloride ion penetration at 180 days.

When penetrated in the concrete, chloride ion starts to diffuse in the deeper zone. The diffusion process is assumed to follow Fick's 2nd law of diffusion. The rate of diffusion depends on many factors such as the mix proportions, curing period, maturity, w/b ratio, presence of supplementary cementitious materials etc. (Roy, Chye & Northwood, 1993; Page, Short & El-Tarras, 1981; Thomas & Matthews, 2004). The chloride diffusion test was carried out on three specimens from each mix at 28 days and 180 days of age. The minimum exposure period as specified in NT Build 443 (1995) is 35 days. However, the exposure period of the specimens were extended considering the presence of fly ash and lower w/b ratio, the curing period and age of the sample. The specimens of 28 days age were exposed to chloride solution for a period of 56 days and the specimens of 180 days age were exposed for about 5 months (147 days).

Table 9. Chloride diffusion test results (Nath & Sarker, 2011b)

Mixture		28 days			180 days		
Series	Mix ID	Diffusion coefficient, D_a ($\times 10^{-12}$ m^2/s)	Surface chloride, C_s (mass % of concrete)	Penetration parameter, k_{cr} (mm/yr$^{1/2}$)	Diffusion coefficient, D_a ($\times 10^{-12}$ m^2/s)	Surface chloride, C_s (mass % of concrete)	Penetration parameter, k_{cr} (mm/yr$^{1/2}$)
A	A00	4.68	0.57	29.7	3.41	0.62	25.3
	A30	3.51	0.37	26.6	0.70	1.38	11.3
	A40	3.42	0.40	26.0	0.90	1.05	12.9
B	B00	1.97	0.55	19.5	1.40	0.69	16.3
	B30	2.52	0.42	22.3	0.59	1.37	10.4
	B40	2.37	0.47	21.3	0.65	1.32	10.9

The diffusion coefficient (D_a), surface chloride concentration (C_s) and chloride penetration parameters are (k_{cr}) presented in Table 9. Figures 10 and 11 show the variation of the diffusion coefficients and surface chloride concentrations of all samples respectively.

It can be seen from Figure 10 that, fly ash concretes of series A (A30 and A40) have shown better resistance to chloride diffusion as compared to the control concrete (A00). The diffusion of the fly ash concretes reduced by about 25% of the control concrete's diffusion at 28-day test. The diffusion coefficient did not vary significantly with the increase of fly ash from 30% to 40%. Nevertheless, mixture A40 has shown slightly lower diffusion coefficient than mixture A30 at 28 day.

In series B, the concretes incorporating fly ash (B30 and B40) B have shown higher chloride diffusion than the control concrete (B00) at 28 days. It is usually observed that chloride diffusion is depended on the binder type and w/b ratio in the concrete (Song, Lee & Ann, 2008). Since the concretes of series B had same w/b ratio in each mix, the fly ash concretes had relatively less hydration as compared to the control concrete at 28 days. However at the 180 days test, the chloride diffusion of the fly ash concrete samples decreased by more than 50% of that of the control concrete samples. It indicates continued hydration of fly ash concretes beyond 28 days, which resulted in enhanced resistance to chloride penetration.

The surface chloride concentration (C_s) over the period of exposure is important in explaining chloride diffusion. From Figure 11, it can be seen that, fly ash concretes have shown less amount of surface chloride at 28 days test (with 56 days of exposure) in comparison to control concrete. While control concrete have shown very small increase in surface chloride concentration at 180 days (with 147 days exposure), fly ash concrete have shown significant rise. This could be due to increased resistance to chloride penetration in fly ash concrete through pore refinement over the age. The result supports the observation of Thomas and Matthews (2004). However it can be varied for exposure other than submersed condition (Costa & Appleton, 1999).

The penetration parameter (k_{cr}) results also indicated the higher resistance to chloride penetration due to fly ash inclusion and followed similar trend to the diffusion coefficient. The resistance to chloride diffusion of fly ash concrete increased at a faster rate over the age as compared to the control concrete.

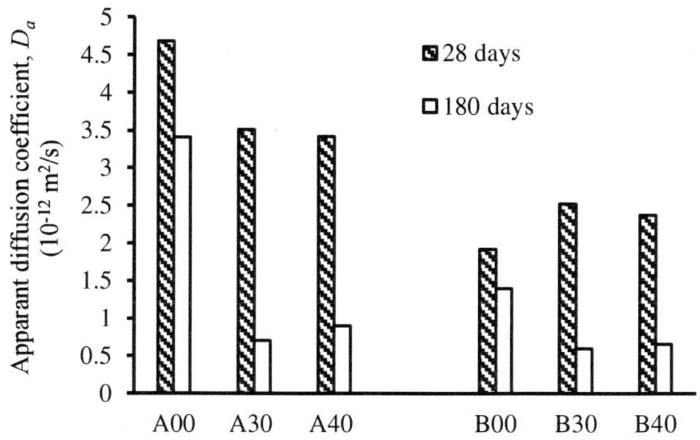

Figure 10. Variation of chloride diffusion coefficient of concretes of series A and B.

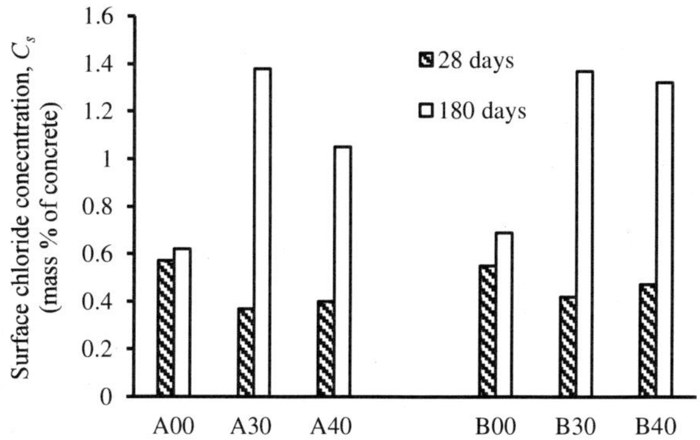

Figure 11. Variation of surface chloride concentration of concretes of series A and B.

The results of chloride diffusion test indicate that, inclusion of fly ash greatly influence the resistance of chloride penetration from the early age of concrete. The resistance to chloride diffusion is more significant when reduced w/b ratio and higher total binder content as compared to control concrete are used in the mixture (series A) to incorporate additional fly ash. On the other hand, RCPT test results indicate enhanced resistance of fly ash concrete of both series A and series B. Similar trends of the effect of fly ash in resisting chloride ion penetration are reported in literature (Malhotra, 1990; Sengul, Tasdemir, C. & Tasdemir, M.A, 2005). The chloride ion penetration depends on the chloride binding capacity of the constituent materials of concrete. Fly ash increases the chloride binding capacity of concrete. While entering into the pores, some chloride ion reacts with the cement compound C_3A (tricalcium aluminate) and forms stable chloro-complexes (Dinakar, Babu & Santhanam, 2008). In normal cement concrete, chloride ions can move deeper in the paste which is facilitated by interconnected capillary pores. Fly ash, due to its pozzolanic reaction, gradually diminishes the interconnectivity of pores. Hence fly ash concretes achieve greater resistance to chloride ion penetration.

CONCLUSION

Fly ashes from different sources yield different chemical composition which influences the properties of concrete. Hence it is necessary to have a clear understanding of the fly ash before using it in concrete mixture. Most of the fly ashes of Australia can be classified as Class F as per ASTM standard. Some of the fly ashes have high content of iron oxide and unburnt carbon. According to the report published by Ash Development Association of Australia (ADAA), fly ashes generated in Australia contain elements of heavy metal below the limit approved by state regulatory agencies. Radioactivity concentrations of these fly ashes are less than allowable limits and similar to other commonly used building materials. Hence Australian fly ashes can be used in the Civil Engineering applications without causing hazards to health and environment.

This chapter presented a study on the use of a class F fly ash as substantial part of the binder for making high strength concrete. Two series of high-strength concretes designed with 30% and 40% cement replacement by a class F fly ash. Mixtures were designed with (series A) and without (series B) adjustments in the w/b ratio and the total binder content, to account for the inclusion of fly ash. The mechanical and durability properties of the concretes were determined up to 180 days of age. The following conclusions are drawn from the test results:

The 28-day compressive strengths of fly ash concretes were similar to that of the control concrete when designed with adjusted w/b ratio and total binder content (series A). The 28-day strength dropped when cement was partially replaced with fly ash without adjustment in the w/b ratio and total binder content (series B). However, fly ash concretes achieved same strength as control concrete at 335 days. Significant strength development of the fly ash concretes occurred during the period between 28 days and 56 days of age. Indirect tensile strength and flexural strength results indicated similar trend as that of compressive strength.

Fly ash concrete achieved reduced drying shrinkage than the control concrete of similar grade when the w/b ratio and the binder content were adjusted to take into account the inclusion of fly ash. Inclusion of fly ash without necessary adjustment in the mixture proportions may not result in reduced drying shrinkage.

Fly ash concretes achieved less sorptivity and water permeability than the control concretes, however significant reduction was observed when adjustments were made in the mixtures. Incorporation of fly ash improved the resistance to water permeation of concrete from an early age and permeability further decreased at six months.

Fly ash concrete designed with adjusted w/b ratio and total binder content achieved similar 28-day compressive strength which eventually resulted in similar carbonation as that of the control concrete. Higher carbonation was observed when no adjustment was made in the mixture proportion to account for fly ash inclusion.

The resistance of fly ash concretes to chloride penetration was better when w/b ratios and binder contents were modified to account for the fly ash. Concretes having fly ash without adjusted mixture proportion caused higher chloride diffusion at 28 days due to inadequate hydration.

In general, substantial inclusion of class F fly ash can increase strength, reduce drying shrinkage and enhance the resistance to water and chloride ion penetration of high strength concrete. This can be achieved by proper proportioning of the mixture with modified w/b ratio and total binder content to obtain superior results from the early age of 28 days. When replacing cement with fly ash, adjustments in the mixture proportions to achieve similar strength of the control Portland cement concrete eventually yields better durability properties.

REFERENCES

ACI Committee 211. (2008). *Guide for selecting proportions for high-strength concrete using Portland cement and other cementitious materials* (ACI 211.4R-08). American Concrete Institute, MI, USA.

AS 1012.10. (2000). *Methods of testing concrete – Method 10: Determination of indirect tensile strength of concrete cylinders ('Brazil' or splitting test)* Standards Australia. Retrieved from http://www.saiglobal.com

AS 1012.11. (2000). *Methods of testing concrete – Method 11: Determination of the modulus of rupture.* Standards Australia. Retrieved from http://www.saiglobal.com

AS 1012.13. (1992). *Methods of testing concrete – Method 13: Determination of the drying shrinkage of concrete samples prepared in the field or in the laboratory.* Standards Australia. Retrieved from http://www.saiglobal.com

AS 1379. (2007). *Specification and supply of concrete.* Standards Australia. Retrieved from http://www.saiglobal.com

AS 2758.1. (1998). *Aggregates and rock for engineering purposes Part 1: Concrete aggregates.* Standards Australia. Retrieved from http://www.saiglobal.com

AS 3600-2009. (2009). *Concrete Structures.* Standards Australia. Retrieved from http://www.saiglobal.com

AS 3972. (1997). *Portland and blended cements.* Standards Australia. Retrieved from http://www.saiglobal.com

Ash Development Association of Australia (2009a). *Assessment of Naturally Occurring Radionuclides in Australian Coal Combustion Products (CCPs). TECHNICAL MEMORANDUM: AM_TN_2009_03_06.* Retrieved from www.adaa.asn.au

Ash Development Association of Australia. (2009b). *Coal Combustion Product (CCP) Environmental Monitoring Program 2008/9.* February. Retrieved from www.adaa.asn.au

Ash Development Association of Australia. (2009c). *Australian experience with fly ash in concrete: Application and opportunities.* FLY ASH TECHNICAL NOTE No. 8. Retrieved from www.adaa.asn.au

Ash Development Association of Australia (2012). *Annual Membership Survey Results, January - December 2011.* Retrieved from www.adaa.asn.au

ASTM Standard C618 – 08a. (2008). *Standard Specification for Coal Fly Ash and Raw or Calcined Natural Pozzolan for Use in Concrete.* ASTM International, West Conshohocken, PA, 2008. Retrieved from http://specs4.ihserc.com

ASTM Standards C1202-07. (2007). *Standard test method for electrical indication of concrete's ability to resist chloride ion penetration.* ASTM International, West Conshohocken, PA, 2007. Retrieved from http://specs4.ihserc.com

ASTM Standards C1585-04. (2004). *Standard test method for measurement of rate of absorption of water by hydraulic cement concretes.* ASTM International, West Conshohocken, PA, 2008. Retrieved from http://specs4.ihserc.com

Atis, C. D. (2003a). Accelerated carbonation and testing of concrete made with fly ash. *Construction and Building Materials*, 17(3), 147–152.

Atis, C. D. (2003b). High-volume fly ash concrete with high strength and low drying shrinkage. *Journal of Materials in Civil Engineering*, 15(2), 153-156.

Australian Coal Association (2013). *Energy*. Retrieved April 2, 2013 from http://www.australiancoal.com.au/energy.html

Bentz, D. P., & Ferraris, C. F. (2010). Rheology and setting of high volume fly ash mixtures. *Cement and Concrete Composites*, 32(4), 265–270.

Berndt, M. L. (2009). Properties of sustainable concrete containing fly ash, slag and recycled concrete aggregate. *Construction and Building Materials*, 23(7), 2606–2613.

Bucea, L., Cao, H. T., & Sirivivatnanon, V. (1996). Durability Properties of High Volume Fly Ash Concretes. *National Symposium on the Use of Recycled Materials in Engineering Construction: Programme & Proceedings*. Barton, ACT, Institution of Engineers, Australia, 141-146.

Burden, D. (2006). *The Durability of Concrete Containing High Levels of Fly Ash*. Masters Thesis. University of New Brunswick, Canada.

Cabrera, J. G., Woolley, G. R., A Study of 25 Year Old Pulverized Fuel Ash Concrete Used in Foundation Structures, in *Proceedings of Instrumentation Civil Engineering*, Part 2; 149–165, March, 1985.

Camoes, A., Aguiar, B. & Jalali, S. (2003). Durability of low cost high performance fly ash concrete. *Proceedings of International Ash Utilization Symposium*, Centre for Applied Energy Research, University of Kentuky.

Carette, G., Bilodeau, A., Chevrier, R. L., and Malhotra, V. M. (1993). Mechanical properties of concrete incorporating high volumes of fly ash from sources in the U.S. *ACI Materials Journal,* 90(6), 535-544.

Castellote, M., & Andrade, C. (2006). Round-robin test on methods for determining chloride transport parameters in concrete. *Materials and Structures*, 39(10), 955-990.

Chindaprasirt, P., Homwuttiwong, S., and Sirivivatnanon, V. (2004). Influence of fly ash fineness on strength, drying shrinkage and sulfate resistance of blended cement mortar. *Cement and Concrete Research*, 34(7), 1087–1092.

Costa, A., & Appleton, J. (1999). Chloride penetration into concrete in marine environment – Part II: Prediction of long term chloride penetration. *Materials and Structures,* 32(5), 354-359.

Dinakar, P., Babu, K. G., & Santhanam, M. (2008). Durability properties of high volume fly ash self-compacting concretes. *Cement and Concrete Composites*, 30(10), 880-886.

Duxson, P., Provis, J. L., Lukey, G. C., & Van Deventer, J. S. J. (2007). The role of inorganic polymer technology in the development of green concrete. *Cement and concrete research*, 37(12), 1590-1597.

Energy White Paper 2012 Australia's energy transformation. (2012). Department of Resources, Energy and Tourism, Australian Government, Canberra. Retrieved from http://www.ret.gov.au/energy/Documents/ewp/2012/Energy_ White_Paper_2012.pdf

Escalante-Garcia, J. I., & Sharp, J. H. (2004). The chemical composition and microstructure of hydration products in blended cements. *Cement and Concrete Composites*, 26(8), 967-976.

fib CEB-FIP. (2006). *Model code for service life design (fib Bulletin 34)*. International Federation for structural concrete (fib), Switzerland.

Fly ash. (2013). Wikipedia, the free encyclopedia. Retrieved 1 April, 2013 from http://en.wikipedia.org/wiki/Fly_ash#cite_ref-Moore.2C_Pantheon_14-0

French, D. & Smitham, J. (2007). *Fly ash characteristics and feed coal properties.* Research report 73, CSIRO Energy Technology, Cooperative research centre for coal in sustainable development, Pullenvale, QLD, Australia. Retrieved from http://www.ccsd.biz /publications/files/RR/RR 73 Fly Ash Char_web.pdf

Ganesh Babu, K., & Rao, G. S. N. (1993). Efficiency of fly ash in concrete. *Cement and Concrete Composites,* 15(4), 223 – 229.

GHD/TEL Method. (n.d.). *Determination of water permeability of concrete*. Obtained by personal communication, 25 August 2009.

Gopalan, M. K. (1993). Nucleation and pozzolanic factors in strength development of class F fly ash concrete. *ACI Materials Journal*, 90 (2), 117–120.

Heidrich, C. (2003). Ash utilisation – an Australian perspective. *International Ash Utilisation Symposium.* Centre for Applied Energy Research, University of Kentucky. Retrieved from http://www.flyash.info

IAEA Safety Guide No. RS-G-1.7. (2004). *Table 1, Application of the Concepts of Exclusion, Exemption and Clearance.*

Jerath, S. & Hanson, N. (2007). Effect of Fly Ash Content and Aggregate Gradation on the Durability of Concrete Pavements, *Journal of Materials in Civil Engineering,* 19 (5), 367–375. DOI: 10.1061/(ASCE)08991561(2007)19:5(367)

Jiang, L., Lin, B., & Cai, Y. (2000). A model for predicting carbonation of high-volume fly ash concrete. *Cement and Concrete Research*, 30(5), 699-702.

Jiang, L. H., & Malhotra, V. M. (2000). Reduction of water demand of non-air-entrained concrete incorporating large volumes of fly ash. *Cement and Concrete Research*, 30 (11), 1785–1789.

Khan, M. I., & Lynsdale, C. J. (2002). Strength, permeability, and carbonation of high-performance concrete. *Cement and Concrete Research*, 32(1), 123–131.

Killingley, J., McEvoy, S., Dokumcu, C., Stauber, J., & Dale, L. (2000). *Trace element leaching from fly ash from Australian power stations.* End of Grant Report, Australian Coal Association Research Program Project C8051, CSIRO Division of Energy Technology, 98.

Kouloumbi, N. & Batis, G. (1992). Chloride Corrosion of Steel Rebars in Mortars with Fly Ash Admixtures. *Cement & Concrete Composites*, 14(3), 199-207.

Kumar, B., Tike, G.K., & Nanda, P.K. (2007). Evaluation of properties of high-volume fly-ash concrete for pavements. *Journal of Materials in Civil Engineering*, 19(10), 906-911.

Lam, L., Wong, Y. L., & Poon, C. S. (2000). Degree of hydration and gel/space ratio of high-volume fly ash/cement systems. *Cement and Concrete Research*, 30 (5), 747– 756.

Langley, W. S., Carette, G. G., Malhotra, V. M., 1989. Structural concrete incorporating high volumes of ASTM class F fly ash, *ACI Materials Journal*, 86(5), 507- 514.

Malhotra, V. M., 1990. Durability of concrete incorporating high-volume of low-calcium (ASTM class F) fly ash, *Cement and Concrete Composites*, 12(4), 271-277.

Malhotra, V. M., (2002). High performance high volume fly ash concrete, an environmentally friendly solution to the infrastructure needs of developing countries. *Concrete International*, 24 (7), July.

Marsh, B.K., Day, R.L. & Bonner, D.G. (1985). Pore structure characteristics affecting the permeability of cement paste containing fly ash. *Cement and Concrete Research*, 15(6), 1027-1038.

McCaffrey, R. (2002). Climate change and cement industry. *Global cement and Lime Magazine (Environmental special issue)*, 15-19.

Mehta, P. K. (1985). Influence of fly ash characteristics on the strength of Portland-fly ash mixtures. *Cement and Concrete Research*, 15(4), 669-674.

Mehta, P. K. (1999). Concrete technology for sustainable development. *Concrete International,* 21(11), November, 47-53.

Montgomery, D. G., Hughes, D. C., & Williams, R. I. T. (1981). Fly ash in concrete - a microstructure study. *Cement and Concrete Research*, 11(4), 591-603.

Munt, V. C. & Munroe, D. C. (1950). Use of East Perth Power Station Fly Ash as a Pozzolan, *The Journal, Institution of Engineers*, Australia - No.1035, September.

Naik, T. R., Ramme, B. W., Kraus, R. N., & Siddique, R. (2003). Long-term performance of high-volume fly ash pavements. *ACI Materials Journal*, 100(2), 150–155.

Naik, T. R., Singh, S. S., & Hossain, M. M. (1994). Permeability of concrete containing large amounts of fly ash. *Cement and Concrete research*, 24(5), 913-922.

Nath, P., & Sarker, P. (2010). Resistance to permeation of high strength concrete containing fly ash. *Proceedings of 2nd International Symposium on Life-cycle Civil Engineering*, Taiwan Tech, Taipei, Taiwan, 27-31 October.

Nath, P. & Sarker, P. (2011a). Effect of Fly Ash on the Durability Properties of High Strength Concrete. *Procedia Engineering*, 14, 1149–1156.

Nath, P and Sarker, P. (2011b). Improvement of Durability and Service Life of Concrete Using Class F Fly Ash. *Proceedings of the Concrete 2011 Conference*, Perth, WA, October 12, 2011, The Concrete Institute of Australia.

Neville, A. M. & Brooks, J. J. (1987). *Concrete technology*. Longman Scientific and Technical, England.

NT Build 443, 1995-11. (1995). *Concrete, Hardened: Accelerated Chloride penetration*. Nordtest Method. Retrieved from www.nordtest.org

Onera, A., Akyuzb, T. S., & Yildiza, R. (2005). An experimental study on strength development of concrete containing fly ash and optimum usage of fly ash in concrete. *Cement and Concrete Research*, 35(6), 1165–1171.

Page, C. L., Short, N. R., El-Tarras A., (1981). Diffusion of chloride ions in hardened cement paste. *Cement and Concrete Research*, 11(3), 395–406.

Papadakis, V.G. (1999). Effect of fly ash on Portland cement systems Part I. Low-calcium fly ash. *Cement and Concrete Research,* 29(11), 1727-1736.

Papadakis, V. G. (2000a). Effect of fly ash on Portland cement systems: Part II. High-calcium fly ash. *Cement and Concrete Research*, 30(10), 1647– 1654.

Papadakis, V. G. (2000b). Effect of supplementary cementing materials on concrete resistance against carbonation and chloride ingress. *Cement and Concrete Research,* 30(2), 291–299.

Papadakis, V.G., and Tsimas, S. (2002). Supplementary cementing materials in concrete: Part I. Efficiency and design. *Cement and Concrete Research*, 32(10), 1525–1532.

Papworth, F., and Grace, W. (1985). Designing for concrete durability in marine environs. *Proc., Concrete 85 Conference*, Brisbane; Concrete Institute Australia.

Poon, C. S., Lam, L., & Wong, Y. L. 2000. A study on high strength concrete prepared with large volumes low calcium fly ash. *Cement and Concrete Research*, 30(3), 447-455.

Puyvelde, D. V. (2013). *An update on CO_2 capture from cement production*. Retrieved April 1, 2013 from http://www.globalccsinstitute.com/insights/authors/dennisvanpuyvelde/2013/02/20/update-co2-capture-cement-production

Ramezanianpour, A.A., and Malhotra, V.M. (1995). Effect of curing on the compressive strength, resistance to chloride-ion penetration and porosity of concretes incorporating slag, fly ash or silica fume. *Cement and Concrete Composites*, 17(2), 125-133.

Roper, H., Baweja, D. & Kirkby, G. A. (1984). Fly Ash in Australian Concrete. *Second international Conference on Ash Technology and Marketing*, Central Electricity Generating Board, London, UK, September.

Roy, S. K., Chye, L. K., & Northwood, D. O. (1993). Chloride ingress in concrete as measured by field exposure test in the atmospheric, tidal and submerged zones of a tropical marine environment. *Cement and Concrete Research*, 23(6), 1289-1306.

Sengul, O., Tasdemir, C. & Tasdemir, M.A. (2005). Mechanical properties and rapid chloride permeabilites of concretes with ground fly ash. *ACI Materials Journal,* 102(6), 414–421.

Shehata, M.H., Thomas, M.D.A., and Bleszynski, R.F. (1999). The effects of fly ash composition on the chemistry of pore solution in hydrated cement pastes. *Cement and Concrete Research*, 29(12), 1915–1920.

Siddique, R. (2004). Performance characteristics of high-volume Class F fly ash concrete. *Cement and Concrete Research*, 34(3), 487-493.

Sisomphon, K., & Franke, L. (2007). Carbonation rates of concretes containing high volume of pozzolanic materials. *Cement and Concrete Research*, 37(12), 1647–1653.

Sivasundaram, V., Carette, G. G., & Malhotra, V. M. (1990). Long-term strength development of high volume fly ash concrete. *Cement and Concrete Composites*, 12(4), 263-270.

Srinivasan, P., Tiwar, A. K., & Banchhor, A. (2004). Stability of HVFA concrete for pavements. *Indian Concrete Journal*, 78(11), 58–61.

Song, H. W., Lee, C. H., & Ann, K. Y. (2008). Factors influencing chloride transport in concrete structures exposed to marine environments. *Cement and Concrete Composites*, 30(2), 113–121.

Tasdemir, C (2003). Combined effects of mineral admixtures and curing conditions on the sorptivity coefficient of concrete. *Cement and Concrete Research*, 33(10), pp. 1637-1642.

Technical Bulletin 2. (2005). *Chemical comparison of fly ash and Portland cement*. Retrieved August 11, 2010 from http://www.flyash.com/data/upimages/press/TB.2 Chemical Comparison of Fly Ash and Portland Cement.pdf

Thomas, M. D. A., Matthews, J. D. (2004). Performance of pfa concrete in a marine environment–10-year results. *Cement and Concrete Composites*, 26(1), 5-20.

Thomas, M. D. A, Matthews, J. D., & Haynes, C. A. (2000). Carbonation of Fly Ash Concrete, *Fifth CANMET/ACI International Conference on Durability of Concrete*, Barcelona, Spain, p. 539-556. Retrieved from http://tris.trb.org/view.aspx?id=657380

Toutanjia, H, Delattec, N., Aggounb, S., Duvalb, R., & Dansona, A. (2004). Effect of supplementary cementitious materials on the compressive strength and durability of short-term cured concrete. *Cement and Concrete Research*, 34(2), 311–319.

Ward, C. R., & French, D. (2006). Determination of glass content and estimation of glass composition in fly ash using quantitative X-ray diffractometry. *Fuel*, 85(16), 2268–2277.

In: Fly Ash
Editor: Prabir Kumar Sarker

ISBN: 978-1-62948-044-2

Chapter 9

USE OF FLY ASH IN COMBINATION WITH BAUXITE RESIDUE IN SUSTAINABLE CONCRETE MANUFACTURE

***Ade Sri Wahyuni*[1,*] *and Hamid Nikraz*[2]**

[1]Department of Civil Engineering, Bengkulu University, Indonesia
[2]Department of Civil Engineering, Curtin University, Western Australia

ABSTRACT

This chapter presents the application of fly ash as a partial replacement of cement together with the use of red sand (bauxite residue) as fine aggregate in concrete manufacture. The experimental results of 5 types of concrete e.g.: control mix, concrete incorporating red sand with 0%, 10% and 30% fly ash, and the second control mix with 10% fly ash are used to study the combined effect of fly ash and red sand on various properties of concrete.These concrete mixtures are compared in terms of mechanical properties and durability by applying the following generic tests: compressive strength, splitting tensile strength, flexural strength, modulus of elasticity, water permeability, and chloride diffusion. The pH dependence leaching test was applied only to the concrete specimen that achieved the highest strength.The concrete strength is not shown to be significantly affected when the fine aggregate of natural sand is replaced completely with red sand. The fly ash addition seems to effect in a positive way to improve the workability and durability of concrete. The leaching test results did not exceed the guidelines given by NSW EPA.

Keywords: Fly ash, red sand, concrete

[*] Corresponding Author address: Jurusan Teknik Sipil, Fakultas Teknik, Universitas Bengkulu, Jl W.R Supratman, Kandang Limun, Bengkulu, 38371.A , Indonesia. Email: ade_sri_w@yahoo.com.

1. INTRODUCTION

Today's trend in technology is the green technology towards sustainability. Research is being conducted globally to utilize more industrial by-products to minimize its impact on the environment and at the same time finding alternative building materials. This chapter looks at utilizing fly ash from the coal industry and the red sand as the by-product from alumina industry to manufacture concrete.

1.1. Fly Ash

Nowadays, fly ash as the by-product of coal-fired power stations is widely used in concrete due to its ability to enhance the properties of concrete. It has been shown that fly ash with its pozzolanic properties can increase the strength of concrete (ACI 2003; Neville 1995; Ryan and Samarin 1992).

In addition, there are some more advantages of using fly ash such as: reducing the temperature rise during initial hydration, improving the resistance of concrete to sulphate, reducing the expansion due to alkali-silica reaction and its contribution to the durability and strength of hardened concrete (ACI, 2003).

Fly ash contains siliceous glass which plays an important role in the pozzolanic reaction. The pozzolanic particles of fly ash will react continuously with the solutions $Ca(OH)_2$ in the pore structure of cement paste to form additional Calsium-Silicate-Hydrate (the binder in concrete) until there is no place for C-S-H to occupy (Philleo, 1991).

Research by Odler (1991) suggested that the maximum content of the fly ash in hardened cement paste is about 30% by mass of the cementitious material. Beyond this point, the hardened cement paste does not get benefit in terms of strength development.

1.2. Red Sand

Alcoa World Alumina Australia has refineries in Western Australia annually produces approximately 15 million tonnes of waste product from bauxite refining, known as red mud. Nguyen and Boger (1998) reported that the disposal of bauxite residue might account for 30 – 50% of the operations cost in an alumina industry. The use of red sand in concrete manufacture could alleviate the residue storage volume, and would provide an alternative source of construction material.

Many researchers have been working to reduce bauxite residue storage by investigating the possible use of the residue in some different areas in construction materials such as: ceramic tiles, bricks, filler for resin (Kurdowski and Sorrentino 1997), as raw materials for cement (Singh et al., 1996; Pan et al., 2002; Tsakiridis et al., 2004) and as corrosion inhibitor (Cabeza et al., 2003 ;Diaz et al., 2004). Alcoa has also been working to develop red sand by a process of separation of the bulk residue, and preliminary investigations have shown that it is similar to crushed rock; hence, there may be many potential uses.

2. MATERIALS USED

2.1. Fly Ash

The fly ash used in the work presented in this chapter was sourced from Western Australia. The chemical oxide composition of the fly ash is shown Table 1.

Table 1. Chemical Analysis of Fly Ash (Hardjito et al. 2004)

Compound	Percentage %
SiO_2	53.36
Al_2O_3	26.49
Fe_2O_3	10.86
CaO	1.34
Na_2O	0.37
K_2O	0.80
TiO_2	1.47
MgO	0.77
P_2O_5	1.43
SO_3	1.70
LOI*	1.39

* LOI = Loss on ignition.

2.2. Fine Aggregate

The fine aggregate was natural sand for the control mix and red sand for other mixes. The red sand was obtained from Alcoa refinery in Western Australia. The chemical composition of the red sand is provided in the Table 2:

Table 2. Chemical composition of coarse bauxite residue (Alcoa 2002)

Chemical compound	Percentage (%)
Silica amorphous	19-44
Ferric oxide	11-45
Aluminium oxide	11-26
Calcium oxide	2-8
Sodium monoxide	1-6
Titanium dioxide	1-5

The physical properties and sieve analysis results of both natural sand and red sand are given in Tables 3 and 4 respectively. The sieve analysis results of fine aggregate are plotted in Figure 1.

Table 3. Physical properties of fine aggregate

Property	Natural sand	Red sand
Specific gravity	2.64	2.81
Fineness modulus*	2.47	1.99

* Fineness modulus: index of the fineness of an aggregate- the higher the FM, the coarser the aggregate.

Table 4. Sieve analysis of fine aggregate

Sieve Aperture (mm)	Percent Passing Natural sand	Percent Passing Red sand	Grading Requirement
2.360	100.00	99.63	60-100
1.180	99.97	96.21	30-100
0.600	86.32	79.36	15-100
0.425	50.15	63.99	
0.300	16.88	46.45	5-50
0.150	1.10	15.45	0-20
0.075	0.07	1.17	0-5

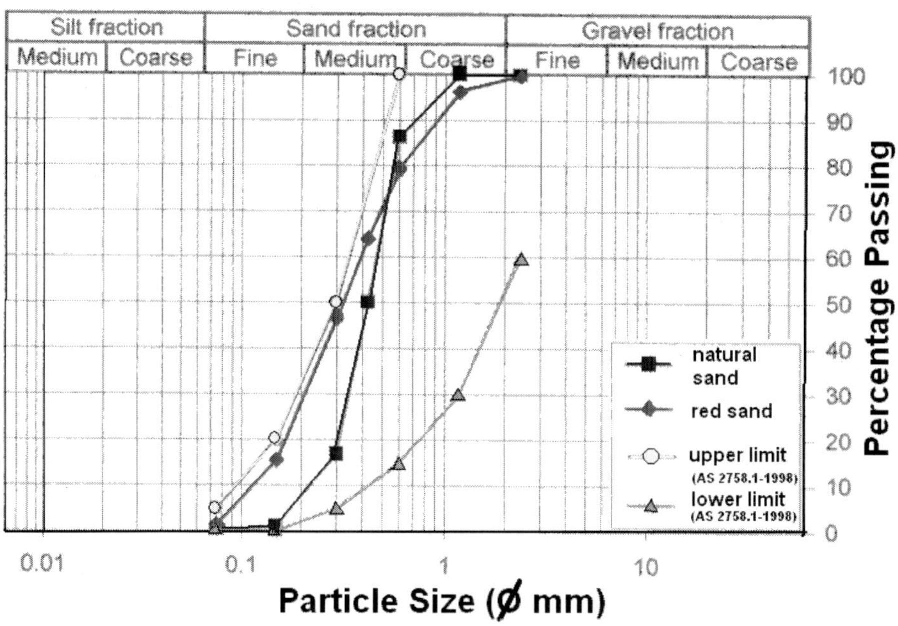

Figure 1. Particle size distribution of fine aggregate.

The particle size distribution indicates that the red sand is reasonably well graded and falls within the limits of the particle size distribution curves given in the Australian Standard AS 2758.1-1998.

2.3. Coarse Aggregate

Coarse aggregate used in this research was crushed granite with nominal size of 20 mm and 10 mm. The physical properties and sieve analysis results of the aggregate are given in Table 5 and 6 respectively.

Table 5. Physical properties of coarse aggregate

Property	Crushed granite
Specific gravity	2.6
SSD absorption	0.40%

Table 6. Sieve analysis of coarse aggregate

Sieve Aperture (mm)	Percent Passing 10 mm Agg	Percent Passing 20 mm Agg	Grading Requirement
19.00	100	91.78	85-100
9.50	39.67	6.31	25-55
4.75	2.09	1.26	0-10
2.36	0.91	1.15	0-5
1.18	0.75	1.13	
0.60	0.53	1.07	
0.30	0.29	0.85	
0.15	0.12	0.57	

The combination of 60% to 40% of aggregate 10 mm and 20 mm respectively meets the grading requirement as recommended by AS 2758.1-1998 as shown in Figure 2.

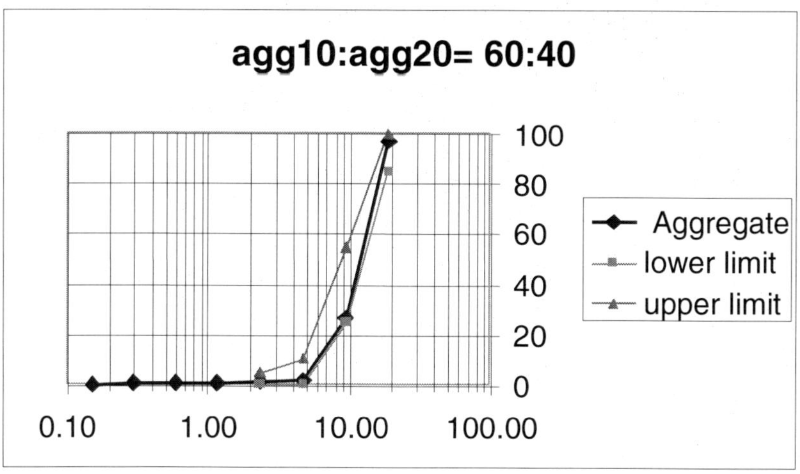

Figure 2. The combination of 10 mm and 20 mm aggregate.

3. Mix Design

Five mixture proportions were made. Mix #1 was the first control mix with natural sand and no fly ash. Mix # 5 was the second control mix to compare with mix #3. The mix proportions are given in Table 7.

Table 7. Concrete mix design

Mix No	Fine Agg Type	Water/ Binder	Water kg/m^3	OPC kg/m^3	Fly Ash %	Fine Agg kg/m^3	Coarse Agg kg/m^3
#1	Nat sand	0.47	169.2	360	0	712.3	1162.13
#2	Red sand	0.47	169.2	360	0	712.3	1162.13
#3	Red sand	0.47	169.2	324	10	712.3	1162.13
#4	Red sand	0.47	169.2	252	30	712.3	1162.13
#5	Nat sand	0.47	169.2	324	10	712.3	1162.13

4. Result and Discussion

4.1. Compressive Strength

The test for determining the compressive strength of concrete was performed in accordance with AS 1012.9-1999. The cylinder specimens were cast in 100 mm in diameter × 200 mm in height moulds, demoulded 24 hours after casting and cured in the curing tank until the day of testing. The concrete compressive strength test was carried out at day 3, day 7, day 14, day 28, day 56, and day 90 of concrete age. Each test consisted of 5 cylinders except for the tests on day 56 and day 90 where 3 cylinders were used. The results were obtained from the average strength of specimens on each testing day. The results on the strength development of five types of concretes are listed in Table 8 and plotted in Figure 3.

Table 8. Concrete strength development with curing age

Mix	Compressive strength (MPa)					
	3 day	7 day	14 day	28 day	56 day	90 day
# 1(control mix)	37±3	45.5±1.5	49.5±1.5	56±2	59.5±1.5	62±2
#2 (RS +0 %FA)	37±1	43.5±1.5	49±2	53.5±1.5	55.5±1.5	58±2
#3 (RS + 10%FA)	37±1	48.5±1.5	53±2	60±2	66.5±2.5	66.5±2.5
#4 (RS + 30%FA)	21.5±1.5	29.5	35±2	40±1	45±3	50.7±0.6
#5(NS + 10%FA)	30.1±0.3	45±2	50±1	58.5±2.5	64±1	66±3

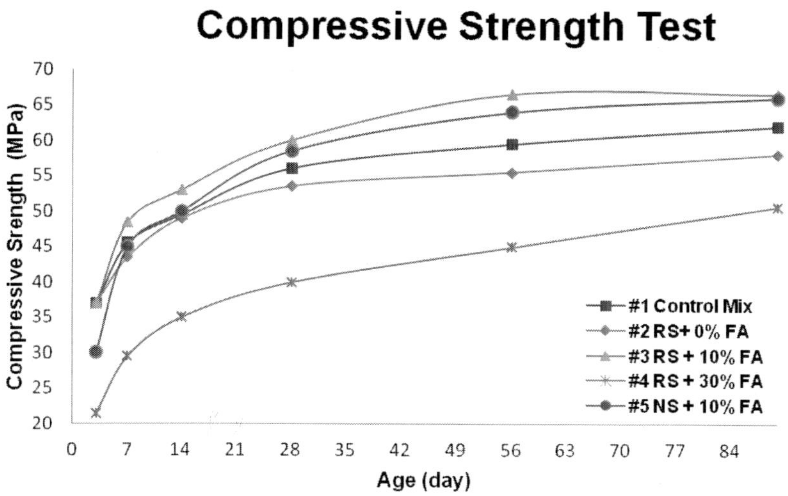

Figure 3. Development of concrete strength with age (Wahyuni 2005).

The strength of concrete using red sand as complete replacement of natural sand is similar to that of the control mix at age 3 to 28 days with a 6% decrease at age 56 and 90 days. The strength of concrete incorporating red sand and10 % fly ash at 3 days is 18% higher than that of concrete mix of natural sand and 10% fly ash. However at later ages the strength development of mix#3 and mix#5 are similar, which indicates that there is little difference between these two mixes.

The replacement of 30% cement by fly ash, shows a decrease of concrete strength. This finding is in line with the previous research by Odler (1991) about the correlation between fly ash content with the relative strength.

4.2. Indirect Tensile Strength

The AS 1012.10-2000 was used to determine the indirect tensile strength of concrete. The specimens for this test were cylinders measuring in 150 mm in diameter and 300 mm in height. The samples were demoulded 24 hours after casting and cured in the curing tank until the testing day. The splitting tensile test of concrete was conducted at day 7, day 14, day 28 of curing. The results on the splitting tensile strength of six types of concretes are listed in Table 9 and shown in Figure 4.

Table 9. Indirect tensile strength of concrete at different ages

Mix	Type of concrete	Splitting tensile strength (MPa)		
		7 day	14 day	28 day
#1	Control mix	3.4±0.1	3.85±0.15	4.55±0.15
#2	Red Sand + 0%Fly ash	4.0±0.2	4.35±0.15	4.45
#3	Red Sand + 10%Fly ash	3.5±0.1	3.6±0.1	3.75±0.35
#4	Red Sand + 30%Fly ash	2.45	2.7±0.1	3.05±0.15
#5	Natural Sand + 10%Fly ash	3.35±0.15	3.75±0.35	4.35

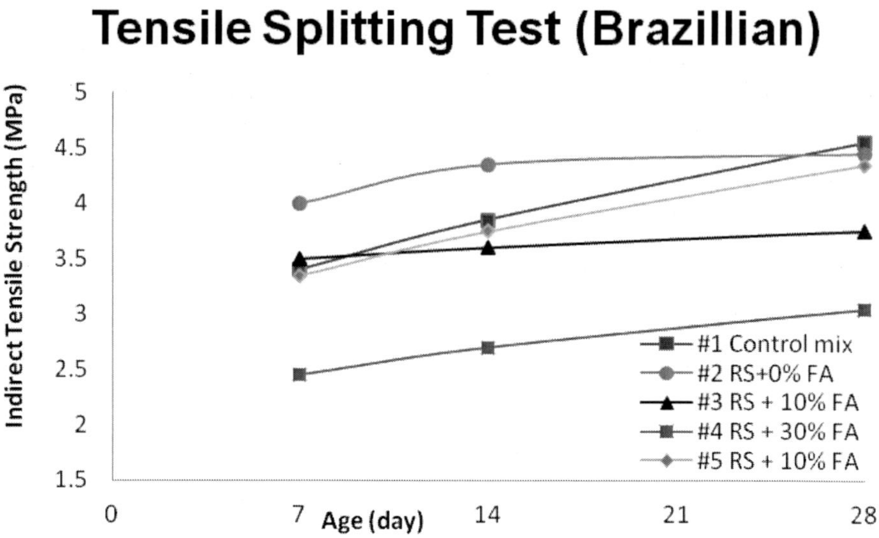

Figure 4. Development of tensile strength of concrete with age (Wahyuni 2005).

As expected, the tensile strength increased with the increase of compressive strength at a decreasing rate (Neville, 1995).It is clear from the results that the addition of fly ash as a cement replacement reduced the splitting tensile strength of concrete up to 28 days.

4.3. Flexural Strength

The test specimen had a dimension of 100mm × 100mm × 400mm. Three beams were cast from each mixture, demoulded after 24 hours and cured in the curing tank. The flexural strength test was carried out at day 28 in accordance with AS 1012.11-2000.

Table 10 shows the flexural strength of five types of concrete.

No significant variation is observed in the flexural strengths of the concrete mixtures, except for the red sand with 30% fly ash, which shows the lowest strength in comparison with others. This is because the compressive strength of this type of concrete is also the lowest.

Table 10. Flexural strength test result

No	Type of concrete	Flexural strength (MPa)
		Day 28
#1	Control mix	5.1±0.4
#2	Red Sand + 0%FA	5.55±0.15
#3	Red Sand + 10%FA	5.15
#4	Red Sand + 30%FA)	4.0±0.3
#5	Natural Sand + 10%FA	4.95±0.15

4.4. Modulus of Elasticity

The modulus of elasticity and Poisson's ratio of concrete were determined in accordance with AS 1012.17-1997. The specimens for this test were cylinders measuring in 150 mm in diameter and 300 mm in height. The samples were demoulded 24 hours after casting and cured in the curing tank until the testing day. Table 11 shows the modulus of elasticity values of the concrete mixtures.

Table 11. Modulus of elasticity and Poisson's ratio of concrete

Mix	Type of concrete	f_c' (MPa)	E (GPa)	μ	Age (days)
#1	Control mix	55.13	28.66	0.15	28
#2	Red sand + 0%FA	53.98	26.78	0.16	28
#3	Red sand + 10%FA	68.47	32.33	0.19	176
#4	Red sand + 30%FA	53.13	26.69	0.20	183
#5	Natural sand + 10%FA	67.25	32.16	0.18	93

The value of the Modulus of Elasticity depends on the concrete compressive strength and the type of aggregates. The results in table 11 are within the range of the diagram shown by Kliszczewics and Ajdukiewicz , in which the modulus of elasticity lies between 23GPa to 37GPa for the characteristic compressive strength between 25 MPa to 84 MPa. This type of concrete used crushed granite as coarse aggregate. In addition, Nassif et al (2005) reported that the addition of 10% of fly ash as cement replacement in concrete mixture increased the modulus of elasticity of concrete by 6-17%.

4.5. Water Permeability

The water permeability test of concrete was in accordance with the GHD water permeability method. (Taywood Engineering Limited,1993)

Two slices of concrete with 45-50 mm thick were cut from a concrete cylinder (100mm x 200mm). The curved surface of samples was coated with epoxy prior to being placed in the water permeability rig. Subsequently high water pressure (about 950 kPa) was applied from the bottom of the concrete. The water permeability test set-up can be seen in Figure 5.

The water permeability test takes seven days and every day the flow rate of water is measured by channeling the water along a calibrated pipette of 2 mL capacity.

The permeability coefficient of concrete is calculated by The Darcy's equation in the following formula:

$$k = \frac{QL}{AH} \quad (1)$$

where:

k = Permeability coefficient (m/sec).
Q = Flow rate (m^3/sec).
A = Cross sectional area (m^2).
L = Depth of specimen (m).
H = Head of water (m).

Figure 5. The water permeability test arrangement.

The water permeability test of concrete was conducted at day 56 of curing.

The results of water permeability of test specimen is summarized in Table 12 and Figure 6. These results are the average values of two specimens from the same cylinder.

Table 12. Water permeability results

Mix	Type of concrete	Permeability coef (Ks/m/sec)	slump
#1	Control Mix	13.7×10^{-12}	50 mm
#2	Red sand + 0%FA	30.0×10^{-12}	0 mm
#3	Red sand + 10%FA	19.9×10^{-12}	10 mm
#4	Red sand + 30%FA	14.6×10^{-12}	25 mm
#5	Natural sand + 10%FA	13.5×10^{-12}	55 mm

In general, concrete incorporating red sand exhibited higher water permeability values in comparison with the reference mix. The chemical testing procedures manual produced by GHD (Taywood Engineering Limited ,1993) suggested the coefficient of water permeability of concrete with water cement ratio between 0.46 – 0.53 will lie between $3 \times 10^{12} - 12 \times 10^{12}$ Ks/m/sec. Based on this manual the water permeability found the mixtures is considered high, even for the control mix.

Theoretically, higher slump will create high permeability concrete. This is because more porosity left since water evaporates after drying. Furthermore, the permeability of mixes containing additives such as fly ash depends on the formation of secondary hydration products, which fill in the pores left by evaporated water. However in this case, low slump value makes the concrete hard to handle and to be compacted even though by vibrating table, resulting in more pores for concrete and easy ingress for water.

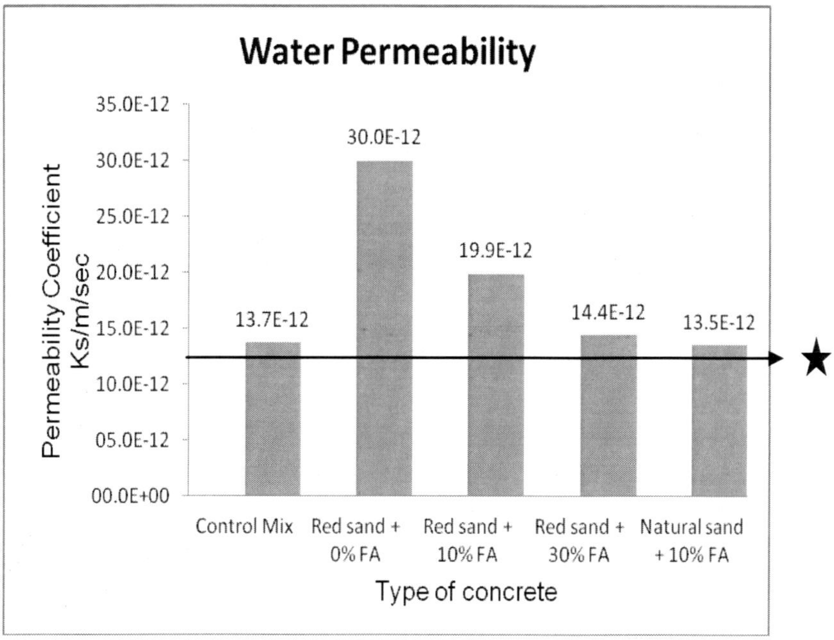

★= The Water Permeability value recommended by Taywood Engineering Limited (1993).

Figure 6. Water permeability results (Wahyuni 2005).

The concrete with low slump value is very hard to handle and not suitable for industry applications. The low slump value of concrete incorporating red sand was possibly due to the finer particle of red sand in comparison with the natural sand. The finer particle has more surface area than the coarser one. Consequently, even though the amount of water used in the mix design for both natural sand and red sand, are the same, concrete incorporating red sand will absorb more water. Thus, addition of superplastisizer is recommended when red sand is used.The fly ash addition seems to affect the mix in a positive way; to improve the workability of concrete and subsequently decrease permeability.

4.6. Chloride Diffusion

The Nordtest Method NT Build 443 (1995) was employed to estimate the resistance against chloride penetration into hardened concrete. The test involves exposing concrete in sodium chloride solution for 35 days. Following the exposure, about eight layers measuring 2 mm thick each were ground from the concrete. The collected concrete powder from grinding was analysed for acid-soluble chloride content by using a potentiometric auto titrator as shown in Figure 7.

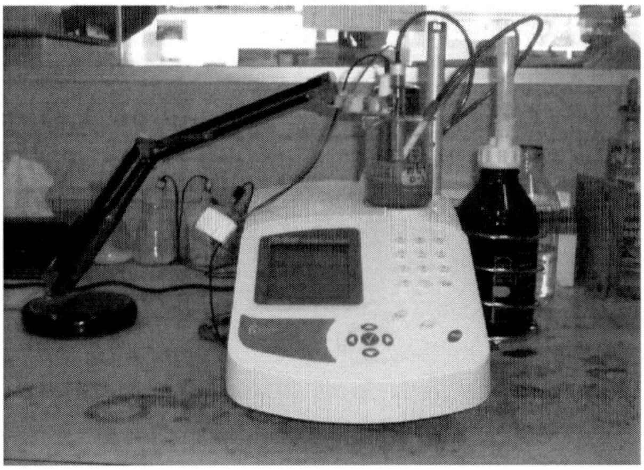

Figure 7. Auto burette titrator used for chloride analysis.

The diffusion coefficient describes a diffusion curve in the form of Fick's Second Law of Diffusion given in the Nordtest Method NT BUILD 443 (1995):

$$C_{(x,t)} = C_s - (C_s - C_i).\ Erf\left(\frac{x}{\sqrt{4.D_e.t}}\right) \tag{2}$$

where :

$C_{(x,t)}$[mass %] = chloride concentration measured at the depth x, at the exposure time t.
$C_{s)}$ [mass %] = boundary condition at the exposed surface.
C_i [mass %] = initial chloride concentration measured on the concrete slice.
D_e [m^2/s] = chloride diffusion coefficient.
x [mm] = depth below the exposed surface (to the middle of a layer).
t [s] = exposure time.
Erf = error function.

Table 13. Chloride diffusion coefficient of concrete specimen

No	Type of concrete	C_s (% mass)	C_o(% mass)	D (m^2/sec)
#1	Control mix	0.72	0.02	7.81×10^{-12}
#2	Red sand + 0% fly ash	0.84	0.02	8.31×10^{-12}
#3	Red sand + 10% fly ash	0.93	0.02	11.79×10^{-12}
#4	Red sand + 30% fly ash	0.93	0.03	15.71×10^{-12}

C_s = Surface chloride concentration.
C_o = Initial chloride concentration.
D = Chloride diffusion coefficient.

The chloride penetration test of concrete was conducted at day 56 of curing. The result of chloride penetration of 4 test specimens is tabulated in Table 13. These results are the average values of three specimens from the same cylinder.

The chemical testing procedures manual produced by GHD suggested the coefficient for chloride diffusion of concrete with water/ cement ratio between 0.46-0.53 will lie in the range of $2.8 - 5.8 \times 10^{-12}$.The chloride diffusion coefficient found in this research is considered high in accordance with GHD manual.

The process in which fluid shifts under a differential in concentration is called diffusion. The important component of concrete is diffusivity. The two gases concerned in diffusion are carbon dioxide and oxygen. The availability of CO_2 leads to carbonation of hydrated cement paste, while the presence of O_2 can progress the corrosion of embedded steel (Neville, 1995).

It is important to understand the flow rate of chloride ions into concrete for predicting the service life of concrete in marine environment (Truc et al. 2000).

Oh et al. (2002) indicated that durability of concrete is generally influenced by chloride permeability. According to their findings, adding fly ash and reducing the water to binder ratio will minimize the chloride permeability of concrete.

Furthermore, Bader (2003) determined other factors that may affect the chloride diffusion such as cement content, level of consolidation and period of curing. Insufficient curing of concrete may increase the permeability of the surface layer by 5 to 10 times.

4.7. Leaching Test

Leaching of some particular elements into the environment is an important environmental issue. Therefore leaching test of concrete incorporating a by-product such as red sand and fly ash is considered important.

It is common to use cement based materials for immobilizing waste containing hazardous compounds because the leaching possibility is often small in materials with a high silica content (EUR 17869 EN, 1997).

However, concrete is a porous material that cannot immobilize substances completely. There is always the possibility of leaching some particular elements to the environment during exposure to the surrounding climate or being crushed for reuse or recycling (Andersson, 2000).

There are two factors that influence leaching (Townsend et al. 2003):

1) Physical factors: particle size, contact time, homogeneity, liquid to solid ratio, porosity, sorption, partitioning, temperature and type of flow.
2) Chemical – Biological factors: pH, redox, complexation, precipitation, organic carbon content, alkalinity and common ion effect.

The pH-leaching test was performed in accordance with European Standard (PrEN 14429) - pH dependence leaching test. This is an appropriate test to determine the influence of pH on the leachability of inorganic constituent from a waste material. The test involves extractions of material at liquid to solid ratio (L/S) = 10 for 48 hours at pH value in the range of 4 – 12.

The leaching procedure consists of three defined stages:

1) Period A (acid/base addition) from t_0 up to t_{0+4} hours for acid/base addition in three steps.
2) Period B (equilibrium period) from t_0 + 4 hours up to t_0 + 44 hours.
3) Period C (verification period) from t_0 + 44 hours up to t_0 + 48 hours for verification of equilibrium condition.

Leaching test result can be expressed either as leachate concentration (mg/L) or as constituent release (mg/kg). To convert mg/L to mg/kg, the leachate concentration is multiplied by the L/S value (normally L/S = 10 ml/gr dry material).

The eluate of the specimen is analyzed by Inductively Coupled Plasma (ICP) Mass Spectrometry and ICP Optical Emission Spectrometry by contract laboratories.

The pH dependence leaching test was carried out at day 28 of the curing. The results of some heavy metals which leached out from two specimens are presented in Table 14, Table 15 and Figure 8.

Table 14. Leaching data of concrete incorporating natural sand + 10% fly ash

Sample	pH	Ag mg/L	As mg/L	Cd mg/L	Cr mg/L	Hg mg/L	Pb mg/L	Se mg/L
1	11.79	0.01	0.05	0.01	0.1	0.005	0.5	0.5
2	10.82	0.01	0.05	0.01	0.5	0.005	0.5	0.5
3	10.51	0.01	0.05	0.01	0.54	0.005	0.5	0.5
4	9.97	0.01	0.05	0.01	0.58	0.005	0.5	0.5
5	9.26	0.01	0.05	0.01	0.60	0.005	0.5	0.5
6	8.56	0.01	0.05	0.01	0.60	0.005	0.5	0.5
7	5.51	0.01	0.05	0.01	0.50	0.005	0.5	0.5
8	3.71	0.01	0.05	0.01	0.60	0.005	0.5	0.5
TCLP(NSW EPA) mg/L		0.5	0.5	0.1	0.5	0.02	0.5	0.1

Table 15. Leaching data of concrete incorporating red sand + 10% fly ash

Sample	pH	Ag mg/L	As mg/L	Cd mg/L	Cr mg/L	Hg mg/L	Pb mg/L	Se mg/L
1	12.35	0.01	0.1	0.05	0.65	0.01	0.5	0.5
2	10.89	0.01	0.1	0.05	0.6	0.01	0.5	0.5
3	9.79	0.01	0.1	0.05	0.63	0.01	0.5	0.5
4	9.13	0.01	0.1	0.05	0.7	0.01	0.5	0.5
5	8.50	0.01	0.1	0.05	0.7	0.01	0.5	0.5
6	7.18	0.01	0.1	0.05	0.7	0.01	0.5	0.5
7	5.30	0.01	0.1	0.05	0.5	0.01	0.5	0.5
8	4.34	0.01	0.1	0.05	0.4	0.01	0.5	0.5
TCLP(NSW EPA) mg/L		0.5	0.5	0.1	0.5	0.02	0.5	0.1

Note: Table A4 of NSW EPA (1999) suggested that "Aluminium, Barium, Boron, Chromium (0 and III oxidation states), Cobalt, Copper, Iron, Manganese, Vanadium and Zinc have deliberately not been listed in the table and need not be tested for".

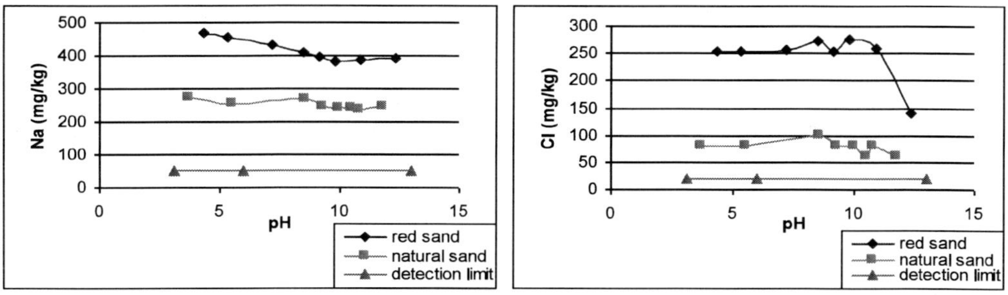

Figure 8. Leaching characteristics of Natrium and Chloride.

It is clear from the Table, that the leaching results do not exceed the guidelines given by NSW EPA. The only exception is for selenium. However the leaching concentration from the two specimens tested does not show any influence by the type of sand used.

Figure 8 shows that the amount of Sodium (Na) and Chloride (Cl) leaching out from red sand is more than that of natural sand. The amount of chloride recommended in reinforced concrete is about 100 mg/L (Rump and Krist, 1992). The amount of Chloride leaching out from red sand is about 280 mg/kg or 28 mg/L). Consequently, even though the amount of Chloride leaching out from red sand is about 60 % higher than that of natural sand, the washed red sand could be used in reinforced concrete.

Conclusion

The strength of concrete using red sand as complete replacement of natural sand is similar to that of the control mix at age 3 to 28 days with a 6% decrease at age 56 and 90 days.

The addition of 10% fly ash increased the strength of concrete incorporating red sand; the compressive strength was consistently greater than the control mix at all ages tested.

The addition of 10% fly ash increase the strength of concrete incorporating red sand and resulted in compressive strengths comparable to the control mix with 10% fly ash (mix #5).

The addition of 30% fly ash decreased the strength of concrete incorporating red sand by 25%, 19%, and 13% at day 28, 56 and 90 respectively. However this finding is in line with the previous research by Odler (1991) about the correlation between fly ash content with the relative strength.

The durability tests indicated high water permeability and high chloride diffusion of concrete containing red sand. This might be due to low slump value which makes the concrete hard to handle and to be compacted even though by vibrating table, resulting in more pores for concrete and easy ingress for water. The fly ash addition seems to effect in a positive way to improve the workability and durability of concrete.

The leaching test results do not exceed the guidelines given by NSW EPA. Concrete incorporating red sand leached chloride 60% higher than that of natural sand. However the chloride content in washed red sand, about 28 mg/L, is tolerable as the amount of chloride recommended in reinforced concrete is about 100 mg/L (Rump and Krist, 1992).

REFERENCES

ACI COMMITTEE 232 (2003). Use of Fly Ash in Concrete. *American Concrete Institute*, Michigan.

Alcoa (2002). *CHEMWATCH Full Report (Review)*, Retrieved December 16, 2003, from http://wao-prodappl.wao.alcoa.com/chemwatch/full.

Andersson, A. (2000). *Leaching of Hazardous Substances from Concrete Constituents*, Chalmers University of Technology.

Bader, M. A. (2003). Performance of Concrete in a Coastal Environment. *Cement and Concrete Composites*, 25(4-5), 539-548.

Cabeza, M., Collazo, A., Novoa, X. R., and Perez, M. C. (2003). Red Mud as a Corrosion Inhibitor for Reinforced Concrete. *Corrosion Science in the 21th Century*, UMIST, UK.

Danish Standard NT BUILD 443 (1995), Concrete Hardened: Accelerated Chloride Penetration NORDTEST".

Diaz, B., Joiret, S., Keddam, M., Novoa, X. R., Perez, M. C., and Takenouti, H. (2004). Passivity of Iron in Red Mud's Water Solutions. *Electrochimica Acta*, 49(17-18), 3039-3048.

EUR 17869 EN, E. C. R. (1997). Development of a Leaching Method for the Determination of the Environmental Quality of Concrete.

Hardjito, D., Wallah, S.E., Sumajouw, D.M.J. (2004). Properties of Geopolymer Concrete with Fly Ash as Source Material: Effect of Mixture Composition. 7th CANMET/ACI International Conference on Recent Advances in Concrete Technology, Las Vegas, U.S.A, 109-117.

Kliszczewicz, Alina. Ajdukiewicz, Andrzej.(2002). Differences in Instantaneous Deformability of HS/HPC According to the Kind Of Coarse Aggregate. *Cement & Concrete Composite* 24(2) 263-267.

Kurdowski, W., and Sorrentino, F. (1997). *Red Mud and Phosphogypsum and their Fields of Application*. Waste Materials Used in Concrete Manufacturing, William Andrew Publishing/Noyes,

Nassif, H. H.,Najm, H., Suksawang, N. (2005). Effect of Pozzolanic Materials and Curing Methods on the Elastic Modulus of HPC. *Cement & Concrete Composites* 27, 661-670.

Neville, A. M. (1995). *Properties of Concrete*, 4th edition, England: Pearson Education Limited, Essex.

Nguyen, Q. D., and Boger, D. V. (1998). Application of Rheology to Solving Tailings Disposal Problems. *International Journal of Mineral Processing*, 54(3-4), 217-233.

NSW EPA. (1999) Environmental Guidelines: Assessment, Classification & Management of Liquid & Non-liquid Wastes." EPA Sydney.

Odler, I. (1991). Final report of Task group 1, 68-MMH Technical Committee on Strength of Cement. *Materials and Structures*, 24(140), 143-157.

Oh, B. H., Cha, S. W., Jang, B. S., and Jang, S. Y. (2002). Development of High-Performance Concrete Having High Resistance to Chloride Penetration. *Nuclear Engineering and Design*, 212(1-3), 221-231.

Pan, Z., Cheng, L., Lu, Y., and Yang, N. (2002). Hydration Products of Alkali-Activated Slag-Red Mud Cementitious Material. *Cement and Concrete Research*, 32(3), 357-362.

Philleo, R. E. (1991). Concrete Science and Reality, Material Science of Concrete II. *American Ceramic Society*, 1-8.

Rump, H.H., and Krist, H. (1992)." Laboratory Manual for the Examination of Water, Waste Water and Soil. 2nd edition, New York : VCH Publisher Inc.,.

Ryan, W. G., and Samarin, A. (1992). *Australian Concrete Technology*, Australia : Addison Wesley Longman.

Singh, M., Upadhayay, S. N., and Prasad, P. M. (1996). Preparation of Special Cements from Red Mud. *Waste Management*, 16(8), 665-670.

Taywood Engineering Limited (1993).Chemical Testing Procedures Manual. Perth Laboratory, Western Australia.

Townsend, T., Jang, Y. -C., and Tolaymat, T. (2003). *A guide to the Use of Leaching Test in Solid Waste Management Decision Making*. Department of Environmental Engineering Sciences, University of Florida.

Truc, O., Ollivier, J. P., and Carcasses, M. (2000). A New Way for Determining the Chloride Diffusion Coefficient in Concrete from Steady State Migration Test. *Cement and Concrete Research*, 30(2), pp 217-226.

Tsakiridis, P. E., Agatzini-Leonardou, S., and Oustadakis, P. (2004). Red Mud Addition in the Raw Meal for the Production of Portland Cement Clinker. *Journal of Hazardous Materials*, 116(1-2), pp 103-110.

Wahyuni, A.S.(2005). *Sustainable Use of Residual Bauxite Tailings Sand (Red Sand) in Concrete.* Masters Thesis. Curtin University of Technology, Australia.

In: Fly Ash
Editor: Prabir Kumar Sarker
ISBN: 978-1-62948-044-2

Chapter 10

THE USE OF ALKALI-ACTIVATED FLY ASH IN THE PRODUCTION OF CEMENT WITH REDUCED ENVIRONMENTAL IMPACT

Alexandre Silva de Vargas*[*,1], *Angela Borges Masuero*[2], *Denise C. C. Dal Molin*[2] *and Antônio C. F. Vilela*[3]

[1]Department of Professional Masters in Materials Technology and Industrial Processes, Feevale University;
postdoctoral student, Federal University of Rio Grande do Sul (UFRGS), Brazil
[2]Department of Civil Engineering,
[3]Department of Metallurgical Engineering,
Federal University of Rio Grande do Sul (UFRGS), Brazil

ABSTRACT

Traditional waste products employed as mineral additions in Portland cement can undergo alkali-activation to produce special clinker-free cements with low environmental impact. One such addition is fly ash (FA), a by-product of coal burning power plants. Because of high transportation costs, FA can only be used by the civil construction sector if power plants are located near cement and concrete production facilities. When this is not the case, FA becomes only another type of waste material whose disposal incurs additional costs and becomes a source of pollution. Alkali-activation technologies offer a possible alternative to increase the use of FA produced by power plants. To contribute to the awareness of this innovative technology, this paper presents a bibliographical review of alkali-activated cements. First, a brief history of these cements is provided alongside nomenclature information and the major techniques used in their analysis. This is followed by a discussion of fly ash-based alkali-activated cements, including the main variables that affect their mechanical properties. Finally, an assessment of solidification/stabilization (S/S) technologies of heavy metals in alkali-activated matrices is provided.

* Feevale University, Campus II, RS 239 n° 2755, Novo Hamburgo/RS, Brazil, CEP 93525-000. Email: alexandrev@feevale.br.

Keywords: Fly ash, alkali-activation, solidification/stabilization

1. INTRODUCTION

Alkali-activation is a technology that opens a wealth of possibilities for a wider use of many by-products traditionally used as mineral additions in Portland cement, such as the fly ash (FA) produced by coal burning power plants.

Acidic FA (CaO < 10%) displays pozzolanic properties[1], whereas alkaline FA (CaO > 10%) offers a combination of pozzolanic and binding properties. These characteristics allow cement and concrete manufacturers to develop a wide range of uses for FA.

FA waste can only be fully used by the civil construction sector when power plants are located in surrounding areas because of the high costs incurred in transporting the ash. For power plants located far from cement and concrete production facilities, FA is simply a waste material whose disposal creates additional costs and becomes a source of pollution.

Mineral additions are often used in Portland cement. ASTM C595/C595M (2010) classifies Portland cements containing mineral admixtures as: i) Type IS—Portland blast-furnace slag cement; ii) Type IP—Portland-pozzolan cement; iii) Type IT—Ternary blended cement. Such additions offer important benefits in the form of mortar and concrete technology and by helping preserve the environment.

For the technological point of view, for most cement pastes with the same mechanical strength, those with mineral additions display greater durability than those with no additions. This can be better understood from a brief discussion of the products generated by Portland cement hydration reactions and by pozzolanic reactions.

The main products of cement hydration reactions are hydrated calcium silicate (C-S-H), hydrated calcium aluminate (C-A-H) and calcium hydroxide ($Ca(OH)_2$). The first two account for the physical and mechanical properties of the matrix and lend it dimensional stability. The latter, a by-product of the hydration reaction of clinker silicates, contributes to maintain the pH of the matrix in the range of 12.5 and 13.5. When mineral additions with reactive silica (such as fly ash) are incorporated in the matrix, chemical reactions with $Ca(OH)_2$ produce increased amounts of C-S-H (pozzolanic reaction), which has a higher Si/Ca ratio. This increase in C-S-H in a matrix with additions helps improve mechanical properties and increase durability, and it provides greater chemical stability when compared to matrices with no mineral additions.

From an environmental perspective, four major benefits result from the partial substitution of mineral additions for Portland cement: (i) waste material that would otherwise be dumped into landfills can be used as a by-product of power plants; (ii) reduced need for mining natural clay and limestone quarries (raw materials used in the production of clinker); (iii) energy saving, because temperatures in the range of 1450 to 1550°C are required for the partial fusion of these raw materials to produce clinker; (iv) lower emissions of gaseous pollutants such as CO_2, created by the decarbonation of lime in rotary kilns and by the

[1] A pozzolan is a siliceous or siliceous and aluminous material which, in itself, possesses little or no cementitious value but which will, in finely divided form and in the presence of water, react chemically with calcium hydroxide at ordinary temperature to form compounds possessing cementitious properties (Metha & Monteiro, 1994, p. 217).

burning of fossil fuels. NOx and CH_4 are also produced in relatively smaller amounts. According to Lima (2010), approximately 840 kg of CO_2 are generated for each metric ton of cement. Given that the world production of cement in 2012 amounted to 3.7 billion metric tons (Van Oss, 2013), the estimated production of CO_2 in that year was approximately 3.1 billion metric tonnes. As CO_2 is one of the major gases that contributed to the greenhouse effect, it is in the interest of the industry to reduce emissions.

The limestone decarbonation reaction that takes place in the kilns during the production of cement is shown in equation 1 (Roy, 1999, p. 250).

$$5CaCO_3 + 2SiO_2 \Rightarrow 3CaO \cdot SiO_2 + 2CaO \cdot SiO_2 + 5CO_2 \quad (1)$$

Although the clinkering process generates huge amounts of CO_2, several technological and environmental developments have occurred since the first patent for Portland cement was filed by Aspdin in 1824.

As the use of Portland cement in buildings and structures became widespread, it was observed that the contact of this material with specific compounds under certain environmental conditions, expansions stages that compromised the mechanical properties of concrete were created. Examples of such problems are alkali aggregate reactions (AAR) and sulfate attack.

The limitations of Portland cement, combined with a growing production of FA by power plants, prompted researchers to find new uses for this material by means of new technologies, such as alkali-activation.

With alkali-activation, special clinker-free binders known in the literature as geopolymers, poly(sialates) or alkaline mineral polymers are created. These binders are made up of Si-O-Al bonds, with a combination of fully amorphous and semicrystalline structures. Geopolymers possess good mechanical properties and chemical stability, and are obtained from the alkali- activation of geological or waste materials (such as FA) that are rich in amorphous silicon (Si) and aluminum (Al). The covalent and ionic reactions generated by these elements create structural units in tetrahedral coordination that combine to create polymer chains. The arrangement of these chains depends primarily on the existing Si/Al ratios found in the combined raw materials. In addition, the presence of Ca^{2+} traces further contributes to the creation of crosslinks with aluminum.

These materials belong to a new family which, among other noteworthy properties, brings together the specific features of cement, the refractory properties of traditional ceramics and chemical adsorption characteristics only found in zeolites (Palomo, Fernández-Jiménez, & Blanco-Varela, 2004).

Roy (1999) states that geopolymers are a new family of high performance binders that led to the development of very high strength materials for cement. Some actually do not fall within the standard definition of traditional cementitious materials and so the term 'chemically bonded ceramics' has been used to describe them. Chemically bonded ceramics are defined as ceramic-like materials that result from chemical reactions that occur close to room temperature. Moreover, these materials can be more durable than hydrated Portland cement.

Poly(sialates) are chain or ring polymers with Si^{4+} and Al^{3+} in tetrahedral coordination with oxygen. An excess negative charge is created because of the four-fold coordination of aluminum with oxygen, which must be balanced with the positive charge of metallic cations (e.g. K^+, Na^+, Ca^{2+}) (Davidovits, 1991, p.1635).

The empirical formula that characterizes the structural arrangement of poly(sialates) was suggested by Davidovits (1982), as follows:

$$Mn\ [-(SiO_2)z - AlO_2]n\ .\ wH_2O \quad (2)$$

where z is 1, 2 or 3, M is a monovalent cation such as sodium (Na^+) or potassium (K^+), whose source is the alkaline activator; n is the degree of polycondensation, and w has a value close to 7. The different names that Davidovits suggested for poly(sialates) are related to the value of z, which in fact represents the atomic ratio of Si/Al in the polymeric structure. For z = 1, the mineral is called poly(sialate) or PS, and it is the K-PS polymer when M is potassium. For z = 2, the mineral is called poly(sialate-siloxo) or PSS. When M is sodium, (sodium)-poly(sialate-siloxo) (Na-PSS) results, or a mixture of sodium and potassium called (sodium, potassium)-poly(sialate-siloxo) or ((Na,K)-PSS). For z = 3, the mineral is called poly(sialate-disiloxo) or PSDS. When M is potassium, it is called (potassium)- poly(sialate-disiloxo) (K-PSDS).

This chapter evaluates the main features of alkali-activation, while presenting a history of major research studies on the production of alkali-activated binders, the differences in their nomenclature adopted in the literature, the variables affecting their final properties as well as the use of alkali-activated matrices in the solidification/stabilization (S/S) technology.

2. Alkali-Activation

2.1. History

One of the first systematic studies on alkali-activation dates back to 1939 and was carried out by René Feret. Féret (1939), in John (1995, p.11), claimed that Portland clinker-slag composite cements were taking over the market of lime-slag composite cements, mainly because the clinker-slag composite showed greater chemical and mechanical strength. At the time, sodium carbonate, sodium chloride, sodium sulfate or calcium sulfate were often added to cement in order to accelerate strength gains in lime-slag cements. Purdon (1940) added that even though ground granulated blast furnace slag (GGBFS) was considered a cementitious material, the reactions required to create products with superior mechanical characteristics were (and still are) extremely slow, which made (and still makes) their use in isolation impractical.

This led Purdon (1940) to study the use of NaOH as a possible alkaline activator for GGBFS. He observed that GGBFS-based mortars activated with solutions of NaOH 5.5% to 7.5% achieved higher initial (32 MPa at 7 day) and final strength (58 MPa at 1 years) when compared with conventional Portland-based mortars, which displayed strengths of 18 MPa and 43 MPa, respectively. A note should be made about the chemical composition of GGBFS used in the preparation of these mortars, with 41% CaO, 30% SiO_2 and 20.5% Al_2O_3. In addition to these alkali-activated mortars, others mortars were prepared using GGBFS supplied by other steel mills.

The international impact of the works of Ferét (1939) and Purdon (1940) caused them to be seen as the pioneering researchers in the filed of alkali-activation of GGBFS. It should be noted that the final product resulting from the activation of GGBFS is made up mostly of hydrated calcium silicate structures.

However, it was only in 1957, in Kiev, Ukraine, in the former Soviet Union, that Professor Victor Glukhosvky started his studies on alkali-activation. He aimed to produce special binders from materials with low calcium and high silicon (Si) and aluminum (Al) content, which were very different from the binders with GGBFS that had been studied until then. Aluminosilicates, when exposed to solutions of alkaline metals (Li, Na, K, Rb, Cs) – the first group in the periodical table – displayed hydraulic binding properties similar to those obtained when alkaline earth metals (Mg, Ca, Sr, Ba) were used (Krivenko, 2002). The first study of Glukhovsky on this material was published in 1959.

Krivenko (2002) shows that the theory proposed by Glukhovsky was based on geological data on sodium-potassium-calcium aluminosilicate compounds found on the Earth's crust and known for their high chemical stability and resistance to weathering. In 1957, Glukhovsky called the new material "soil-cement", after his theory on the formation of new rocks.

Although 1957 is seen as a milestone in the studies of alkali-activation of aluminosilicates for the production of binders, many structures erected by ancient civilizations are believed to have been built using this technology, which was lost in time and rediscovered in the 20th century. Some studies on the topic were made by Davidovits (1987), Davidovits et al. (1981), Demortier (2004), Davidovits and Davidovits (1999). This latter study used magic angle spinning nuclear magnetic resonance (MAS-NMR) of the ^{29}Si resonance to identify the chemical shift at -86 ppm in the spectra of samples of ancient Roman cements. These authors claim that this suggests that the hardening of these samples was not the result of the carbonation of lime alone, as archeologists have often suggested, but it is rather the result of chemical reactions between lime and aluminosilicates from calcined clay and volcanic tuffs. This chemical reaction yields an aluminosilicate structure with a major resonance peak at -86 ppm, which suggests the presence of $Si(Q_3Si, 1OH)$[2] and $Si(Q_4)$ structures. These authors believe that this chemical reaction could represent a geopolymer. They also claim that any unreacted lime underwent slow carbonation and formed calcite, which would explain why the cement matrix of the mortar contains calcium carbonate and aluminosilicates.

This theory is supported by the works of Langton and Roy (1984) and Roy and Langton (1989), as quoted in Roy (1999), who claimed that they found analcime[3] in mortars and concretes made with Roman pozzolanic cement. According to Roy (1999), the presence of zeolites in different samples of ancient cements suggests that they represent a final, stable stage in a series of conversions of amorphous phases into semicrystalline zeolites that took place over long periods of time. This suggests that cement mixes with a similar composition could also enjoy long term durability.

---

[2] The degree of polymerization of SiO_4 tetrahedra is represented by Q_n ($n = 0, 1, 2, 3$), a distinction can be made between the branched chains of silicates (Q_3) and silicates forming the three-dimensional structure (Q_4).

3 Hydrated sodium aluminosilicate of the zeolite family ($AlSi_2O_6Na\ .H_2O$)

2.2. Nomenclature and Main Factors Affecting the Final Characteristics of Alkali-activated Products

After it was observed that alkaline metals modified the structure of the bonds of potential raw materials used in the production of binders, a variety of names have been proposed in the literature.

Krivenko (2002) lists some of the names used since the early days of alkali-activation studies: soil-cement (Glulhovsky, 1957); geopolymers (Davidovits, 1981); alkali-activated cements (Narang *et al.*, 1983); F-cement (Forss, 1983); geopolymeric cements (Davidovits et al. 1995); alkaline cements (Krivenko & Skurchinskaya, 1991); geocements (Changgo, 1991).

In most cases, these binders are produced in an alkaline environment, with the optional use of clinkers. They differences are related to the raw materials used, the types and concentrations of activators, reaction kinetics, chemical reactions and, most important of all, the structural arrangement of the bonds that make up the matrix.

Palomo, Grutzeck and Blanco (1999) propose the following classification of the study of alkali-activation:

- Alkali-activation of blast furnace slags, which have been studied and used for a long time but still cause a great deal of controversy related to the nature of their reaction products;
- alkali-activation of some calcined clays (mainly metakaolin) and fly ash. This is a more recent development that has attracted great attention lately.

Palomo et al. (1999) also showed that the main differences between both alkali-activation cases would be:

- composition of the activated material, which in the first case is mostly Si and Ca, and Si and Al in the latter case;
- activator concentrations: low to moderate for the former and high for the latter.

This chapter deals with the latter case, with a focus on the alkali-activation of fly ash.

Palomo et al. (1999) describe alkali-activation as a chemical process that can be used to convert glassy structures (partially or totally amorphous and/or metastable) into highly compact cement composites. Van Jaarsveld, Van Denverter and Lorenzen (1997) add that a strongly alkaline medium is necessary for alkali-activation (or geopolymerization) to occur so that some silicon and alumina will dissolve and the surface of the raw material particles can undergo hydrolysis[4]. This can be achieved by means of a single or by a combination of alkaline solutions known as activators. According to Roy (1999), another characteristic of alkali-activators is that they boost the latent hydraulic properties of the materials used in the process. In addition, the concentration of dissolved silica can be increased and controlled by the addition of soluble silicates.

It seems that this surface reaction is responsible for binding waste particles as well as immobilizing any toxic metals that they might contain (S/S). The eventual strength

[4] Hydrolysis is the cleavage of chemical bonds by the addition of water.

development therefore greatly depends on the extent to which this surface reaction takes place (Van Jaarsveld et al. 1997, p. 665).

For Kruger (1990), in Van Jaarsveld et al. (1997), the surface reaction on waste particles and its contribution to the synthesis reaction depends on four main factors: (1) mineralogy; (2) silica and alumina content; (3) fineness or reactive surface area and (4) morphology.

In addition to these factors, Palomo, Blanco-Varela, Alonso and Granizo (2003) show that the alkali-activation of FA is a result of the interaction between the activator found in the mix water and the glassy aluminosilicates found in the FA. The main reaction product obtained in the alkali-activation of FA is an amorphous aluminosilicate gel with a short range three-dimensional structure called 'zeolitic precursor'. In addition, small quantities of zeolite types such as hydrosodalite[5] and herchelite were also detected using x-ray diffraction (XRD). Their presence indicates that zeolitic crystals may represent a thermodynamically stable stage towards which the system would evolve over time.

According to Van Jaarsveld et al. (1997), alkali-activated cements, often called geopolymers, are better understood as amorphous equivalents of certain synthetic zeolites[6] and generally display a more or less similar chemical composition, in spite of the absence of a distinctive zeolitic structure, which makes them amorphous to X-rays. The formation of geopolymers follows almost the same route observed in most zeolites. According to Breck (1974), in Var Jaarsveld et al. (1997), the three main steps in the formation of zeolites are: (1) dissolution, with the formation of mobile precursors through the complexing action of hydroxide ions; (2) partial orientation of mobile precursors as well as the partial internal restructuring of alkaline polysilicates and (3) reprecipitation, where the whole system hardens into an inorganic polymeric structure.

However, Van Jaarsveld et al. (1997) point out that there are some marked differences between zeolite formation and geopolymerization, most of which are related to the composition of the initial reaction mixture. Therefore, an amorphous to semicrystalline poly(sialate) is formed by the copolymerization of aluminosilicate samples that result from the dissolution of the Si and the Al found in waste materials that have been exposed to a strong alkaline medium created by the presence of soluble alkaline metals.

To sum up, the alkali-activation reaction of fly ash takes place through an exothermic process of dissolution, in which the breakdown of the Si O Si and Al-O-Al covalent bonds in the glass takes place and ions (silicon and aluminum) are transferred to the solution. The products of the structural destruction of the raw materials build up over a period of time (induction period), in which heat release is minimal.

Finally, a condensation of the structure occurs (strongly exothermal step), in which a cementitious material with a poorly ordered structure but high mechanical strength appears (Palomo et al., 1999, p.1327).

---

[5] Hydrosodalite (Si_2O_4,Al_2O_4,2Na) is the synthetic equivalent of naturally occurring sodalite ($Si_3Al_3O_{12}$,Na_3,NaCl)

[6] Zeolites are cystalline aluminosilicates thatdisplay molecular size pores and have wide application in heterogeneous catalysis and in mixture sepation processes. Zeolites can be synthetized by the conventional hydrothermal method or by microwave heating.

2.3. Techniques Used in the Analysis of Alkali-activated Samples

Among the techniques suggested by Xu and Van Deventer (2002) for the analysis of products generated from the polymerization of the original structure of raw materials, the following can be mentioned: scanning electron microscopy (SEM) and transmission electron microscopy (TEM) combined with energy dispersion spectroscopy probe (EDS), X-ray diffraction (XRD) and magic angle spinning solid state nuclear magnetic resonance (MAS-NMR) of the ^{27}Al and ^{29}Si nuclei. With MAS-NMR the nature of chemical bonds and the spatial configurations of coordinated SiO_4 and AlO_4 tetrahedrons can be identified, given that the alkali-activated matrix can range from amorphous to semicrystalline. This is corroborated by Van Jaarsveld and Van Deventer (1999), who showed that some FA particles are only partly dissolved, which means that the final product will display some degree of crystallinity.

Figure 1 illustrates the appearance of a FA sample after partial attack by an alkaline activator. The surface of the pleosphere was partially solubilized by the activator, which also reached part of the cenospheres inside (Vargas, Dal Molin, Vilela, Jalali & Gomes, 2007, p. 467).

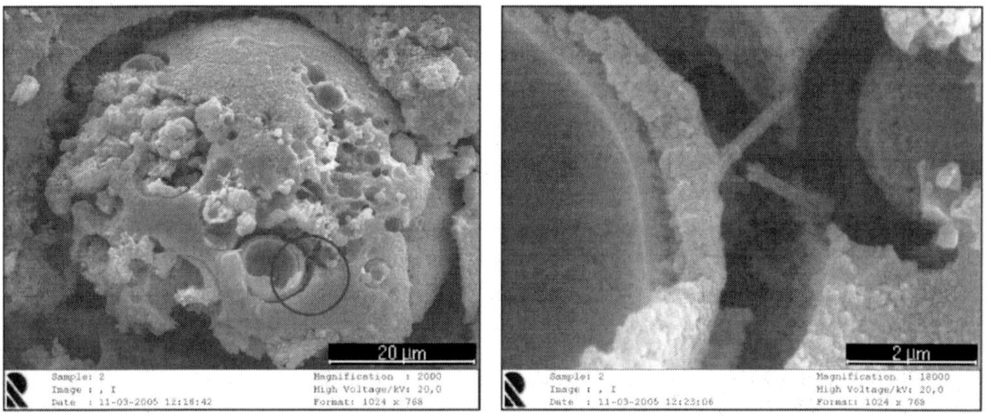

(a) (b)

Source: Vargas et al., 2007, p. 467.

Figure 1. Morphological appearance of an alkali-activated fly ash paste at 28 days; (a) partially solubilized plerospheres, and a view of microspheres inside; (b) appearance of the external layer of the microsphere after attack by an alkaline activator. Region highlighted in Figure 2a.

Another important support technique used in the analysis of alkali-activated samples is Fourier Transform Infrared Spectroscopy (FTIR). Barbosa (1999) shows spectroscopy tools are mainly used to determine local structural properties such as coordination numbers, symmetric sites, local variations, imperfections and impurities. In addition, they provide information on the nature of the chemical bonds between the atoms in the network and on the surface structural properties, as well as on the interactions that do not contribute to the formation of the structural network.

2.4. Mechanical Assessment of Alkali-activated Binders

In this section, the mechanical properties of alkali-activated products are assessed. The properties of the materials and activators and their respective concentrations, curing conditions used, etc., are presented in order to assess the influence of these variables on the final characteristics of alkali-activated products.

2.4.1. Temperature and Curing Times

The use of thermal curing for samples of alkali-activated fly ash (AAFA) is a convenient method to increase the polycondensation rate of aluminosilicates. Otherwise, long periods of time would be required to achieve strength increases in the samples, which would compromise the technical and economic feasibility of the project. Shi (1996) and Vargas (2006) observed that FA-based samples cured at 20^{0}C displayed no strength at 28 days. It should be highlighted that the effects of temperature need to be assessed in combination with those of other variables, as described below.

Hardjitoet al.(2005) assessed the influence of temperature and curing periods in the compressive strength of AAFA-based concrete, as shown in tables 1 and 2, respectively.

Table 1. Effect of curing on compressive strength of AAFA-based concrete

Temperature (°C)	**Compressive strength (MPa)***	
	Concrete:	
	A (n°9)	**B (n° 10)**
35	20	30
60	57	68
90	66	70

Source: Adapted from Hardjito et al. ,2005, p. 5.

Table 2. Effect of curing time on compressive strength of AAFA-based concrete

Compressive strength A (n° 9)							
Time							
hs	*4*	*8*	*12*	*24*	*48*	*72*	*96*
(MPa)	22	30	41	60	72	78	82

Source: Adapted from Hardjito et al. ,2005, p. 5.

Table 1 shows compressive strength values for concrete samples at 7 days, subjected to different curing temperatures in the first 24h (35, 60 and 90°C). A substantial increase in compressive strength was observed when the curing temperature was increased from 35 to 60°C, both for concrete sample A (8M NaOH) and for concrete sample B (14M NaOH). However, the same behavior was not observed in the 60°C to 90°C temperature range: for concrete sample B, the increase in strength was much smaller when compared to the behavior observed for the temperatures of 35 and 60°C; for this sample, no significant gain in compressive strength was observed when the temperature was increased from 60°C to 90°C.

This indicates that the temperature range that improves the mechanical performance of alkali-activated samples is related to the concentration of the activator used.

Table 2 shows the compressive strength behavior of concrete sample A over time. These samples were stored at a curing temperature of 60°C until the test age. According to Hardjito et al. (2005), longer curing times improve the polymerization process of aluminosilicates, resulting in higher compressive strength. They point out, however, that the rate of increase in strength was higher in the first 24 hours of curing. After 24 hours, the gain in strength was only moderate, which means that for practical purposes heat-curing times need not exceed 24 hours.

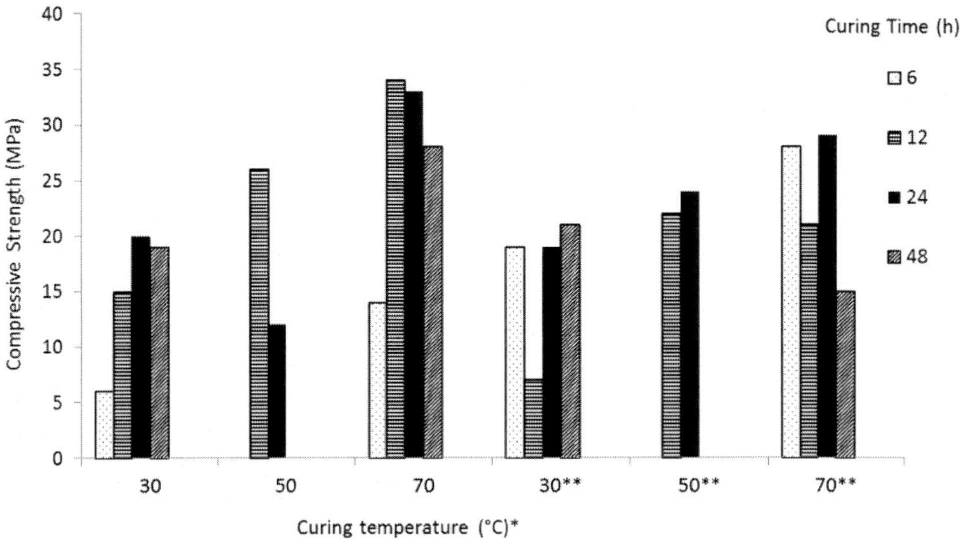

* Oven-cured samples, under atmospheric conditions.
** Same as *, molds coated in plastic to create hydrothermal cure.
Source: Adapted from Van Jaarsveld et al. (2002, p. 68).

Figure 2. Effect of curing conditions on the compressive strength (MPa) of AAFA matrices.

In line with the findings of Hardjito et al. (2005), it was observed that for AAFA-based samples the ideal temperature was 60°C and the ideal curing time was 24h. This shows that it is important to determine the optimal values of these variables in the alkali-activation process.

Criado at al. (2007) also assessed the compressive strength behavior of AAFA-based samples. In their study, FA samples were activated with different alkaline solutions (sodium hydroxide and sodium silicate) containing similar concentrations of sodium oxide (8%) but with different concentrations of soluble silica. All samples were cured at 85°C (humidity>90%) until the test age (8h, 20h; 7, 28, 90 or 180 days). The samples with the lowest soluble silica content (0% and 1.62%) showed a faster increase in compressive strength between the ages of 8h and 20h, thus corroborating the findings of Hardjito et al. (2005). However, even though the increase in strength of those AAFA-based samples with the highest concentrations of soluble silica (5.4% and 9.07%) was not as fast, the highest final compressive strength values were observed in the samples with 5.4% soluble silica at the ages of 90 and 180 days (98MPa). Another important point is that the compressive strength of all samples showed an upward trend to the age of 180 days at 85°C.

Van Jaarsveld, Van Deventer and Luckey (2002) also assessed the effect of curing conditions on the compressive strength of AAFA matrices, as shown in Figure 2.

The results in Figure 2 indicate that the samples cured at 70°C achieved the highest strength values when compared to the samples cured for the same intervals at 30°C and at 50°C. However, Van Jaarsveld et al. (2002) point out that when samples are cured at high temperatures for longer periods, the alkali-activated matrix seems to be weakened. They suggest that small amounts of water must be retained to reduce cracking and maintain structural stability. This could be attributed to the breakup of the aluminosilicate gel, resulting in dehydration and excessive shrinkage. A drop in strength is noticeable at curing temperatures of 70°C between periods of 24 and 48h, both for samples with and without plastic coating.

Van Jaarsveld et al. (2002) suggested that the drop in strength between the curing times of 24h and 48h was related to the curing temperature of 70°C, which is considered high. However, in a study by Criado et al. (2007), samples with AAFA that were cured at 85°C to the age of 180 days showed an upward trend in compressive strength. It should be noted that, in addition to the temperature of 85°C, the samples were cured in an environment with humidity >90%, which may have contributed to the gain in strength. Moreover, the increase in strength observed in the two studies may have been affected by the type of waste materials used and the type and concentration of activators.

However, it should be remembered that for practical applications, the optimal situation would be to produce alkali-activated products with improved mechanical properties for civil engineering applications without the need for heat curing. If heat curing is required, the process should not exceed 24h to reduce additional production costs.

2.4.2. Activators and Concentrations

Palomo et al. (1999) claim that the key factor affecting the development of polycondensation reactions in aluminosilicates is the choice of alkaline activator. They observed faster increases in compressive strength in alkali-activated samples that used a combination of hydroxide (K or Na) and soluble silicate (K or Na) solutions – 68.0 MPa – than with alkali-activated solutions with hydroxides alone – 27.3 MPa – kept at 85°C until the test age of one day.

Slota (1987) adds that the effect of the activator depends significantly on the physical and chemical nature of the waste material used in the process. Moreover, Van Jaarsveld et al. (2002) show that the relationship between the mineralogy and the individual reactivity of minerals is highly complex, and that further studies on the subject are needed. It is reasonable to assume that the type and nature of the waste material used has a direct effect on the chemical and physical properties of alkali-activated samples.

Shi (1996) studied the effect of solutions of 4% Na_2SO_4 or 4% $CaCl_2.2H_2O$ on the compressive strength of samples with 80% FA and 20% $Ca(OH)_2$, as shown in Figure 3. Control samples were prepaed with 80% FA and 20% $Ca(OH)_2$, but no activators were used.

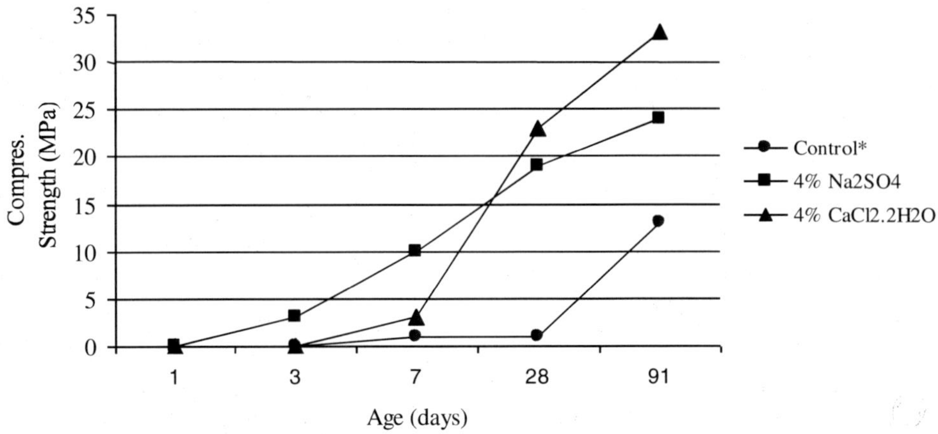

Source: Adapted from Shi (1996, p. 1354)

Figure 3. Compressive strength of alkali-activated samples of FA and $Ca(OH)_2$ with 4% Na_2SO_4 or 4% $CaCl_2 \cdot 2H_2O$, at different ages. Samples cured at 23°C.

It can be observed in Figure 3 that the addition of Na_2SO_4 resulted in significant increases in the strength of samples at both the initial and the final age when compared to the strength of control samples. The use of $CaCl_2$ was shown to be less effective at the ages of 3 and 7 days, but it resulted in greater strength at 28 days when compared to the control and to the Na_2SO_4 samples.

Another important variable in alkali-activation technology is the concentration of alkaline activators. Hardjito and Rangan (2005) show that an increase in the molar concentrations of NaOH from 8M (17 MPa) to 14M (48 MPa) resulted in higher compressive strengths of AAFA samples when the same curing conditions wre used (60°C/24 h).

A similar behavior was observed by Kats (1998) – figure 4 – and by Bakharev (2005) – table 3. These authors observed that higher NaOH concentrations contributed to an increase in the compressive strength of AAFA samples.

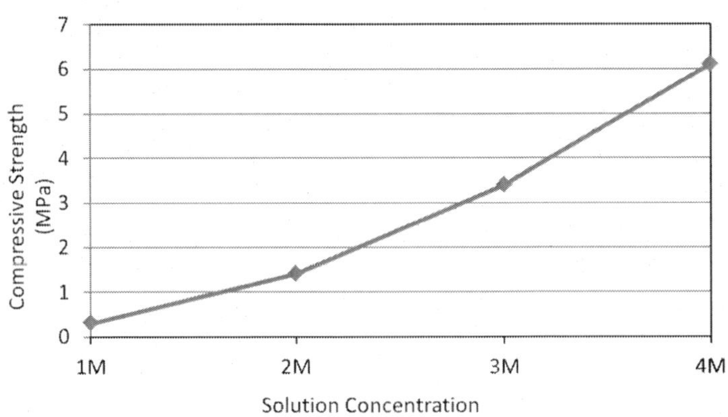

Source: Modified from Katz (1998, p. 200).

Figure 4. Effect of solution concentration on the compressive strength at 7 days.

Table 3. Effect of NaOH concentrations (6% and 8%) on the compressive strength of AAFA-based pastes

Age (days)	Compressive strength (MPa)*	
	Samples activated with NaOH at concentrations of:	
	6%	8%
1	6	10
28	18	45

* Samples subjected to the same curing conditions: 2h at room temperature and at 75°C until test age.
Source: adapted from Bakharev (2005, p. 1226).

Similarly, Vargas et al. (2011) concluded that the molar ratio Na_2O/SiO_2 (N/S) played the most important role in the compressive strength of AAFA. For very low molar ratios, such as N/S 0.2, compressive strength was only 2 MPa. In this case, because of the reduced formation of aluminosilicate gels, factors such as temperature and age had no significant effect on mechanical properties. On the other hand, a molar ratio N/S 0.4 resulted in the highest compressive strength in the samples, particularly after the age of 91 days. In this case, when sufficient amounts of aluminosilicate gels are formed, factors such as temperature and age start to have a significant effect on compressive strength, as illustrated in figure 5.

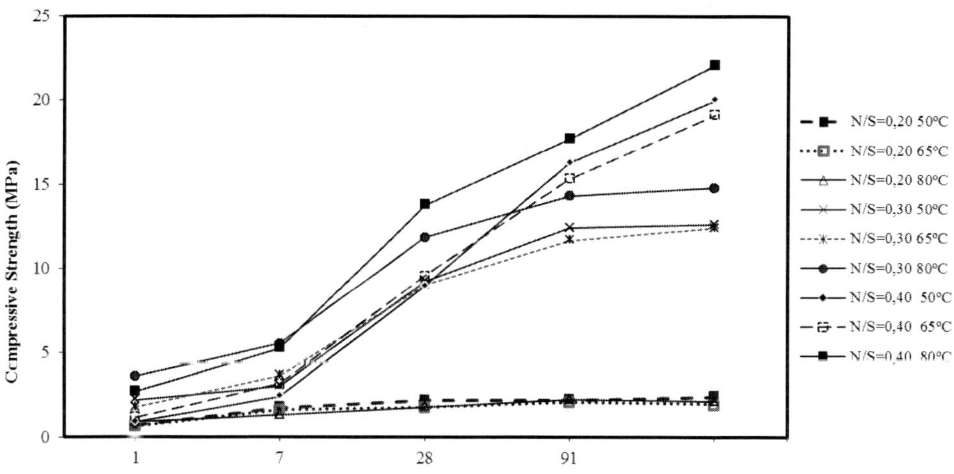

Source: Modified from Vargas et al. (2011, p. 656).

Figure 5. Effect of interactions between age, curing temperature and Na_2O/SiO_2 (N/S) molar ratio on the compressive strength (MPa) of alkali-activated mortars.

Fernández-Jiménez and Palomo (2004) studied the influence of the atomic ratio Si/Na on the compressive strength of AAFA-based mortars, as shown in figure 6.

Figure 6 shows, there is an optimal Si/Na atomic ratio in which the maximum strength resulted (≅ 90 MPa). As we move away from this optimal ratio, the strength is reduced.

In Table 4, Silva and Thaumaturgo (1998) show the compressive strength behavior of AAFA with reference to the percentage of Na_2O (NaOH solution).

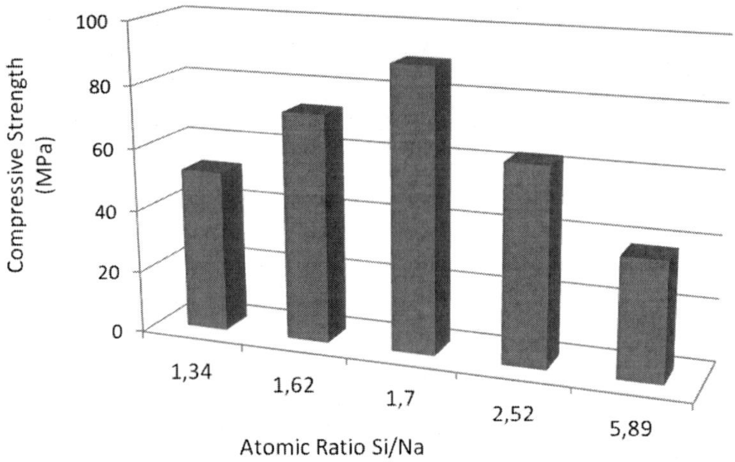

Source: Modified from Fernández-Jiménez et al. (2004, p. 9).

Figure 6. Compressive strength x atomic ratio Si/Na of AAFA-based mortars, cured at 85°C for 20h.

Table 4. Compressive strength of AAFA-based mortars with reference to curing temperature and Na_2O concentrations

% (Na_2O)	Compressive strength (MPa)*
10	1.8
20	6.2
30	12.1
50	8.2
70	5.9

* Curing temperature 65°C.
Source: Modified from Silva and Thaumaturgo (1998, p. 4).

Table 4 shows an optimal concentration of 30% Na_2O (12.1 MPa). Concentrations above or below this value show a decrease in the compressive strength of alkali-activated samples, which confirms the behavior described by Fernández-Jiménez et al. (2004) shown in figure 6.

Silva and Thaumaturgo (1998) explain that for concentrations above the optimal value, such as 50 and 70%, a supersaturation of the activating solution occurs and causes the alkali to crystallize in the hardened matrix and reduces strength when compared with the behavior of samples activated with 30% Na_2O. This is in agreement with the findings of Palomo et al. (1999), who show that the optimal concentrations of the alkali activator are related to the materials used as geopolymeric precursors. These authors claim that for activator concentrations above the optimal level, free OH^- ions found in the matrix can modify the structure of the activated matrix and reduce its strength.

On the other hand, if activator concentrations are below optimal (lower solution pH), the dissolution of FA will be affected and produce less aluminosilicate gel, which is responsible for the mechanical properties of the material. This results in lower strength when compared with the samples prepared with the optimal concentration of the alkali activator.

It follows that the optimal ratios of the different components (waste materials and activators) in the reactive mixtures must be determined for an effective polycondensation of aluminum silicates. This way, no excess or shortage of alkalis if found in the matrix and an alkali-activated product with mechanical properties and chemical stability properties that meet the requirements of technical standards results.

Davidovits (1982) suggests molar ratios for the mixture of reactants used to produce alkali-activated poly(sialate) M – PSS binders, where M can be Na, K or both, and which can be used as a starting point for production, as follows:

$0.20 < M_2O/SiO_2 < 0.48$
$3.50 < SiO_2/Al_2O_3 < 4.50$
$10.0 < H_2O/M_2O < 25.0$
$0.80 < M_2O/Al_2O_3 < 1.60$

The optimal combination of ratios depends largely on the waste material(s) and activator(s) used. The ratios above serve as a starting point that can and should be extrapolated in order to create improved alkali-activated materials.

2.4.3. Other Variables in the Alkali-activation Process

Fernándes-Jiménez and Palomo (2003) claim that the most important physical property that affects the reactivity of FA is its particle size distribution. The fineness of FA is influenced mostly by the fineness of coal, the thermal process and the type of electrostatic filters fitted in the power plants. Higher compressive strength values were obtained with the use of AAFA samples with particles smaller than 45μm than with samples with larger particles. In addition, these researchers found that the highest strength values were obtained in the samples prepared with FA that had the highest concentrations of silica fume. Van Jaarsveld, Van Deventer and Luckey (2003) corroborate these findings and show that one of the parameters that affect the structure and the physical properties of alkali-activated materials is the thermal history of the waste material and its degree of crystallinity. Puertas and Fernándes-Jiménez (2003) reported similar findings and underscored the importance of knowing the silica fume content and the particle size distribution of FA used in alkali-activation. Silica fume is the FA component that reacts with alumina and the alkali to produce cementitious materials, while the particle size distribution is the physical property that most affects FA reactivity.

Van Jaarsveld et al. (2003) mention additional parameters that may affect the structure and physical properties of alkali-activated materials, such as water content, calcium content in FA and water/FA ratio.

Finally, Fernándes-Jiménez and Palomo. (2003) list some desirable characteristics that FA samples should have to be considered for use in the alkali-activation process: loss on ignition lower than 5%; concentration of Fe_2O_3 below 10%; low concentrations of CaO; concentration of silica fume between 40 and 50%; between 80 and 90% of particles smaller than 45μm; large concentration of glassy phase.

Therefore, in general terms, the production of alkali-activated cements depends on the waste material(s) used, their mineralogy, morphology, particle size distribution, chemical composition, concentration of silica fume, type and concentration of activator(s) as well as the curing process used with the samples. These variables are interconnected and if only a single aspect is considered, this can lead to incorrect conclusions.

3. Solidification/Stabilization Technology

The term "solidification/stabilization" (S/S) refers to a general category of processes that are used to treat a number of wastes, including solids and liquids. Solidification and stabilization are distinct technologies, as follows (EPA, 1997):

3.1. Solidification

Solidification refers to processes that encapsulate a waste to form a solid material and to restrict contaminant migration by decreasing the surface area exposed to leaching and/or by coating the waste with low-permeability materials. Solidification can be accomplished by a chemical reaction between a waste and binding (solidifying) reagents or by mechanical processes. Solidification of fine waste particles is referred to as microencapsulation, while solidification of a large block or container of waste is referred to as macroencapsulation (EPA, 1997)

3.2. Stabilization

Stabilization refers to processes that involve chemical reactions that reduce the leachability of a waste. Stabilization chemically immobilizes hazardous materials or reduces their solubility through a chemical reaction. The physical nature of the waste may or may not be changed by this process (EPA, 1997).

3.3. Solidification / Stabilization Technology Using Portland Cement Matrices

Cocke et al. (1992) recall that the cornerstone of S/S technology was the Resource Conservation and Recovery Act (RCRA), enacted by the U.S. Congress in 1976, after which S/S became an essential element in the treatment of hazardous materials that cannot be eliminated by recovery, incineration and/or reduction sources.

The RCRA regulated and gave power to Environmental Protection Agencies to act and provide guidelines on the generation, classification, transportation, treatment and disposal of home and industrial waste, hazardous or not. The first handbook on the use of these regulations was published in 1990 and has since undergone several changes and corrections that reflect developments in public and private waste management policies (EPA, 2011).

For Santhanam and Lunt (1981), the state of development of S/S processes ranges from laboratory tests to marketing. There are basically three methods by which S/S can be used to improve waste disposal:

- improvements in the physical properties of waste materials can make handling easier. This makes better control/management of the disposal area possible and lowers the impact associated with the physical stabilization and the contamination of surface and underground waters;
- reducing waste exposure by reducing its surface area and/or permeability, or by encapsulating waste, which limits the contact between water and waste;
- chemical reactions with the waste that reduce the solubility of chemical components and prevent widespread contamination through leaching or solubilization in water.

Wiles (1987) and Cocke et al. (1992) point out that S/S technology has been in use since the 1970s to manage industrial waste. In this technology, selected materials are used (Portland cement, fly ash, lime, etc.) to change the physical and chemical properties of hazardous waste so that it can be disposed in landfills or converted into new products.

According to Cocke (1990), the complexity of S/S processes is a result of the combined effects of the chemistry of Portland cement with equilibrium solutions, kinetic processes and near-surface phenomena. The interaction between cement and waste must be determined and incorporated into the detailed models of the process.

Daniali (1990), Vempati et al. (1995) and Cohen et al. (1997) describe the use of Portland cement in S/S techniques as a technology that is particularly suitable for the treatment of hazardous wastes as it is easy to use and its cost compares favorably to pyrometallurgical or incineration processes. To Wiles (1987), this is associated with the widespread availability of low-cost waste materials (such as fly ash, cement and lime) that can be used in well-known processes with equipment that is easy to operate and readily available in the construction industry.

Wiles (1987) claims that the role of S/S technology, whose aim is to dispose of hazardous waste in an environmentally sound manner, is largely dependent on regulations and legal interpretation, as well as on the availability of technological resources that can address specific performance indicators that may be set.

The alkaline conditions of cementitious matrices can greatly influence the behavior of heavy metals. By increasing alkalinity, it is possible to reduce the solubility of hazardous cations. According to the works of Cocke et al. (1992) and Li et al. (2001), the S/S technology can be seen as pH-dependent, and therefore, on the solubility of the metallic hydroxides present.

Moreover, high concentrations of heavy metals, such as zinc, in waste, may interfere with the hydration reactions of Portland cement and hinder its use in civil construction. Vargas (2002) studied the effect of electric air furnace dust (EAFD), with 13% Zn on concrete blocks of Portland cement. Although cement blocks with EAFD showed higher compressive strength at 7 days when compared to samples of the same age without this addition, the use of EAFD slowed down the hardening of Portland cement. Brehm (2004) added 10% ZnO to Portland cement pastes and observed that, in some cases, hardening was completed only 7 days after mixing, which demonstrates the detrimental effect of ZnO on hardening times. Gineys, Aouad, Damidot (2011) and Palomo and Palacios (2003) observed that additions of 0.7% Zn and 3.12% Pb, respectively, delayed the hydration reaction of Portland cement.

This shows how the chemical composition of waste can affect the hardening times of Portland cement matrices.

According to Vempati et al. (1995), in some cases it may be necessary to develop special cement compositions that can immobilize or block toxic substances, given that the chemical environment can be manipulated to synthesize new classes of minerals that affect the solubility of heavy metals. Therefore, alkali-activated cements may be an alternative to the use of S/S technology.

3.4. Solidification / Stabilization Technology Using Alkali-activated Fly Ash Matrices

Vargas et al. (2009) investigated the compressive strength– figure 7, and Vargas et al. (2009[a]) investigated the leaching (Brazilian Standard NBR 10005, 2004) – table 5 – and solubilization (Brazilian Standard NBR 10006, 2004) of AAFA mortars prepared with 0%, 5%, 15% and 25% of electric air furnace dust (EAFD) by weight of FA. Samples were cured at 70°C for 24 h and then stored at room temperature until the test age. In the leaching test, the granular material (<9.5 mm) is stirred for 18h in a solution of acetic acid and deionized water (pH=5) with a liquid/solid ratio (L/S) of 20. In the solubilization test, a sample of 250g of the granular material (<9.5mm) is placed in 1000g of deionized water for 7 days. After that, the leachate and the solubilized extract are filtered and analyzed.

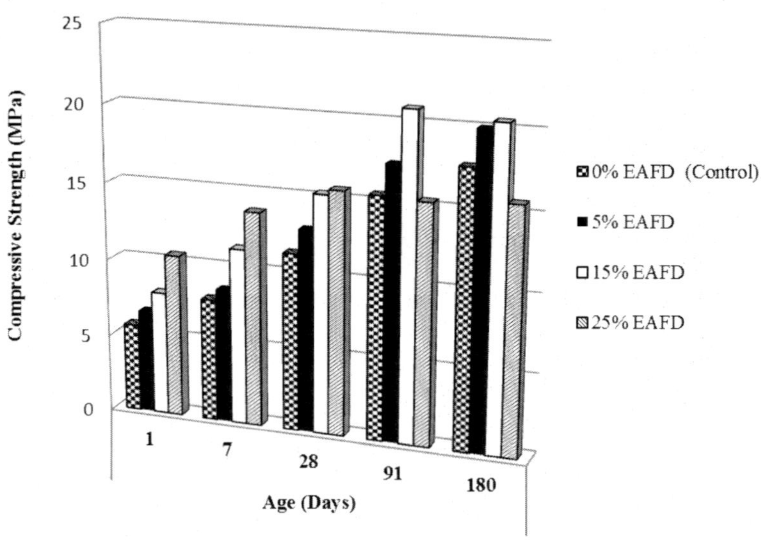

Source: Modified from Vargas et al. (2009, p. 5).

Figure 7. Compressive strength (MPa) of AAFA samples with 0%, 5%, 15% and 25% EAFD at 1, 7, 28, 91 and 180 days. Samples were heat cured at 70°C for 24h, and then stored at room temperature.

It should be noted that the same samples subjected to the compressive strength tests were used in the leaching and solubilization tests, which made it possible to establish relationships between strength and the encapsulation properties of the cementitious matrix.

According to Vargas et al. (2009), the use of EAFD did not interfere in the polycondensation reactions of aluminosilicates of the alkali-activated samples. On the contrary, additions of up to 15% of EAFD contributed to increase the compressive strength of samples, as figure 7 shows.

Increases in strength ranging from 10% to 17% were observed in the sample with 5% EAFD when compared to the control sample (0% EAFD). The increase in strength in this composition remains almost constant to the age of 180 days, when it reaches a value that is similar to that of the sample with 15% EAFD.

The sample with 25% EAFD illustrates the case when the optimal limit for the addition of this waste was exceeded. At the ages of 1 and 7 days, the strength of this composition is higher, but then strength values stabilize and there is no significant increase after 28 days.

The results of EAFD and alkali-activated leachates (ages of 1, 28 and 91 days) are shown in table 5 (Vargas et al., 2009[a])

Table 5. Concentration of hazardous metals in EAFD and alkali-activated leachates (mg/l) with 0%, 5%, 15% and 25% EAFD at 1, 28 and 91 days. Leaching tests were carried out according to Brazilian Standard NBR 10005 (2004)

Element	Sample													
	EAFD	0%			5%			15%			25%			NBR
(mg/L)		Age (days)												
		1	28	91	1	28	91	1	28	91	1	28	91	*10.004**
Cd	6,9	0,03	0,04	0,02	0,02	0,08	0,03	0,1	0,16	0,16	0,26	0,11	0,08	*0,5*
Cr	0,002	0,02	0,003	0,02	0,02	0,02	0,02	0,01	0,16	0,02	0,06	0,02	0,02	*5*
Pb	61	0,28	0,32	0,2	0,02	0,61	0,32	0,95	0,9	0,78	1,3	0,78	0,77	*1*

* Brazilian standard NBR 10004 (2004) threshold values.
Source: Modified from Vargas et al. (2009[a], p. 6).

The concentrations of cadmium (Cd) and lead (Pb) found in the unprocessed EAFD leachate exceeded the threshold values of Brazilian Standard NBR 10004 (2004) by 12.8 and 60 times, respectively. This waste is then classified as a Class I hazard.

This strategy of creating monoliths or microliths with good mechanical strength, which can encapsulate heavy metals and prevent leaching and carry-over to the environment, summarizes the aims of S/S technology. The results in table 7 show that the heavy metals were encapsulated in the alkali-activated matrix for all samples and ages studied, except for the samples with 25% EAFD at 1 day.

However, after some time it was possible to identify a drop in leachate concentrations, which resulted from the increase in strength and the stronger encapsulation of the cementitious matrix. For the sample with 25% EAFD, Cr concentrations decreased and the waste that had been classified as hazardous (at 1 day) was then reclassified as non-hazardous at 28 and 91 days (Brazilian Standard NBR 10004, 2004). In order to classify a material as inert or non-inert, solubilization tests were carried out (Brazilian Standard NBR 10006, 2004). The results of solubilization tests show that the samples with 5% EAFD were classified as inert and the samples with 15 and 25% EAFD were classified as non-inert, as their concentrations of Cr and Cd exceeded the threshold values prescribed by Brazilian Standard NBR 10004 (2004) – 0.05 and 0.005 mg/L, respectively (Vargas et al., 2009[a]).

However, Pereira et al. (2009) warn that the leaching test method affects the results of the immobilization of heavy metals. According to these authors, the pH of the leachate is one of the most important variables in S/S technology. This is in line with the leaching results presented by Galiano, Pereira, and Vale (2011). These authors assessed two different leaching methods used to characterize AAFA samples with different concentrations of fly ash produed from incinerated municipal solid waste (MSW). When the UNE EM 12457-4 method was used, some alkali-activated samples prepared with the waste were classified as inert, non-hazardous (for their Cr and Sb content) or hazardous (for their Sb content). However, when the TCLP method was used, all samples were classified as inert.

Xu et al. (2006) show that other variables in S/S technologies that use alkali-activated fly ash and metakaolin include the concentration of activator and the type of metal that is to be encapsulated. A comparison of the leaching behavior of Cd, Cu, Pb and Cr confirmed that the immobilization of Pb and Cr was more effective than the immobilization of Cu and Cd. Palomo e Palacios (2003) found distinctive behaviors of heavy metals in AAFA-based samples. While it was possible to solidify and stabilize lead (Pb) particles in alkali-activated samples, when Cr^{6+} was present the material did not develop mechanical strength, which suggests that this chromium interfered in the alkali-activation mechanism of FA.

Zhang et al (2008) also mention that AAFA matrices did not perform well in the encapsulation of Cr^{+6}. However, they observed that additions of Na_2S helped to reduce the concentration of Cr^{+6} in leachates as Cr^{+6} is reduced to Cr^{3+} in the presence of sulfides.

Van Jaarsveld and Van Deventer (1999) concluded that the immobilization of metallic contaminants was not only due to the physical effect of the alkali-activated matrix. They observed that different metallic ions affected the chemistry and the morphology of the alkali-activated matrix during the polycondensation process. They also showed that the metal is chemically bound in the structure, even though this bond does not affect the tetrahedral Al and Si bonds that make up the alkali-activated matrix. Fernandez-Jiminez (2004) also concluded that small amounts of arsenic compounds were chemically bound in the alkali-activated matrix, rather than physically encapsulated. This is in agreement with the findings of Xu et al (2006), who observed that heavy metals seem to have a relatively important effect on the chemical and physical properties of alkali-activated matrices.

Therefore, the use of S/S technology using alkali-activated FA by itself or in combination with other precursor materials such as metakaolin and GGBFS may improve the immobilization of heavy metals in the cementitious matrix. However, this depends on the type of waste material(s), the activator(s) used and its/their concentration and the curing method for the alkali-activated samples. It is also important to examine the interaction of the heavy metal (or heavy metals) with the cementitious matrix, which, as it was observed in some cases, affected the alkali-activation process of the fly ash. Finally, the method adopted to assess the performance of the immobilization process can also be an important variable. The chosen method should be the one required by the applicable legislation in the country where the assessment is made.

Conclusion

Alkali-activation technologies show important results in terms of mechanical and environmental aspects. However, variables such as the type of waste material, the type and concentration of the activator, the curing temperature and the particle size distribution affect the final properties of the product. Therefore, it is necessary to determine these variables in order to optimize alkali-activation processes.

The presence of heavy metals in the alkali-activated matrix resulted in different behaviors depending on how the cementitious matrix was obtained. In some cases, Cr^{6+} interfered with aluminosilicate reactions. Therefore, the effect of heavy metals on alkali-activation processes should be carefully assessed before this process can be used in S/S technologies.

Cementitious matrices obtained from clinker-free waste were shown to be an alternative to Portland cement.

REFERENCES

American Society for Testing and Materials –Standard Specification for Blended Hydraulic Cements (ASTM C595/C595M - 2010).

An XRD study of the effect of the SiO2/Na2O ratio on the alkali activation of fly ash.*Cement and Concrete Research 37, 671-679.*

Bakharev, T. (2005) Geopolymeric materials prepared using Class F fly ash elevated temperature curing. *Cement and Concrete Research*, 35, 1224-1232.

Barbosa, V. F. F. Síntese e caracterização de Polissialatos. (1999). PhD Thesis. Instituto Militar de Engenharia (IME), Rio de Janeiro, Brasil.

Breck, D.W. (1974). Zeolite Molecular Sieves. London: John Wiley and Sons.

Brehm, F. A. (2004). Adição de óxido de zinco (ZnO) em pastas de cimento visando viabilizar a reciclagem de pós de aciaria elétrica (PAE) na construção civil. PhD Thesis. Escola de Engenharia – Universidade Federal do Rio Grande do Sul (UFRGS), Brazil.

Cocke, D. L. (1990). The binding chemistry and leaching mechanisms of hazardous substances in cementitious solidification/stabilization systems. *Journal of Hazardous Materials*, 24 (2-3), 231-253.

Cocke, D. L., Mollah, M.Y.A., Parga, J.R., Hess, T.R., Ortego, J.D. (1992). An XPS and SEM/EDS characterization of leaching effects on lead – and zinc – doped Portland cement. *Journal of Hazardous Materials*, 30, 83-95.

Cohen, B., Cilliers, J.J., Petrie, J.G. (1997). Optimization of solidification/stabilization treatment of ferro-alloy waste products through factorial design. *Journal of Hazardous Materials*, 54 (3), 175-188.

Criado, M., Fernandez-Jiménez, A., La Torre, A.G. de., Aranda, M.A.G., Palomo, A. (2007).

Daniali, S. (1990) Solidification/stabilization of heavy metals in latex modified Portland cement matrices. *Journal of Hazardous Materials*, 24, 225-230.

Davidovits, J. (1982). *Mineral polymers and methods of making them.* U.S. Patent 4349386 A

Davidovits, J. (1987). Ancient and modern concretes: What is the real difference? *Concrete International*, 9, 23-35.

Davidovits, J. (1991). Geopolymers: inorganic polymeric new materiasl. *J. Thermal Analysis*, 37, 1633-1656.

Davidovits, J. (1994). Properties of Geopolymer Cements. Paper presented at the 1994 First International Conference on Alkaline Cements and Concretes , Kiev, Ukraine: Kiev State Technical University.

Davidovits, J. (1994). Properties of Geopolymer Cements. Paper presented at the 1994 First

International Conference on Alkaline Cements and Concretes , Kiev, Ukraine: Kiev State Technical University

Davidovits, J. (1994a). Geopolymers: man-made rock geosynthesis and the resulting development of very early high strength cement. *J. Materials Education*, 16 (2&3), 91-139.

Davidovits, J. (1999, 30 jun-2 jul). *Chemistry of geopolymeric systems, therminology.* Paper presented at the 1999 2ª International Conference Geopolymer, University of Picardie, Saint-Quentin, France.

Davidovits, J., & Davidovits, F.(1999, 30 jun-2 jul). Archaeological analogues and long-term stability of gepolymeric materials. Results from the European Research Project Geocistem. Paper presented at the 1999 2ª International Conference Geopolymer, University of Picardie, Saint-Quentin, France.

Davidovits, J., Legrand, J. J. (1981). *Expanded minerals based on potassium poly(sialates) and/or sodium, potassium poly(sialates-siloxo) types*. FR Patent 2490626A18203328

Davidovits, J., Sawyer, J. L. (1985). *Early high-strength mineral polymer.* U. S. Patent 4509985.

Demortier, G. (2004). PIXE, PIGE and NMR study of the masonry of the pyramid of Cheops at Giza. *Nuclear Instruments and Methods in Physics Research B, 226*. 98-109

Environmental Protection Agency (EPA) (1997). Innovative Site Remediation Technology, Solidification/Stabilization. Design & Application, 4. *Retrieved April 30, 2013, from http://www.epa.gov/epawaste/inforesources/online/index.htm*

Environmental Protection Agency (EPA) (2011). Orientation Manual, 2011. Retrieved April 30, 2013, from http://www.epa.gov/tio/download/remed/ss_sfund.pdf

Feret, R. (1939, July). Slags for the manufacture of cement. *Rev Mater Constr Trav Publish.*

Fernándes-Jiménez, A., Palomo, A. (2003). Characterization of fly ashes. Potencial reactivity as alkaline cements. *Fuel*, 82, 2259-2265.

Fernándes-Jiménez, A., Palomo, A. (2003ª, May 11-16). Alkali activated fly ashes: propreties and characteristics. Paper presented at the TH International Congress on the Chemistry of Cement ICCC), 11, 2003, Durban, África do Sul.

Fernándes-Jiménez, A., Palomo, A.; (2004). *Alkali-Activated Fly Ash Concrete: Alternative Material for the Precast Industry*. Paper presented at the 2004 Seminário Geopolimeros – desenvolvimento recentes e suas aplicações na engenharia, Vila Real, Portugal.

Forss, B. (1983). Experiences from the use of F-cement-a binder based on alkali-activated blastfurnace slag, in: G. M. Idorn, Steen Rostam (Eds.), Alkalis in Concrete, Danish Concrete Association, Copenhagen, Denmark, 101-104.

Galiano, Y. L., Pereira, C. F., Vale, J. (2011). Stabilization/solidification of a municipal solid waste incineration residue using fly-ash based geopolymers. *Journal of Hazardous Materials.* 185, 373-381

Gineys, N., Aouad, G., Damidot, D. (2011). Managing trace elements in Portland cement – Part II: Comparison of two methods to incorporate Zn in a cement. *Cement & Concrete Composites* 33, 629–636.

Glukhovsky, V. D. (1959) Soil Silicates. Kiev:Gostroiizdat (in Russian)

Hardjito, D., Rangan, B. V. (2005). Development and properties of low-calcium fly ash-based geopolymer concrete. Paper presented at the 2005 Research Report GC 1. Faculty of Engineering Curtim University of Technology, Perth, Australia.

Hardjito, D., Wallah, S.E., Sumajouw, D.M.J., & Rangan, B.V. (2005). Fly ash-based geopolymer concrete. Australian Journal of Structural Engineering, 6 (1), p. 1-9

Katz, A. (1998). Microscopic study of alkali-activated fly ash. *Cement and Concrete Research*, 28, 2, 197-208.

Krivenko, P. (1994). Alkaline Cements, in: P. V. Krivenko (Ed), Alkaline Cements and Concretes, Proceedings of the Ist International Conference. VIPOL Stock Co., Kiev, Ukraine, 11-130.

Krivenko, P. (2002, 29-29 out). Alkaline Cements: From Research to Application. Paper presented at the 2002 Geopolymers Turn potential into profit. Hotel Sofitel, Melbourne, Austrália, 28-29 Out. 2002a.

Krivenko, P., Petropavloskii, P. (2002, 29-29 out). Industrial uses of Geocement-based materials in construction and others industries. Paper presented at the 2002 Geopolymers Turn potential into profit. Hotel Sofitel, Melbourne, Austrália, 28-29 Out. 2002a.

Krueger, R. A. The chemistry of fly ash the pozzolanic reaction. (1990). ChemSA, 301-303.

Langton, C.A., Roy, D.M. (1984). Longevity of borehole and shaft sealing materials: characterization of ancient cement-based building materials. In: G. McVay (Ed.). *Scientific Basis for Nuclear Waste Management* (pp. 543-549). North Holland, New York.

Li, X. D., Poon, S.S., Sun, H., Lo, I.M.C., Kirk, D.W. (2001). Heavy Metal Speciation and Leaching Behaviors in Cement Based Solidified/Stabilized Waste Materials. Journal of Hazardous Materials, 82 (3), 215-230.

Lima, J.A.R. (2010). *Avaliação das conseqüências da produção de concreto no Brasil para mudanças climáticas.* PhD Tesis. São Paulo University.

Mehta, P. K., & Monteiro, P.J.M. (1994). *Concrete: Microstructure, Properties and Materials.* PINI. São Paulo, Brazil (In Portuguese).

Palomo, A., Blanco-Varela, T., Alonso, S., & Granizo, L. (2003, may 11-16). NMR of alkaline activated "calcium hidroxide – metakaolin" solid mixture. Paper presented at the 2003 of the 11TH International Congress on the Chemistry of Cement (ICCC), Durban, África do Sul.

Palomo, A., Fernández-Jiménez, A., & Blanco-Varela, M.T. (2004, june 15-17). Polímeros Minerales Alcalinos. Paper presented at the 2003 of the 7° National Congress of the Materials, Valência, Spain.

Palomo, A., Grutzeck, M., Blanco, M. T. (1999). Alkali-acivated Fly Ashes. A Cement for Future. *Cement and Concrete Research*, 29, 1323-1329.

Pereira, C. Fernández., Luna, Y., Querol, X., Antenucci, D., Vale, J. (2009) Waste stabilization/solidification of na electric arf furnace dust using fly ash-based geopolymers. *Fuel, 88, 1185-1193.*

Puertas, F., Fernández-Jiménez, A. (2003). Mineralogical and microstructural characterisation of alkali-activated fly ash/slag pastes. *Cement & Concrete composites*, 25, 287-292.

Purdon, A. O. (1940). The action of alkalis on blast-furnace slag. *Journal Society Chemical Industries,* 59, 191-202.

Roy, D. M. (1999). Alkali-activated cements opportunities and challenges. *Cement and Concrete Research, 29, 249-254.*

Roy, D.M., Langton, C.A. (1989). Studies of Ancient Concrete as Analogs of Cementious Sealing Materials for a Repository in Tuff, LA-11527-MS, Los Alamos National Laboratory, Los Alamos, NM.

Santhanam, C. J.; Lunt, R. R. (1981). Flue gas cleaning wastes disposal and Flue Gas Cleaning Disposal and Utilization. Noyes Data Corporation, New Jersey – USA.

Shi, C. (1996). Early Microstructure Development of Activated Lime-Fly Ash Pastes. *Cement and Concrete Research, 26 (9), 1351-1359.*

Silva, F. J., Thaumaturgo, C. (1998, October). Use of Enviromental Scanning Electron Microscopy (ESEM) to Study Alkali-Activated Pulverized Fly-Ash. Paper presented at the 1998 6 TH Brazilian Conference on Microscopy of Materials, Micromat 98 – ACTA Microscópica, São Paulo, BR.

Slota, R. (1987).Utilization of water glass as an activator in the manufacturing of cementitious materials from waste by-products. *Cement and Concrete Research,*17 (5), 703-708

Standards Brazil. (2004). Procedure *for obtention leaching extract of solid wastes.* Associação Brasileira de Normas Técnicas (ABNT) (AS NBR 10005-2004). Rio de Janeiro, Brazil.

Standards Brazil. (2004). *Procedure for obtention leaching extract of solid wastes.* Associação Brasileira de Normas Técnicas (ABNT).(AS NBR 10006-2004). Rio de Janeiro, Brazil

Standards Brazil. (2004). *Solid waste – Classification.* Associação Brasileira de Normas Técnicas (ABNT) (AS NBR 10004-2004). Rio de Janeiro, Brazil.

Van Jaarsveld, J. G. S., Van Deventer, J. S. J. (1999). Effect of the alkali metal activator on the properties of fly ash-based Geopolymers. *Industrial & Engineering Chemistry Research, 38 (10), 3932-394.*

Van Jaarsveld, J. G. S., Van Deventer, J. S. J., Lorenzen, L. (1997). The potential use of Geopolymeric materials to immobilise toxic metals: part I. Theory and applications. *Minerals Engineering, 10 (7), 659-669.*

Van Jaarsveld, J. G. S., Van Deventer, J. S. J., Luckey, G. C. (2002). The effect of composition and temperature on the properties of fly ash - and kaolinite - based geopolymers. *Chemical Engineering Journal, 89, 63-73.*

Van Jaarsveld, J. G. S., Van Deventer, J. S. J., Lukey, G. C. (2003) The characterization of source materials in fly ash-based geopolymers. *Materials Letters, 57, 1272-1280.*

Van Jaarsveld, J. G. S., Van Deventer, S. A. (1999). The potential use of Geopolymeric materials to immobilize toxic metals: part II. Materials and leaching characteristics. *Minerals Engineering, 12 (1), 75-91.*

Van Oss, H. G. *(2013). Cement. U.S.Geopoligical Survey.* Retrieved April 15, 2013, from http://minerals.usgs.gov/minerals/pubs/commodity/cement/mcs-2013-cemen.pdf

Vargas, A. S. , Pavão, B., Masuero, A. B., Dal Molin, D. C.C. , Vilela, Antonio C. F. (2009a). Influence of electric arc furnace dust (EAFD) in alkali-activated fly ash binder. Paper presented at 2009 11th International Conference on Non-conventional Materials and Technologies (NOCMAT), 2009, Bath, United Kingdom

Vargas, A. S. ; Pavão, B., Dal Molin, D. C.C. , Masuero, A. B. , Vilela, A. C. F. (2009). Cinzas volantes álcali-ativadas: efeito da adição do pó de aciaria ao ligante. Paper presented at the 2009 64th Congresso da ABM, 2009, Belo Horizonte, Brazil.

Vargas, A. S., Dal Molin, D. C.C., Vilela, A. C. F., Jalali, S., Gomes, J. C. (2007). Cinzas volantes álcali-ativadas com solução combinada de NaOH e Ca(OH)2. *Matéria (UFRJ), 12, 462-469.*

Vargas, A.S. (2002). Estudo da viabilidade do uso do pó de aciaria elétrica a arco na confecção de blocos de concreto para pavimentação. Dissertation. Escola de Engenharia – Universidade Federal do Rio Grande do Sul (UFRGS), Brazil.

Vargas, A.S. (2006). Cinzas volantes álcali-ativadas para obtenção de aglomerantes especiais. PhD Thesis. Escola de Engenharia – Universidade Federal do Rio Grande do Sul (UFRGS), Brazil.

Vargas, A.S., Dal Molin, D.C.C., Vilela, A.C.F., Silva,F.J., Pavão, B., Veit, H. (2011), The effects of Na2O/SiO2 molar ratio, curing temperature and age on compressive strength, morphology and microstructure of alkali-activated fly ash-based geopolymers. *Cement and Concrete Composites, 33(6), 653-660.*

Vargas, A.S., Masuero, A.B., Vilela, A.C.F. (2006). Investigations on the Use of Electric-Arc Furnace Dust (EAFD) in Portland Cement Pastes. *Cement and Concrete Research, 36, 1883-1841.*

Vempati, R.V., Mollah, M.Y.A., Chinthala, A.K., Cocke, D.L., Beeghly, H. (1995). Solidification/stabilization of toxic metal wastes using coke and coal combustion by-products. *Waste Management*, 15 (5–6), 433-440.

Wiles, C. C. (1987). A Review of Solidification/Stabilization tecnology. *Journal of Hazardous Materials*, 14 (1), 5-21.

Xu, H., Van Deventer, J.S.J. (2002). Microstructural characterisation of geopolymers synthesised from kaolinite/stilbite mixtures using XRD, MAS-NMR, SEM/EDX, TEM/EDX, and HREM. *Cement and Concrete Research, 32, 1705-1716.*

Xu, J. Z., Zhou, Y.L., Chang, Q., Qu, H.Q. (2006). Study on the factors of affecting the immobilization of heavy metals in fly ash-based geopolymers. Materials Letters, 60 (6), 820-822.

Zhang, J., John L., Provis, J.L., Feng, D., Jannie S.J., Van Deventer, J.S.J. (2008). *Cement and Concrete Research*, 38, 681–688.

In: Fly Ash
Editor: Prabir Kumar Sarker
ISBN: 978-1-62948-044-2

Chapter 11

GEOPOLYMER CONCRETE USING FLY ASH

Prabir K. Sarker and B. Vijaya Rangan
Department of Civil Engineering, Curtin University,
Perth, Western Australia

ABSTRACT

A summary of the extensive studies conducted on geopolymer concrete is presented; most of the Chapter is devoted to fly ash-based geopolymer concrete. The effects of factors that influence the properties of the geopolymer concrete in the fresh and hardened states are discussed. The properties include compressive strength, tensile strength, modulus of elasticity, drying shrinkage, creep, resistance to sulfate and acid, and fire resistance. The last part of the Chapter describes the behavior of large-scale reinforced geopolymer concrete members and illustrates the application of the geopolymer concrete in the construction industry. Suitability of the available Standards and Codes for design of fly ash based geopolymer concrete beams and columns are presented. Some more recent studies on geopolymers and the economic merits of the geopolymer concrete are also included.

Keywords: Alkali activation; beam; column; durability; geopolymer concrete; fly ash; strength, structures

1. INTRODUCTION

The global use of concrete is second only to water. As the demand for concrete as a construction material increase, so does the demand for Portland cement. It was estimated that the production of cement would increase from about from 1.5 billion tonnes in 1995 to 2.2 billion tonnes in 2010 (Malhotra, 1999). However, the global cement production in 2012 was about 3.7 billion tonnes (Van Oss, 2013). This shows that the global cement production is increasing at a higher rate than what was predicted. Thus, the increasing trend of global demand of cement is expected to continue in order to meet the demand of the construction industry.

On the other hand, the climate change due to global warming has become a major concern. The global warming is caused by the emission of greenhouse gases, such as carbon dioxide (CO_2), to the atmosphere by human activities. Among the greenhouse gases, CO_2 contributes about 65% of global warming (McCaffery, 2002). The cement industry is held responsible for some of the CO_2 emissions, because the production of one tonne of Portland cement emits approximately one tonne of CO_2 into the atmosphere (Davidovits, 1994; McCaffery, 2002).

Several efforts are in progress to supplement the use of Portland cement in concrete in order to address the global warming issues. These include the utilization of supplementary cementing materials such as fly ash, silica fume, granulated blast furnace slag, rice-husk ash and metakaolin, and the development of alternative binders to Portland cement.

In this respect, the geopolymer technology shows considerable promise for application in concrete industry as an alternative binder to the Portland cement (Duxson et al, 2007). In terms of global warming, the geopolymer technology could significantly reduce the CO_2 emission to the atmosphere caused by the cement industries as shown by the detailed analyses of Gartner (2004).

Coal is used by power stations because its high efficiency as a fuel which generates fly ash after combustion. The world's reserve of coal in the year 2002 was estimated as 985 billion tonnes and the global annual fly ash production in the year 2003 was estimated as 414 million tonnes (Mukharjee et al, 2008). The present global reserve of coal is estimated as 860 billion tonnes (World Energy Council, 2012). Thus at present, a large volume of fly ash is generated globally and substantial amount of it remains unused. Utilisation of the fly ash in new value-added applications will save substantial amount of landfill space and help sustainable development by reducing the demand for virgin materials. This chapter describes the utilization of fly ash to make geopolymer binder for manufacturing of concrete. The properties of fly ash based geopolymer concrete and the behavior of structural members are described.

2. GEOPOLYMERS

Davidovits (1988; 1994) proposed that an alkaline liquid could be used to react with the silicon (Si) and the aluminum (Al) in a source material of geological origin or in by-product materials such as fly ash and rice husk ash to produce binders. Because the chemical reaction that takes place in this case is a polymerization process, he called the term 'Geopolymer' to represent these binders.

Water is released during the chemical reaction that occurs in the formation of geopolymers. This water, expelled from the geopolymer matrix during the curing and further drying periods, leaves behind nano-pores in the matrix, which provide benefits to the performance of geopolymers. The water in a low-calcium fly ash-based geopolymer mixture, therefore, plays no direct role in the chemical reaction that takes place; it merely provides the workability to the mixture during handling. This is in contrast to the chemical reaction of water in a Portland cement concrete mixture during the hydration process.

There are two main constituents of geopolymers, namely the source materials and the alkaline liquids. The source materials for geopolymers based on alumina-silicate should be rich in silicon (Si) and aluminium (Al). These could be natural minerals such as kaolinite. Alternatively, by-product materials such as fly ash, silica fume, blast furnace slag, rice-husk ash, etc could be used as source materials. The choice of the source materials for making geopolymers depends on factors such as availability, cost, type of application, and specific demand of the end users. The alkaline liquids are from soluble alkali metals that are usually Sodium or Potassium based. The most common alkaline liquid used in geopolymerisation is a combination of sodium hydroxide (NaOH) or potassium hydroxide (KOH) and sodium silicate or potassium silicate.

This Chapter deals primarily with low-calcium fly ash-based geopolymer concrete. Low-calcium (ASTM Class F) fly ash is preferred as a source material than high-calcium (ASTM Class C) fly ash. The presence of calcium in high amounts may interfere with the polymerization process and alter the microstructure (Gourley, 2003; Gourley and Johnson, 2005). However, the presence of calcium bearing compounds in controlled amount can help reduce the setting time and formation of calcium silicate hydrate in the geopolymer matrix.

3. Constituents of Geopolymer Concrete

Geopolymer concrete can be manufactured by using the low-calcium (ASTM Class F) fly ash obtained from coal-burning power stations. The chemical composition and the particle size distribution of the fly ash must be established prior to use. An X-Ray Fluorescence (XRF) analysis may be used to determine the chemical composition of the fly ash.

Low-calcium fly ash has been successfully used to manufacture geopolymer concrete when the silicon and aluminum oxides constituted about 80% by mass, with the Si-to-Al ratio of about 2. About 80% of the particles of the fly ash used in previous research were smaller than 50 μm (Gourley, 2003; Gourley and Johnson, 2005; Hardjito and Rangan, 2005; Wallah and Rangan, 2006; Sumajouw and Rangan, 2006; Fernandez-Jimenez et al, 2006a; Sofi et al, 2006a; Siddiqui, 2007). The reactivity of low-calcium fly ash in geopolymer matrix has been studied by Fernandez-Jimenez, et al (2006b). Usually, the finer the particle, higher is the reactivity of the fly ash.

Coarse and fine aggregates used by the concrete industry are suitable to manufacture geopolymer concrete. The aggregate grading curves currently used in concrete practice are applicable in the case of geopolymer concrete (Hardjito and Rangan, 2005; Wallah and Rangan, 2006; Sumajouw and Rangan, 2006; Gourey, 2003; Gourley and Johnson, 2005; Siddiqui, 2007).

A combination of sodium silicate solution and sodium hydroxide (NaOH) solution can be used as the alkaline liquid. The sodium silicate solution is commercially available in different grades. The sodium hydroxide with 97-98% purity, in flake or pellet form, is commercially available. The solids must be dissolved in water to make a solution with the required concentration. The concentration of sodium hydroxide solution can vary in the range between 8 Molar and 16 Molar. Though an 8 Molar solution may be adequate to achieve the required strength for many applications, higher strength can be achieved by increasing the molarity of the solution. Note that the mass of water is the major component in both the alkaline solutions.

In order to improve the workability, a high range water reducing super plasticizer and extra water may be added to the mixture. No negative effect of the addition of Napthalene based superplasticiser was observed on the strength development of geopolymer concrete when used up to a dosage of 2% of the fly ash (Hardjito et al., 2005).

4. Influence of the Main Mixture Parameters on the Properties of Geopolymer Concrete

The primary difference between geopolymer concrete and Portland cement concrete is the binder. The silicon and aluminum oxides in the low-calcium fly ash reacts with the alkaline liquid to form the geopolymer paste that binds the loose coarse aggregates, fine aggregates, and other un-reacted materials together to form the geopolymer concrete.

As in the case of Portland cement concrete, the coarse and fine aggregates occupy about 75 to 80% of the mass of geopolymer concrete. This component of geopolymer concrete mixtures can be designed using the principle used for Portland cement concrete. The compressive strength and the workability of geopolymer concrete are influenced by the proportions and properties of the constituent materials that make the geopolymer paste.

Experimental results (Hardjito and Rangan, 2005) have shown the following effects of the main mixture parameters on the properties of geopolymer concrete:

- Higher concentration (in terms of molarity) of sodium hydroxide solution results in higher compressive strength of geopolymer concrete.
- Higher the mass ratio of sodium silicate solution-to-sodium hydroxide solution, higher is the compressive strength of geopolymer concrete.
- The addition of naphthalene sulphonate-based super plasticizer, up to approximately 4% of fly ash by mass, improves the workability of the fresh geopolymer concrete; however, there is a slight degradation in the compressive strength of hardened concrete when the super plasticizer dosage is greater than 2%.
- The workability of the fresh geopolymer concrete measured by the slump value increases when the water content of the mixture increases.
- As the H_2O-to-Na_2O molar ratio increases, the compressive strength of geopolymer concrete decreases.

These experimental results show the interaction of various parameters on the compressive strength and the workability of geopolymer concrete. In order to assist the design of low-calcium fly ash-based geopolymer concrete mixtures, a single parameter called 'water-to-geopolymer solids ratio' by mass was devised. In this parameter, the total mass of water is the sum of the mass of water contained in the sodium silicate solution, the mass of water in the sodium hydroxide solution, and the mass of extra water, if any, present in the mixture. The mass of geopolymer solids is the sum of the mass of fly ash, the mass of sodium hydroxide solids used to make the sodium hydroxide solution, and the mass of solids in the sodium silicate solution (i.e. the mass of Na_2O and SiO_2).

Tests were performed to establish the effect of water-to-geopolymer solids ratio by mass on the compressive strength and the workability of geopolymer concrete (Hardjito and

Rangan, 2005). The test specimens were 100×200 mm cylinders, heat-cured in an oven at various temperatures for 24 hours. The results of these tests showed that the compressive strength of geopolymer concrete decreases as the water-to-geopolymer solids ratio by mass increases. It was shown that the average compressive strength of concrete with water-to-geopolymer solids ratio of 0.174 varied between 40 and 70 MPa for a 24 hours curing at temperatures varying between 30 and 90 ^{o}C. The strength reduced by about 15 to 20% with the increase of water-to-geopolymer solids ratio from 0.174 to 0.197. The compressive strength decreased by about 30 to 35% for increasing water-to-geopolymer solids ratio from 0.174 to 0.22. The trend of this strength variation is analogous to the well-known effect of water-to-cement ratio on the compressive strength of Portland cement concrete. Obviously, as the water-to-geopolymer solids ratio increased, the workability increased as the mixtures contained more water. Similar trend was also observed by Siddiqui (2007) in the studies conducted on steam-cured reinforced geopolymer concrete culverts.

5. Mixing, Casting and Compaction of Geopolymer Concrete

Geopolymer concrete can be manufactured by adopting the conventional techniques used in the manufacture of Portland cement concrete. In the laboratory, the fly ash and the aggregates were first mixed together dry in a pan mixer for about three minutes. The aggregates are usually prepared to saturated-surface-dry (SSD) condition.

The alkaline liquid was mixed with the super plasticiser and the extra water, if any. The liquid component of the mixture was then added to the dry materials and the mixing continued usually for another four minutes. The fresh concrete could be handled up to 120 minutes without any sign of setting and without any degradation in the compressive strength. The fresh concrete was cast and compacted by the usual methods used in the case of Portland cement concrete (Hardjito and Rangan, 2005; Wallah and Rangan, 2006; Sumajouw and Rangan, 2006).

Fresh fly ash-based geopolymer concrete is usually cohesive. The compaction of small specimens can be done by using a vibrating table or by rodding while electrically operated concrete vibrator is more suitable for casting of larger members. The workability of the fresh concrete was measured by means of the conventional slump test.

Figure 1(a) shows a photo of typical geopolymer concrete made from fly ash and mixed by using the usual mixing facilities used for traditional cement concrete. Figure 1 (b) shows that fly ash based geopolymer concrete can be designed for a high slump which indicates a reasonable workability of the mixture.

Figure 2 shows typical fly ash-based geopolymer concrete specimens after hardening. A vibrating table was used for compaction of the specimens. The compressive strength of geopolymer concrete is influenced by the wet-mixing time. Test results show that the compressive strength increased as the wet-mixing time increased (Hardjito and Rangan, 2005).

(a) Geopolymer concrete after mixing. (b) Slump test of geopolymer concrete.

Figure 1. Typical freshly mixed fly ash-based geopolymer concrete.

Figure 2. Fly ash-based geopolymer concrete cylinder specimens for compressive strength test.

6. Curing of Geopolymer Concrete

Heat-curing substantially assists the chemical reaction that occurs in the geopolymer paste. Both curing time and curing temperature influence the compressive strength of geopolymer concrete. The rate of increase in strength was rapid up to 24 hours of curing time; beyond 24 hours, the gain in strength is only moderate. Therefore, heat-curing time need not be more than 24 hours in practical applications.

The required heat-curing regime can be manipulated to fit the needs of practical applications. In laboratory trials (Hardjito and Rangan, 2005), precast products were manufactured using geopolymer concrete; the design specifications required steam-curing at 60 ^{O}C for 24 hours. In order to optimize the usage of formwork, the products were cast and steam-cured initially for about 4 hours. The steam-curing was then stopped for some time to allow the release of the products from the formwork. The steam-curing of the products then continued for another 21 hours. This two-stage steam-curing regime did not produce any degradation in the strength of the products.

A two-stage steam-curing regime was also used by Siddiqui (2007) in the manufacture of prototype reinforced geopolymer concrete box culverts. It was found that steam curing at 80 ^{O}C for a period of 4 hours provided enough strength for de-moulding of the culverts; this was then followed by steam curing further for another 20 hours at 80 ^{O}C to attain the required design compressive strength.

Also, the start of heat-curing of geopolymer concrete can be delayed for several days. Tests have shown that a delay in the start of heat-curing up to five days did not produce any degradation in the compressive strength. In fact, such a delay in the start of heat-curing substantially increased the compressive strength of geopolymer concrete (Hardjito and Rangan, 2005). This may be due to the geopolymerisation that occurs prior to the start of heat-curing.

The temperature required for heat-curing can be as low as 30 ^{O}C. In tropical climates, this range of temperature can be provided by the ambient conditions, as illustrated by recent studies (Nuruddin, et al., 2010; Barber, 2010). Geopolymers using Class F fly ash alone may take very long time such as more than 24 hours to set if cured at temperatures below 20 ^{O}C. However, the setting time can be significantly reduced by the addition of 5 to 20% of blast furnace slag with class F fly ash (Nath and Sarker, 2012). Addition of a small proportion of other Calcium bearing compounds such as cement can also reduce the setting time. Thus, the required setting time at normal temperature can be obtained by adjusting the proportion of the calcium bearing component of the binder.

The above flexibilities in the curing regime of geopolymer concrete can be exploited in practical applications.

7. Design of Geopolymer Concrete Mixtures

Concrete mixture design process is generally based on performance criteria. Geopolymer concrete mixtures can be designed based on the information given in Sections 3 to 6 above. The role and the influence of aggregates are considered to be the same as in the case of Portland cement concrete. The mass of combined aggregates may be taken to be between 75% and 80% of the mass of geopolymer concrete.

The performance criteria of a geopolymer concrete mixture depend on the application. For simplicity, the compressive strength of hardened concrete and the workability of fresh concrete are selected as the performance criteria. In order to meet these performance criteria, the alkaline liquid-to-fly ash ratio by mass, water-to-geopolymer solids ratio by mass, the wet-mixing time, the curing temperature, and the curing time are selected as parameters. With regards to alkaline liquid-to-fly ash ratio by mass, values in the range of 0.30 and 0.45 are recommended. Sodium silicate solution is usually cheaper than sodium hydroxide solids.

Examples of two fly ash-based geopolymer concrete mixture proportions are given in Table 1 (Chang et al., 2009). The mass fractions of the major components of the fly ash were 48% SiO_2, 29% Al_2O_3, 12.7% Fe_2O_3 and 1.76% CaO. The sodium hydroxide was 14 M solution made by dissolving NaOH flakes in water. The Na_2SiO_3 was purchased in liquid form and contained 14.7% Na_2O, 29.4% SiO_2 and 55.9% H_2O, by mass. The two mixtures are similar except higher extra water in GPC1 than in GPC2. The higher amount of water provided in mixture 1 resulted a higher workability that mixture 2. The average slumps of the Mixtures GPC1 and GPC2 were 250 and 210 mm respectively.

Table 1. Examples of mixture proportions of fly ash-based geopolymer concrete (kg / m^3) (Chang et al., 2009)

Ingredient	Mixture GPC1	Mixture GPC2
Aggregate 10mm	551	556
Aggregate 7mm	643	650
Sand	643	650
Fly ash	406	410
Sodium hydroxide solution (14M)	41	41
Sodium silicate solution	103	103
Superplasticiser	6	6
Extra water	26	17

8. Mechanical Properties of Geopolymer Concrete

Experimental studies have shown that the aggregate-binder interfaces are stronger in geopolymers than in the case of Portland cement (Lee and van Deventer, 2004). This may lead to superior mechanical properties and long-term durability of geopolymer concretes (Provis et al, 2007).

The mean compressive strengths of mixtures GPC1 and GPC2 given in Table 1 after curing for 24 hours at 60 ^{0}C were 32 and 51 MPa respectively. Mixture GPC2 had a rest period of 3 days before start of steam curing while steam-curing of mixture GPC1 started immediately after casting of the specimens. While the higher extra water improved the workability of the mixture GPC1, it reduced the compressive strength. Also, the 3 days of rest period improved the compressive strength of mixture GPC2.

The behaviour and failure mode of fly ash-based geopolymer concrete in compression is similar to that of Portland cement concrete, as shown in Figure 3. It can be observed from the figure that the failure plane passed through the aggregates indicating the strong bonding between the aggregates and the geopolymer matrix. The coarse aggregates were granites in the specimens. Test data show that the strain at peak stress is in the range of 0.0024 to 0.0026 (Hardjito and Rangan, 2005). The mean values of modulus of elasticity for mixtures GPC1 and GPC2 were 20 GPa and 24.5 GPa respectively.

The Poisson's ratio of fly ash-based geopolymer concrete with compressive strength in the range of 40 to 90 MPa falls between 0.12 and 0.16. These values are similar to those of Portland cement concrete.

The splitting tensile strength of geopolymer concrete is only a fraction of the compressive strength, as in the case of Portland cement concrete (Hardjito and Rangan, 2005). Tensile strength of geopolymer concrete increases with the increase of compressive strength. However, it was observed that the tensile strength of fly ash-based geopolymer concrete is higher than that of conventional cement concrete of the same compressive strength. For the mixtures GPC1 and GPC2 in Table 1, the mean splitting tensile strengths obtained from tests of 150 mm × 300 mm cylinders were 3.17 MPa and 4.40 MPa respectively.

The unit-weight of concrete primarily depends on the unit mass of aggregates used in the mixture. Tests show that the unit-weight of the low-calcium fly ash-based geopolymer concrete is similar to that of Portland cement concrete. When granite-type coarse aggregates were used, the unit-weight varied between 2330 and 2430 kg/m^3 (Hardjito and Rangan, 2005).

Figure 3. Failure of fly ash-based geopolymer concrete specimens in compression.

9. Creep and Drying Shrinkage

The creep and drying shrinkage behaviour of heat-cured low-calcium fly ash-based geopolymer concrete was studied for a period of one year (Wallah and Rangan, 2006). Test results show that heat-cured fly ash-based geopolymer concrete undergoes very little drying shrinkage in the order of about 100 micro strains after one year. This value is significantly smaller than the range of values of 500 to 800 micro strains experienced by Portland cement concrete.

The creep coefficient, defined as the ratio of creep strain-to-elastic strain, after one year of loading for heat-cured geopolymer concrete with compressive strength of 40, 47 and 57 MPa is between 0.6 and 0.7, while for geopolymer concrete with compressive strength of 67

MPa this value is between 0.4 and 0.5. These values are about 50% of the values recommended by the Australian Standard AS3600 for Portland cement concrete.

The drying shrinkage strains of geopolymer concrete cured in ambient conditions are larger than those experienced by the heat-cured specimens. As mentioned earlier, water is released during the chemical reaction process of geopolymers. In the specimens cured in ambient conditions, this water may evaporate over a period of time causing large drying shrinkage strains especially in first two weeks (Wallah and Rangan, 2006).

10. Durability Properties

Permeability is considered as the most important property of concrete to describe its durability. This is because a higher permeability allows the aggressive agents to penetrate concrete and reduce its durability. Thus, concrete with low permeability is generally considered to be more durable. Olivia et al. (2008) studied the permeation properties of geopolymer concrete and compared with those of conventional OPC concrete of similar strength grade. Permeability tests were conducted under a water head of 95 m. It was found that the permeability coefficients of geopolymer concrete were similar to those of OPC concrete of similar strength grade.

Tests were performed to study the sulfate resistance of heat-cured low-calcium fly ash-based geopolymer concrete. Test results (Wallah and Rangan 2006) showed that heat-cured low-calcium fly ash-based geopolymer concrete has an excellent resistance to sulfate attack. There was no damage to the surface of test specimens after exposure to sodium sulfate solution up to one year. There was no sign of surface erosion, cracking or spalling on the specimens. There were no significant changes in the mass and the compressive strength of test specimens after various periods of exposure up to one year. The change in length was extremely small and less than 0.015% (Wallah and Rangan, 2006).

The deterioration of Portland cement concrete due to sulfate attack is attributed to the formation of expansive gypsum and ettringite which causes expansion, cracking, and spalling in the concrete. Low-calcium fly ash-based geopolymer concrete undergoes a different mechanism to that of Portland cement concrete and the geopolymerisation products are also different from hydration products. The main product of geopolymerisation is not susceptible to sulfate attack like the hydration products. Because there is generally no gypsum or ettringite formation in the main products of geopolymerisation, there is no mechanism of sulfate attack in heat-cured low-calcium fly ash-based geopolymer concrete. However, presence of high calcium either in the fly ash or in the aggregates could cause the formation of gypsum and ettringite in geopolymer concrete.

Tests were performed to study the sulfuric acid resistance of heat-cured low-calcium fly ash-based geopolymer concrete. The concentration of sulfuric acid solution was 2%, 1% and 0.5%. The sulfuric acid resistance of geopolymer concrete was evaluated based on the mass loss and the residual compressive strength of the test specimens after acid exposure up to one year.

The visual appearance of specimens after exposure to sulfuric acid solution showed that acid attack slightly damaged the surface of the specimens (Wallah and Rangan, 2006). The maximum mass loss of test specimens of about 3% after one year of exposure is relatively small compared to that for Portland cement concrete as reported in other studies. Exposure to sulfuric acid caused degradation in the compressive strength; the extent of degradation depended on the concentration of the acid solution and the period of exposure (Wallah and Rangan, 2006).

The acid resistance of geopolymer concrete must be considered in relation to the performance of Portland cement concrete in a similar environment. Past research data have shown that geopolymeric materials performed significantly better in acid resistance as compared to Portland cement (Davidovits, 1994; Gourley and Johnson, 2005). The superior performance of geopolymeric materials in acidic environment is attributed to the lower calcium content of the source material.

11. Geopolymer Concrete Structural Members

In order to demonstrate the application of heat-cured low-calcium fly ash-based geopolymer concrete, twelve reinforced columns and twelve reinforced beams were manufactured and tested (Sumajouw and Rangan, 2006). The columns were tested for axial load and uniaxial bending. Twelve other geopolymer concrete columns were tested for combined axial load and biaxial bending (Rahman and Sarker, 2011). The test columns were 175 x 175 mm in cross-section and 1500 mm in length. The concrete compressive strength at the age of testing of the columns varied from 37 to 63 MPa. The longitudinal reinforcement ratio was either 1.47 % or 2.95%. A fly ash-based geopolymer concrete column under test is shown in Figure 4.

The behavior and failure modes of reinforced geopolymer concrete columns were similar to those observed in the case of Portland cement concrete columns. As expected, the load capacity of columns was influenced by the load-eccentricity, the concrete compressive strength, and the longitudinal reinforcement ratio. When the load eccentricity decreased, the load capacity of columns increased. The load capacity also increased when the compressive strength of concrete and the longitudinal reinforcement ratio increased.

The load-carrying capacity of reinforced geopolymer concrete columns was calculated using both a simplified stability analysis proposed by Rangan (1990) and the moment-magnifier method incorporated in the Australian Standard for Concrete Structures (AS 3600) and the American Concrete Institute Building Code (ACI 318). Strength of the geopolymer concrete columns under combined axial load and biaxial bending can be calculated by using the well-known Bresler's reciprocal load formula used for OPC concrete columns. The calculated failure loads correlated well with the test values. These results demonstrate that the methods of calculations used for reinforced Portland cement concrete columns are applicable for reinforced geopolymer concrete columns.

Test of a fly ash-based geopolymer concrete beam is shown in Figure 5. The behaviour and failure mode of reinforced geopolymer concrete beams were similar to those observed in reinforced OPC concrete beams. The flexural capacity of beams was influenced by the concrete compressive strength and the tensile reinforcement ratio. The flexural strength of reinforced geopolymer concrete beams was calculated using the conventional flexural strength theory of reinforced concrete beams as described in Standards and building Codes such as the the Australian Standard, AS 3600 and the ACI Building Code, ACI 318. For beams with tensile reinforcement ratio of 1.18%, 1.84%, and 2.69%, the test and calculated values agreed well. In the case of beams with tensile steel ratio of 0.64%, as expected, the calculated values were conservative due to the neglect of the effect of strain hardening of tensile steel bars on the ultimate bending moment.

Figure 4. Load-test of a fly ash-based geopolymer concrete column.

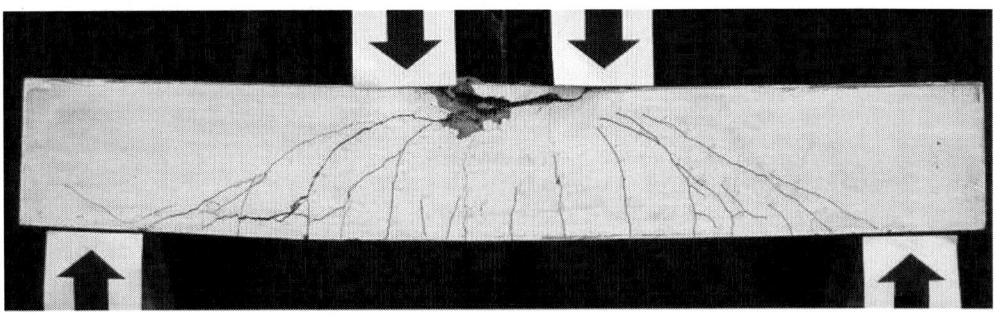

Figure 5. A fly ash-based geopolymer concrete beam after load test.

Mid-span deflection at service load of reinforced geopolymer concrete beams was calculated using the elastic bending theory and the serviceability design provisions given in the Australian Standard, AS 3600. According to AS3600, the calculation of short-term deflection of reinforced concrete beams should include the effects of cracking, tension stiffening, and shrinkage properties of the concrete. In these calculations, the service load was taken as the test failure load divided by 1.5; measured values of modulus of elasticity and drying shrinkage strain of geopolymer concrete were used. Good correlation of test and calculated deflections at service load was seen (Sumajouw and Rangan, 2006).

Sarker, et al (2007a) and Sofi, et al (2007b) showed by pull-out tests of beam-end specimens that adequate bond exists between fly ash-based geopolymer concrete and reinforcing steel. The bond strength of geopolymer concrete is found to be higher than Portland cement concrete of similar compressive strength. Chang et al (2007, 2009) studied the shear and bond strength of reinforced geopolymer concrete beams. The test beams were 200 mm × 300 mm in cross-section. Twelve 2500 mm long beams were tested for the bond strength of spliced bars with concrete. Nine 2000 mm long beams were tested for failure under shear loads. The failure modes and crack patterns observed for reinforced geopolymer concrete beams were similar to those reported in the literature for reinforced OPC concrete beams. The design provisions contained in the Australian Standard for Concrete Structures AS3600 and American Concrete Institute Building Code ACI318 are found to give conservative predictions for the shear strength and bond strength of reinforced geopolymer concrete beams. These design provisions are, therefore, applicable to design of reinforced geopolymer concrete beams.

In all, the above results demonstrate that reinforced low-calcium (ASTM Class F) fly ash-based geopolymer concrete structural members can be designed using the design provisions currently used in the case of reinforced Portland cement concrete members.

12. Fire Resistance

The fire resistance of fly ash-based geopolymers has been studied by Zhu et al (2009). The geopolymer paste (i.e. with no coarse aggregates and sand) gained strength after exposure to high temperature. Geopolymer mortars (geopolymer + sand) sometimes increased in strength and other times decreased in strength after exposure to elevated temperature of 800 ^{O}C. The behaviour of the geopolymer mortar appears to be related to two opposing processes in action at high temperature exposures. That is, sintering and/or further geopolymarisation process at high temperature increases the strength, whereas the thermal incompatibility may cause a decrease in strength. In the case of geopolymer mortars with low strength, the loss in strength due to thermal incompatibility may be minimal with the result that there is a gain in strength. On the other hand, in the case of high strength geopolymer mortars, the loss of strength due to thermal incompatibility is larger than the strength gained by the other process, and hence there is an overall strength loss.

Sarker and deMeillon (2007) studied the residual strength of fly ash-based geopolymer concrete cylinder specimens exposed to fire up to 1000 ^{O}C for two hours. It was found that there was no strength loss up to 400 ^{O}C and strength started decreasing for exposures beyond this temperature. Similar to OPC concrete, strength loss increased with the increase of the fire temperature. Geopolymer concrete was found to retain higher proportion of the original strength than OPC concrete up to 600 ^{O}C. However, at higher temperature such as 1000 ^{O}C, the proportion of strength loss was same in both types of concrete. This is because the damage occurred in the concrete by the very high thermal shock in the cooling period was similar in both types of concrete.

13. Geopolymer Concrete Precast Products

High-early strength gain is a characteristic of geopolymer concrete when cured by dry-heat or steam, although ambient temperature curing is possible for geopolymer concrete. It has been used to produce precast railway sleepers, sewer pipes, and other prestressed concrete building components. The early-age strength gain is a characteristic that can best be exploited in the precast industry where steam curing or heated bed curing is common practice and is used to maximize the rate of production of elements. Recently, geopolymer concrete has been tried in the production of precast box culverts with successful production in a commercial precast yard with steam curing (Siddiqui, 2007; Cheema et al, 2009).

Geopolymer concrete has excellent resistance to chemical attack and shows promise in the use of aggressive environments where the durability of Portland cement concrete may be of concern. This is particularly applicable in aggressive marine environments, environments with high carbon dioxide or sulphate rich soils. Similarly in highly acidic conditions, geopolymer concrete has shown to have superior acid resistance and may be suitable for applications such as mining, some manufacturing industries and sewer systems.

Gourley and Johnson (2005) have reported the details of geopolymer precast concrete products on a commercial scale. The products included sewer pipes, railway sleepers, and wall panels. Reinforced geopolymer concrete sewer pipes with diameters in the range from 375 mm to 1800 mm have been manufactured using the facilities currently available to make similar pipes using Portland cement concrete. Tests performed in a simulated aggressive sewer environment have shown that geopolymer concrete sewer pipes outperformed comparable Portland cement concrete pipes by many folds. Gourley and Johnson (2005) also reported the good performance of reinforced geopolymer concrete railway sleepers in mainline tracks and excellent resistance of geopolymer mortar wall panels to fire.

Concrete Institute of Australia (2011) has produced a Recommended Practice Note on Geopolymer Concrete. Aldred and Day (2012) discussed their experience of the use of geopolymer concrete in various field applications such as pavements, retaining walls, water tanks, boat ramps and precast bridge decks. The study showed that their geopolymer concrete in these applications satisfied the relevant performance requirements of the Australian Standards. Recently, geopolymer concrete has been used to produce 33 precast floor beams of the size up to 10.8 m span and 2.4 m wide in construction of the Global Change Institute at the University of Queensland, Australia. This is considered as a major step forward in the large scale structural applications of geopolymer concrete.

14. High Performance Geopolymer Concrete and Geopolymer Ferrocement

Recently, Ng (2011), working under the direction of Professor Steve Foster of the University of New South Wales Australia, has reported the details of a comprehensive study into the development, properties, and applications of high performance geopolymer concrete. The source materials comprised both low-calcium fly ash from a power station, ground granulated iron blast furnace slag, and Kaolite high performance ash (HPA) from a power station. The alkaline liquid was a combination of sodium silicate solution and sodium hydroxide solution, in a ratio of 2.5:1 by mass.

Both sand and 10 mm basalt were used as aggregates. Some mixtures also included steel fibres. The specimens were heat-cured in a hot water bath at 80-90 ^{O}C for three to seven days. The compressive strength of the geopolymer concrete varied between 44 and 63 MPa. Using these mixtures, several geopolymer beams were cast and tested under load to study the shear strength of such beams. The study also included some preliminary work on the development of low-density high performance geopolymer mortars and structural behavior of beams made using such mixtures.

Sreevidya and her colleagues (2011) have reported studies on geopolymer ferrocement slabs. Low-calcium fly ash from a power station and sand were used to manufacture the geopolymer mortar. The alkaline liquid was a combination of sodium silicate solution and sodium hydroxide solution. The test specimens were cured in ambient conditions, and the 28-day compressive strength ranged between 35 and 62 MPa. Using this geopolymer mortar, several geopolymer ferrocement slab specimens were made and load-tested.

15. Economic Benefits of Geopolymer Concrete

Heat-cured low-calcium fly ash-based geopolymer concrete offers several economic benefits over Portland cement concrete. The price of one tonne of fly ash is only a small fraction of the price of one tonne of Portland cement. Therefore, after allowing for the price of alkaline liquids needed to make the geopolymer concrete, the price of fly ash-based geopolymer concrete is estimated to be about 10 to 30 percent cheaper than that of Portland cement concrete.

In addition, the appropriate usage of one tonne of fly ash earns approximately one carbon-credit that has a redemption value of about 10 to 20 Euros. Recently, Australian Government has proposed to implement carbon tax to a value of Aus$20 per tonne. Based on the information given in this paper, one tonne low-calcium fly ash can be utilized to manufacture approximately three cubic meters of high quality fly ash-based geopolymer concrete, and hence earn monetary benefits through carbon-credit trade.

Furthermore, the very little drying shrinkage, the low creep, the excellent resistance to sulfate attack, and good acid resistance offered by the heat-cured low-calcium fly ash-based geopolymer concrete may yield additional economic benefits when it is utilized in infrastructure applications.

Conclusion

The Chapter presented information on concrete using fly ash-based geopolymer binder. Low-calcium fly ash (ASTM Class F) is used as the source material, instead of the Portland cement, to make concrete.

Low-calcium fly ash-based geopolymer concrete has excellent strength properties and is suitable for structural applications. The mixture design parameters that influence the properties of the fresh and the hardened concrete have been discussed. Typical mixture proportions of fly ash-based geopolymer concrete and their mechanical properties are presented.

The basic properties of fly ash-based geopolymer concrete and the behaviour of structural members made using this concrete are discussed. The properties of hardened geopolymer concrete and the behaviour and strength of reinforced geopolymer concrete structural members are similar to those observed in the case of Portland cement concrete. Therefore, the design provisions contained in the current Standards and Codes can be used to design reinforced low-calcium fly ash-based geopolymer concrete structural members.

Heat-cured low-calcium fly ash-based geopolymer concrete also shows excellent resistance to sulfate attack, good acid resistance, and good fire resistance, undergoes low creep, and suffers very little drying shrinkage. The Chapter has identified some possible economic benefits of using geopolymer concrete. Some recent studies on geopolymers and large scale structural applications of geopolymer concrete have also been included in the discussion.

REFERENCES

ACI Committee 318 (2008), *Building Code Requirements for Structural Concrete*, American Concrete Institute, Farmington Hills, MI.

Aldred, J. and Day, J. (2012), "Is Geopolymer Concrete A Suitable Alternative to Traditional Concrete?", *Proceedings of the 37th Conference of Our World in Concrete & Structures*, 29 – 31August, Singapore. Rerieved on 12 June 2013 from: http://www.cipremier.com/e107_files/downloads/Papers/100/37/100037002.pdf

Barber, S. (2010), Final Year Project Report, Curtin University (*private communication*).

Bresler, B. (1960), "Design Criteria for Reinforced Columns under Axial Load and Biaxial Bending", *ACI Journal*, 57(5), 481-490.

British Petrolium (2012), BP Statistical Review of World Energy, June, Retrieved on 04 June 2013 from http://www.bp.com/statistical_review

Chang, E. H. , Sarker, P. K., Lloyd, N. and Rangan, B.V. (2007), "Shear behaviour of reinforced fly ash-based geopolymer concrete beams", *Proceedings of the23rd Biennial Conference of the Concrete Institute of Australia*, Adelaide, Australia, 679 – 688.

Chang, E. H. Sarker, P. K. Lloyd, N. Rangan, V. B. (2009), Bond Behaviour of Reinforced Fly Ash Based Geopolymer Concrete Beams, *Concrete Solutions 09, 24th Biennial Conference of the Concrete Institute of Australia*, Sydney, 17 – 19 September.

Cheema, D.S., Lloyd, N.A., Rangan, B.V. (2009), "Durability of Geopolymer Concrete Box Culverts- A Green Alternative", *Proceedings of the 34th Conference on Our World in Concrete and Structures*, Singapore.

Committee BD-002 Standards Australia (2009), *Concrete Structures: Australian Standard AS3600-2009,* Standards Australia.

Concrete Institute of Australia (2011), "Recommended Practice Note on Geopolymer Concrete".

Davidovits, J. (1988) "Soft Mineralogy and Geopolymers", *Proceedings of the Geopolymer 88 International Conference*, the Université de Technologie, Compiègne, France.

Davidovits, J. (1994) "High-Alkali Cements for 21st Century Concretes. in Concrete Technology, Past, Present and Future", *Proceedings of V. Mohan Malhotra Symposium*, Editor: P. Kumar Metha, ACI SP- 144, 383-397.

Duxson P., Provis J. L., Lukey G. C. and van Deventer J. S. J. (2007), "The Role of Inorganic Polymer Technology in the Development of Green Concrete", *Cement and Concrete Research*, 37(12), 1590-1597.

Fernández-Jiménez A. M., Palomo A., and López-Hombrados C (2006a), "Engineering Properties of Alkali-activated Fly Ash Concrete", *ACI Materials Journal*, 103(2), 106-112.

Fernández-Jiménez A. M., de la Torre A. G., Palomo A., López-Olmo G, Alonso M. M., and Aranda M. A. G. (2006b), "Quantitative Determination of Phases in the Alkali Activation of Fly Ash, Part I, Potential Ash Reactivity", *Fuel*, 85(5-6), 625-634.

Gartner E. (2004), "Industrially Interesting Approaches to 'Low-CO_2' Cements", *Cement and Concrete Research,* 34(9), 1489-1498.

Gourley, J. T. (2003), "Geopolymers; Opportunities for Environmentally Friendly Construction

Materials", Paper presented at the *Materials 2003 Conference: Adaptive Materials for a Modern Society,* Sydney.

Gourley, J. T., & Johnson, G. B. (2005), "Developments in Geopolymer Precast Concrete", Paper presented at the *International Workshop on Geopolymers and Geopolymer Concrete*, Perth, Australia.

Hardjito, D. and Rangan, B. V. (2005), *Development and Properties of Low-Calcium Fly Ash-based Geopolymer Concrete,* Research Report GC1, Faculty of Engineering, Curtin University of Technology, Perth, available at *espace@curtin* or www.geopolymer.org.

Hardjito, D., Wallah, S., Sumajouw, D. & Rangan B. V. (2005), fly Ash-Based Geopolymer Concrete, Australian Journal of structural Engineering, 6(1), 77-85.

Lee W. K. W. and van Deventer J. S. J. (2004), "The Interface between Natural Siliceous Aggregates and Geopolymers", *Cement and Concrete Research*, 34(2) 195-206.

Malhotra, V. M. (1999), "Making concrete 'greener' with fly ash", *ACI Concrete International*, 21, 61-66.

Malhotra, V.M. (2006), "Reducing CO2 Emissions", *ACI Concrete International*, 28, 42-45.

McCaffrey, R. (2002), "Climate Change and the Cement Industry", *Global Cement and Lime Magazine (Environmental Special Issue)*, 15-19.

Mukherjee, A. B., Zevenhoven, R., Bhattacharya, P., Sajwan, K. S., Kikuchi, R. (2008). Mercury flow via coal and coal utilization by-products: A global perspective. Resources, *Conservation and Recycling*. 52, 571–591.

Nath. P. and Sarker, P. (2012), "Geopolymer Concrete for ambient Curing Condition", *Proceedings of Australasian Structural engineering Conference (ASEC 2012)*, July 11-13, Perth, Australia.

Ng, Tian Sing (2011), "An Investigation into the Development of High Performance Geopolymer Concrete", PhD thesis, The University of New South Wales, Sydney, 551pp.

Nuruddin M. F., Qazi S. A., Kusbiantoro A., and Shafiq N. (2011), "Utilization of Waste Material in Geopolymeric Concrete", *Proceedings of the ICE – Construction Materials*, 164 (6), 315 – 327.

Olivia, M., Sarker, P., Nikraz, H. (2008), Water penetrability of low calcium geopolymer concrete. Proceedings of the International Conference on Construction and Building Technology, June 16 – 20, Kuala Lumpur, Malaysia. 517 – 530.

Provis J. L., Muntingh Y., Lloyd R. R., Xu H., Keyte L. M., Lorenzen L., Krivenko P. V., and van Deventer J. S. J. (2007), "Will Geopolymers Stand the Test of Time?", *Ceramic Engineering and Science Proceedings*, 28(9), 235-248.

Rahman, M., and Sarker, P. K. (2011), "Geopolymer Concrete Columns under Combined Axial Load and Biaxial Bending", Concrete 2011 Conference, Oct 12-14, Perth, Western Australia.

Rangan, B.V. (1990), "Strength of Reinforced Concrete Slender Columns", *ACI Structural Journal*, 87(1) 32-38.

Rangan, B.V. (2008) "*Low-Calcium Fly Ash-based Geopolymer Concrete*", Chapter 26 in Concrete Construction Engineering Handbook, Editor-in Chief: E.G. Nawy, Second Edition, CRC Press, New York.

Rangan, B.V. (2009), "*Engineering Properties of Geopolymer Concrete*", Chapter 11 in Geopolymers: Structures, Processing, Properties, and Applications, Editors: Provis J. and van Deventer J., Woodhead Publishing Limited, London.

Sarker P. K., Grigg A., and Chang E. H. (2007), "Bond Strength of Geopolymer Concrete with Reinforcing Steel", Proceedings of *Recent Developments in Structural Engineering, Mechanics and Computation*, CD ROM, Editor: A. Zingoni, Millpress, the Netherlands, 1315-1320.

Sarker P. K., and deMeillon, T. (2007), "Residual Strength of Geopolymer Concrete After Exposure to High Temperature", Proceedings of *Recent Developments in Structural Engineering, Mechanics and Computation,* CD ROM, Editor: A. Zingoni, Millpress, the Netherlands, 1566-1571.

Siddiqui, K.S. (2007), "Strength and Durability of Low-Calcium Fly Ash-based Geopolymer Concrete", *Final Year Honours Dissertation*, The University of Western Australia, Perth.

Sofi, M., van Deventer J. S. J., Mendis, P. A. and Lukey, G. C. (2007a), "Engineering Properties of Inorganic Polymer Concretes (IPCs)", *Cement and Concrete Research*, 37(2), 251-257.

Sofi, M., van Deventer, J. S. J., Mendis, P. A. and Lukey, G. C. (2007b), "Bond Performance of Reinforcing Bars in Inorganic Polymer Concrete (IPC)", *Journal of Materials Science*, 42(9), 3007-3016.

Sreevidya, V., Anuradha, R., and Venkatasubramani, R. (2011), "Experimental Study on Ferrocement Slab using Various Wire Meshes", under publication (private communication).

Sumajouw, M. D. J. and Rangan, B. V. (2006), *Low-Calcium Fly Ash-Based Geopolymer Concrete: Reinforced Beams and Columns,* Research Report GC3, Faculty of Engineering, Curtin University of Technology, Perth, available at *espace@curtin* or *www.geopolymer.org*.

Van Oss, H. G. (2013), *Cement. U.S.Geoligical Survey.* Retrieved 04 June, 2013, from http://minerals.usgs.gov/minerals/pubs/commodity/cement/mcs-2013-cemen.pdf

Wallah, S. E. and Rangan, B. V. (2006), *Low-Calcium Fly Ash-Based Geopolymer Concrete: Long-Term Properties*, Research Report GC2, Faculty of Engineering, Curtin University of Technology, Perth, available at *espace@curtin* or *www.geopolymer.org*.

van Jaarsveld, J., van Deventer, J. S. J., and Lorenzen, L. (1997), "The Potential Use of Geopolymeric Materials to Immobilise Toxic Metals: Part I. Theory and Applications", *Minerals Engineering* 10(7), 659-669.

Zhu, P., Sanjayan J. G. and Rangan B. V. (2009), "An Investigation of the Mechanisms for Strength Gain or Loss of Geopolymer Mortar after Exposure to Elevated Temperature", *Journal of Material Science,* Vol.44, 1873-1880.

In: Fly Ash
Editor: Prabir Kumar Sarker
ISBN: 978-1-62948-044-2

Chapter 12

COMPOSITE MATERIALS BASED ON POLYPROPYLENE (PP) REINFORCED WITH FLY ASH: RHEOLOGICAL, MECHANICAL AND THERMAL PROPERTIES

***Santiago García Pardo*[1], *Ana Ares*[1], *María José Abad*[1] *and Celina Bernal*[2*]**

[1]Grupo de Polímeros, Universidad de A Coruña, Ferrol, España
[2]Departamento de Ingeniería Mecánica, Facultad de Ingeniería, Universidad de Buenos Aires, Buenos Aires, Argentina

ABSTRACT

In the present chapter, recent results about the development and characterization of composite materials based on Polypropylene (PP) reinforced with fly ash are reviewed. Several investigations on different PP/ash composites including those reinforced with raw ash and others with ash treated with different silane based modifiers are presented. The incorporation of maleic anhydride functionalized iPP (MAPP) as coupling agent or an olefin block copolymer (OBC) as compatibilizer in the composites formulation was also analyzed in these research works. The rheological, mechanical (deformation and fracture) and thermal behaviors were studied for different composites. The effect of filler content on these behaviors was also analyzed. The failure mechanisms responsible for the fracture response of the different composites are also discussed. A comprehensive analysis of the most relevant published results is presented. Finally, future trends on the development of composites based on PP and fly ash are discussed.

Keywords: Polypropylene based composites, fly ash, fracture, failure, thermal behavior, rheological behavior

---

* Corresponding Author address: Email: cbernal@fi.uba.ar.

1. Introduction

Over the recent years, an increasing pressure has been driving the reuse of industrial by-products that have been discarded in the past. Hence, reusing industrial wastes is the objective of research and development and many industries have been making significant efforts to reutilize their wastes and to produce low-cost composites and ecofriendly designs. Coal-burning power stations generate large quantities of fly ash, so its handling and storage imply challenging tasks that affect the cost effectiveness of the relevant industry (Nath et al., 2009). Fly ash is generally disposed in landfills or other areas prescribed by public institutions. However, several problems such as selection of damping sites and increasing costs of disposal have made it difficult to dispose. For this reason, different methods to beneficially use ash have been developed (Setsuda et al., 2012). Research on recycling and reusing of fly ash has recently become an interesting field of study in the emerging markets for green and eco-friendly manufacturing processes and products (Nath et al., 2009). Fly ash has been found to have value-added usage in the construction industry. Large volumes of fly ash are currently being used in structural fills, concrete, blended cement and mining applications (Setsuda et al., 2012)).

In addition, one of the main challenges of current industries is to obtain new plastics with improved properties and processability at relatively low cost. Hence, a common practice in industry is the use of inorganic fillers into plastic items.

Fly ash seems to be an attractive alternative for mineral fillers as plastic products are frequently compounded with fillers with similar chemical composition than fly ash and it is cheaper and more environmentally friendly. Both ashes and fly ashes are produced from the combustion of carbon and other fossil fuels. Inorganic fly ashes are mainly composed of a significant amount of SiO_2 and lower contents of Al_2O_3, Fe_2O_3, Na_2O, MgO, K_2O, etc.

Polymer composites reinforced with different types of ash have been already the subject of many investigations (Bose and Mahanwar, 2004; Stefani et al. (2005); Stefani et al. (2006); Chaudhary et al. (2004); Wang et al. (2004); Siriwardena et al. (2001); Siriwardena et al. (2003); Ismail et al. (2002); Kruger et al. (1999); Chand and Vashishtha (2000); Chand and Khare (1999); Chand and Khare (2000); Chand and Pandey (2000); Setsuda et al. (2012); Iraola-Arregui et al. (2011); Alfaro et al. (2013)). Usually, ash fillers with a particular chemical composition, reduced size and shape are used; however, this approach generates residues and entails a multi-stage process.

Furthermore, when inorganic fillers are introduced into a polymeric matrix poor interaction and/or interfacial adhesion between polymer and filler is generally observed. To improve interfacial adhesion, and therefore obtain enhancements in several desirable properties, different additives having reactive groups compatible with the chemical nature of the polymer and the filler are frequently added in the formulations (Pionteck et al. (2005); Thongsang and Sombatsompop (2006)).

Silane-type coupling agents are widely used (Chaudhary and Jollands (2004)). They are silicon-based chemicals that contain both inorganic and organic reactive groups in the same molecule.

In addition, interfacial adhesion has been also successfully improved by using polymeric coupling agents such as grafted polyolefins, especially maleic anhydride grafted PP (MAPP).

On the other side, elastomers are also used to compatibilize polymer composites, leading to improved morphological and rheological properties of polymers while increased toughness is simultaneously achieved.

In this chapter, recent results about the development and characterization of composite materials based on Polypropylene (PP) reinforced with fly ash are reviewed. A comprehensive analysis of the most relevant published results about rheological, mechanical and thermal behaviors is presented. Finally, future trends on the development of composites based on PP and fly ash are briefly discussed.

2. Polypropylene (PP) and Its Composites: General Characteristics

The most commonly used thermoplastics in industry are polyolefins. Among them, a useful commodity polymer having outstanding properties is polypropylene (PP). It is characterized for having low density, sterilizability, good surface hardness, very good abrasion resistance, excellent electrical properties, as well as good mechanical and barrier properties to water. It also has low cost, worldwide production, simplicity of processing, capability to burn without producing toxic emissions, working security and recyclability.

Polypropylene, its blends and composites find wide applications in home appliances, automotive parts, extruded profiles, packaging industry, construction, etc. (Rohlmann et al. (2006); Zhang et al. (2004)). Although PP is relatively ductile at room temperature, it has the disadvantage of having insufficient toughness at low temperatures and in the presence of notches. Therefore, its use is still limited in many engineering applications.

The incorporation of rigid particles into PP may induce improvements in mechanical properties such as stiffness, strength, fracture resistance, impact toughness, etc. (Fu et al. (2008); Kiss et al. (2007)). In particular, this is a promising approach to obtain simultaneous enhancements in stiffness and toughness (Zhang et al. (2004); Fu and Wang (1992); Bartczak et al. (1999); Thio et al. (2002); Zuiderduin et al. (2003); Argon and Cohen (2003); Chan et al. (2002)).

It is well known that the mechanical properties of composites are determined by component properties, composition, structure and interfacial interactions (Moczó and Pukánszky (2008)). In particular, they are strongly related to type, size and shape, content and surface treatment of the fillers (Moczó and Pukánszky (2008); Dangtungee et al. (2005)). Generally, commercial fillers present a relatively broad particle size distribution hence, most of them are expected to form aggregates in some extent.

In addition, as stated before, interfacial adhesion between inorganic fillers and polyolefins is often rather poor and this can be detrimental to several material properties. Therefore, different additives are frequently added in the formulations (Pionteck et al. (2005); Thongsang and Sombatsompop (2006)). While stiffness is independent of coupling, strength and fracture toughness are expected to be greatly affected by it. It is well known (Zuiderduin et al. (2003)) that changes in interfacial interactions between filler and matrix influence the debonding process, the material failure behavior and hence, the overall composites performance. Uniform dispersion of the filler in the polymer matrix, on the other side, avoids the creation of crack-initiating large agglomerates (Thio et al. (2002)) hence, it is critical to the material impact toughness (Fuad et al. (2010)).

3. Fly Ash Particles As Reinforcement of Polymers

It has been reported in the literature (Nath et al. (2009)) that several factors determine the properties of fly ash such as the nature of the coal, coal preparation before firing, the firing system, the method of collection and any subsequent processing. Knowledge of the surface morphology and the elemental composition of fly ash is very important when fly ash is incorporated into metals, ceramics or polymers. Many experimental techniques such as optical and electron microscopy, atomic force microscopy as well as wide-angle X-ray diffraction have extensively been used to characterize fly ash. Some examples of this will be presented later.

Table 1. Chemical composition of fly ashes used as reinforcement of PP

Compounds	Pardo et al. (2009)	ICP-AES Nath et al. (2009)	XRF Nath et al. (2009)	Setsuda et al. (2012)
SiO_2 (%)	30.3 ± 0.8	47	47.27	62.4
CaO (%)	16.5 ± 0.5	1.5	5.03	0.75
Al_2O_3 (%)	12.9 ± 0.1	31.1	28.96	23.5
Fe_2O_3 (%)	8.8 ± 0.3	15.1	11.46	3.3
K_2O (%)	6.9 ± 0.1	0.12	0,8	0.61
TiO_2 (%)	3.1 ± 0.1	2.2	1.48	1.43
SO_3 (%)	2.83 ± 0.07	-	-	-
MgO (%)	2.58 ± 0.02	1,1	1.48	0.51
P_2O_5 (%)	1.575 ± 0.035	0.09	1	-
Na_2O (%)	1.195 ± 0.007	0.13	0.77	0.35
MnO (%)	0.856 ± 0.017	0.24	0.17	-
Cl (%)	0.500 ± 0.033	-	-	-
ZnO (%)	0.131 ± 0.006	-	-	-
CuO (%)	0.097 ± 0.003	-	-	-
BaO (%)	0.094 ± 0.005	0.02	-	-
SrO (%)	0.085 ± 0.003	0.01	-	-
ZrO_2 (ppm/%)	391 ± 8	-	-	-
Rb_2O (ppm/%)	276 ± 7	-	-	-
$Cr2O_3$ (ppm/%)	215 ± 7	0.05	-	-
NiO (ppm/%)	162 ± 17	-	-	-
CoO (ppm/%)	74 ± 6	-	-	-
PPC (%)	11.5 ± 0.2	-	-	-
LOI (%)	-	0.92	1.65	-

Fly ash particles are classified into two types: the solid spherical particles which are called precipitators and the hollow particles with density lower than 1.0 g cm^{-3} which are called cenospheres. A common type of fly ash is generally composed of crystalline compounds such as quartz, mullite and hematite, glassy compound such as silica glass, and other oxides. The precipitator fly ash (density in the range 2.0–2.5 g cm^{-3}) can improve several properties of selected matrices, including stiffness, strength, and wear resistance and reduce the density. Cenosphere fly ash can be used for the synthesis of ultra-light composite materials due to its significantly low density (in the range 0.4–0.7 g cm^{-3}) compared with the densities of metal matrices (in the range of 1.6–11.0 g cm^{-3}) (Matsunaga et al. (2002)).

Chemically, fly ash itself is mainly a mixture of alkali and transition metal oxides principally of silicon, aluminum and iron, and a small percentage of oxides of calcium, magnesium, potassium, sodium, titanium depending on the processing and coal composition (Nath et al. (2009)). Table 1 lists the composition of several fly ashes used by different authors as reinforcement of PP.

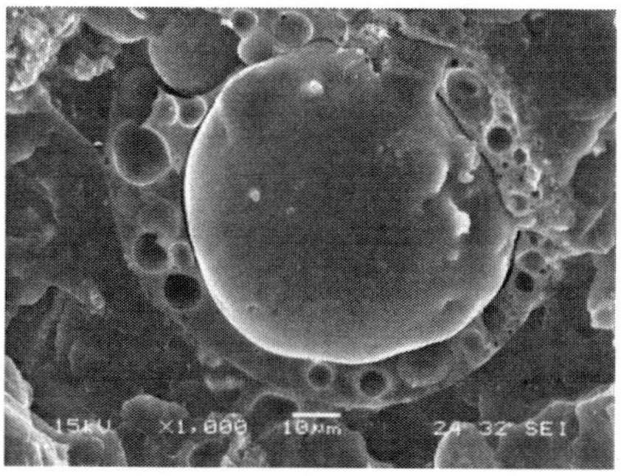

Figure 1. SEM micrographs showing the morphology of cenospheres used as reinforcement of polymers by Pardo et al. (2009), Ares et al. (2010) and Stocchi et al. (2013).

The chemical composition of fly ash reported by Nath et al. (2009) was determined by two different methods: Inductively Coupled Plasma Atomic Emission Spectroscopy (ICP-AES) and X-Ray Fluorescence Spectroscopy (XRF). The ratio of SiO_2, Al_2O_3, Fe_2O_3 and CaO were observed to be slightly different in the two methods (Table 1) whereas the percentage of the minor compounds in fly ash was almost the same.

Pardo et al. (2009), Ares et al. (2010) and Stocchi et al. (2013) used the same fly ash as reinforcement of polymers. Their ash particles were obtained when part of the waste generated during fiberboard manufacturing was burned to produce electrical energy. They were cenospheres which consist of hollow particles containing a central porosity as well as porosity in the walls (Matsunaga et al. (2002); Stocchi et al. (2013)).

The same authors (Pardo et al. (2010); Stochi et al. (2013)) also observed that their ash was not well separated and uneven particle sizes, many coarse particles and irregularly shaped impurities were found.

Pardo et al. (2009) reported a monomodal particle size distribution with a mean diameter value of 30.88 $\pm$ 16.42 μm and a few particles having diameter higher than 100 μm.

On the other hand, Nath et al. (2009) used solid spherical fly ash particles and reported a bimodal peak in the particle size distribution showing a wide range (5-70 μm) of particle size. Most of their fly ash particles were in the size around 40 μm. Although some small particles and interparticle agglomeration were observed, irregularly shaped amorphous particles were not present in their fly ash. This type of irregularly shaped amorphous particles is usually generated in rapid cooling and combustion temperature (Nath et al. (2009)).

Finally, Nath et al. (2009) also determined the constituent phases of fly ash by Wide angle-X-ray diffraction. They found that their fly ash had 68.7% orthorhombic mullite (aluminium silicate, $3Al_2O_3.2SiO_2$) (essentially a type of glass), 23.2% hexagonal quartz (silicon oxide, SiO_2), and 8.1% of hematite (Fe_2O_3). By simulation analysis they also revealed that their fly ash contained 70% aluminum-silicate mullite and 10% silica phase, with various metal oxides. Although mullite is not a major ingredient in coal, it forms in the process of thermal decomposition of natural mineral kaolinite present in coal by a series of chemical reactions (Nath et al. (2009)).

The fly ash used by Pardo et al. (2009) and Stocchi et al. (2013), on the other hand, presented a crystalline phase composed by 2.7 vol% of quartz (SiO_2) and 17.0 vol% of kyanite (Al_2SiO_5).

From the results presented above it is clear that the physical, chemical and mineralogical characteristics of fly ash can vary widely and this complicates the analysis of the effect of the incorporation of fly ash into polymeric matrices and, in particular, into Polypropylene.

4. PP/Ash Composites

Several authors have used fly ash as filled reinforcing material in metals and polymers such as Nylon-6 (Bose and Mahanwar (2004)), epoxy (Stefani et al. (2006)), vinylester (Stocchi et al. (2013)), PP (Siriwardena et al. (2003)), etc.

The resulting mechanical properties of composites generally depend on the nature of the filler, its size and distribution, aspect ratio, volume fraction, and the intrinsic adhesion between component surfaces. High aspect ratio fillers generally increase the material strength because the filler is able to attain high local stress transferred from the polymer matrix, whereas lower aspect ratio fillers such as fly ash do not produce this effect. The reduced tensile strength in particulate fillers such as fly ash may arise from: i) poor adhesion at the polymer/filler interface, and ii) the initiation and growth of cracks in the particle and subsequent failure of the composites (Nath et al. (2009)).

Although PP/fly ash composites present the above-mentioned drawbacks, they can still find several interesting engineering applications, particularly at elevated temperatures, where the fly ash becomes a source of strength while the polymer provides improved ductility. One major environmental value of using the already available huge quantities of fly ash is that it can lead to conservation of other fillers widely used in polymers in significant proportions (Nath et al. (2009)).

In the following, a brief discussion about some commonly used coupling agents is presented. Then, the most relevant recently published results on the rheological, thermal and mechanical behavior of PP/fly ash composites are reviewed.

4.1. Coupling Agents

As it has been mentioned before, interfacial adhesion between inorganic fillers and polymers is often rather poor. Hence, different additives having reactive groups compatible with the chemical nature of the polymer and the filler are frequently added in the formulations.

Silane-type coupling agents are widely used (Chaudhary et al. (2004)). They are silicon-based chemicals that contain two kinds of reactivity (inorganic and organic reactive groups in the same molecule). Their typical structure is $YSi(OR)_3$, where OR is a hydrolyzable group such as methoxy, ethoxy or acetoxy, and Y is an organofunctional group such as amino, methacryloxy, epoxy, vinyl, etc.

A silane coupling agent will act as a link between an inorganic substrate (such as ash) and an organic material (such as commercial polypropylene) to bond, or couple, the two dissimilar materials together. Although it is usual to pretreat the surfaces of mineral fillers with the silane coupling agent, it may also be added to the blend of filler and polymer directly. Polycondensation of the trimethoxy-silyl group also proceeds rapidly; it is already completed in the starting phase of the reaction. The silane coupling agent is selected according to the organic functionality of the target polymer.

The methoxysilyl group is subject to hydrolysis in water or water/alcohol solutions. The initial product of hydrolysis is a silanetriol: RSi(OCH...)... + 2HLORSi(OH)... + 3CH...OH.

Silantriols are moderately stable at dilute concentrations in polar solvents such as water and alcohols.

Silanols groups are capable of condensing with hydroxyl groups at the surface of ash.

After condensing with the ash surface, the remaining silanol groups are capable of hydrogen bonding or condensing with adjacent silanol groups. By this combination of covalent and hydrogen bonding, the coupling agent is bonded to the inorganic surface, and modifies it so that it is organo-reactive.

Interfacial adhesion has been also successfully improved in polymer based composites by using polymeric coupling agents such as grafted polyolefins, especially maleic anhydride grafted PP (MAPP).

The presence of maleated PP in PP/ash composites (Hristov et al. (2004)) is expected to greatly improve interfacial adhesion contributing to the formation of a strong interfacial layer. It is believed that the maleic anhydride groups of MAPP are able to react with the surface hydroxyl groups of ash particles. Therefore, maleic anhydride which is a rigid five-membered ring with a permanent dipole moment is expected to act as a coupling agent, breaking up agglomerates and increasing the degree of particles dispersion in the matrix and interfacial adhesion between both phases (Bikiaris et al. (2005)). The polypropylene segments of MAPP form miscible blends with the bulk PP through cocrystallization, and the polar part of MAPP form a chemical bond with ash, or a polar filler in general. The surface hydroxyl groups of ash particles react as a nucleophile with the maleic anhydride groups of the grafting agent.

In the case of ash, the presence of particles with high surface roughness can also contribute to a strong interaction offering sites for mechanical anchoring. Due to reactions taking place between hydroxyl groups of the filler and maleic anhydride groups of copolymer (Bikiaris et al. (2005)), as well as the mechanical anchoring mentioned above, the formation of a strong bonding between ash particles and PP macromolecules can be achieved.

4.2. Rheological Behavior

It is well known that filler inclusion in a plastic matrix causes changes in its processability. Similarly, the addition of coupling agents affects the flow properties of the polymer molecules. Generally, the viscosity of the system rises as the interactions between the filler and the polymer increase. However, when coupling agents are incorporated in the formulation, the melt viscosity can be reduced due to the higher dispersion of the filler particles into the matrix. The effect of different coupling agents on composite rheology has been already studied for several materials (Suhara and Kutty (1998); Takahashi et al. (1992); Agrawal et al. (2007); Joseph et al. (2007); Filipe et al. (2005)). These rheological studies (high shear) are extremely important to analyze composite processing by injection molding or extrusion. In the following paragraphs, the rheological behavior of different PP/ash composites is reviewed.

PP/Raw Ash Composites

Ares et al. (2010)) studied the dynamic viscoelastic behavior of PP composites with different amounts of untreated ash (10, 20 and 30 wt%).

An increase in the storage modulus (G') and loss modulus (G'') was reported with the addition of filler, being this enhancement higher for the composite with 30 wt% fly ash. Compositions with a lower amount of filler were found to produce a small increase in moduli and the curves followed the unfilled PP values closely with increasing frequency. The polymer melt viscosity of the composite increased initially due to the restriction to deformation imposed by filler particles. In the 10 wt% fly ash composite, the amount of filler added did not seem to be sufficient to provoke this effect. With the addition of a higher percentage of filler, the moduli enhancement observed was attributed to the agglomeration of the filler particles within the matrix. The modulus-enhancement effect of filler in PP composites was significantly higher at lower frequencies than at high frequencies (the typical frequency of conventional manufacturing techniques). For both storage and loss moduli, the slope of the curves was observed to decrease with increasing filler concentrations.

Ares et al. (2010) also observed an enhancement in viscosity with the increase in filler concentration. This effect was more noticeable in composites with 30 wt% ash. However, at high frequency, the viscosity difference for the composites with different filler concentrations was very small and the curves seemed to converge at high frequencies, which explained the successful exploitation of this filler as reinforcement for plastics in conventional processing technologies.

PP/Silane Treated Ash Composites

Pardo et al. (2010) evaluated the compatibilizing effect of three silane coupling agents containing different reactive functional groups (functional groups with different size): amine (GF96), vinyl (XL10) and vinyl-benzyl amine (Z-6032) in PP/fly ash composites. They observed that the presence of fly ash particles perturbed normal polymer flow and hindered the mobility of chain segments, therefore, the viscosity of the filled polymer increased with the amount of filler. All composites studied by Pardo et al. (2010) showed a shear thinning behavior which they attributed to a greater degree of polymer-filler interaction, which required higher shear stress and longer relaxation times for the composites to flow. They also observed the same trend for dynamic moduli as a result of the restriction to deformation due to the presence of filler particles: G´ and G¨ increased with increasing fly ash content.

Furthermore, Pardo et al. (2010) also observed that the treatment with different coupling agents caused diverse effects depending on the type of silane. For the silane with the vinyl reactive group, an increase in composite viscosity has been attributed to a greater interaction between PP and fly ash and to a better dispersion of the filler in the matrix. For the silane with the amine functional group, on the other hand, a decrease in viscosity was observed and attributed to the better dispersion of the filler into the Polypropylene matrix. For the silane with the vinyl-benzyl amine reactive group, higher viscosity values were reported by these authors due to the restrictions to flow caused by this reactive group.

PP/OBC/Treated Ash Composites

More recently, Ares et al. (2010) studied the rheological behavior of ternary systems based on Polypropylene, an olefin block copolymer (OBC) and different contents of ash with an amino silane coupling agent.

The OBC consists of crystallizable ethylene–octene blocks (hard) with very low comonomer content and high melting temperature, alternating with amorphous ethylene–octene blocks (soft) with high comonomer content and low glass transition temperature. Therefore, in thermal studies OBCs show a higher melting temperature than random olefin copolymers with comparable densities. The long ethylene sequences in the hard blocks of OBC could crystallize in lamellar crystals with less defects and higher melting temperature. OBCs exhibit such highly elastomeric performance with polypropylene (PP) blends (Dias et al. (2008); Kamdar et al. (2009); Liu et al. (2011); Chum and Swogger (2008); Liu et al. (2012)).

The main effect of the simultaneous incorporation of the silane coupling agent and OBC on the rheological properties of PP/ash composites was a decrease in G' caused by an improvement in flow behavior. The addition of OBC copolymer with lower viscosity than PP improved flow properties, as reflected in a drastic decrease in G' that increased with the amount of filler (Ares et al. (2010)).

4.3. Thermal Behavior

The behavior of PP based composites in engineering applications critically depends on the extent of crystallinity and the nature of the crystalline morphology of PP (Xu et al. (2003)). In addition, rigid particles are expected to act as nucleating agents making the spherulite size decrease and the crystallinity increase (Yu and He (2000)) with the subsequent important effects on the material performance. Reported results about the effect of fly ash on the thermal and, in particular, on the crystallization behavior of PP are presented below.

PP/Raw Ash Composites

Nath et al. (2009) performed thermal analysis of PP/fly ash composites by Differential Scanning Calorimetry (DSC) and observed that neat PP mainly showed the mono-modal peak of monoclinic α-crystalline phase whereas the incorporation of fly ash led to the appearance of the pseudo hexagonal β-phase. Hence, in their composites, fly ash seemed to be a heterogeneous β nucleator in agreement with the results reported in the literature for PP reinforced with other fillers (Zhao and Li (2005); Zhang et al. (2004), Pérez et al. (2013)). However, Igarza et al. (2013) developed DSC and X-ray diffraction (XRD) analysis for PP/ash and PP/MAPP/ash composites and they did not observe any change in the crystalline structure of PP as a consequence of maleinization or reinforcement.

In PP/ash binary composites Pardo et al. (2013) found that the crystallization temperature (T_c) increased with filler addition. The crystallization peak of PP was shifted to higher temperatures, suggesting that the crystallization of PP started earlier and ash acted as nucleating agent increasing the crystallization rate. This was in agreement with the results of Bartczak et al. (1999) who reported the influence of the filler on the nucleation of crystallites of PP by increasing the crystallization temperature. This effect was also noticeable in the PP/ash composites of Pardo et al. (2013) in the degree of crystallization. It increased proportionally with ash content and except in the sample having 10 wt% ash, crystallinity was higher than that of raw PP. Hence, the presence of solid surfaces seemed to reduce free enthalpy for critical nucleation and promoted PP crystallinity in these composites.

In addition, Pardo et al. (2013) found that the PP melting temperature slightly decreased with respect to the value of raw PP. Therefore, the obtained PP crystallites were smaller or less perfect, probably because they were formed around the filler particles and as a result altered the structure of the sample. The size of ash particles can also influence the morphology of the crystalline state (Yuan et al. (2005)). In the case of Pardo et al. (2013), the size distribution of filler particles was quite broad, probably promoting a heterogeneous crystalline state in the PP with different sizes and perfection in the crystallites.

PP/Silane Treated Ash Composites

Pardo et al. (2010) studied the thermal behavior of different PP/fly ash composites through Thermogravimetric Analysis (TGA). They found that the presence of ash led to improved thermal stability of PP, being this effect greater in the composites containing treated ash. In addition, composites with ash treated with vinyl and amino silanes presented higher degradation temperatures (higher thermal stability) than the composite with ash treated with vinyl-benzylamine silane as a result of the stronger filler/matrix adhesion (stronger bonds) in the former composites.

The nucleation effect observed for PP/raw ash composites by Pardo et al. (2013) mentioned above, was similar when ash treated with silane was incorporated into the composites: crystallization temperature (T_c) increased with filler addition. However, for the composites with treated ash, the degree of crystallinity was similar to that of the PP matrix. In spite of the nucleating effect of the ash surface as a result of the treatment with silane, the excess silanols in the samples may be crosslinked, hindering PP chain motion and reducing its crystallinity. The two effects, the nucleating effect of ash particles and the reduction of chain motion due to crosslinking, are competitive and overcome one or another depending on ash content. It was observed that with 10 and 20 wt% of treated ash, the crystallinity degree was similar to that of raw PP. But with 30 wt% ash, the nucleating effect of the filler increased the crystallinity degree, although the ash was treated with silane.

PP/MAPP/Ash Composites

The presence of MAPP in the formulation of PP based composites has been reported to change the crystallization behavior of PP leading to a decrease in the crystallinity and hence, in the matrix stiffness (Kim et al. (2007)). The decrease in the regularity of the chains as a result of the modification by maleic anhydride units leads in this case to reduced crystallization tendency (decreased crystallinity and melting point) (Menyhárd and Varga (2006); Seo et al. (2000); Cho et al. (1999); Li et al. (2003). On the other side, ash particles are expected to act as nucleating agents making the spherulite size decrease and the crystallinity increase (Yu and He (2000)).

However, Igarza et al. (2013) did not found a clear influence of MAPP over the crystallization process of PP. PP/ash composites with and without coupling agent exhibited similar values of crystallization temperature (T_c), melting temperature (T_m) and degree of crystallinity (x_c). The expected behavior was only observed in the case of the crystallization temperature: an increase in T_c in the composites prepared using both neat PP or PP/MAPP as matrix was found.

PP/OBC/Treated Ash Composites

For PP/OBC/ash composites, Pardo et al. (2013) observed that T_c remained without variations with respect to the value for PP. The nucleating effect of ash seemed to be counteracted by the presence of OBC. Other authors had previously shown that the crystallization temperature of PP decreased in blends with olefinic-copolymers (like OBC) (Jang et al. (1985)). The reason is that the olefinic block tends to diffuse into the PP matrix and can hinder access of PP chains to the arrangement and growth of nuclei, resulting in a less crystalline matrix. The effect was more pronounced in the PP/OBC/ash composites of Pardo et al. (2013) when the OBC:ash ratio was 1:1, because the amount of ash was low and there was more free OBC, not associated to ash particles, forming microdomains in the PP matrix. For the other OBC:ash ratios investigated (1:2 and 1:3), the OBC was mainly placed around ash particles and affected crystallinity of PP at the interfacial regions. The decrease in the T_c values was minor or even disappeared in those cases.

It is also worth noting that Pardo et al. (2013) observed that the DSC thermogram for the composite with 10 % of OBC and 30 % ash showed a shoulder centered at 150 °C, which indicated that the crystalline structure of PP had a small β polymorphous portion, as also observed by others (Liang et al. (2000); Liang (2007)). These authors proved that the particle surfaces of rigid fillers can act as nucleation site for β crystals. As it was mentioned before, Nath et al. (2010) has also confirmed how coal fly ashes have acted as heterogeneous β-nucleating agents inducing a secondary β-crystalline phase in PP/ash composites using a higher amount of ash.

In summary, the ash inclusion in PP matrix changes its crystalline state and improves the thermal stability of the polyolefin. The modifications in thermal properties of ash composites mainly depend on the ash amount and the surface treatment applied to the ash particles.

4.4. Deformation Behavior

In many applications of polymers, mechanical properties are very important and, particularly, the optimization of the material toughness is often required. For this purpose, knowledge of the relationship between morphology and deformation behavior is needed. In the following, the results obtained on the deformation behavior of different PP/fly ash composites with and without coupling agents are presented.

PP/Raw Ash Composites

Pardo et al. (2010) studied the deformation behavior of PP/fly ash composites with different ash content. They observed that the incorporation of ash to PP led to more brittle and weaker materials, as both tensile strength and elongation at break were found to significantly decrease with ash content. However, the material stiffness was observed to increase from the incorporation of a stiffer second phase (Siriwardena et al. (2003)) and this effect was found to be more pronounced as the ash content increased (Beltran and Mijangos (2000); Martí-Ferrer et al. (2006)).

The observed loss of strength has been attributed by Pardo et al. (2010) to the lack of interfacial adhesion between PP and ash. It had been previously reported in the literature (Wong and Mai (1999)) that poor tensile strength and low failure strain were caused by particle debonding from the matrix prior to yielding as a result of poor interfacial adhesion. The presence of big particles with a wide particle size distribution found in the composites developed by Pardo et al. (2010) has been considered by these authors as a contributing factor to the observed embrittlement effect.

Nath et al (2009) studied the tensile behavior for different PP/fly ash composites with varying filler content at different test temperatures. They found that, generally, tensile strength values were lower than that of neat matrix and that the drop increased with increase of fly ash content. They attributed this to the fact that as fly ash content increased, larger defects were created, which increased in the material linear elastic fracture mechanics type of failure. At higher temperatures, they observed a small increase in strength between 0 and 20% fly ash content due to the enhanced free volume of the thermoplastic polymer at the elevated temperature. This was filled up by the fly ash apparently causing preferable wetting between the PP and the fly ash, thereby increasing the strength.

PP/Silane Treated Ash Composites

Pardo et al. (2010) also analyzed the effect of different silane-type coupling agents on the deformation behavior of PP/fly ash composites. They observed that coupling agents containing amine and vinyl functional groups gave higher stiffness and strength values compared to composites without coupling agent. In contrast, the system with the coupling agent having the vinyl-benzylamine group led to a more ductile composite. Vinyl and amino reactive groups seemed to have shorter and more rigid couplings with the polyolefin than the vinyl-benzilamine groups whose molecular structure produced more flexible bonds with the polypropylene chains. Based on this fact, Pardo et al (2010) attributed their results on Young´s modulus more to the presence of the stiffer second phase and to the stiffness of the coupling rather than to interfacial adhesion which was not expected to have a significant influence on material stiffness (Lauke et al. (1994)).

In the case of the strength, on the other hand, the degree of filler-matrix interaction promoted by the coupling agent was found to play an essential role (Pardo et al. (2010)). Composites with coupling agents containing amine and vinyl functional groups were found to be stronger than the unmodified system or the system with the vinyl-benzilamine group, as a result of the better interaction between PP and ash in the former systems.

Pardo et al. (2009) also studied the deformation behavior of PP composites reinforced with different contents of ash treated with a silane-type coupling agent having an amine functional group (GF96 Wacker-Chemie Italia S.r.L.). They observed that, independently of the presence of the coupling agent, all composites displayed less ductility and strength than PP matrix as a result of a poor interfacial adhesion between both components and the subsequent large particle debonding which induced premature failure. The incorporation of the silane was found to lead to composites with slightly better properties. This was attributed by the authors to an improvement in the ash particle dispersion and/or to a better interaction between PP and ash.

PP/MAPP/Ash Composites

Igarza et al. (2013) compared the uniaxial tensile behavior of PP/ash composites with different ash content without and with maleic anhydride grafted PP. They observed that, irrespectively of the presence of MAPP, all composites exhibited semiductile tensile behavior displaying some amount of plastic deformation after maximum stress (yield point). For the composites without coupling agent, the amount of plastic deformation was found to decrease as the ash content increased. Composites samples containing MAPP, on the other hand, failed almost immediately after the maximum or even at this point, showing no significant effect of ash content.

The addition of ash to PP led to an increase in stiffness as expected from the incorporation of a much stiffer second phase (Móczó and Pukánszky (2008); Fu et al. (2008)), whereas a significant reduction in both tensile strength and strain at break values was observed. This was attributed to the debonding of ash particles from the PP matrix (Wong and Mai (1999)).

In addition, the incorporation of MAPP in the composites formulation led to a slight increase in stiffness, suggesting better dispersion of ash particles in the PP matrix. Furthermore, tensile strength and elongation at break for the PP/MAPP/ash composites exhibited a significant change in their behavior as they were roughly independent of ash content. While tensile strength maintained a relatively high value within the composition range investigated by Igarza et al. (2013), elongation at break greatly decreased for 10 wt% ash and limited ductility was observed for all composites independently of filler loading.

A negative effect from the addition of MAPP on the elongation at break had been previously reported by others (Bikiaris et al. (2005)) and explained in terms of a greater interaction between filler and matrix. This improved adhesion, facilitates the transfer of properties between the two materials. Therefore, rigid particles which are unable to deform hinder composites elongation, making them less ductile.

In the composites of Igarza et al. (2013), the reduction of the PP cross-section surface area and its substitution by the rigid ash particles unable to elongate, as well as the presence of big particles, acted as stress raisers and failure points. All these factors, in addition to the improved interfacial adhesion promoted by MAPP, had a concomitant negative effect on the material ductility, independently of ash content.

PP/OBC/Silane Treated Ash Composites

Ares et al (2010) compared the tensile response of PP/fly ash composites with and without an olefin block copolymer (OBC). They found that composites without OBC presented complete break immediately after maximum stress and the specimens broke without necking. Nevertheless, load-displacement curves showed a small nonlinear region before reaching maximum stress causing whitening of the fractured surface. PP/OBC/fly ash composites, on the other hand, displayed a more ductile behavior than PP matrix or composites without OBC. These specimens presented clear stress whitening and slight necking before the break point.

Composite stiffness was found to grow with respect to the matrix proportionally to ash content. However, the incorporation of OBC led to a decrease in composite modulus. Tensile strength values, on the other hand, were found to decrease slightly with ash content but they kept reasonably acceptable taking into account that a rigid filler was incorporated into PP. The presence of silane in the composites developed by Ares et al. (2010) contributed to this result by increasing the interaction between the ash particles and the PP molecules. Strain at break values exhibited a drastic drop (less than half the value of the matrix) from the incorporation of ash into PP. However, the addition of OBC allowed enhancing ductility although the improvement depended on the OBC/ash ratio.

4.5. Fracture Behavior under Different Loading Conditions

Quasi-Static Loading Conditions

PP/Ash Composites with Raw Ash and Silane-Treated Ash

Pardo et al. (2009) studied the fracture of PP/raw ash composites. They observed that the PP matrix (considering the blend of PP and a processing additive as the PP matrix) and the composites displayed ductile fracture with stress-whitened fracture surfaces.

Pardo et al. (2009) also analyzed the effect of a silane-type coupling agent containing amine functional groups on the quasi-static fracture behavior of PP/fly ash composites. Irrespectively of the presence of the coupling agent, they found a decreasing trend of the fracture initiation parameters with increasing ash content. They attributed this result to the increase in the number of critical-size flaws that induced premature failure. However, all composites displayed higher critical initiation values than neat PP or the matrix as a result of the toughening induced by the presence of small ash particles. In addition, a significant improvement in the crack propagation resistance was found by Pardo et al (2009) for the composites in comparison to the matrix, as more energy was absorbed by the composite specimens to fracture. Poor adhesion between PP and ash was considered by Pardo et al (2009) as responsible for debonding at the PP/ash interface and hence, for inducing matrix stretching around small particles. This mechanism required a large amount of energy (Vu-Khanh and Fisa (1986); Vu-Khanh and Fisa (1990)). Furthermore, composite samples with treated ash displayed slightly higher values of fracture parameters, suggesting a positive effect of the incorporation of the silane with amine functional groups to ash.

PP/MAPP/Ash Composites

Igarza et al. (2013) observed in PP/MAPP/ash composites that, irrespective of the ash content used, all materials in quasi-static fracture tests displayed non-linear load-displacement behavior with stable crack growth and their fracture surfaces were found to be highly stress-whitened.

The presence of ash led to improved fracture properties with respect to the matrix due to the development of the toughening mechanisms of particle debonding and subsequent matrix ductile tearing induced by the presence of ash particles mentioned before (Pardo et al. (2009)). However, the incorporation of MAPP in the composites formulation led to reduced fracture toughness values which, in the case of the energy release rate, were even lower than those of the matrix and were much lesser dependent on ash content. Igarza et al. (2013) suggested that the increased adhesion between ash and PP achieved from the addition of MAPP, hindered the toughening mechanisms induced by the ash particles independently of ash content.

PP/OBC/Silane Treated Ash Composites

Pardo et al. (2013) reported the load–displacement curves obtained in quasistatic three-point-bending tests for PP/OBC/ash composites. They observed that all PP/OBC/ash composites displayed ductile fracture with stress-whitened fracture surfaces. In fact, the area under load–displacement curves was found to be significantly larger than that of their binary PP/ash composites (Pardo et al. (2009)).

Values of the fracture parameters (K_{IQ}, G_{IQ} and W_f) obtained by Pardo et al. (2013) indicated that the incorporation of the OBC copolymer to PP led to improved fracture behavior in comparison with the PP matrix or the binary PP/ash composites previously investigated (Pardo et al. (2009)).

In addition, a decreasing trend of the fracture toughness parameters with increasing ash content was reported by Pardo et al. (2013) and it was attributed to the increase in the number of critical-size flaws that induced premature failure. However, all composites displayed higher critical initiation values than the matrix as a result of the toughening mechanism induced by the presence of ash and OBC.

Also note that the improvement observed by Pardo et al. (2013) in the case of the fracture energy of the PP/OBC/ash composites (critical initiation value (G_{IQ}) and work of fracture value (W_f) was remarkable, as much more energy was absorbed by these composites. To induce matrix stretching around filler particles and debonding at the interfaces, a weak adhesion between polymer matrix and filler is needed. As the same authors had mentioned in a previous article (Ares et al. (2010)), the copolymer transformed rigid ash particles into soft particles with a hard core and a soft shell. Because of this, the particles could better absorb the energy and the material toughness increased. On the other hand, a part of the OBC segregated as almost spherical inclusions within the continuous PP matrix and around the ash particles, which have been taken also as contributing factors to toughening. These two effects made ternary composites to absorb significantly higher deformation energy than binary composites.

Impact Loading Conditions

Force-displacement records obtained on SENB specimens in impact for PP/ash composites with and without MAPP by Igarza et al. (2013) showed that fully brittle behavior was exhibited for all materials, irrespectively of ash content. Under impact loading, plastic deformation of the matrix is suppressed by the high strain rate as well as the constraint imposed by the rigid filler. In addition, the material in front of the crack tip is subjected to plane-strain conditions and the crack propagates through the matrix with little or no plastic deformation (Histrov et al. (2004)). Igarza et al (2013) observed that the addition of MAPP led to more repetitive records, suggesting improved dispersion of ash particles in the PP matrix.

Macroscopically, fracture surfaces of samples broken in impact fracture tests did not exhibit stress whitening confirming the absence of the matrix plastic deformation and being in agreement with the fully brittle load-displacement records observed.

For all composites, critical initiation energy release rate values obtained in impact were higher than those of PP and they also displayed a maximum at about 20 wt% ash. The important decrease of the scatter of experimental data for the PP/MAPP/ash composites investigated by Igarza et al. (2013) was also evident in critical initiation energy release rate values.

The maximum in the impact energy observed for the composites with 20 wt % ash, was related to the crystallinity of PP. A maximum in impact strength had been previously reported in the literature (Xu et al. (2001)) for PP modified with different contents of an effective nucleating agent. This maximum agreed in the materials of Igarza et al. (2013) with a minimum in the degree of crystallinity. It is thought that crystallites act as stress concentrators, intensifying the stress locally far above the applied stress and hence, they probably reduce the material susceptibility to multiple crazing and shear yielding (Galeski (2003)).

Furthermore, the presence of MAPP in the formulation of the composites of Igarza et al. (2013) had a detrimental effect on the materials impact fracture behavior in agreement with Bikiaris et al. (2005).

4.6. Failure Mechanisms

It has been established in the literature (Zhao and Li (2005)) that there are two main approaches to study the materials fracture behavior. One is to develop a quantitative calculation of fracture toughness values by fracture mechanics. Another approach is to disclose and analyze the fracture morphology directly by microscopy. In the following, a brief review of recent studies about the failure behavior of different PP/ash composites is presented.

PP/Raw Ash Composites

From SEM micrographs of lateral views of PP/ash samples tested up to subcritical displacement levels, Pardo et al (2009) observed that small ash particles (particle size smaller than 1 μm) debonded from the matrix, which also elongated around them. The difference in the elastic properties between PP and ash made small ash particles act as stress concentrators during deformation processes and the triaxial stress around them to build up. Interfacial debonding occurred, changing the stress state in the surrounding matrix. This mechanism of toughening had been proposed by Kim and Michler (Kim and Michler (1998)) for particle-filled semicrystalline polymers and later observed by others (Zuiderduin et al. (2003); Lazzeri et al. (2005)) for different polypropylene composites.

Hence, in the PP/ash composites developed by Pardo et al. (2009), as a result of poor adhesion, the debonding of small ash particles induced the deformation mechanism in the matrix, which led to higher energy absorption in both crack initiation and propagation steps and composites displayed higher values of fracture parameters than the PP matrix. On the other hand, debonding of large particles did not seem to contribute to the deformation mechanism but led to the formation of critical-sized flaws. Consequently, the increase in ash content led to a decrease in the tensile and fracture properties due to the greater number of large particles able to induce premature failure.

PP/Silane Treated Ash Composites

Pardo et al. (2010) obtained SEM micrographs of fracture surfaces of specimen broken in tensile tests of PP/fly ash composites using three different silane-based coupling agents. They observed that the composite with coupling agent containing vinyl-benzylamine groups exhibited ductile tearing of the PP matrix widely spread throughout the material whereas composites with amino-silane and vinyl-silane, in contrast, showed more restricted and localized matrix yielding. Poor interfacial adhesion between ductile polymers and rigid fillers promotes debonding of the particles from the matrix, which subsequently causes ductile tearing around these particles.

Pardo et al. (2009) also analyzed by SEM the fracture surfaces of composites SENB samples with different contents of ash treated with a silane containing amine functional groups. They clearly observed matrix ductile tearing, confirming the ductile behavior exhibited by these composites in quasistatic fracture tests. These authors also performed energy-dispersive X-ray analysis from a line that crossed an ash particle and the interface. They observed that several elements other than carbon such as Al, Si, Ca, etc. existed, indicating that the treatment with silane led to the formation of an interface around ash particles. Therefore, filler-matrix interaction seemed to be improved by this treatment. The coupling agent containing a primary amine as organofunctional group reacted instantaneously with the carboxyl groups of the polymer forming secondary amide bonds. During processing, the secondary amide groups took part in further reactions with the continuously forming carboxyl groups, resulting in imide groups. These fast reactions of the silane and the polymer might lead to the enhanced interaction between the components of PP/ash composites (Pardo et al. (2009)).

It has been established in the literature (Thio et al. (2004)) that weaker interaction between rigid particles and the polymer matrix leads to higher toughness, as the debonding process is favored. Hence, the slight improvement in the materials fracture properties observed by Pardo et al. (2009) mentioned before from the incorporation of silane have been attributed by these authors more to an improvement of the dispersion of ash particles in the PP matrix rather than to the observed increased interaction between PP and ash.

PP/MAPP/Ash Composites

In SEM micrographs of fracture surfaces of PP/ash composites without and with MAPP, Igarza et al. (2013) clearly observed that ash particles debonded from the PP matrix which underwent ductile tearing around these particles. However, the composites with MAPP exhibited a lower degree of ductile tearing of the matrix material as PP fibrils appeared shorter. This was in agreement with the important decrease in toughness observed for PP/MAPP/ash composites respect to the composites without coupling agent.

No signs of fibrils of PP indicative of the matrix ductile tearing were observed in the SEM micrographs of fracture surfaces of samples broken in impact tests (Igarza et al. (2013)), thus confirming that the amount of plastic deformation on the fracture plane was rather limited. Therefore, a small amount of energy was dissipated within the damage zone and hence, the measured toughness was low (impact fracture energy values were significantly lower than quasi-static values) (Zebarjad et al. (2003)).

Many particles were present on the fracture surface of the PP/ash composites without coupling agent (Igarza et al. (2013)). Most of these particles were nearly clean, suggesting that during crack propagation, the fracture front preferentially passed from the interface rather than going through ash particles. This indicated a rather poor adhesion between ash and PP in this case.

On the other hand, the incorporation of MAPP in the composites formulation led to a more homogeneous dispersion of ash particles in the PP matrix and also improved interfacial adhesion between ash and PP, as ash particles were not distinguishable.

PP/OBC/Silane Treated Ash Composites

From scanning electron micrographs of the fracture surfaces of PP/OBC/ash specimens Pardo et al. (2012) observed that composites with OBC exhibited ductile behavior; with ductile tearing of the PP matrix. This ductile tearing was greater than that observed in their PP/ash composites without OBC (Pardo et al. (2009)). In addition, particles debonded from the PP matrix, which elongated around them. Furthermore, these authors found evidences of OBC inclusions into PP. This failure mechanism had been previously reported for binary PP/ash composites as mentioned before (Ares et al. (2010)). In addition, ash particles appeared encapsulated by OBC and fibrillation occurred through the OBC layer that covered ash particles.

More detailed views of SEM images of PP/OBC/ash composites (Figure 2) (Pardo et al. (2012)) showed how fibrils of OBC were formed around ash particles. Despite prior initiation of particle debonding, these fibrils allowed stress transfer between the matrix and the filler. Thus, the copolymer created a soft interface for stress transfer. The ash particles were coated with a thin and soft copolymer layer that could be deformed plastically, and OBC fibrils were stretched between filler particles and the PP matrix. These fibrils seemed to have been formed at some stage of the deformation process, and their orientation created a plastic deformation zone. In addition, SEM fractographs revealed how PP and OBC acted like a single matrix. Although it was known that OBC inclusions existed within the PP matrix, they could not be distinguished due to the extensive ductile tearing and stretching.

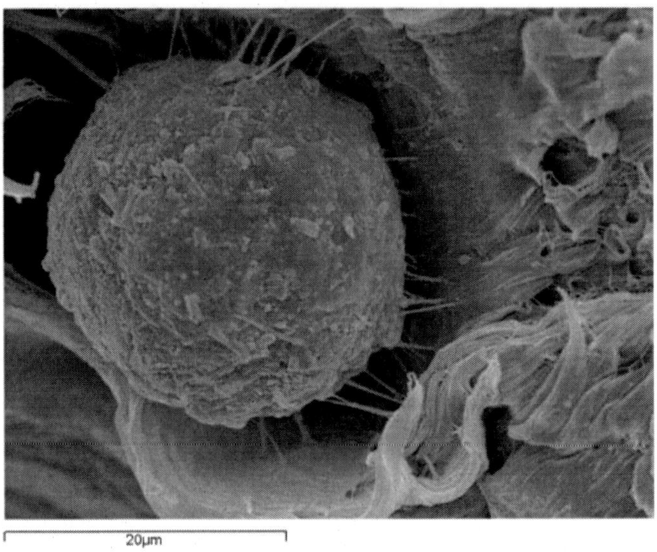

Figure 2. Detailed view of a SEM image of a PP/OBC/ash composite investigated by Pardo et al. (2012).

The main aim of OBC incorporation was to enhance toughening as a result of a better composite morphology. Some adhesion between the silane-modified ash particles and the matrix seemed to exist, deformation occurring through a debonding process with fibrillization at the interface between OBC-encapsulated ash particles and matrix. In the first stage, stress concentration occurred around these particles and they were slightly stretched or deformed. Due to the existence of some interfacial adhesion, fibrils were formed within the OBC layer, which covered the ash particles and PP and simultaneously shear bands appeared in the matrix. A rich OBC matrix region was formed around the ash particles and its deformability stimulated the increase in fibrils. By increasing the deformation process, the fibrils formed at the polar regions of the particles break down, and only a few fibrils remained at the equator. In the last stage, further shear flow of the matrix was considerably enhanced. Hence, a fibrillized debonding process at the interface has been considered by Pardo et al. (2012) as one of the major toughening mechanisms (Ares et al. (2010); Kim and Michler (1998)). Even so, the presence of critical large ash particles still led to the decrease in the fracture parameter values with ash content (Pardo et al. (2009); Liang et al. (2000); Liang (2007)).

CONCLUSION

The most relevant recently published results about the development and characterization of composites based on Polypropylene reinforced with fly ash have been reviewed.

The results reported in the literature on the morphological and chemical characteristics of fly ash are really diverse hence, a generalization of the influence of different parameters on the performance of PP/ash composites is impossible. An individual analysis of each system must be carried out.

Regarding the deformation behavior of PP/ash composites, in general, the incorporation of fly ash into PP leads to stiffer but weaker materials and this is more noticeable as filler content increases. However, the changes in interfacial adhesion could alter the latter trend.

Under both quasi-static and impact fracture tests, enhancements in fracture toughness are generally observed with respect to the PP matrix due to the toughening mechanisms derived from the presence of ash particles. Nevertheless, a decreasing trend of toughness is often found with increasing fly ash content as a result of the increased number of critical-sized flaws that induce premature failure.

Coupling agents and compatibilizers are frequently added in the composites formulations in order to improve ash dispersion in the PP matrix, interfacial adhesion between phases and/or to enhance composites processability. Nevertheless, their presence complicates the scenario and the evaluation of the different influencing factors on the composites performance becomes extremely difficult.

The influence of fly ash on the crystallization behavior of PP is also a key factor and a source of complexity that must be taken into account.

FUTURE TRENDS

Finally, much more work is still needed to take full advantage of fly ash as reinforcement or toughener of polymers and particularly, of Polypropylene. Low cost processing techniques to improve the quality of fly ash (control of particle size and its distribution, morphology) and the distribution and dispersion of the filler in the polymer, optimization of interfacial properties as a function of final requirements, etc. are some of the main routes to reach that goal.

The capability of fly ash particles as nucleating agents for the β phase of PP should also be fully exploited when materials with better ductility, toughness or weldability are required.

Finally, only once the performance of PP/ash composites can be optimized, their use as reinforcement of polymers will be extended to a wide range of applications and hence, massive utilization of ash will be possible.

REFERENCES

Agrawal P., Oliveira S.L., Araujo E.M., & Melo T.J.A., (2007). Effect of different polypropylenes and compatibilizers on the rheological, mechanical and morphological properties of nylon 6/PP blends, *J. Mater. Sci.*, 42, 5007-5012.

Ares A, Santiago GP, Abad MJ, Jesus C & Luis B. (2010) Effect of aminomethoxy silane and olefin block copolymer on rheomechanical and morphological behavior of fly ash-filled polypropylene composites. *Rheol Acta*; 49: 607–618.

Argon AS & Cohen RE. Toughenability of polymers. (2003) *Polymer*; 44(19): 6013–6032.

Bartczak Z, Argon AS, Cohen RE, Weinberg M. (1999) Toughness mechanism in semi-crystalline polymer blends: I. High density polyethylene toughened with rubbers. *Polymer*; 40:2331.

Bartczak Z, Argon AS, Cohen RE & Weinberg M. (1999) Toughness mechanism in semicrystalline polymer blends: II. High-density polyethylene toughened with calcium carbonate filler particles. *Polymer*; 40: 2347–2365.

Bikiaris DN, Vassillou A, Pavlidou E & Karayannidis GP. (2005) Compatibilisation effect of PP-g-MA copolymer on iPP/SiO2 nanocomposites prepared by melt mixing. *European Polymer Journal*, 41: 1965-1978.

Bose S. & P.A. Mahanwar, (2004) Effect of Flyash on the Mechanical, Thermal, Dielectric, Rheological and Morphological Properties of Filled Nylon 6. *J. Miner. Mater. Character.* Eng., 3, 65-72.

Chan CM, Wu JS, Li JX, Cheung YK. (2002) Polypropylene/calcium carbonate nanocomposites. *Polymer*; 43(10): 2981–2992.

Chaudhary D.S., Jollands M.C., & F. Cser, (2004) Recycling rice hull ash: A filler material for polymeric composites. *Adv. Polym. Tech.*, 23, 147.

Chand N. & S.R. Vashishtha, (2000) Utilization of fly ash from coal-fired power plants in China *Clay. Res.*, 19, 49.

Chand N. & N. Khare, (1999) Effect of flyash addition on dielectric properties of polypropylene, *Indian J. Eng. Mater. Sci.*, 6, 342-345.

Chand N. & N. Khare, (2000) Effect of flyash loading on dielectric relaxation of polypropylene/low density polyethylene blends. *Met. Mater. Process.*, 12, 319-326.

Chand N. and A. Pandey, (2000) High stress abrasive wear study on fly ash filled polypropylene/polystyrene blend. *Met. Mater. Process.*, 12, 91-95.

Cho K, Li F & Choi J. (1999) Crystallization and melting behavior of polypropylene and maleated polypropylene blends. *Polymer*, 40: 1719-1728.

Chum Steve P & Swogger Kurt W. (2008) Olefin polymer technologies history and recent progress at the Dow chemical company. *Prog Polym Sci*; 33(8): 797–819.

Dangtungee R, Yun J, & Supaphol P. (2005) Melt rheology and extrudate swell of calcium carbonate nanoparticle-filled isotactic polypropylene. *Polym Test*; 24(1):2–11.

Department. U.S. Department Energy. Available at: http://www.netl.doe.gov/.

Dias P., Lin Y.J., Poon B., Chen H.Y., Hiltner A. & Baer E. (2008) Adhesion of statistical and blocky ethylene–octene copolymers to polypropylene. *Polymer*. Vol. 49, 2937-2946.

Filipe S., M. Joao, A. Duarte, C.R. Leal, & M.T. Cidade, (2005) *J. Appl. Polym. Sci.*, 98, 694.

Fu S-Y, Feng X-Q, Lauke B & Mai Y-W, (2008) Effects of particle size, particle/matrix interface adhesion and particle loading on mechanical properties of particulate–polymer composites. *Composites: Part B.*, 39: 933-961.

Fu Q & Wang G. (1992) Polyethylene thoughened by rigid inorganic particles. *Polym Eng Sci*; 32(2): 94–97.

Fuad MYA, Hanim H, Zarina R, Mohd Ishak ZA & Hassan A. (2010). Polypropylene/calcium carbonate nanocomposites – effects of processing techniques and maleated polypropylene compatibiliser. *Exp Polym Lett*; 4(10): 611–620.

Galeski A. Strength and toughness of crystalline polymer systems. (2003) *Progress in Polymer Science*, 28: 1646-1699.

Gensler R, Plummer CJG, Grein C & Kausch HH. (2000) Influence of the loading rate on the fracture resistance of isotactic polypropylene and impact modified isotactic polypropylene. *Polymer*; 41: 3809–3819.

Hristov VN, Lach R & Grellmann W. (2004) Impact fracture behavior of modified polypropylene/wood fiber composites. *Polymer Testing*, 23: 581-589.

Hwang J.Y., Beneficial Use of Fly Ash, National Energy Technology Department, US Department of Energy, USA, http://www.nctl.doc.gov/

Ismail H.& H.B. Hong, C.Y. Ping, and H.P.S. Abdul Khalil, (2002) *J. Reinf. Plast. Compos.*, 21, 1685.

Jang BZ, Uhlmann DR & Sande JBV. (1985) Rubber-toughening in polypropylene. *J Appl Polym Sci*; 30: 2485–2504.

Joseph S., A.R.R. Menon, J. Aju, & S. Thomas, (2007) Melt rheology and extrudate morphology studies of polystyrene/polybutadiene blends in the presence and absence of compatibilisers *J. Mater. Sci.*, 42, 2054.

Kamdar A.R., Ayyer R.K., Poon B.C., Marchand G.R., Hiltner A., & Baer E. (2009) Effect of tie-layer thickness on the adhesion of ethylene–octene copolymers to polypropylene, 50, 3319-3328.

Karger-Kocsis J, (2000) Reinforced Polymer Blends, in Polymer Blends, Volume 2: Performance, ed by Paul DR and Bucknall CB, New York, Wilcy.

Kim D, Fasulo PD, Rodgers WR, Paul DR. (2007) Structure and properties of polypropylene-based nanocomposites: Effect of PP- g-MA to organoclay ratio. *Polymer*, 48: 5308-5323.

Kim G. M. and G. H. Michler, (1998) Micromechanical deformation processes in toughened and particlefilled semicrystalline polymers: Part 1. Characterization of deformation processes in dependence on phase morphology, *Polymer* 39, 5689–5697.

Kim G. M. and G. H. Michler, (1998) Micromechanical deformation processes in toughened and particlefilled semicrystalline polymers. Part 2: Model representation for micromechanical deformation processes, *Polymer* 39, 5699–5703.

Kiss A, Fekete E, Pukánszky B. (2007) Aggregation of CaCO3 particles in PP composites: effect of surface coating. *Compos Sci Technol*; 67(7–8):1574–83.

Kruger R.A., M. Hovy, & D. Wardle, The Use of Fly Ash. Fillers in Rubber, in International Ash Utilization Symposium, Paper 72 Lexington, Kentucky (USA) (1999).

Lauke B., W. Beckert, & K. Schneider, (1994) Interlaminar shear strength evaluation of curved composite samples *Appl. Compos. Mater.*, 1, 267-271.

Lazzeri A., S. M. Zebarjad, M. Pracella, K. Cavalier & R. Rosa, (2005) Filler toughening of plastics. Part 1. The effect of surface interactions on physico-mechanical properties and rheological behavior of ultrafine $CaCO_3$/HDPE nanocomposites, *Polymer* 46, 827–844.

Li J, Zhou C, & Gang W. (2003) Study on nonisothermal crystallization of maleic anhydride grafted polypropylene/montmorillonite nanocomposite. *Polymer* Testing, 22: 217-223.

Liu, G., Zhang, X., Liu, C., Chen, H., Walton, K. & Wang, D. (2011) Morphology and mechanical properties of binary blends of polypropylene with statistical and block ethylene-octene copolymers, Journal of Applied Polymer Science. 119, 3591-3597.

Liang JZ, Li RKY & Tjong SC. (2000) Impact fracture behavior of PP/EPDM/glass bead ternary composites. *Polym Eng Sci*; 40: 2105–2111.

Liang J-Z. (2007) Impact fracture toughness of hollow glass bead filled polypropylene composites. *J Mater Sci*; 42: 841–846.

Liu G, Zhang X, Li X, Chen H, Walton K & Wang D. Correlation of miscibility and mechanical properties of polypropylene/olefin block copolymers: effect of chain composition. *J Appl Polym Sci,* 2012; 125: 666–675.

Liu G, Zhang X, Liu C, Chen H, Walton K & Wang D. (2011) Morphology and mechanical properties of binary blends of polypropylene with statistical and block ethylene-octene copolymers. *J Appl Polym Sci*; 119: 3591–3597.

Matsunaga T., J. K. Kim, S. Hardcastle & P. K. Rohatgi, (2002) *Mat. Sci. Engng* 325, 333.

Menyhárd A, & Varga J. (2006) The effect of compatibilizers on the crystallisation, melting and polymorphic composition of β-nucleated isotactic polypropylene and polyamide 6 blends. *European Polymer Journal*, 42: 3257-3268.

Móczó J, & Pukánszky B. (2008) Polymer micro and nanocomposites: Structure, interactions, properties. *J. Ind. Eng. Chem.*, 14: 535-563.

Nath DCD, Bandyopadhyay S, Yu A, Zeng Q, Das T, Blackburn D & White C. (2009) Structure-property interface correlation of fly ash-isotactic polypropylene composites. *J Mater Sci* 44: 6078-6089.

Nath DCD, Bandyopadhyay S, Yu AB, Blackburn D & White C. (2010) Novel observations on kinetics of nonisothermal crystallization in fly ash filled isotactic polypropylene composites. *J Appl Polym Sci*; 115: 1510–1517.

Nath DCD, Bandyopadhyay S, Yu A, Blackburn D, White C & Varughese S. (2010) Isothermal crystallization kinetics of fly ash filled iso-polypropylene composite and a new physical approach. *J Therm Anal Calorim*; 99: 423–429.

Oksman K and Lindberg H. (1998) Influence of thermoplastic elastomers on adhesion in polyethylene wood flour composites. *J Appl Polym Sci*; 68: 1845–1855.

Pardo SG, Bernal C, Abad MJ, Cano J, & Barral Losada L. (2009) Deformation and Fracture Behavior of PP/Ash Composites. *Composite Interfaces, Special Issue*, 16: 97-114.

Pardo SG, Bernal C, Abad MJ, Cano J, & Ares A. (2012). Fracture and thermal behaviour of biomass ash polypropylene composites. *J Thermoplast Compos Mater*. doi: 10.1177/0892705712452740

Pionteck J, Sadhu VB, Jakisch L, Pötschke P, Häubler L, & Janke A. (2005) Crosslinkable coupling agents: synthesis and use for modification of interfaces in polymer blends. *Polymer*; 46 (17):6563–74.

Rohlmann CO, Failla MD, & Quinzani LM. (2006) Linear viscoelasticity and structure of polypropylene-montmorillonite nanocomposites. *Polymer*; 47(22): 7795–804.

Seo Y, Kim J, Kim KU, & Kim YC. (2000) Study of the crystallization behaviors of polypropylene and maleic anhydride grafted polypropylene. *Polymer*, 41: 2639-2646.

Siriwardena S., Ismail H., and Ishiaku U.S., (2001) Effect of Mixing Sequence in the Preparation of WRHA Filled EPDM/PP Blends. *Polym. Test.*, 20, 105-139.

Siriwardena S., Ismail H., & Ishiaku U.S., (2003) A Comparison of the Mechanical Properties and Water Absorption Behaviour of White Rice Husk Ash and Silica Filled Polypropylene Composites *J. Reinf. Plast. Compos.*, 22, 1645-1666.

Stefani, P.M. Cyras V., Tejeira Barchi A., & Vazquez A., (2006) Mechanical Properties and Thermal Stability of Rice Husk Ash Filled Epoxy Foams. *J. Appl. Polym. Sci.*, 99, 2957-2965.

Stefani P.M., Garcia D., Lopez J., & Jimenez A., (2005) Thermogravimetric analysis of composites obtained from sintering of rice husk-scrap tire mixtures. *J. Therm. Anal. Cal.*, 81, 315-320.

Stocchi A., E. Rodríguez, Vázquez A. & C. Bernal, (2013) Deformation and fracture behavior of vinylester/fly ash composites, *J. Applied Polym Sci.*, in press.

Suhara F. & Kutty S.K.N., (1998) Rheological Properties Of Short Polyester Fiber-Polyurethane Elastomer Composite With Different Interfacial Bonding Agents, *Polym. Plast. Technol. Eng.*, 37, 57-70.

Takahashi M., Hayashi J., & Suzuki S., (1992) Improvement of the rheological properties of the zirconia/polypropylene system for ceramic injection moulding using coupling agents, *J. Mater. Sci.*, 27, 5297-5302.

Thio Y. S., A. S. Argon & R. E. Cohen, (2004) Role of interfacial adhesion strength on toughening polypropylene with rigid particles, *Polymer* 45, 3139–3147.

Thio YS, Argon AS, Cohen RE, & Weinberg M. (2002) Toughening of isotactic polypropylene with CaCO3 particles. *Polymer*; 43(13):3661–74.

Thongsang S, & Sombatsompop N. (2006) Effects of NaOH and Si69 treatments on the properties of fly ash/natural rubber composites. *Polym Compos*; 27(1):30–40.

Vu-Khanh T & Fisa B. (1990) Effects of fillers on fracture performance of thermoplastics: strain energy density criterion. *Theor Appl Fracture Mech*; 13: 11–19.

Vu-Khanh T & Fisa B. (1986) Fracture behavior of mica-reinforced polypropylene: effects of coupling agent, flake orientation, and degradation. *Polym Compos*; 7: 219–226.

Wang M., Z. Shen, C. Cai, S. Ma, & Y. Xing, (2004) Experimental investigations of polypropylene and poly(vinyl chloride) composites filled with plerospheres. *J. Appl. Polym. Sci.*, 92, 126-131.

Williams JG. (1984) *Fracture mechanics of polymers*. London, UK: Ellis Horwood Limited.

Wong SC & Mai YW. (1999) Effect of rubber functionality on microstructures and fracture toughness of impact-modified nylon 6,6/polypropylene blends: 1. Structure property relationships. *Polymer*; 40: 1553–1566.

Xi XL, Fung KL, Li RKY, Tjong SC & Mai YW. (2002) Structural and mechanical behavior of polypropylene/ maleated styrene-(ethylene-co-butylene)-styrene/sisal fiber composites prepared by injection molding. *J Polym Sci Part B Polym Phys*; 40: 1214–1222.

Xu W, Liang G, Zhai H, Tang S, Hang G, & Pan W-P, (2001) Preparation and crystallization behaviour of PP/PP-g-MAH/Org-MMT nanocomposite. European Xu T, Yu J, Jin Z. Effects of crystalline morphology on the impact behavior of polypropylene. *Materials and Design*, 22: 27-31.

Yu J, & He J. (2000) Crystallization kinetics of maleic anhydride grafted polypropylene ionomers. *Polymer*, 41: 891-898.

Yuan Q, Jiang W, An L & Li RKY. (2005) Effects of filler size and heat treatment on the crystallization behavior of glass bead-filled polypropylene. *J Polymer Sci B Polym Phys*; 43: 306–313.

Yuan Q, Jiang W, An L, Christiansen JD & Li RKY. (2005) Competing effect between filled glass bead and induced b crystal on the tensile properties of polypropylene/glass bead blends. *J Appl Polym Sci*; 96: 1729–1733.

Zebarjad SM, Lazzeri A, Bagheri R, Seyed Reihani SM, Frounchi M. (2003) Fracture mechanism under dynamic loading of elastomer-modified polypropylene. *Materials Letters*, 57: 2733-2741.

Zhang QX, Yu ZZ, Xie XL, & Mai YW. (2004) Crystallization and impact energy of polypropylene/CaCO3 nanocomposites with nonionic modifier. *Polymer*; 45(17):5985–94.

Zhao H, & Li RKY. (2005) Crystallization, mechanical, and fracture behaviors of spherical alumina-filled polypropylene nanocomposites. *J Polym Sci Pol Phys*; 43:3652–64.

Zuiderduin WCJ, Westzaan C, Huetink J, & Gaymans RJ. (2003) Toughening of polypropylene with CaCO3 particles. *Polymer*; 44(1):261–75.

Zuiderdein W. C. J., C. Westzaan, J. Huétink & R. J. Gaymans, (2003) Toughening of polypropylene with calcium carbonate particles, *Polymer* 44, 261–275.

In: Fly Ash
Editor: Prabir Kumar Sarker

ISBN: 978-1-62948-044-2

Chapter 13

TECHNOLOGICAL INITIATIVES AND STATUS OF FLY ASH IN INDIA

Anjan K. Chatterjee*
Conmat Technologies Private Limited
Kolkata, India

ABSTRACT

The concern for promoting fly ash as an industrial resource effectively started in India in 1988 through the National Housing Policy providing financial incentives for use of fly ash bricks. The generation of fly ash at that time in the country was only 30 million tonnes and the disposal mode was wet. In the following two and a half decades, there has been a phenomenal increase in the generation of fly ash to about 200 million tonnes with large scale adoption of the dry disposal system. The environmental issues in dealing with such large volumes of fly ash are likely to become alarming with the projected generation of 450 million tonnes in 2021-22. At present, with all the engineering and scientific back-up the country is at a utilization of about 50% of the quantity generated primarily in cement, concrete and construction. Hence, the research community is now engaged in more intense characterization of fly ash and establishing more innovative and diversified application areas such as mechano-chemical activation of fly ash, development of geopolymeric materials, polymeric fillers, composite products, extraction of high value yielding constituents, conversion to nano-particulate substance, etc. This chapter is aimed at culling out a historical track of fly ash usage and research in India over the last few decades.

1. INTRODUCTION

At present India ranks third amongst the coal producing countries in the world. The known coal reserves as on 01 April 2012 were 293.5 billion tonnes, 6.66% of the World total.

* Corresponding Author address: CF 318, Sector – I, Salt Lake, Kolkata – 700 064, India. E-mail : technologiesconmat@gmail.com.

Most of the coal in India is of non-coking type with high to very high ash content, but has other quality parameters such as low sulfur content and low concentration of trace elements. The production level of coal in 2011-12 was 426 million tonnes per annum and about 88% of coal mined was consumed by the power (electricity) producing units in the country. On the other hand, the lignite reserves in the country are rather low, only 6.5 billion tonnes, and quite restricted in geographical occurrence in the Southern and Western parts of the country. Due to the abundance of coal reserves it is natural that the Indian Power Industry is overwhelmingly dependent on this fuel. According to the statistics of Central Electricity Authority (2013), the operative wing of the Ministry of Power, Government of India, the installed capacity of power generation as on 31 May 2013 was as given in Table 1.

Further according to the plans laid by the Central Electricity Authority (2012 a), the installed capacity of coal-based thermal power is likely to increase by about 116 GW by 2022. Therefore, it is obvious that there is awesome abundance of coal ash in India as a process by-product and its quantity will continue to increase as the generating capacity of coal and lignite based power will increase to meet the ever-increasing demand and supply gap of electricity in the country. This chapter is aimed at taking a stock of the generation and utilization of coal ash in a historical perspective and the future technological strategies of dealing with this refuse-turned-resource material.

Table 1. Installed capacity of power generation in India

Fuel /Type	Capacity (MW)	Percentage of the Total
Coal	131628	58.47
Diesel	20360	9.04
Gas	1200	0.53
Total Thermal	153188	68.04
Nuclear	4780	2.12
Hydro	39623	17.60
Renewable	27452	12.18
Grand Total	225133	100.00

2. Power Generation Process and Terminological Parlance of Coal Ash

For generation of power the traditional grate boilers with lump coal feed have been replaced by Pulverised Coal Bed (PCB) and Fluidized Bed Combustion (FBC) boilers. Although the coal quality in the system is low, as already mentioned, the final quality of coal ash depends on the milling system used for the preparation of coal for injection to the combustion chamber. Due to the process involved in firing of the boilers (Figure 1) about 75-80% of the total ash is finely divided and moves out of boilers along with the flue gases and is collected either by the mechanical means or by electrostatic precipitators. This ash is generally termed as “fly ash” or “pulverized fuel ash” (pfa). The remaining 20-25% of the ash is collected at the bottom of the air preheater, economizer and the boiler itself and is taken out by suitable facilities.

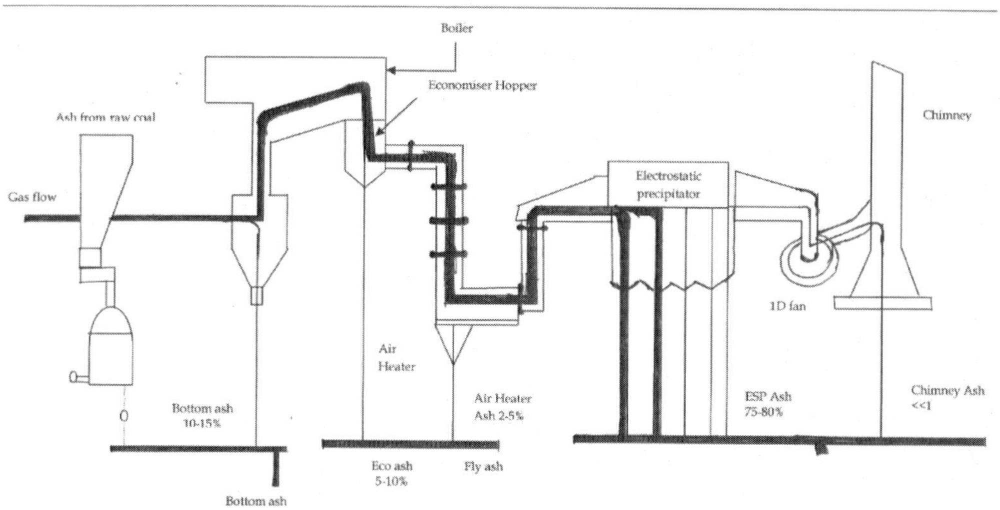

Figure 1. Ash accumulation points in a dry bottom furnace.

The coal ash collected from the bottom of the boiler is referred to as "bottom ash". When fly ash alone or with bottom ash is carried to a lagoon storage in the form of a slurry, it is termed as "pond ash", while the same material stored in a relatively dry form of deposition after conditioning is termed as "mound ash". Of all the above varieties of coal ash, the dry fly ash is obviously the most important from the perspective of usefulness and recycling. It is estimated that about 1000 million tonnes of coal ash are being dumped in ash ponds in India, while the quantity of fresh coal ash generation continues to rise every year in the country. It may be pertinent to mention here that pulverized lignite fired boilers and pressurized fluidized bed combustion (PFBC) system are in operation in the country to a limited extent and, hence, small but, significant volumes of high-lime lignitic ash and PFBC ash are also available in the country, although the predominant generation of coal ash is from bituminous coal-fired boilers.

3. Historical Trend of Fly Ash Generation

3.1. Parameters of Significance in Generation and Approximation in Quantitative Estimates

Generation of coal ash and consequently of fly ash is dependent on a large number of parameters, the more important ones being:

- coal quality in terms of ash content and its gross calorific value
- mode of preparation of the pulverized fuel
- type of boiler
- installed capacity of the plant
- operating hours
- plant load factor

- specific heating rate (heat for one unit of power)
- dust collection system and its efficiency

As many of the above parameters keep on varying within a plant and from plant to plant, the total ash generation cannot be expected to be steady, causing difficulties in the estimation of ash generation. From the statistics of ash generation, it is also difficult to understand if the bottom ash quantity gets duly excluded from the total quantity in a measured manner. On the whole the statistical presentation for fly ash and coal ash appears as synonymous. It may also be pertinent to mention here that in power generation, the coal consumption which used to be in the range of 0.3 – 0.4 million tonnes per 100 MW appears to be in the range of 0.22 million tonnes for every 100 MW in some plants in recent times and the ash content in coal is seen varying from 34 to 45%, although the attempt has been to keep it close to 34% by using de-ashed coal (see Table 2) (Senapati, 2011; CEA 2012).

Table 2. Data relating to the generation of coal ash

Year	Installed capacity, MW	Coal consumption, Million tonnes	Specific coal consumption, Million tonnes/ 100 MW	Coal Ash generated, Million tonnes	Ash % in coal
1995	54000	200	0.37	75	37.5
2000	70000	250	0.36	90	36.0
2010	98000	300	0.31	110	34.0
2011* (April– Sep)	55840	120	0.22	43	35.6

*Data based on feed back of a fairly representative number of plants (CEA, 2012 b).

Table 3. Fly ash generation and utilization in India

Year	Generation, Mt	Utilization, Mt (% of generation)	Remarks
1987-88	30	Not known	Wet disposal and brick making
1993-94	40	1.2 (3%)	Commencement of the National Fly Ash Mission
2004-05	112	42 (38%)	Fly Ash Mission continuing as Fly Ash Utilization Programme
2006-07	130	60 (46%)	Same as above
2010-11	200	100 (50%)	Total utilization planned but not achieved
2016-17	300	300 (100%)	Projection made
2021-22	450	450 (100%)	Projection made
2031-32	900	?	Yet to be planned

Based on comparable norms of say, 0.30 million tonnes of coal consumption per 100 MW installed capacity and average ash percentage of 40 in coal, for the current installed capacity of 131628 MW the coal ash generation may not exceed 150 million tonnes per annum. With similar norms, if the projected additional installed capacity of 116 GW is realized by 2022, the coal ash generation may touch more than 400 million tonnes per annum then. Although not identical, more or less similar set of ash generation and utilization data have emanated from the Department of Science and Technology, Government of India and other sources, which are given in Table 3 (Kumar, 2011).

3.2. Promotional Efforts for Utilization of Coal Ash in a Historical Perspective

In late nineteen eighties and early nineties the growing quantum of coal ash and its negligible utilization created great concern amongst the policy making authorities in the country (BMTPC, 1993). The National Housing Policy (1988) and the subsequent policy documents laid adequate stress on promoting low-cost building materials based on agricultural and industrial wastes. In order to enhance the utilization of fly ash in the manufacture of building products proposals were made to the Department of Revenue to provide financial incentives by way of exemption of duties to fly ash bricks containing 50% or more of fly ash. This was followed by proposals to cover other building materials using different percentages of fly ash. Some of these proposals were implemented. The Research and Development efforts at different institutions were financially supported from the country's Science and Technology budget (BMTPC, 1993). Entrepreneurs were encouraged to set up fly ash brick production units.

In order to evolve an integrated system and provide thrust to technology transfer efforts, the Government of India set up the Building Materials Technology Promotion Council (BMTPC) in 1990 as an interministerial apex organization. With the efforts of BMTPC, use of fly ash bricks was promoted in the construction undertaken by Central Public Works Department (CPWD). Other public sector organizations like Housing and Urban Development Corporation (HUDCO) and National Housing Board extended financial support to promote industrial units for production of fly ash based building materials. Subsequently, the Department of Power, Ministry of Environment and Bureau of Indian Standards were involved more effectively in furthering the utilization of fly ash. All these efforts of different wings of Government of India ultimately led to the creation of a Fly Ash Mission in 1994 under the Department of Science and Technology (DST), which had a tenure up to 2002 to give real push to application development. This mission thereafter converted itself into Fly Ash Utilisation Programme (FAUP), which continued up to 2007. Thereafter this activity is continuing as Fly Ash Unit of the DST (DST website).

Almost simultaneously with the formation of Fly Ash Mission the retired executives of West Bengal Power Development Corporation in Kolkata formed a not-for-profit registered society under the banner Coal Ash Institute of India, devoted singularly to the promotion of fly ash in the industry and agriculture as well as to ensure the benign modes of coal ash disposal (CAII website). In a similar manner, in June 2007 a not-for-profit company named as Centre for Fly Ash Research and Management (C-FARM) was registered in Delhi under Section 25 of the Indian Companies Act, 1956 (C-RARM website). With all these promotional efforts the fly ash as a material has progressively risen in its technological and application status as depicted in Figure 2. Based on the known and practiced technologies, a projection has been made about the future potential of fly ash utilization (Kumar, 2011) (Table 4).

Table 4. Projected future scope of utilization of fly ash (Million Tonnes)

Sl. No.	Areas	2011	2012	2017	2022
1.	PPC Production	50	85	130	200
2.	Concrete	8	15	20	25
3.	Road embankment	8	15	20	20
4.	Low lying area filling	9	20	20	20
5.	Ash bund raising	8	5	5	10
6.	Mine fill	12	20	70	120
7.	Brick and blocks	4	10	15	25
8.	Agriculture & others	1	2	20	30
	TOTAL:	100	172	300	450

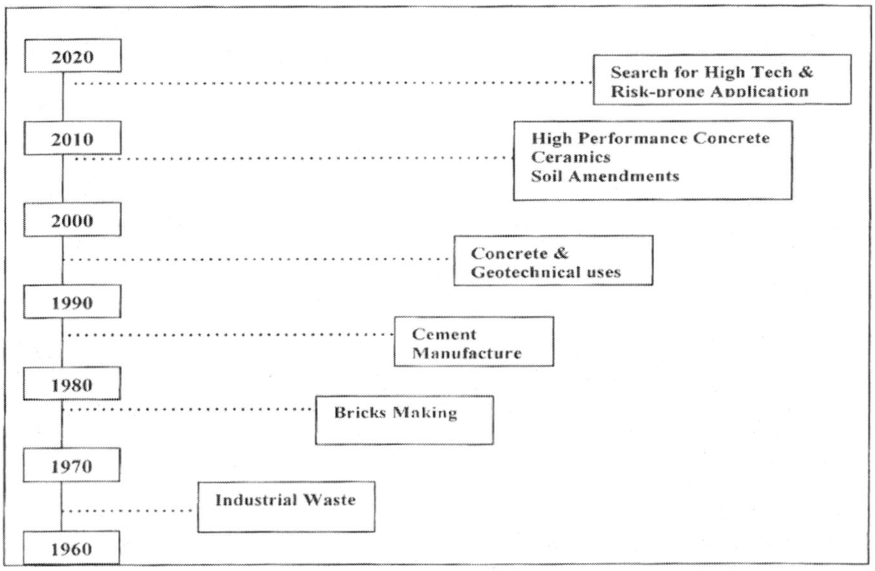

Figure 2. Rise of fly ash in application horizons in six decades.

As a result of concerted efforts in the country, fly ash and its products have been included as a regular construction item in the Specification and Schedule of Rates of Central and State Public Works Departments. About 10 new Standards for fly ash have been formulated or thoroughly revised and 60 Standards have been amended for fly ash use. The Gazette Notification issued by Ministry of Environment & Forests (MoEF) issued first in September 1999 and amended later in August 2003 for conservation of soil especially highlights the use of fly ash in almost all types of construction. It makes it mandatory to produce and use fly ash and fly ash based products (viz., bricks, blocks, tiles, etc.) within 100 km radius of a thermal power plant. It further prohibits excavation of soil for construction of any roads and embankments within that area. The Ministry has also made it mandatory to use fly ash in roads and embankments within 100 km of a thermal power plant. The required guidelines for such applications of fly ash including the construction of rural roads have been issued by the concerned authorities. To further facilitate the availability of fly ash of required quality, the thermal power plants have been directed to install dry fly ash collection and delivery systems.

During the course of the last decade the fly ash has been recognized as a resource material and its demand is also on the rise in certain parts of the country. Currently, it has also been declared as a tradeable commodity by the thermal power plants, which earlier had the directive of supplying fly ash free of cost to the intending users. Notwithstanding all these efforts, and advantages, the overall use of fly ash is stagnating at around 50% of the quantity generated. The target of 100% use of fly ash likely to be generated is, by itself, a daunting task. If one considers the expected generation of fly ash over the next two decades, the volume is gigantic and its use plan will have to be far more challenging than what is perceived today. No niche application strategy would work and one will have to look for newer technological avenues of bulk usage. For this purpose, the characteristics of the Indian fly ash and its processing for specific application should to be more comprehensively examined.

4. Characteristics of Indian Fly Ash

Characterization of coal ash in general and fly ash in particular can be conducted or observed broadly from three angles:

(i) Determination of intrinsic physico-chemical and phase characteristics for the basic understanding of the materials.
(ii) Detecting the special attributes, which are essential for specific fields of applications, e.g., road sub-base, embankment, cement, concrete, mine filling, masonry products, etc.
(iii) Determining the amenability of fly ash for further processing for the new and advanced fields of application.

For (i) above, the standard operating procedures are widely adopted, although there are scopes for improvement in the quantitative determination of phases, trace elements, etc. For (ii), generally the practices of certain laboratories are taken recourse to while for (iii), the approaches are essentially non-standard. In the last case, the inter-relationship of process-equipment-properties becomes very significant.

4.1. Bulk Compositional Ranges of Different Types of Coal Ash

In the context of geo-technical applications the bulk composition of fly ash, pond ash and bottom ash has been compared with soil in Table 5 (Pandian, 2004). The table confirms the compositional closeness of the coal ash varieties with soil. Amongst the three types of ashes, the pond and bottom ash may sometimes have higher concentration of silica. The alumina percentage in the bottom ash appears to be lower than the other two varieties. The presence of soluble solids is an important parameter from the application considerations. The investigations carried out by Pandian (2004) showed that the soluble solids ranged from 400 to 17600 ppm in the fly ash, 800 to 3600 ppm in the pond ash and 1400 – 4100 ppm in the bottom ash. Further, all the coal ash samples tested were found to be nearly alkaline. In relative measure, fly ash showed higher pH values than pond and bottom ash.

Table 5. Ranges of chemical composition of coal ash and soil

Oxide %	Fly Ash	Pond Ash	Bottom Ash	Soil
SiO_2	38.0 – 63.0	37.7 – 75.1	23.0 – 73.0	43.0 – 63.0
Al_2O_3	27.0 – 44.0	11.7 – 53.3	13.0 – 26.7	12.0 – 39.0
TiO_2	0.4 – 1.8	0.2 – 1.4	0.2 – 1.8	0.2 – 2.0
Fe_2O_3	3.3 – 6.4	3.5 – 34.6	4.0 – 10.9	1.0 – 14.0
MnO	0.0 – 0.5	Traces – 0.6	Traces – 0.3	Traces – 0.1
MgO	0.1 – 0.5	0.1 – 0.8	0.1 – 0.7	0.2 – 3.0
CaO	0.2 – 8.0	0.2 – 0.6	0.1 – 0.8	0 – 7.0
K_2O	0.04 – 0.9	0.1 – 0.7	Traces – 0.56	0.3 – 2.0
Na_2O	0.07 – 0.43	0.03 – 0.31	Traces – 0.30	0.2 – 3.0
LOI	0.2 – 0.34	0.01 – 20.9	0.61 – 51.6	5.0 – 16.0

Coming specially to fly ash, one may note that the Indian low-lime fly ash (ASTM Class F variety) is characterized by relatively higher concentration of silica and lower contents of ferric oxide and calcium oxide than the high-lime (ASTM Class C variety) as shown in Table 6 (Chatterjee, 2010). The chemical composition of the low-lime type implies higher fusion temperature of this fly ash, and, consequently, the chances of lower glass formation. However, this type shows lower variability than the high-lime variety as studied by Das (2003) and Das and Yudhbir (2003). A study has also been carried out to classify the oxide compositions of fly ashes using radial basis function neural network (Nataraj et al, 2006). The study parameters are shown in Figure 3 and the attributes and group centroids are shown in Table 7.

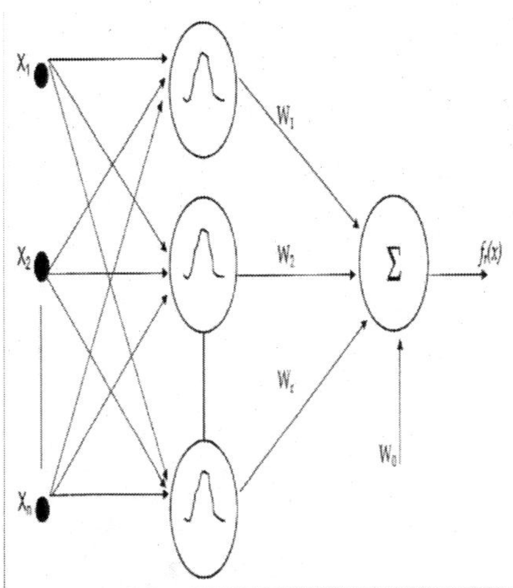

- Training Data Set : 45
- Testing Data Set : 25
- No. of input neurons : 11
- No. of neurons in hidden layer : 3
- The basic function : Gaussian
- No. of output neuron : 1
- Determination of cluster centres by K-means clustering algorithm
- Determination of the width of RBF
- Weight estimation by LMS algorithm

Figure 3. Parameters of the radial basis function neural network study (RBF : Radial Basis Function; LMS : Least Mean Square).

Table 6. Typical comparison of low-lime and high-lime fly ash samples

Oxide %	Low-lime fly ash	High-lime fly ash
SiO_2	63.6	26.4
Al_2O_3	27.2	25.89
TiO_2	0.5	0.6
Fe_2O_3	4.4	25.8
CaO	2.4	16.9
MgO	0.4	1.0
Na_2O	0.25	1.8
K_2O	1.2	0.3
SO_3	0.3	1.1
LOI	2.5	1.5

Table 7. Fly ash attributes and group centroids

Attributes Centroids	SiO_2	Al_2O_3	Fe_2O_3	CaO	MgO	Na_2O	K_2O	SO_3	LoI	Particles %, 45 μm	Pozzolanic Activity Index at 28 days
υ1	59.5	28.4	4.62	0.97	0.58	0.25	1.2	0.86	2.11	87.65	109.00
υ 2	46.2	2.05	21.1	3.52	1.29	1.19	1.83	1.77	1.87	82.80	95.22
υ 3	39.3	2.06	5.82	27.7	2.50	0.90	0.31	2.66	0.26	82.32	101.80

GROUP RANGES

Group Index	SiO_2 %	CaO %	SO_3 %
1 (Pozzolanic)	53.0 – 63.1	0.6 – 3.0	Traces
2 (Moderately Pozzolanic)	39.0 – 50.0	3.0 – 16.0	0.8 – 2.5
3 (Cementitious)	30.3 – 38.0	16.0 – 29.0	2.3 – 3.0

4.2. Heavy Metal and Radionuclides in Fly Ash

From the perspective of environmental concerns the concentration of heavy metals and radionuclides have been studied by several researchers with materials drawn from different thermal power stations. The heavy metal concentrations are shown in Table 8 (Kumar, 2009).

Table 8. Heavy metals in fly ash (in ppm)

Metals	Total	Available
Se	0.6 – 2.6	0.1 – 0.4
Cr	50.0 – 150.0	0.3 – 0.6
Pb	10.0 – 70.0	Not detectable
Co	18.0 – 50.0	0.05 – 0.15
Ni	50.0 150.0	0.15 0.25
Cd	5.0 – 10.0	0.03 – 0.07
As	1.0 – 4.0	Not detectable
Hg	Not detectable	Not detectable

Table 9. Radioactivity levels in fly ash

Radionuclides	Radiation range, $BqKg^{-1}$
^{226}Ra	30 – 110
^{228}Ac	30 – 110
^{40}K	180 – 500

In another study (Senapati, 2011) the heavy metal values were reported as, Ni: 77.6 ppm, Cd: 3.4 ppm, Sb: 4.5 ppm Cr:136.0 ppm, Pb:56.0 ppm. The values are fairly in the range of Table 7 barring the lower content of Cd and presence of Sb. Kumar (2009) also reported the range of radioactivity as given in Table 9. According to the data obtained by Kumar ((2009), the trace and heavy metals as well as radionuclides in fly ash samples are negligible and comparable with most of the field soils. It may be relevant to mention here that the radionucleides of primary concern are K, U, Th and their decay products like radium and radon. An average terrestrial radiation from U, Th and others naturally found in soil is 30 mrem/year.

Estimation of atmospheric release and inhalation risk from radionuclides have been studied by Pandit et al (2011) with coal, bottom ash and fly ash samples drawn from three different power plants. The results showed that the fly ash and bottom ash contained two to five times more natural radionuclides than the feed coal. None of the fly ash and bottom ash samples had radium equivalent activities and external hazard index values more than 370 $BqKg^{-1}$ and unity respectively. However, the absorbed dose rate at 1m above the ash pond was 79.19 $nGyh^{-1}$ (average of three plants) higher than the global average value of 55 $nGyh^{-1}$. The corresponding annual external effective dose was estimated at 0.68 $mSvy^{-1}$, which is also more than that in areas of natural background radiation (0.46 $mSvy^{-1}$). The 5th percentile, 95th percentile and mean values for total inhalation risk arising from radionuclides (^{226}Ra, ^{228}Th, ^{210}Pb and ^{Nat}U) were found to be 3.83 x 10^{-9},6.50 x 10^{-8} and 2.08 x 10^{-8} respectively.

The radioelemental characterization of fly ash from a Super Thermal Power Plant situated in the Western part of India has also been reported by Menon et al (2011). The mean natural radioactivity measured in the coal sample was 35 $BqKg^{-1}$ (17 – 60 $BqKg^{-1}$ range) for ^{238}U, 30 $BqKg^{-1}$ (1 – 64 $BqKg^{-1}$ range) for ^{232}Th and 400 $BqKg^{-1}$ (140 – 850 $BqKg^{-1}$ range) for ^{40}K. The contents of U and Th in the coal sample were 1.7 and 3.2 ppm respectively. Because of the mass loss of coal on firing the enhanced concentration of radionuclides was expected and the results obtained for finer fraction of particles are given in Table 10. Data furnished by different investigators on radioelemental characterization of fly ash samples do show variance, which can be related to varying methodologies, sample size, material properties, etc. Comparison of the results of different investigators does not seem possible at this stage. However, there is an overall comfort level in India that radioactivity of fly ash is not a deterrent in its use, as the source coal in general is very low in radionuclides.

Table 10. Radioactivity levels of the finer fly ash particles ($BqKg^{-1}$)

Particle size	^{40}K		^{238}U		^{232}Th	
(- 45 μm)	Economizer	ESP	Economizer	ESP	Economizer	ESP
Fly ash (magnetic fraction)	68.04	640.08	28.06	207.03	44.04	176.44
Fly ash (non- magnetic)	171.36	312.48	79.97	97.64	132.11	693.99

4.3. Phase Composition of Fly Ash

It is widely known that the reactivity of fly ash is dependent on its glass content and other mineral phases present. It has been observed that the Indian low-lime fly ash is more crystalline than those obtained in other countries, the glass content ranging from 47.0 to 60.9% essentially from the 2nd and 3rd chambers of ESP's in selected power plants in the southern part of India (Fournier et al., 2004), and ranging from 27 to 32 % in mixed fly ash samples in certain other plants (Table 11) (Chatterjee, 2001).

Table 11. Comparative phase composition of the Indian fly ash

Phase %	Fly ash from				
	UK	USA	Canada	Japan	India
Quartz	1.0 – 6.5	0 – 4.0	5.3	5.4 – 11.8	11.2 – 31.4
Mullite	9.0 – 35.0	0 – 16.0	5.6	8.0 – 18.0	8.9 – 31.4
Magnelite	25.0	0 – 30.0	0.3	-	0.1 – 1.3
Hematite	25.0	1 – 8.0	-	0.5 – 5.3	-
Glass	50.0 – 90.0	50.0 – 90.0	81.5	69.0 – 84.0	47.0 – 60.9*

*Values as low as 27 to 32 % measured in mixed fly ash samples of other plants (Chatterjee, 2001).

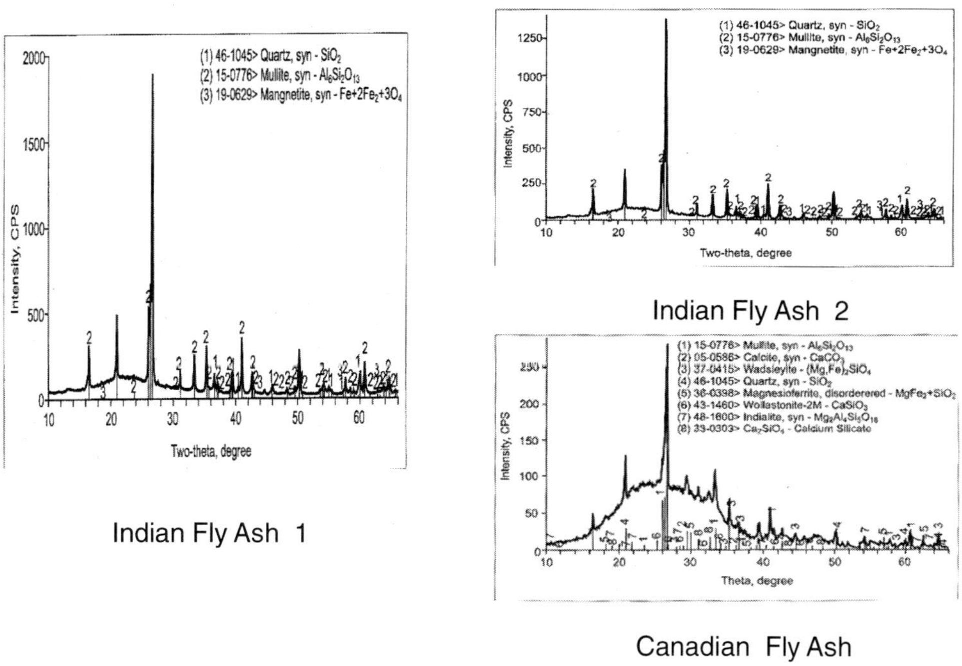

Figure 4. Comparative crystallinity of the Indian and Foreign fly ash samples.

The network theory of glass formation (Zachariassen, 1932) provides a theoretical basis to explain the relatively poor glass content in the Indian fly ash. As the ratio of network formers ($SiO_2 + Al_2O_3 + Fe_2O_3$) to network modifiers ($Na_2O + K_2O + CaO + MgO$) is very high and imbalanced, the glass content is low. The crystallinity of the low-lime Indian fly ash can be easily observed from the x-ray diffractogram (Chatterjee, 2008) (Figure 4). The hump around 22 ° – 24° two-theta indicates the presence of glass in the X-ray diffractogram of the low-lime fly ash, which may however shift to higher two-theta position depending on the lime content. The intensity and precise position of the tip of the hump, therefore, signify the quantity and likely composition of the glass phase. In Figure 4 one may observe variations in the hump characteristics between the two Indian fly ash samples and also between the Indian and the Canadian fly ash samples.

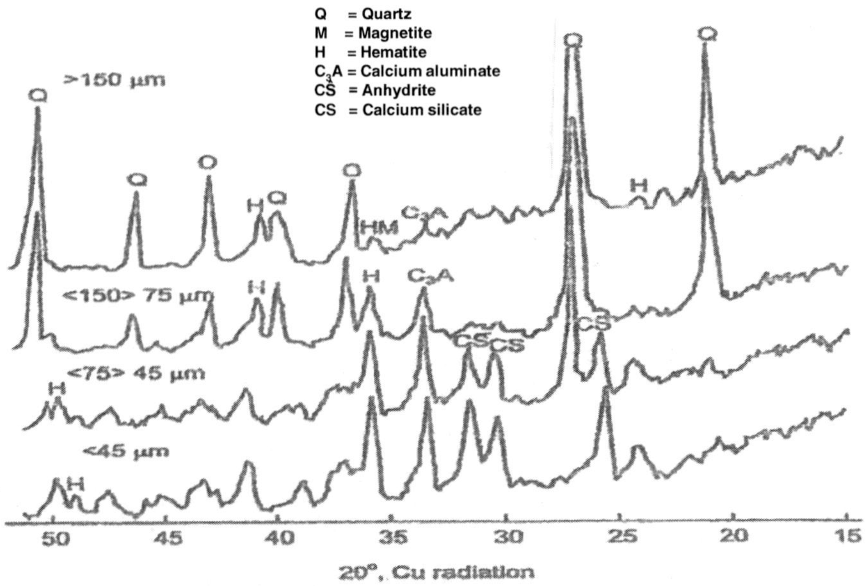

Figure 5. XRD pattern for different fractions of Neyveli (a high-lime) fly ash.

The x-ray diffractograms of different fractions of an Indian high-calcium fly ash sample are shown in Figure 5 (Das and Yudhbir, 2003). The patterns showed the presence of quartz and magnetite in the coarser fraction (>75 µm) whereas phases like tri-calcium aluminate, calcium silicate and anhydrite were present in the finer fractions without any quartz. It may be relevant to mention that such variations in phase composition with particle size distribution was not observed in the case of low-lime fly ash samples. It may also be interesting to note that no mullite phase was detected in the X-ray diffractogram of this high-lime fly ash sample.

4.4. Morphology of Fly Ash

While the fly ash particles are widely believed to be spherical in nature, the Indian fly ash samples display both spherical and angular particles. The spherical particles can either be hollow and thin-walled (known as Cenospheres) or encapsulated ones containing numerous cenospheres (called Plerospheres). In addition, the Indian samples also show abundance of opaque spheres, angular glass fragments and devitrified spherical particles (Figure 6).

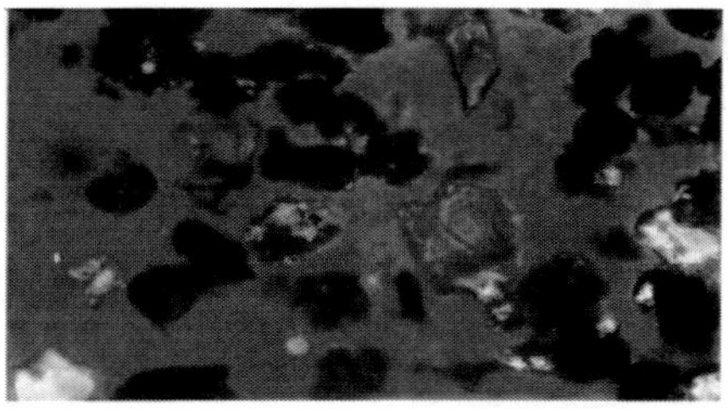

(a) Angular glass fragments

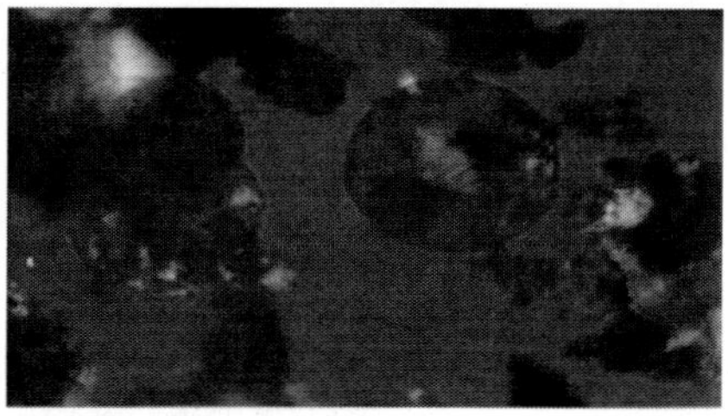

(b) Acicular mullites in a spherical glassy particle

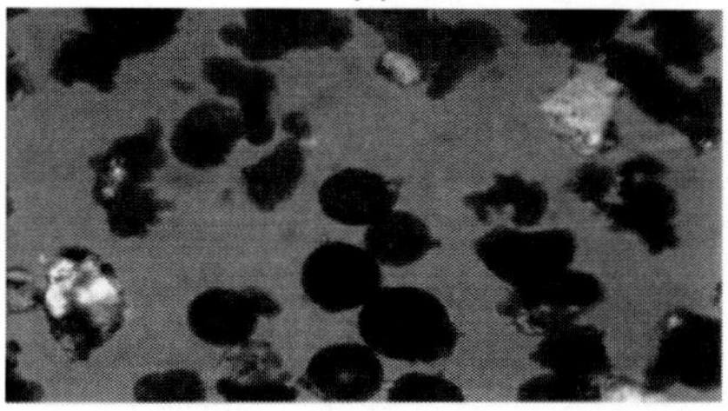

(c) Darker glassy spherical particles

Figure 6. Particle morphology of the Indian fly ashes (magnification 200X).

4.5. Physical Properties of Coal Ash

Coal ash as a particulate material can be classified as sandy silt to silty sand. The pond ash is essentially silty in character with some sand size fraction. The bottom ash is coarser consisting predominantly of sand size fraction with some silty particles. Fly ash is predominantly a silty non-plastic material. In general, the specific gravity of coal ash lies around 2.0 but can vary to a large extent (1.6 to 3.1). The variation of specific gravity of coal ash is the result of the variation of particle size, particle shape and chemical composition. The lower specific gravity of some fly ash samples could be due to larger proportions of hollow spheres from which the entrapped air could not be removed. The bulk density generally varies from 0.85 to 1.2 and the overall porosity lies in the range of 45 to 55%.

5. Overview of the Applications of Fly Ash

It is widely accepted that there could be many areas of application for fly ash (subject to their economic viability). In terms of increasing technology intensity and value addition the fields of use can be tentatively grouped as shown in Figure 7. The current fields of application of fly ash in India are broadly limited to the lower and middle blocks shown in Figure 7. It is also evident from Table 3 that the cement industry will continue to be the major consumer of fly ash from the thermal power plants and it also appear that mine filling may turn out to be the second largest application area in the country. In the following sections some of the important segments of fly ash usage have been reviewed.

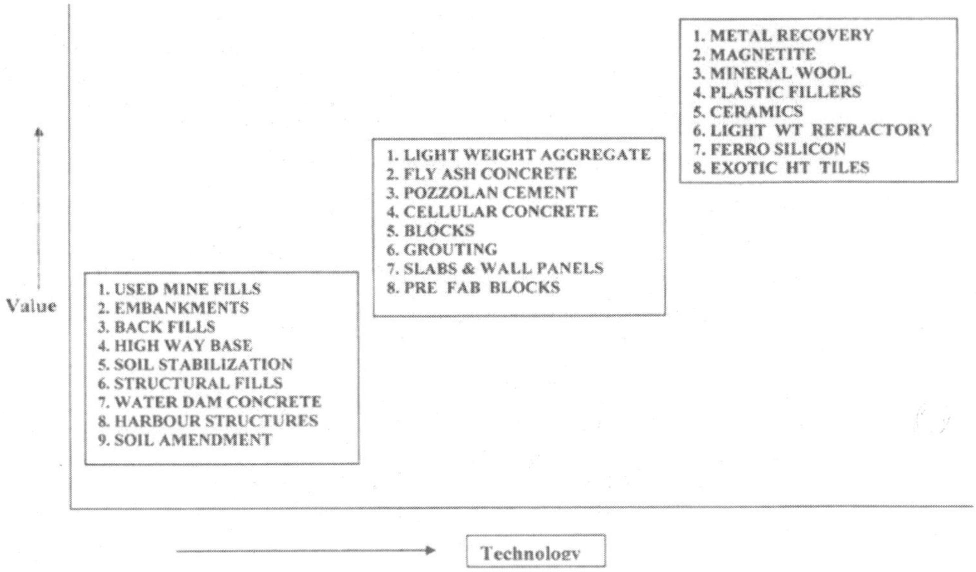

Figure 7. Fly ash utilization by class of application.

5.1. Fly Ash As a Supplementary Cementitious Material (SCM)

The pozzolanic nature of coal ash was highlighted in an anonymous report in 1914 but the pioneering research was done by Davis and his team at the University of California, Berkeley about 75 years ago (Davis et al, 1937). The first significant use of fly ash as SCM was in the construction of Hungry Horse Dam in Montana, USA (USBR, 1948). There has been immense proliferation of its use thereafter. In India, the use of fly ash in cement making started in the nineteen eighties but picked up in late nineteen nineties with sharp increase in the production of Portland Pozzolana Cement (PPC) (Figure 8).

The Indian Standard Specification of pulverized fuel ash (IS 3813: 2003) was first published in 1966 in three parts to cater the requirements of fly ash for three specific uses: Part 1 covering use of fly ash as a pozzolana; Part 2 covering the use of fly ash as an admixture for concrete and Part 3 covering its use as a fine aggregate for mortar and concrete. It was subsequently realized that the performance of fly ash as a pozzolana or an admixture or a fine aggregate was complementary and inseparable and, hence, requirements of fly ash for these uses should be covered by a single Standard. As a result, the Standard was revised in 1981 by combining the three parts into a one single Standard. This revision classified fly ash in two grades: Grade 1 essentially for manufacture of Portland Pozzolana Cement and Grade 2 for the incorporation in cement mortar and concrete as well as in lime – pozzolana mixture. Needless to mention that Grade 1 fly ash was obviously deemed suitable for Grade 2 applications as well.

With improvements in coal combustion technologies as well as ash collection and disposal systems in thermal power plants and also due to further diversified uses of fly ash it has been felt necessary that

(a) There should be a two- part standard for cement, cement mortar and concrete.
(b) Separate Standards should be drawn up for fly ash in lime-pozzolana mixture applications, sintered products making, geotechnical applications and agricultural applications.

While the two-part Standard for cement, cement mortar and concrete is in vogue, Standards for other applications are in the process of formulation. Standard IS 3812 (Part I): 2003 is for use of fly ash as "pozzolana" in cement, cement mortar and concrete, the second part of the same Standard is for use as "admixture" in cement mortar and concrete. The chemical and physical requirements of fly ash as a pozzolanic materials as given in the above Standard are furnished in Tables 12 and 13 respectively.

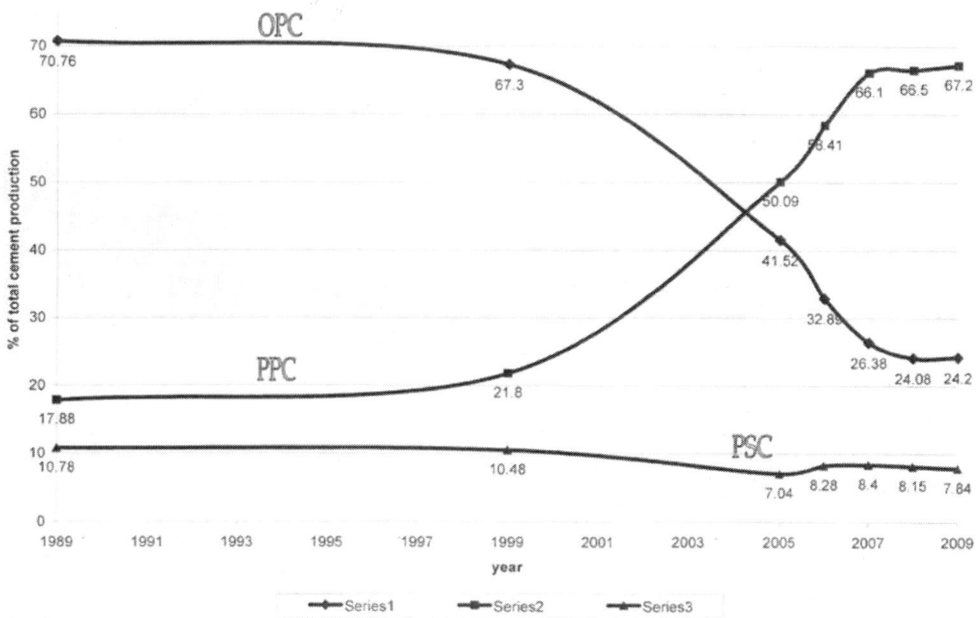

Figure 8. Trend of production of blended cements in India in the last two decades.

Table 12. Chemical requirements of fly ash as a pozzolanic material [IS 3812 (Part I): 2003]

Sl.No.	Composition	Pulverised Fuel Ash		Remarks
		Siliceous	Calcareous	
1.	$SiO_2 + Al_2O_3 + Fe_2O_3$	70	50	Mass % min
2.	SiO_2	35	25	Mass % min
3.	Reactive SiO_2, percent, min	20	20	Optional test Mass %. Min
4.	MgO	5.0	5.0	Mass % max
5.	Total S as SO_3	3.0	3.0	Mass % max
6.	Available alkalis as Na_2O	1.5	1.5	Mass % max
7.	Total chloride percent	0.5	0.05	Mass % max
8.	Loss on ignition	5.0	5.0	Mass % Max

Table 13. Physical requirements of fly ash as a pozzolanic material [IS 3812 (Part): 2003]

Serial No.	Properties	For both siliceous and calcareous pulverized fuel ash	
1.	Specific surface by Blaine's air permeability method	330	m^2/kg, min
2.	Particles retained on 45 μm IS sieve (wet sieving)	34	percent, max (optional test)
3.	Lime reactivity – average compressive strength	4.5	N/mm^2, min
4.	Compressive strength at 28 days (Cement Reactivity)	Not less than 80 percent of the strength of corresponding plain cement mortar cubes	
5.	Soundness by autoclave test – expansion of specimen	0.8	per cent. Max

The aforesaid Indian Standard also permits the use of fly ash with minimum Blaine's specific surface of 250 m^2/kg as a pozzolanic material in the manufacture of PPC by intergrinding with Portland cement clinker, if the fly ash, when ground to 320 m^2/kg Blaine's specific surface or to the fineness of the resultant PPC, whichever is lower, meets all the requirements of Tables 12 and 13. The chemical requirement is different from Table 7 only in respect of total sulfur as sulfur trioxide, which is 5.0% instead of 3.0% but, in terms of physical requirements the minimum specific surface area acceptable has been stipulated as 200 m^2/kg and the maximum particles retention on 45 μm as 50%.

5.2. Variability of Fly Ash As SCM

Notwithstanding the stipulations provided in the Indian Standard as explained above, the fly ash supplied from different sources to the cement and concrete industry show significant variability in respect of the following parameters:

- Particle size distribution as represented by

  (a) Blaine's specific surface area.
  (b) Residue on 45 μm sieve.

- Unburnt carbon as quantified by loss on ignition.
- Glass content or proportions of reactive/soluble silica.
- Moisture content, in case of wet supply of the fly ash.

5.2.1. Particle Size Distribution and Size Classification Technology

As fly ash is generally collected with the help of multi-field ESPs, the specific surface area typically varies from approximately 250 m^2/kg in the first field to approximately 850 m^2/kg in, say, the sixth field. As the final collection hopper contains materials from all the fields, the specific surface area of the mixed fly ash is often coarse and variable (Chatterjee,

2008). The typical spread of the particle size distribution is shown in Figure 9. The mean particle diameter of the fly ash analyzed was found to be 21.71 μm while the particle diameter at 50% was 13.73 μm. One of the most prevalent technologies to derive fly ash fractions of narrow range of particle size distribution is to take recourse to using high efficiency cyclone separators for classifying fly ash. Some illustrations have been provided earlier by Chatterjee (2008) for the improvement in the properties obtained for fly ash and fly ash containing cement mortars by such processing technology. Another example of the effects of classified fly ash on concrete properties can be seen in Table 14 (Mullick 2005).

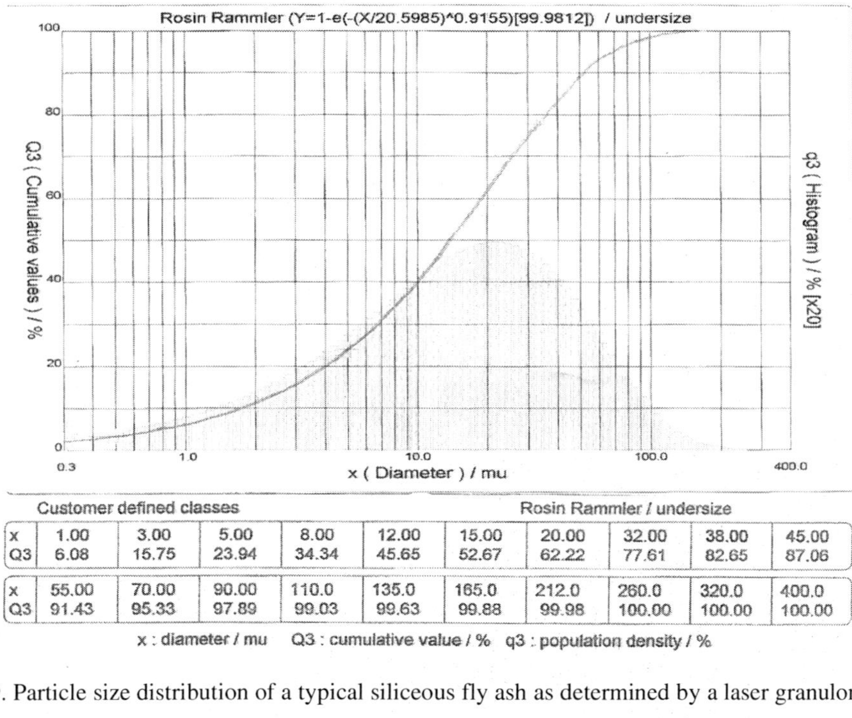

Customer defined classes | Rosin Rammler / undersize

x	1.00	3.00	5.00	8.00	12.00	15.00	20.00	32.00	38.00	45.00
Q3	6.08	15.75	23.94	34.34	45.65	52.67	62.22	77.61	82.65	87.06

x	55.00	70.00	90.00	110.0	135.0	165.0	212.0	260.0	320.0	400.0
Q3	91.43	95.33	97.89	99.03	99.63	99.88	99.98	100.00	100.00	100.00

x : diameter / mu Q3 : cumulative value / % q3 : population density / %

Figure 9. Particle size distribution of a typical siliceous fly ash as determined by a laser granulometer

Table 14. Improvement in quality of fly ash by processing

Item and properties	Fly ash as received	Processed fly ash
Fly ash		
% Retention on 45 μm	45.0	6.0
Lime reactivity, MPa	4.57	4.9
Cement Reactivity %	90.0	103.0
1 : 3 mortar		
28-day strength, MPa	50.2	58.4
M 20		
7-day strength, MPa	19.6	31.5
28-day strength, MPa	29.0	44.2

In another instance, when 30% of OPC was substituted by the processed fly ash, the concrete characteristics (slump, 28-day strength) were observed to be the same as in the control OPC concrete. Much higher slump retention and increased 28-day strength were obtained when additionally superplasticisers were used (Bhattacharjee and Dhuri, 1999).

5.3. Improvements in Concrete Properties with the Use of Fly Ash

It is widely known that in fly ash containing concrete the improvements in properties primarily come from three factors:

(i) continued hydration of cement-pozzolana mixture
(ii) reduced water demand
(iii) enhanced cohesion of paste matrix.

The microstructure of dense hydrated cement paste in a low w/c fly ash concrete is shown in Figure 10 and the overall performance of siliceous fly ash concrete is schematically depicted in Figure 11 with reference to Portland cement concrete.

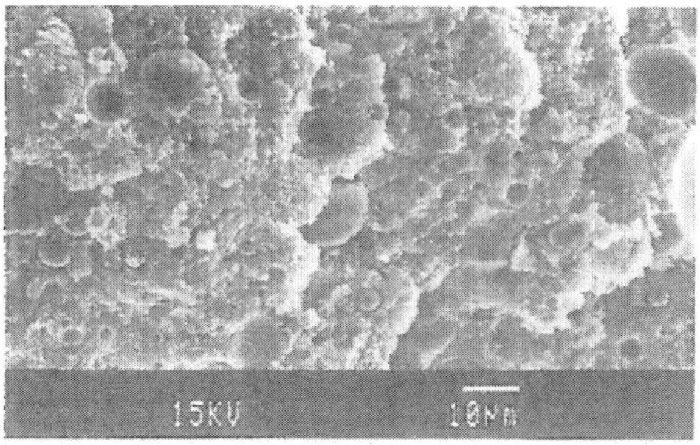

Figure 10. Dense hydrated cement paste in a low w/c fly ash concrete.

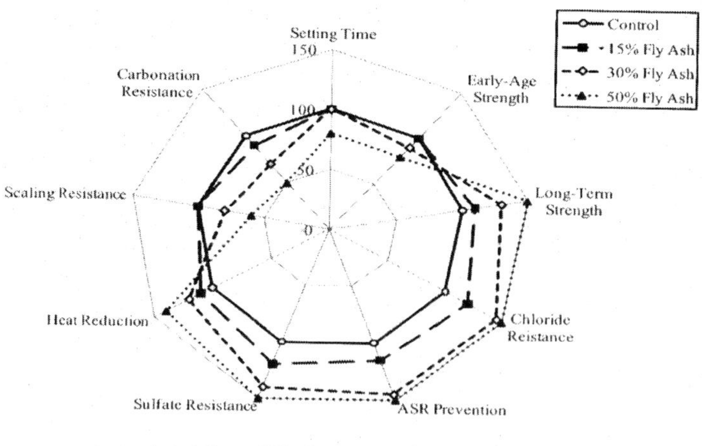

Figure 11. Performance of Fly Ash (Class 'F') Concrete.

The specific effects can be broadly summarized as follows:

a) For improvement in workability, water reduction of about 3% can be achieved for approximately 10% addition of fly ash. This relationship may not hold good if the carbon content in the fly ash is closer to 5%.
b) The low-lime siliceous fly ash is more effective in reducing the heat of hydration.
c) Reduction of early age strength (especially at 1 day) is greater for the siliceous fly ash and for its increased usage. Use of accelerating admixtures, high early strength Portland cement or silica fume appears to help compensate the strength reduction.
d) In order to derive the benefit of long-term strength gain, the acceptance criteria for concrete mix design can be considered to be 56 days in place of 28 days.
e) Without adequate curing the benefits of impermeability and lower chloride penetration cannot be achieved in fly ash based concrete.
f) The low-lime siliceous fly ash is quite effective on its own in reducing the risk of alkali silica reaction but if a high-lime or calcareous fly ash has to be used, it may need to be combined with either slag or silica fume.
g) As in ASR, for sulfate resistance as well the low-lime siliceous fly ash has proved to be quite effective. Use of high-lime fly ash in such situations is better avoided.
h) The risk of deleterious expansion of heat-cured concrete due to delayed ettringite formation can also be prevented by using a low-lime siliceous fly ash in sufficient quantities (more than 25%) as cement replacement.
i) Since fly ash based concretes are more prone to carbonation, adequate curing and sufficient cover requirements have to be taken recourse to. It seems, however, that the carbonation effect is more related to the grade of concrete than the SCM content in it as borne out by some test results obtained under identical laboratory conditions (Table 15).
j) Although it has been observed that the high-lime fly ash (Class C) demonstrates slightly better resistance to deicer salt scaling than the low-lime (Class F) variety in some laboratory experiments, the Indian experience in this regard is very little because of the country's essentially tropical climate

Table 15. Carbonation test results after 28 days curing

Low strength concrete with different levels of cement substitution by fly ash	Average carbonation depth from the surface, mm	High strength concrete with different levels of cement substitution by fly ash	Average carbonation depth from the surface, mm
Reference OPC concrete (M20)	13.2	Reference OPC concrete (M50)	1.6
20% fly ash	7.5	20% fly ash	Nil
30% fly ash	6.3	30% fly ash	Nil
50% fly ash	4.2	50% fly ash	Nil
PPC based (20% fly ash)	8.5	PPC based (20% fly ash)	Nil

5.4. Comparative Properties of Factory Made Fly Ash Based PPC Concrete and Site-Mixed Fly Ash Based OPC Concrete

An experimental investigation was carried out to evaluate the effectiveness of performance characteristics of two concretes: one incorporating fly ash in the form of interground PPC and another as a separate addition of fly ash to the site-mixed OPC concrete. This study was conducted with M20 and M25 grades of concretes as these grades are widely used at general construction sites in India (Banchor and Krishnan, 2006). The fly ash used in this study was a marginally processed one as the residue on 90 μm sieve was rejected before use to bring up the Blaine's specific surface area of the fly ash to 330 m^2/kg from the as-received value of 290 m^2/kg.

The total cementitious material in the M20 grade concrete was 325kg/m^3, while for the M25 grade concrete it was 357kg/m^3. The fly ash content in the total cementitious material in all combinations was 25 per cent and the maximum size of the granite aggregate used was 20 mm. Some of the illustrative results are given in Table 16.

From Table 16 it is evident that there could be marginal benefits of using fly ash in concrete in the form of factory-made PPC. However, the benefits do not appear to be glaring but the advantages of using fly ash in concrete in any mode are clearly visible with respect to the properties and microstructure of OPC concrete.

Table 16. Comparative properties of PPC based and OPC+ fly ash based concretes prepared in a drum mixer

Grade of concrete	Cementitious mix	pH 7-Day	pH 28-Day	MoE Gpa	Water permeability mm	Initial Surface Absorption Test at 10 min, ml/m2/s
M 20	OPC	12.3	12.5	30	12	0.050
M 20	OPC+FA	12.1	12.1	33	10	0.045
M 20	PPC	12.1	12.2	35	9	0.035
M 25	PPC	-	-	35	8	0.030
M 25	OPC+FA	-	-	33	10	0.040
M 25	OPC	-	-	33	10	0.045

5.5. Status of High-Volume Fly Ash Concrete in India

It is now in the domain of public knowledge that the original developmental work on High-Volume Fly ash Concrete (HVFAC) was carried out at the Canadian Centre for Mineral and Energy Technology (CANMET) in 1985. The technology in essence involved the replacement of OPC up to 55% by Class F fly ash with simultaneous use of an effective superplasticizer in making structural grade concretes. The HVFAC technology has been used in constructing several important buildings and roads in Canada and the USA on trial basis. After about two decades these buildings have been observed to display satisfactory performance. An initiative for the transfer of this technology to India was taken by CANMET in 2001.

The project was financially supported by the Canadian Climate Change Development Fund and was administered by the Canadian International Development Agency (CIDA). While the know-how was transferred by CANMET, the lead role at the receiving end was played by the Confederation of Indian Industries (CII). The project comprised numerous activities such as awareness creation through seminars, information dissemination through distribution of technical literature, undertaking demonstration projects, etc.

As a part of the confidence building measure, two Portland cement samples and 20 or so fly ash samples, collected from the Indian sources, were characterized and evaluated in the CANMET laboratories in Canada. The findings have been reported by Fournier et al (2004) and Sivasundaram (2004).

It may be pertinent to mention here that the fly ash samples were collected from six different thermal power stations, but practically all located in the Southern part of the country and the majority of the samples were drawn from the second and third chambers of ESP's, although there were a few all-chambers mixed samples as well. Thus, with somewhat inadequate representativeness, the suitability of the evaluated fly ash samples for making HVFAC containing 50% fly ash was established in the laboratory. The 1-day compressive strength ranged from 8 to 22 MPa and correspondingly the 28-day strength from 45 to 55 MPa. All the lab concretes had RCPT values (ASTM C 1202) below 1000 coulomb.

In order to establish the on-site feasibility of HVFAC in India, a road construction project was undertaken at a place called Ropar in the Northern state of Punjab by a Cement Company. The construction of the 2-lane 7-m wide 300-mm thick and 0.75-km long road was completed in two and half months. The concrete made with 50% replacement of cement by fly ash was of M 40 grade (the actual compressive strength being 41.6 MPa) with flexural strength of 7.6 MPa (Desai, 2004). Thereafter several other pavement projects, both in the private and public sectors, have been completed, some of which were of much longer stretches. Attempts were also made to make a comparison of the first construction costs of OPC and HVFA concrete, which showed an economy of 20% at the prices prevailing about eight years ago (Jain, 2006). Later on, a more comprehensive life-cycle costing was done by the Cement Manufacturers Association of India (Kodiyali et al, 2009) with a view to promoting concrete pavements in preference to flexible types (Table 17). Due to the longer life span and lower maintenance demands, the concrete pavement in general and that with fly ash in particular showed distinct economic advantage.

Table 17. Relative economic advantage of concrete pavement with fly ash

Basic Assumptions *Pavement of 1 km length, 7 m wide, and 2 lanes * CBR of soil 5 *Life of 25 years for concrete pavement and 15 years for flexible pavement *Maintenance every 5 years			
NPV of the Whole Life Cycle (Rs. Million)			
Cost component	Pavement types		
	Flexible	Concrete without fly ash	Concrete with fly ash
Initial	17.2	16.5	15.6
Maintenance	4.7	0.3	0.3
Total	21.9	16.8	15.9

Apart from the economic advantages under normal circumstances, it has also been observed from the technical investigations that the benefits can be enhanced if one uses high strength Portland cements with 28-day compressive strength of more than 53 MPa, fly ash of higher reactivity as reflected in its lime reactivity (> 6.0 MPa) and cement reactivity of more than 90% and if the dosage of the superplasticiser of SNF base is duly optimized, as under those conditions the cement replacement of up to 70% by fly ash may be achieved. Notwithstanding such potentials, the adoption of HVFAC technology in India is some distance away as the present provisions in the relevant Indian codes require review and amendment (Basu and Saraswati, 2006).

5.6. Coal Ash in Geotechnical Applications

As previously indicated, the utilization of fly ash for structural fills, base course construction, subgrade improvements, backfills, grouting and coal mine applications is rather limited. Needless to mention that, apart from cement and concrete the coal ash can be utilized in bulk only in such geotechnical engineering applications but such uses require a different course of characterization as presented by Sridharan et al (1996), Pandian (2004), Dewangan et al (2010), Saha and Pal (2012) and many other investigators. Some important geotechnical properties of a few coal ash samples are given in Table 18. The textural classification range of coal ash in general is summarized in Table 19.

Due to the low value for the specific gravity of coal ash compared to soils, ash fills tend to result in low dry density. The reduction in unit weights is of advantage in the case of its use as a backfill material for retaining walls, since the pressure exerted on the retaining structure as well as the foundation structure will be less. The other application areas include embankments, specially on weak foundation soils, reclamation of the low-lying areas, etc. Based on the grain size distribution, coal ash can be classified as sandy silt to silty sand. They are poorly graded with coefficient of curvature ranging between 0.61 and 3.70 and the coefficient of uniformity in the range of 1.59 – 14.0. Amongst the index properties, liquid limit is an important physical property for use in classification and correlation with engineering properties. Since coal ash is practically non-plastic, no plastic limit can be determined. It is also not possible to carry out shrinkage limit tests since ash pats crumble upon drying. Since the amount of shrinkage is negligible, it is not a constraint in geotechnical applications. This leaves only the possibility of measuring the liquid limit, which is also difficult by the normally adopted Percussion Cup or Fall Cone methods. Nevertheless, the liquid limit values have been measured for the Indian coal ash samples and normalized with respect to those of soil. The values are given in Table 18.

Free swell index is an important parameter in soil engineering to identify swelling soils. Large number of measurements carried out at the Institute of Science in Bangalore, India indicated that 70% of coal ash samples showed negative free swell index due to flocculation. As the clay size fraction in coal ash is generally small, its free swell index is expected to be negligible.

For an effective and efficient use of coal ash, as already mentioned, in geotechnical applications, its classification is important, which is shown in Table 18. As far as the compaction behavior is concerned, the variation of dry density with moisture content for fly ash is less, compared to that of a well graded soil, even when both have the same median size.

Table 18. Geotechnical properties of some Indian coal ash samples.

Coal ash source and composition	Ash type	Specific gravity G	Liquid limit (W_L) [Normalized]	% passing 75 µm sieve	Group index	Coefficient of curvature C_c	Coefficient of uniformity C_u
1, Low-lime	Pond ash	1.93	50.5 [36.8]	16.8	0	1.25	9.00
2. Low-lime	Fly ash	1.98	37.8 [28.2]	72.2	7	1.14	6.00
	Pond ash	1.91	60.9 [43.2]	15.0	0	1.80	3.60
	Bottom ash	2.15	77.0 [62.5]	8.6	0	2.10	4.70
3. Low-lime	Fly ash	1.95	32.1 [23.6]	76.7	8	0.61	5.70
	Pond ash	2.01	56.5 [42.8]	29.5	0	0.79	3.00
	Bottom ash	1.82	53.5 [36.7]	6.0	0	1.36	6.55
4. Low-lime	Fly ash	1.97	38.7 [28.8]	61.6	5	2.47	5.50
	Pond ash	2.00	64.4 [48.6]	25.6	0	1.64	9.40
	Bottom ash	1.95	47.4 [34.9]	13.4	0	1.03	9.09
5. Low-lime	Fly ash	2.18	39.2 [22.3]	92.5	8	1.09	1.59
	Pond ash	2.15	29.9 [48.6]	2.2	0	1.41	3.37
	Bottom ash	2.08	65.1 [34.9]	14.0	0	1.60	8.02
6. High-lime	Fly ash	2.55	60.4 [49.3]	73.4	9	1.04	3.16
	Pond ash	2.50	49.9 [47.1]	16.4	0	1.93	9.67
	Bottom ash	2.08	104.1 [81.7]	8.5	0	2.05	5.20

Note : The normalized water content refers to the water content multiplied by the specific gravity of coal ash divided by 2.65, the specific gravity of most of the soils. This is done to make the value comparable to that of a soil.

Table 19. Classification range of coal ash

Sl. No.	Ash type	Unified system	IS	PRA	Textural	Group index	C_c	C_u
1.	Fly ash	ML	ML	A-4 to A-5	Silty loam to sandy loam	5 – 9	0.61 – 2.47	1.59 – 6.00
2.	Pond ash	SM -SP	SM – SP	A-2 to A-4	Sandy loam to sand	0 – 3	0.79 – 1.93	2.96 – 9.67
3.	Bottom ash	SW - SM	SW – SM	A-2 to A-4	Sandy loam to sand	0	1.03 - 3.70	4.70 - 14.00

Table 20. Compaction characteristics of some coal ash samples

Serial No. & ash type	G	Loose dry density γ_L(kN/m^3)	Compacted air dried state		Lowest compacted state		Maximum density		Compacted saturated state	
			W(%)	γ_d(kN/m^3)	W(%)	γ_d kN/m^3)	W(%)	γ_d(kN/m^3)	W(%)	γ_d(kN/m^3)
1. Pond ash	2.01	9.70	1.0	10.8	24.0	9.7	33.0	10.1	36.0	10.0
			(0.7)	(14.2)	(18.2)	(12.8)	(25.0)	(13.8)	(27.3)	(13.2)
2. Fly ash	2.01	9.10	0.5	12.8	8.8	12.5	20.6	13.0	25.7	12.9
			(0.4)	(16.5)	(6.9)	((16.0)	(16.0)	(16.8)	(20.0)	(16.6)
Pond ash	2.10	9.40	1.8	10.2	24.4	9.6	39.9	10.2	4.2	10.1
			(1.4)	(12.9)	(19.3)	(12.1)	(31.7)	(12.9)	(33.3)	(12.7)
Bottom ash	1.90	--	2.0	7.4	25.0	7.2	64.9	7.7	69.8	7.4
			(1.4)	(10.3)	(17.9)	(10.1)	(46.5)	'(10.7)	(50.0)	(10.4)
3. Fly ash	2.11	9.6	0.1	12.4	6.4	11.2	25.0	12.7	25.4	12.5
			(0.1)	(15.6)	(5.1)	(14.1)	(19.9)	(16.0)	(20.2)	(15.8)
Pond ash	2.01	--	2.0	10.2	15.0	9.3	36.7	20.4	38.3	10.3
			(1.5)	(13.4)	(11.4)	(12.3)	(27.8)	(13.7)	(29.1)	(13.6)
Bottom ash	2.04	8.3	0.1	9.5	18.4	8.5	38.3	9.3	42.3	9.1
			(0.1)	(12.4)	(14.2)	(11.1)	(29.5)	(12.0)	(32.6)	(11.9)
4. High lime										
Fly ash	2.55	7.7	3.9	8.8	30.9	8.0	62.3	8.9	63.9	8.9
			(3.8)	(9.2)	(29.8)	(8.4)	(59.9)	(9.2)	(61.5)	(9.2)
Pond ash	2.3	6.9	4.4	10.7	24.9	9.8	40.4	10.2	44.5	10.3
			(3.9)	(12.2)	(21.9)	(11.1)	(35.5)	(11.6)	(39.1)	(11.7)
Bottom ash	2.05	5.9	1.0	6.8	47.9	5.6	75.1	5.8	80.8	6.1
			(0.8)	(11.1)	(37.0)	(7.3)	(58.1)	(7.6)	(62.5)	(7.9)

This is explained by the higher air void content in fly ash (5 to 15%) than in soil (1 to 5%). The higher void content tends to limit the build-up of pore pressures during compaction, thus allowing the fly ash to be compacted over a larger range of water content. Table 20 shows the actual dry densities and corresponding water content as well as their normalized values obtained for some of the coal ash samples.

As the dry densities in coal ash are less sensitive to compaction water content, the field compaction can be carried out at the dry side of the optimum, taking care of the dust problem. With good shearing resistance available even at low densities, the field compaction need not be constrained to achieve the maximum dry density. The shear strength of freshly compacted fly ash comes primarily from the internal friction, although some apparent cohesion has been observed in low-lime pozzolanic fly ash. The shear strength of fly ash is affected by its density and moisture content with maximum shear strength occurring at optimum moisture content. The low-lime siliceous fly ash has a friction angle in the range of 26 to 42^0. Many research workers have conducted consolidated drained or undrained tests with pore water pressure measurements and reported the mean value of the angle of shearing resistance to be 34^0 (Dewangan et al, 2010).

A major application area for geotechnical use of fly ash is as a sub-base material in the construction of pavements. The California Bearing Ratio (CBR) method is widely used in the design of pavements. The CBR results of the Indian fly ashes in unsoaked conditions ranged from 10.1 to 20.6%. When the pond ashes and the bottom ashes were taken into consideration the lower end of the range got extended to 6.8%. In soaked condition the CBR values of all coal ashes tested were found to range from 0.2 to 8.5%. It seems that fly ash mixed with additives such as lime, cement, sand and soil can be effectively used as sub-base materials even under soaked conditions. As already mentioned, silt content and porous surface of fly ashes reduce the compressibility of the soil. If soil – fly ash – soil or fly ash – soil – fly ash layers are provided, there could be situations with high strength and low compressibility simultaneously with the advantage of lower dry unit weights. It may also be borne in mind that the pond ash and bottom ash varieties may often exhibit good CBR values both under soaked and unsoaked conditions without any additives. It may be pertinent to mention here that the hydraulic conductivity of well-compacted fly ash ranges from 10^{-4} to 10^{-6} cm/s, which is roughly equivalent to that of a silty sand to silty clay soil. This property helps to minimize the seepage of water through a fly ash embankment.

It may finally be borne in mind that all the above geotechnical characteristics of coal ash have to be seen in conjunction with its chemical, mineralogical and reactivity parameters to finally decide on the mode and scope of use of coal ash in geotechnical applications. Needless to emphasize that the overall potential in this respect is quite large and varied, which needs to be fully realized.

5.7. Pond Ash as Mine fill

The Research and Development efforts in respect of using coal ash as mine-fill have been reported several years back (Kumbhakar et al, 2005). There is a huge gap between demand and supply of stowing material for underground mines due to which the production rates from underground coal mines have gone down and a large amount of coal remains locked in standing pillars. River sand being the traditional material for filling underground voids is getting depleted due to its overexploitation by the construction industry and low rate of replenishment due to construction of dams on rivers at the upstream side. Moreover, sand has the disadvantage of being heavy, highly abrasive and consumer of high energy in its transportation in slurry form. Coal ash particularly from the pithead thermal power plants can be a good substitute for sand as a stowing material.

The detailed laboratory investigations carried out with more than 10 different coal ash samples demonstrated the following properties in the context of their use as stowing materials:

i. Specific gravity	: 2.0 – 2.4
ii. Bulk density	: 0.85 – 1.3 t/m^3
iii. Average grain size	: 23 – 150 µm
iv. Porosity	: 40 – 60 %
v. Percolation rate	: 10 – 30 cm/h

vi. Settlement time of solids : 30 – 120 min
vii. Unburnt carbon : < 5%
viii. Moisture content : < 1 %
ix. Crossing Ignition Point Temperatures : > 200^0 C
x. Compressibility : < 2% at a pressure of 90 kg/cm^2
xi. Uniaxial compressive strength (28 days curing): 10 – 160 kPa
xii. Concentration of heavy metals and trace elements in ash leachate within permissible limits for drinking water (IS 10500:1996) and for industrial effluents (IS 2490 : 1982)

After several field trials with the above coal ashes the following observations were made:

a) Pond ash is lighter and energy saving in transportation and filling.
b) Since the percolation rate is lower and the settlement rate is slower, the fines will reach the barricade, choke its pores or escape along with the drainage water. The escaping fines are likely to create certain operational problems, which have to be overcome by improving the barricade design with fines arresting arrangements.
c) The requirement of water to acquire the flow velocity in pipeline is greatly reduced as compared to sand. It has been estimated that the pond ash slurry with 1:1 (by volume) water-to-solid ratio is flowable through 155mm ID pipeline as a pseudo-plastic solid slurry, whereas the water-to-solid ratio for the sand slurry can be as high as 2.5:1.
d) The low levels of unburnt carbon and moisture content are necessary to ensure that the ignition point temperatures remain below 200^0 C so that there is no propensity for spontaneous heating.
e) The ash pack develops strength within a short time and behaves like a monolithic fill, unlike sand packs.
f) The moist pond ash is sticky with high cohesive forces that restricts the free flow of ash. This leads to discontinuous feeding of ash for thorough mixing, resulting in the formation of air pockets and prevention of full bore stowing. However, the problem can be overcome by an arrangement of a central shower, water injection through nozzles in the hopper portions of the storage vessels and the provision of vibrators on the hoppers.
g) With increase in coarseness of ash, the water requirement increases. Therefore, stowing with bottom ashes requires higher amounts of water.
h) The water-to-solid ratio is also dependent on the hydraulic gradient conditions.
i) Because of better compaction the volume shrinkage of 20 – 30% may be noticed when pond ash is placed inside the void against only 8 – 10% in the case of sand.

Coal ashes are proven through numerous laboratory and field trials to be adequately suitable for stowing and any existing stowing plant can be used for ash stowing with certain modified arrangements. However, the future of ash stowing, to a large extent, depends on the development of paste filling technology involving the technology of high concentration slurry in the form of a paste that will be transported through a pipeline and discharged into the goaf to provide roof support. This will induce faster settlement of solids, have better control over the slurry density and escape of fines, and provide better strata control.

Incidentally, the technology of high concentration fly ash slurry transportation has already been adopted in India. One of the power generating stations has been making use of this technology to transport fly ash slurry containing 40 tonnes of dry fly ash per hour for a novel disposal system to create landscape features without releasing water at the site (Roy, 2005). The adoption of this technology for mine filling is likely to substantially increase the application of coal ash in this mining field.

6. Value-Added Fly Ash Products

Bricks and pavement blocks are the common fly ash products being made and used in the country. They are mostly made with fly ash of cement grade to an extent of 40-45%. Either cement or lime is used as the binder in the range of 8-12% and the construction sand is used as the fine aggregate. The processing technology is either moist casting and pressing, or slurry casting, compaction and prolonged water curing or steam curing of shorter duration. Alternatively, fly ash – lime – gypsum combinations are made into bricks by some small producers. A third production technology of producing burnt clay – fly ash bricks has also been practised in the country. Apart from such bricks and blocks, there are certain other value-added products, the know-how for which has either been indigenously developed or imported from elsewhere but the actual production of which is either very limited or yet to commence. Some illustrations of such products are highlighted below.

6.1. Lightweight Fly Ash Aggregates (LFA)

There are three basic technologies available in the country: sintering, cold-bonding and foaming, and proprietary chemical foaming (Neopor). The sintering technology was worked upon in many Research Centres. However, it has been claimed to have been finally perfected for adoption in India by the Regional Research Laboratory in Bhubaneswar in the state of Odisha.

The process involves pelletizing or nodulizing 75- 90% fly ash with some binder and sintering them at 1000 – 1300^0 C. The sintered aggregates are spherical in shape, hard but porous in structure and brownish grey in colour with bulk density of 600 – 700kg/m^3. It is found to be suitable for making lightweight structural concrete. In the second technology of making LFA, lime is used as a binding agent for about 90% fly ash for further processing. This technology was imported by Tata Power Company from M/S Aardelite of the Netherlands in order to set up their LFA plant in the city of Mumbai. Although the technology was proven at this plant, it is reported not to have performed satisfactorily due its high investment and economic unviability. The third technology is provided by M/S Neopor Richter System, by which any type of fine raw material including fly ash could be converted into aggregate.

In this technology a cement – based slurry containing fly ash is mixed with pre-formed stable foam to yield foamed concrete in the density range of 300 – 1200 kg/m^3. This foamed concrete is poured into 500 mm wide and 500 mm deep open channels and allowed to have initial set. An aggregate processor with cutting edges moving over the channels gives an output of 10^3/h of raw aggregate in granular size of 8 – 20 mm with bulk density of 250 – 800 kg/m^3. The import of foam-forming chemical has been a hindrance in popularizing this technology. The indigenous development of equivalent chemicals is being claimed by certain companies, which may open up opportunities for commercializing a similar process.

6.2. Fly Ash Based Cellular Concrete

The operating experience shows that for making cellular concrete it is desirable to have a cement-grade low-lime fly ash with about 300 m^2/kg Blaine's specific surface, a lime of minimum 70% purity (magnesia not exceeding 3 per cent) and a gypsum of 80% purity. Taken in optimum proportions, these ingredients are wet mixed in a high-speed mixer to form a thin homogeneous slurry. A small amount of aluminium metal powder, passing through 75 μm sieve, is added to the slurry. The hydrogen gas liberated by the reaction of aluminium and lime aerates the slurry, which is then poured into a steel mould to about its two-thirds depth and left undisturbed to set in 4-5 h. The stiffened mass is cut into blocks of desired size, which are subsequently autoclaved at a steam pressure of 11-12 kg/cm^2. After autoclaving the blocks are cooled down to the ambient conditions before use. The finished product has bulk density of 700kg/m^3 or more and it is suitable for load-bearing walls in multi-storied buildings. Apart from being lightweight, it has good insulation and fire-resistance properties. It is stable under varying temperature and humidity conditions. Being more economical than cement-sand blocks, the commercialization of this technology appears to be on the rise. Another version of cellular lightweight concrete is also produced under ambient conditions from cement, sand, fly ash, water and preformed foam (Neopor type) stable for 6 – 8 h. The products of this technology can have densities ranging from 400 to 1800 kg/m^3 (Singh, 2011). Concrete of higher density range (1200 – 1800 kg/m^3) is reckoned as structural grade material.

6.3. Fly Ash Zeolites

Synthesis and characterization of fly ash zeolites have drawn the attention of researchers globally over the last few decades and India has not been an exception to this direction of research. Use of fly ash as a raw material for production of zeolites would provide an inexpensive alternative to the normal process of making zeolites from more expensive alumina and silica. Special emphasis has been given to synthesis of detergent grade zeolite – 4A as this zeolite is the most suitable substitute for the phosphatic component of the detergents.

In general, the zeolite synthesis from fly ash involves its alkaline treatment with caustic soda solution. The direct treatment requires prolonged hydrothermal processing ranging from 24 to 48 h and is an. energy intensive step. This is a major disadvantage and has been overcome by optimizing a fusion step. The modified fusion step results in the formation of sodium silicate and sodium aluminate, thus ensuring the probability of formation of zeolite phase with high purity. Comparison of the properties and analysis of FAZ-A and a commercial zeolite of similar type is shown in Table 21. (Das and Segaran, 2005). It is evident from the results that in composition and properties the synthesized FAZ-A compares quite well with the commercially procured sample.

Table 21. Comparative characteristics of FAZ-A and a commercial sample of similar type

Characteristics	FAZ – A	Commercial sample
Calcium binding capacity (meg/100g)	540	500
Average particle size D_{50}, μm	4.0	<4.0
pH (5% aqueous slurry)	<12	<11
Bulk density, g/cm^3	0.45	0.40
Crystal system	Cubic	Cubic
Chemical analysis (%)		
LOI	20.0	21.0
SiO_2	33.0	32.0
Al_2O_3	28.0	28.0
Na_2O	17.0	19.0

A study of FAZ synthesis by the hydrothermal process with fly ashes from seven different thermal power plants has recently been reported by Prasad et al (2012). All experiments were carried out at 100^0 C but with different solid/liquid ratios, different concentrations of alkalis, and different incubation times. The products were characterized in detail. CEC and ammonium adsorption capacity of FAZ up to 250 meq/100g and 22.93 mgNH^{4+}/g respectively were observed and their potential to reduce heavy metals and other environmental pollutants were established. A study pertaining to the synthesis of Na-Y zeolite from coal ash was reported by Kondru et al (2011). Fly ash and NaOH powder were taken in the ratio of 0.8 to 1.5 by volume, ground and mixed to form a homogeneous mixture. The mixture was slowly heated to 550° C and then fired at that temperature for 1 – 2 h. The fired substance was cooled, reground, remixed in about 500 ml distilled water and kept stirred for 10h. The resultant sodium aluminate slurry was subjected to hydrothermal crystallization at 90 – 100° C for 17h. After crystallization, it was cooled, filtered and washed with distilled water to remove excess alumina and dried in oven at 50 – 60° C. the zeolite formed was found to be effective in removing dye from aqueous solution by wet peroxide oxidation.

From the above discussion, one may find the following advantages in making zeolites from fly ash:

a) It may provide a more economical alternative to commercial grade zeolites.
b) It has the potential of providing an effective substitute for environmentally harmful phosphatic component of detergents or it may work as an agent for eliminating heavy metals and other pollutants from disposal streams.
c) It adds substantial value to fly ash as a raw material.

Due to these advantages the FAZ technology merits more attention for scale-up and commercialization than what it has received so far.

7. STUDIES ON IMPROVING THE PROPERTIES AND CHARACTERISTICS OF FLY ASH

From the previous sections of this chapter it is evident that coal ash and more particularly the fly ash has been quite effective as a resource material in certain fields of applications and yet there is a big gap between generation and utilization. Hence, studies are continuing in the country to improve the properties and characteristics of fly ash in particular. Some of these studies are highlighted below.

7.1. Nodulization and Sintering of Fly Ash

One of the problems in using fly ash in bulk is its dust pollution during transportation and handling, for which one has to make special arrangements. An alternative technological approach is the option of producing sintered fly ash nodules (Venkatesh and Harish, 1997, Venkatesh and Srikanth, 2004). In this process the fly ash is first converted into nodules or granules or briquettes by preparing a feed mix with 4 – 7% plastic clay as a binder and using a suitable forming machine for this purpose. Nodules are of 6 – 10 mm size and the sintering temperature is 1100 – 1150° C. Shaft kilns, rotary kilns or stationary or moving grate furnaces can be used for this purpose. The process can use up the unburnt carbon present in the fly ash, supplemented with additional solid fuel, as required. The sintered nodules offer the advantages of non-polluting transportation, handling and storage of fly ash with somewhat improved properties due to the removal of carbon particles on burning. Such sintered nodules can also be used as lightweight aggregates in the building industry.

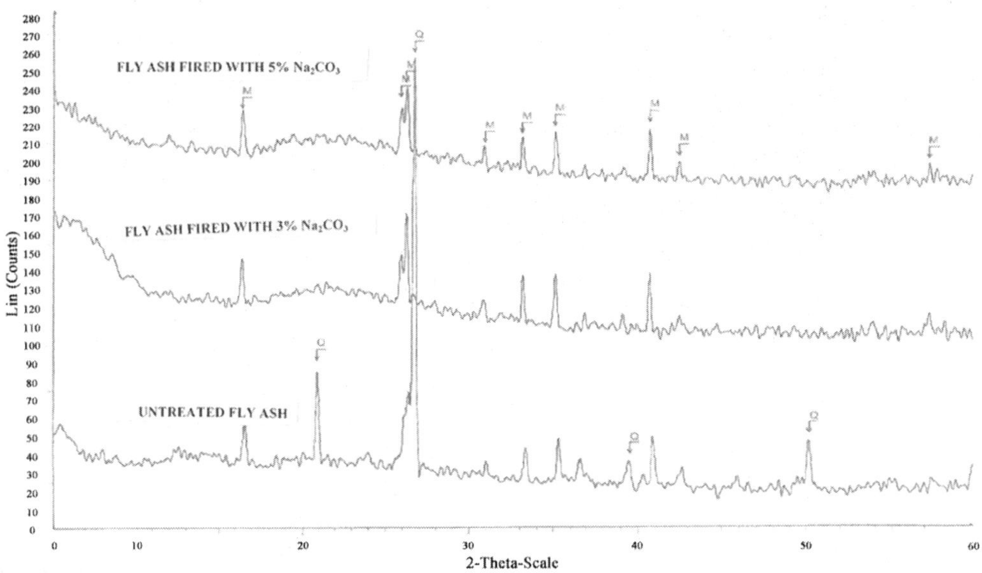

Figure 12(a). XRD patterns of a siliceous fly ash treated with sodium carbonate vis-à-vis the untreated sample.

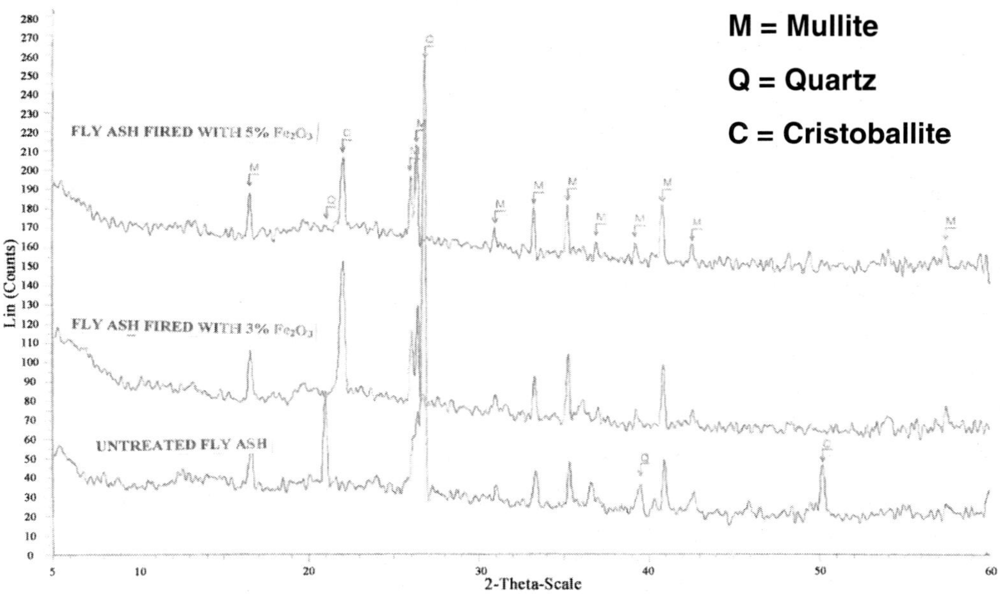

Figure 12(b). XRD patterns of a siliceous fly ash treated with ferrous oxide vis-à-vis the untreated sample.

7.2. Prospects of Revitrification

The basic chemistry shows that the glass content in fly ash is a critical requirement for its reactivity and at the same time the Indian fly ashes, as shown earlier, are predominantly crystalline. In order to find out a solution to this problem the present author attempted to revitrify the low-lime siliceous fly ashes with the help of sodium carbonate and ferrous oxide as fluxes. The fly ash samples were mixed with the fluxes individually up to 5% and fired at different temperatures up to 1525° C. The X-ray diffractograms of the samples fired at 1525°C are shown in Figures 12 (a & b). While there were no phase changes at lower temperatures, at 1525°C with sodium carbonate as the flux, the quartz peaks disappeared but the mullite peaks remained unchanged. But with ferrous oxide as the flux, the quartz phase converted into cristobalite with no change in the mullite phase. The Chapelle test was conducted with the fired and unfired fly ash samples and it was observed that the fired sample with dissolved quartz did not show any visible improvement in the fly ash reactivity. It seemed therefore that, apart from the quartz phase, the dissolution of the mullite phase was necessary for the revitrification process, which required a very high temperature and made the process unworkable. As an alternative, the feasibility of the remelting process by making a clinker of alkali aluminate chemistry is still being examined.

7.3. Chemical Activation

The chemical activation of fly ash usually refers to the processes of alkali and/or sulfate activation of this material. The alkali activation involves dissolution of the glassy phase in an increased alkaline environment, thereby accelerating the pozzolanic reaction. It is obvious that for glass-deficient fly ashes this process of activation may not be very effective. On the other hand, the sulfate action involves its ability to initiate reactions with the aluminous phases, thereby generating the strength-imparting hydrate phases in cement hydration. This also has a relation with the order of crystallinity of the alumina-bearing phases and their dissolution kinetics. Geopolymer cements and concretes have been developed on the basis of the reactions occurring between slag or fly ashes and alkaline fluids. Resulting in three-dimensional polymeric chain and ring structure consisting of Si-O-Al-O bonds (Davidovits, 1994). Here also the glass chemistry and its dissolution characteristics would play an important role. The crystalline nature of the Indian fly ashes is not expected to be very conducive for such reactions under normal circumstances.

Ganguly and Chatterjee, (2011) studied the alkali reactivity of some Indian fly ashes. The studies were carried out with five siliceous fly ash samples collected from five different sources having silica and alumina contents ranging from 55.83 to 61.65 % and 20.67 to 30.06 % respectively. The XRD patterns showed very shallow humps at 24 – 26^0 two-theta indicating the glassy phase to be quite low, tentatively in the range of about 20 % or so.

The activation trials were carried out in three modes:

a) The first set of experiments was conducted with 5% binders prepared individually from the analytical grade sodium hydroxide, potassium hydroxide, calcium hydroxide and their carbonate and sulfate forms dissolved in 200ml of mixing water. The casting mixes were prepared in a mechanical mixer with 500g of each of the fly ash samples and 1500g of standard graded sand. The mixes were cast in 50 mm cube moulds and were cured at 27^0 C and 50^0 C till the age of breaking. In this set of experiments the cubes could not be demoulded as no handling strength was gained even at elevated temperature.
b) In another set of experiments two fly ash samples differing in their mullite contents out of the five mentioned earlier were selectively taken and mixed with 1N sodium hydroxide solution and nodulized. The nodules were heat-treated at 400^0 C for about 20 min. After cooling to the room temperature, the hardened nodules were crushed and the gel-like substance was mixed with standard graded sand in the ratio of 1:3. The mixes were cast in cube moulds and cured at 50^0 C till the age of crushing. In this set of experiments, unlike the previous one, activation was triggered and both the fly ash samples showed 1-day strength of 4.8 and 6.7 MPa and 3-day strength of 7.2 and 9.7 MPa respectively.
c) The third set of experiments was carried out with silicates of sodium, potassium and calcium respectively as binders, the dosage being 5% of the fly ash mass. While the sodium silicate was of the composition Na_2SiO_3. $9H_2O$, the potassium silicate was pure K_2SO_4 powder and the calcium silicate was in the metasilicate form. The rest of the experimental procedure including high-temperature curing remained unchanged. The results obtained are shown in Figure 13. In continuation of this set of experiments the molarity of sodium silicate was altered with with addition of sodium hydroxide to the basic sodium silicate and the effective binder dosage could be reduced for the similar level of strength gain.

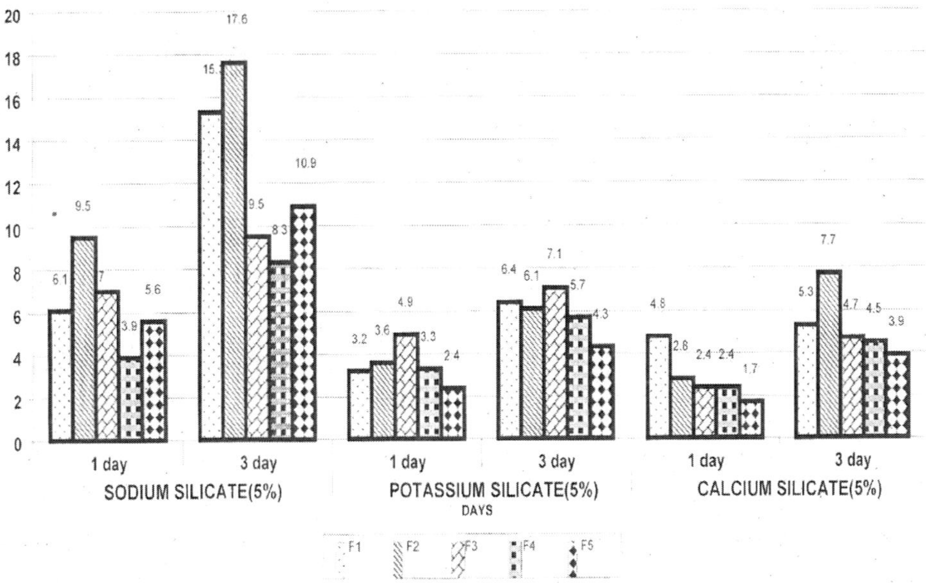

Figure 13. Compressive strengths of fly ash – sand mortars with alkali silicate binders.

The relationship of molarity of the alkaline fluid, fluid-to-fly ash ratio and compressive strength development was published in another study earlier, (Radhakrishna et al, 2006), which is shown in Table 21.

Table 21. Relationship of the molarity of the alkaline binder with the compressive strength of activated siliceous 1:1 mortar in MPa

Fluid/Fly ash ratio	Fly ash/ fluid ratio	4 M Binder		8 M Binder		10 M Binder	
		3-d	7-d	3-d	7-d	3-d	7-d
0.30	3.33	4.8	6.0	16.0	20.0	24.0	26.0
0.40	2.50	3.0	4.0	11.8	14.5	16.0	19.0
0.50	2.00	1.5	2.4	8.0	11.0	12.0	14.0
0.60	1.67	0	0	6.0	8.6	10.0	11.0

Table 22. Strength properties of a siliceous fly ash with 5% alkali fluid of enhanced molarity

Age	Compressive strength, MPa	Flexural strength, MPa
3 – day	14.5	5.5
7 – day	34.0	10.8

It may also be relevant to mention here that in the laboratory of the present author some exploratory work has been carried out to ascertain the relationship between compressive and flexural strengths of alkali-activated siliceous fly ashes without any sand addition through prism mould tests. With alkali binder dosage of 2.5% $Na_2SiO_3.9H_2O$ +2.5% NaOH the strength properties obtained were as shown in Table 22.

It is interesting to note that the fly ash by itself develops good strength and more interestingly the flexural strength is 30 to 40% of the compressive strength. It appears that in the case of crystalline fly ash the dissolved alumina seems to react with the silicate supplied by the activating solution and this leads to the formation of alumina-silicate diagomers. Perhaps this may be the reason why the use of sodium silicate solution has a better activating property than sodium hydroxide. It seems as well that the supplementation of sodium hydroxide to sodium silicate apparently helps in the growth of N-A-S-H gels.

7.4. Ultrafine Grinding for Mechanical Activation

For mechanical activation various milling options are being explored in various laboratories and pilot plants in India. The main objective in such experimental studies has been to ascertain if elastic, plastic and shear deformations could be introduced into the fly ash particles, leading to their fracture and amorphization.

Mechanical milling is known to be the most productive method of producing large quantities of nanocrystalline powders of different types of materials such as metals, alloys, intermetallics, ceramics and composites. Milling and mechanical alloying are carried out using high-energy planetary, ball and vibration mills, where the mean size of the powders produced may vary from 200 to 5-10 nm. Some examples furnished by Gusev and Rempel (2008) are reproduced here. When milling beta form of iron boride in a ball mill it is possible to produce its alpha form with a mean crystallite size of about 8nm. Mechanical treatment of barium titanate in a planetary mill can produce a nanocrystalline powder with a mean particle size of 5-25nm. Mechanical alloying of the powders of borides, carbides, silicides, oxides and sulfides of transition metals has been carried out by the "explosive" method in vibration mills. The powders of transition metal nitrides with particle size of several nanometers are synthesized by milling in vibromills in nitrogen atmosphere. Mechano-chemical synthesis in ball mills of nanocrystalline carbides of titanium, zirconium, vanadium and niobium from powder mixtures of metal and carbon has also been reported. Although carbides form after 4 – 12h milling, the particle size develops after 48h milling to a size range of 6 – 8 nm. Ball milling can be taken recourse to make a variety of new carbon types including carbon nanotubes. Unlike the above experience, the mechanical milling of fly ash has given a different trend of grindability. The typical results are shown in Table 23. Based on a large number of experiments of fly ash grinding conducted by the author in the laboratory scale the comparative results of ball milling and vibro-milling in dry mode and attrition milling in wet mode can be summarized as given in Table 24.

Table 23. Results of extended ball milling of a siliceous fly ash

Grinding time, h	Blaine's surface, m^2/kg	Density, g/cm^3	Lime reactivity, MPa
0	300	2.10	5.8
1	490	2.30	7.0
2	550	2.35	-
3	630	2.39	-
4	710	2.42	9.5
5	780	2.48	-

Table 24. Lab scale grinding trials of fly ashes by different milling systems

Mill system and mode of milling	Milling time, min	Factor of increase in Blaine surface	Reduction ratio of Median particle size	Factor of increase in Lime Reactivity
Ball Mill (Dry)	300 – 360	3.0 - 3.5	3.0 – 3.5	2.0 – 3.0
Vibratory mill (Dry)	50 – 60	3.5 – 3.8	5.0 – 6.0	2.5 – 3.0
Attrition mill (Wet)	30 – 40	3.5 – 4.0	8 - 10	2.5 – 3.0

Table 25. Residual calcium hydroxide in the cement and mineral admixture blends after 3 days of hydration as determined by TG-DSC at 1000^0 C

Serial No.	Blend composition	$Ca(OH)_2$ %
1.	OPC + 25% as received fly ash	10.90
2.	OPC + 25% ball-milled ground fly ash	16.60
3.	OPC + 25% vibratory milled fly ash	7.73
4.	OPC + 25% silica gel	Nil

From the table it is evident that in the case of dry milling the vibratory mills displayed significantly lower milling time and better product characteristics. But the attrition mill in wet mode showed even better performance. However, the average particle size of fly ash ground in different milling systems ranged from 3.0 to 9.0 μm and the enhancement of fly ash reactivity was, for all practical purposes, proportional to the levels of fineness achieved. No discontinuity or sharp change in the product reactivity was observed. The XRD patterns and scanning electron micrographs did not show any noticeable difference. The XRD patterns of differently milled fly ashes and that of a silica gel are compared in Figure 14 and their pozzolanic behavior is compared in Table 25.

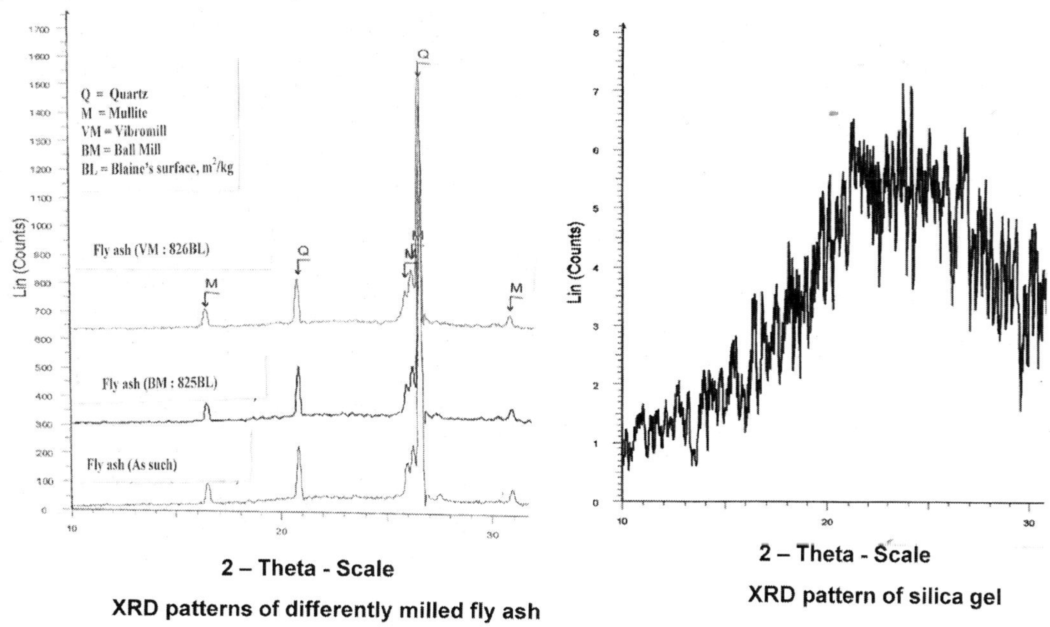

Figure 14. Comparison of XRD patterns.

All these observations, prima facie, indicated the absence of any significant role of milling in the mechanical activation per se. Increase in the specific surface area due to milling was the only factor to enhance the reactivity of fly ash. As one could observe that invibro-milling and attrition milling there is some potential to reduce the milling time and consequently the energy consumption, the scale-up possibilities of these systems were examined.

7.4.1. Scale-Up Possibilities of the New Milling Systems

Two specific attempts made in the recent past are briefly mentioned here. A pilot vibratory mill of 1.5 tph rated throughput with twin tube of 600 mm dia and 3500 mm length had been installed in the premises of one of the cement plants in India. The rated mill drive of 90kW was provided with 62 per cent volumetric filling of cylpebs of 30x30, 25x25 and 20x20 mm sizes. For this installation the problems encountered were the non-achievement of the rated throughput, desired reduction ratio and the expected specific power consumption. Another endeavour has been with a tower mill as a variant of the attrition mill but in dry mode, keeping in view the fact that tower mills or vertimills of throughput of up to 100 tph were in industrial operations but in wet mode. The pilot scale trial runs with the Indian fly ashes having feed particle size range of 1 – 100 μm did not yield the d_{50} value of even 10 μm with specific power consumption of up to 110 kWt/t of fly ash. Use of coarse media (6mm), low tip velocity of the stirrer (3m/s) and dry mode were considered responsible for not achieving the expected performance. Other milling systems like the stirred mills in general and more particularly the horizontal IsaMill type systems merit some attention, as the power intensity in the latter mills is quite high (say, 300 – 1000 kW/m^3 compared to 50 – 100 kW/m^3 in vertical Sala mills). However, for grinding of metallic ores in such systems the particle size reduction below 5 – 10 μm has not been reported. Further, the milling being intrinsically wet, the drying issue becomes critical for fly ashes.

8. Threshold Fineness of Fly Ash for Mechano-Chemical Activation

From the preceding section, it is evident that with the known technologies of fine grinding and high-efficiency cyclone classification the particle size of fly ashes remains above 1 and below 10 μm in general which is much coarser than the field designated as "submicrocrystalline" having mean grain size in the range of 300 to 40nm (Gusev and Rempel, 2008). In Materials Science, it is generally observed that decrease in the size of the crystals below some threshold value results in a large change of properties. These effects occur, as it is believed today in Nanoscience, when the mean size of crystalline grains does not exceed 100nm. Whether this threshold value holds good for fly ashes in general and the highly crystalline fly ash in particular is an issue to be resolved. However, the performance of silica fume having mean particle size of 100nm in general (Table 26) has given an indication that the quantum jump in the properties of fly ashes cannot be expected as long as the mean particle size remains above 1 μm, which has been seen so far as the size reduction limit of the known and potentially up-scaleable fine grinding and classification technologies.

Table 26. Performance comparison of classified fine fly ash with silica fume in concrete

Serial No.	Composition/Properties	Control	Fly ash	Silica fume
1	d_{50}, μm	--	3.0	<0.1
2	Addition, %	0	10.0	10.0
3	w/c ratio	0.40	0.40	0.40
4	Cement, kg/m^3	400	357	356
5	Slump, mm	100	80	80
6	Compressive strength, MPa			
	7 – day	56.2	54.5	71.0
	28 - day	67.9	68.7	80.7
7	Rapid chloride permeability, coulomb			
	7 – day	2922	1083	429
	28 - day	2340	758	297

In search of the threshold particle size of fly ash to effect large change in properties it may be worthwhile to compare with certain other mineral admixtures and fillers used in the cement and concrete industry in terms of their particle characteristics and reactivity. As already mentioned, silica fume is an effective mineral admixture. Depending on the source, the silica content may vary between 94 – 98% in the case of silicon metal and between 86 – 90% in the case of alloys.

The average particle size is about 100 – 150nm and the BET surface is 15 -25 m^2/g (Taylor, 1990). The comparative effect of the use of silica fume has already been shown in Table 26. Compared to silica fume, the performance of aqueous silica colloid suspensions (known as "cembinder") is also worth noting (Chandra, 2002). This admixture contains silica colloid particles to the extent of 8 – 60% by weight of the solution. The specific surface area of the particles is in the range of 50 – 200m ^{2}/g and the mean size of the particles ranges from 5 to 200nm. It is reported that 4 per cent Silica Colloid consumed 60% more calcium hydroxide than 4 percent Silica Fume in the same period of 7 days. Even in one day the Silica Colloid reacted more with calcium hydroxide than the Silica Fume. One may therefore observe that the higher reactivity of Silica Colloid is on account of its nanometric particle size and its purity.

Another important concrete admixture is metakaolin containing 51 – 55% silica and 40 – 45% alumina. The BET surface area is in the range of 14 – 22 m^2/g with particle sizes of 60-90 per cent under 2 μm and less than 15 per cent above 5 μm. This type of admixture shows a pozzolanic reactivity of 840 mg CaO g^{-1} of metakaolin by the Chapelle test (Jones, 2002). A fly ash of similar particle size dimension would also display fairly comparable values. Finally, it may be relevant to compare the behavior of the finely ground limestone filler in cement. It is generally interground with clinker and because of its softness becomes finer than the clinker. For an overall specific surface area (Blaine's) of 420 m^2/kg, the particle size of about 50 per cent of the filler can be below 700nm as compared to 3 μm of the clinker component.

Because of the fineness, it accelerates the hydration of the alite and aluminate phases. Chemically it reacts with the aluminate phase, producing a hydrated carboaluminate phase, thus competing with gypsum (Taylor, 1990).

From the above examples one may observe that the submicrocrystalline materials behave quite differently from materials coarser than 300nm as depicted in Figure 15. Hence, it is certainly worth looking at the feasibility of bringing down the mean particle size of the crystalline fly ashes in the submicrocrystaline range and observe the corresponding improvement in reactivity either by the Chapelle test or the standard lime reactivity test. Using this as the reference point, an exercise needs to be undertaken to optimize the top size of the ultrafine fly ash particles along with their size distribution pattern so as to achieve the tangible benefits of mechano-chemical activation. The scope of this optimization exercise is to answer the following questions:

a) Is there a sharp and distinctive boundary between the bulk and the submicrocrystalline states ?
b) Is there any critical grain size below which the characteristic properties of submicrocrystalline or more particularly of nanocrystalline material become noticeable ?

In other words, one will have to examine if the transition from the bulk to nanocrystalline state is the phase transformation of the first order for the fly ashes under consideration from the thermodynamic point of view. In this exercise the development of an appropriate conversion technology for the Indian fly ashes from the bulk to nanocrystalline state will turn out to be the most important step.

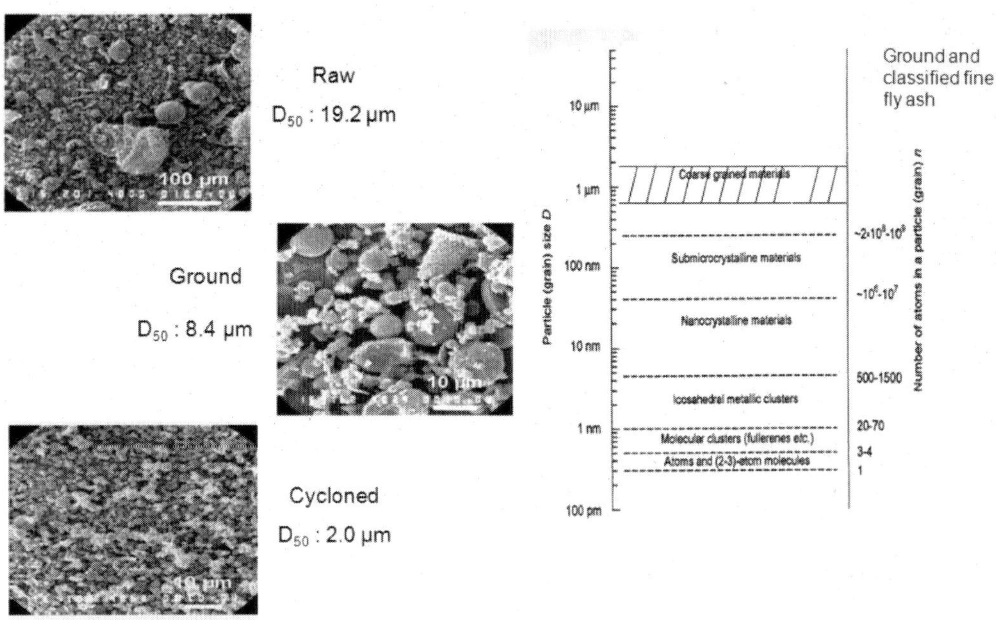

Figure 15. Scale classifying the materials on the basis of their particle size and the position of cyclone and ground fly ash particles.

9. Extraction of Metals and High-Value Materials from Fly Ash

Over the past several decades various attempts have been made to extract the metal values contained in fly ash and also to separate the cenospheres and carbon present in it. Of the major elements available in fly ash, iron, aluminium and carbon extraction have been examined more extensively. Of the minor and trace elements the feasibility of extracting gallium, germanium, molybdenum and uranium have been explored. The objective behind these efforts was more environmental than commercial. Initially the idea was to remove the elements that would otherwise be available for release to the environment after disposal. Along with this perceived environmental need came the commercial interest of transforming the significantly recovered quantities of elements saleable.

Decades back two processes had been tried: direct leaching with hydrochloric acid and pressure digestion acid leaching (Gillium et al, 1982). Later, the applicability of extracting trace elements in a eutectic mix of LiCl+KCl containing 1 mol% of $AlCl_3$ was attempted (Blander, 1984). In the nineteen nineties the technology of biohydrometallurgy was applied (Brombacher, 1998). Trials have also been taken to subject fly ash samples to forced leaching with water, citrate, oxalate, EDTA and carbonate solutions (Nugteren et al, 2002). Technically speaking, all these technologies could remove almost all the metals in substantial proportion. For example, the acid leaching had succeeded to remove all the metals in totality, while the alkali chloride melt technology with reaction time of 3h with no stirring resulted in removal of 98% silver, 82% copper, 98% cadmium, 67% gallium, 97% manganese, 94% lead, 84% titanium and 99% zinc. Chromium and vanadium could not be removed significantly. In the biohydro-metallurgical approach the extents of removal were 81% for zinc, 52% for aluminium, 100% for cadmium, 89% copper, 64% for vanadium and 12% for chromium. In the technology involving forced leaching with water and other extraction agents it was observed that with water alone free lime and substantial quantities of molybdenum (30%), selelium (20-40%), and sulfate (40-50%) were removed. Using the extraction agents these percentages increased to 60-90 and in addition, 15-20% of chromium, vanadium and antimony were removed.

While the technical possibilities of metal extraction could be seen from these and other studies not reported here, the viabilities for up-scaling and industrial adoption of these technologies have apparently been not well established. The industrial practices of such extraction processes are rarely reported even on a global platform. Hence, one can find hardly any adoption or development of technologies in this direction in India at this stage. Further, it should be borne in mind that coal ash in general and fly ash in particular is a multi-oxide substance with decreasing predominance of silica, alumina and iron oxide with, in all likelihood, less than one per cent of high-value metals, especially the rare-earth elements. In India with abundance of metal-grade bauxite, it is obvious that fly ash as a source of alumina may not turn out to be economically viable. The presence of magnetite in the Indian fly ashes is quite insignificant and can hardly justify any recovery attempts under normal circumstances. Similarly, the carbon content in the low-lime fly ashes obtained from the new and modernized large thermal power plants is below two per cent level and could not justify any expensive recovery technologies of flotation and triboelectric type for commercial purposes.

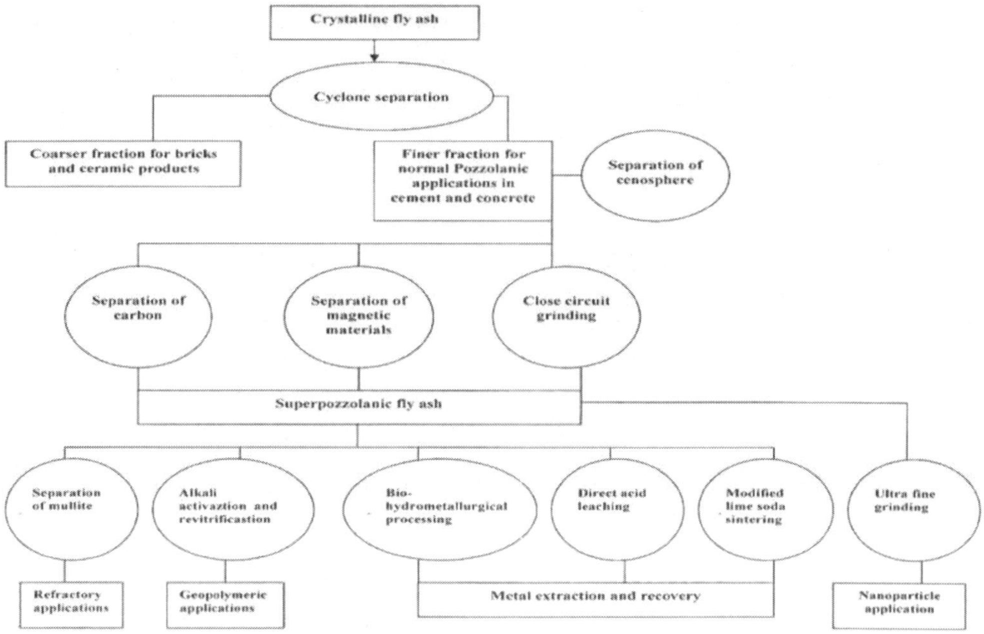

Figure 16. A tentative application-oriented treatment scheme for the crystalline fly ashes.

Finally, the separation of cenospheres by skimming the ash ponds, float-and-sink separation and ultimate classification is in practice in India with trading companies being in operation since 2004. Cenospheres of 0.6-0.9 g/cm^3 density are in trade in India, unlike certain other countries, where spheres of densities up to 2.0 g/cm^3 are collected, separated and used. Under these circumstances, studies are continuing in India for metal separation primarily from the environmental angle, for example, like the single step extraction of metals from coal ash (Lokeshappa and Dikshit, 2012), in which attempts were made to separate water-soluble, ion-exchangeable, acid-soluble and reducible metals with the help of ultrapure water, 1.0M ammonium nitrate, 0.11M acetic acid and a solution of 0.128M of sodium dithionite, 0.3M sodium citrate and 0.1M sodium bicarbonate respectively. For the Indian situations, the environmental focus for removal of heavy metals from fly ash may be a useful approach along with other strategies for enhancing the utilization of coal ash. A tentative application-oriented treatment scheme for the crystalline Indian fly ashes is proposed in Figure 16.

CONCLUSION

The following conclusions can be drawn on the technological initiatives and status of fly ash in India:

- At present over 67% of 214 GW of power generation in India is of thermal type and 85% of the total thermal capacity is based on pulverized fuel firing. As a result the coal ash generation is somewhere in the range of 150 – 200 million tonnes per year, which is likely to reach 300 million tonnes per annum by 2017, when another 100 GW of power generating capacity is added as planned. Thus, in future, in every five-year block period the annual ash generating capacity is likely to increase by 100-150 million tonnes in accordance with the power generation plans. The present and future availability of coal ash, therefore, is an area of concern in the country from the perspective of both environment and resource management.

- The history of transforming the coal ash from an environmental pollutant to an industrial resource started in all seriousness in late nineteen eighties as a simple substitute of silico-aluminous conventional raw materials in bricks and similar products. But the real breakthrough came, when the cement industry realized its potential as a pozzolanic material and increased the production of fly ash based Portland Pozzolana Cement phenomenally. The processes of intergrinding fly ash with clinker, introduction of classification technology for improving the quality and uniformity of fly ash, installing direct collection and transportation of fly ash from the ESPs of power plants with dense phase conveying, etc are the technological initiatives that made the cement industry the largest consumer of fly ash. This preeminent position of the cement industry in this context is likely to continue in the next decade as well.
- More or less simultaneously with the use of fly ash in cement started its use in concrete and in a host of geotechnical applications. But with slow progress in these fields, the total use of coal ash in India at present hovers around 50 per cent of the quantity generated. What is being projected is the leap in the use of coal ash as a mine-fill on one side and in agriculture on the other, which, however, has not been dealt with in this chapter.
- Needless to emphasize that newer applications of coal ash need to be developed, if its hundred percent usage is to be ensured within a reasonable time frame. Large scale application possibilities in geopolymeric materials, high-volume fly ash concrete, metal-matrix composites , polymer fillers, paint and enamel extenders, substitution of silica fume or similar nano-materials, etc. are still awaited. But for such newer applications the properties of the Indian fly ashes that are highly crystalline with variable particle size distribution in the micron range and showing lower reactivities need to be substantially improved.
- The necessity of reliably characterizing and processing the coal ash for separation and recovery of minor, trace and toxic heavy elements is emerging as important for environmental safeguards and resource conservation.
- Presently the improvement in properties of fly ash is being achieved by cyclone separation and re-grinding. With these technologies the size of the fly ash particles remains well above one micron and cannot yield uniform particles of even submicrocrystalline range, which is defined as 300-40nm spread, not to speak of the nano-particulate substance. Along with this shortcoming, the comminution technologies presently adopted have strong engineering limitations of up-scaling, which are awaiting faster solutions.
- The newer comminution technologies will have to answer the following questions: (i) Is there a sharp and distinctive boundary between the bulk and submicrocrystalline states as understood in the modern Materials Science ? (ii) Is

there a critical grain size below which the characteristic properties of nanocrystalline fly ashes become observable with phase transformation of the first order from the thermodynamic viewpoint ?

- As far as the chemical activation is concerned, it has been experimentally observed that the Indian low-lime fly ashes cannot be easily revitrified due to the stable occurrence of the mullite phase. The dissolution chemistry of mullite and its impact on the fly ash properties are still being examined.
- Overall, on the basis of the discussion made in this chapter it appears that a more elaborate treatment scheme needs to be adopted in order to enlarge the application areas of the fly ash component of the coal ash.

REFERENCES

Anonymous, (1914), An investigation of the pozzolanic nature of coal ashes, *Engineering News*, 71 (24), 1334 – 1335.

Banchor, A. and Krishnan, S., (2006), Evaluating the effectiveness of fly ash incorporation as PPC vis-à-vis separate addition to the site mixed OPC concrete, *The Indian Concrete Journal,* 80(4), 17 – 23.

Basu, P.S. and Saraswati, S., (2006), Are existing IS codes suitable for engineering of HVFAC ? *The Indian Concrete Journal* , 80(8), 17 – 21.

Bhattacharjee, K.N. and Dhuri, S.S. (1990), Experiences of utilizing mineral admixtures in concrete under Indian conditions, International Conference on Concrete Technology for Developing Countries, New Delhi, November, Vol. 1, pp. II-3-11.

Blander, M., (1984), Extraction of trace elements from fly ash, US Patent 4475993A.

BMTPC, (1993), State-of-art Report on coal ash utilization in India – Building materials sector, Building Materials & Technology Promotion Council, Ministry of Urban Development, Govt. of India, New Delhi, 1 – 25.

Brombacher, C., Bachofen, R.. and Brandt, H., (1998), Development of a laboratory scale leaching plant for metal extraction from fly ash by Thiobacillus strains, *Applied and Environmental Microbiology*, April 64 (19), 1237 – 1241.

Cement Research Institute of India, (1979), Utilization of Indian fly ash, 5P-3-71, National Council of Cement and Building Materials, New Delhi.

Central Electricity Authority, (2013), Website www.cea.nic.in/installedcap_allindia.pdf.

Central Electricity Authority, (2012a), National electricity plan, Vol.1 (generation), Government of India, New Delhi.

Central Electricity Authority, (2012b), Report on fly ash generation at coal/lignite based thermal power stations and its utilization in the country (April – September 2011), CEA, New Delhi.

Centre for Fly Ash Research and Management, Website http://c-farm.org.

Chandra, S., (2002), Properties of concrete with mineral and chemical admixtures, in : Structure and Performance of Cements, (Eds.) *J. Bensted and P. Barnes*, Spon Press, London, 140 – 186.

Chatterjee, A.K., (2001), Manufacture of PPC in India – effective use as a blending material, Proceedings of National Seminar on Utilization of Fly Ash in Water Resources Sector, New Delhi, April, 29 – 41.

Chatterjee, A.K (2008), Enhancing the potential of industrial use of the Indian fly ashes through mechano-chemical activation – prospects and problems, Proceedings of the VI International Conference on Mechano-Chemistry and Mechanical alloys, National Metallurgical Laboratory, Jamshedpur, India, Published in 2011, pp. 287 – 293.

Chatterjee, A.K., (2010), The Indian fly ashes, their characteristics and potential for mechano-chemical activation for enhanced usability, Proceedings of the 2nd International Conference on Sustainable Construction Material and Technologies, Ancona, Italy, 28 – 30 June, Vol. 1, 41 – 51.

Coal Ash Institute of India, Website http://coalashinstitute.com.

Das, A.K. and Segaran, R.G. (2005), Opportunities of high value application of fly ash, Journal of CAII, Vol. VII, Coal Ash Institute of India, Kolkata, 1 – 11.

Das, S.K., (2003), Morphology of some Indian fly ashes, *The Indian Concrete Journal*, 77(9), 1300 – 1303.

Das, S.K. and Yudhbir, (2003), Chemistry and mineralogy of some Indian fly ashes, *The Indian Concrete Journal*, 77(12), 1491 – 1494.

Dattatreya, J.K., Neelamagam, M. and Rajamane, N.P., (2006), A comparison of effects of ultrafine fly ash and silica fume in concrete, *The Indian Concrete Journal*, 80(2), 44 – 50.

Davidovits, J., (1994), High – alkali cements for 21st century concretes, Proc., V. Mohan Malhotra Symp. On Concrete Technology – Past, Present and Future, sp – 144, P. Mehta ed., ACI, Detroit, 383 – 397.

Davis, R.E., Carlson, R.W., Kelly, J.W. and Davis, H.E. (1937), Properties of cements and concretes containing fly ash, *Journal of the American Concrete Institute*, Vol. 33, 577 – 611.

Department of Science and Technology, Government of India, Website www.dst.gov.in/ scientific program/fly_ash.html.

Desai, J.P., (2004), Construction of HVFA concrete pavements in India : four case studies, *The Indian Concrete Journal*, 78 (11), 67 – 71.

Dewangan, P., Pradhan, M. and Kishore, N. (2010), Utilization of fly ash as a structural fill material for safe and sustainable development : need of the hour, *The Indian Mining and Engineering Journal*, 134 – 139.

Fournier, B., Lu, D., Charland; J.P., and Li, J. (2004), Evaluation of Indian fly ashes for use in HVFA concrete, Part 1; Characterization, *The Indian Concrete Journal*, 78 (11), 22 – 30.

Ganguly, P.P. and Chatterjee, A.K. (2011), Alkali activation of highly crystalline Indian fly ashes towards developing clinkerless binders, Abstracts & Proceedings, XIII ICCC, (Eds.) Palomo, A., Zaragova, A. and Agni, J.C.L., Madrid, Italy, p. 169.

Gillium, T.M., Canon, R.M., Egan, B.Z., Kolmers, T.M., Seelay, F/G/ and Watson, J.S., (1982), Economic metal recovery from fly ash, *Resources and Conservation*, Vol. 9, August, 155 – 168.

Gusev, A.I. and Rempel, A.A., (2008), Nanocrystalline Materials, Viva Books Pvt Ltd, New Delhi.

Jain, A.K., (2006), Development and application of high volume fly ash concrete, Indian Cement Review, December, 12 – 16.

Jones, T.R., (2002), Metakaolin as a pozzolanic addition to concrete, in : Structure and Performance of Cements, (Eds.) J. Bensted and P. Barnes, Spin Press, London, 372 – 399.

Kadiyali, M.R., Rahidekar, V.H. and Sati, R., (2009), Techno-economic feasibility of concrete roads in India's road modernization programme, NCB – CMA Special Publication, 11th NCB International Seminar on Cement and Building Materials, NCB, New Delhi, 75 – 81.

Kondru, A.K., Kumar, P., Teng, T.T., Chand, S., Wasewar, K.L., (2011), Synthesis and characterization of Na-Y Zeolite from coal ash and its effectiveness in removal of dye from aqueous solution by wet peroxide oxidation, *Arch. Environ.Science*, 5, 46 – 54.

Kumar, V., (2009), Fly ash : a rewarding soil ameliorant, Technical Papers, National Conference on the use of fly ash in agriculture and Forestry, Coal Ash Institute of India, Kolkata, India, 1 – 10.

Kumar, V. (2011), A holistic approach for fly ash management. Proceedings of the National Conference on Fly Ash, Centre for Fly Ash research and Management, New Delhi, pp 1.1.1 – 1.1.7.

Kumbhakar, D., Pandey, J.K., Prashant, and Mangal, P.K. (2005), Coal ash as minefill – an innovative approach towards eco-friendly ash utilization, *Journal of CAII*, Coal Ash Institute of India, Vol. VII, March, 23 – 29.

Lokeshappa, B. and Dikshit, A.K., (2012), Single step extractions of metals in coal fly ash, *Resources and Environment*, 2(2), 1 – 8.

Menon, R., Raja, P., Malpe, D., Subramaniyam, K.S.V. and Balaram, V., (2011), Radioclemental characterization of fly ash from Chandrapur Super Thermal Power Station, Maharashtra, India, *Current Science*, 100 (12), p. 1880.

Mullick, A.K., (2005), Use of fly ash in structural concrete : Part 1 – why ? *The Indian Concrete Journal*, 79(5), 13 – 22.

Nataraj, M.S., Jayaram, M.A., and Ravikumar, C.N., (2006), Group indexing fly ashes through radial basis function network. *The Indian Concrete Journal*, 80(7), 39 – 45.

Nugteren, H.W., Janssen – Jurkovicova and Scarlett, B., (2002), Removal of heavy metals from fly ash and the impact on its quality, *Journal of Chemical Technology and Biotechnology*, Vol. 77, Issue 3, 389 – 395.

Pandit, G.G., Sahu, S.K. and Puranik, V.D.,(2011), Natural radionucleides from coal-fired thermal power plants – estimation of atmospheric release and inhalation, risk, *Radioprotection*, 46(6), pp. S 173 – S 179.

Pandian, N.S., (2004), Fly ash characterization with reference to geotechnical applications, *Journal of Indian Institute of Science*, Nov – Dec, 84, 189 – 216.

Prasad, B., Maity, S., Sangita, K., Mahato, A.K., Mortimer, R.J., (2012), Studies on synthesis and characterization of zeolite prepared from fly ash, *Environ. Technol.*, 33 (1 – 3), 37 – 50.

Radhakrishna, Shashishankar, A. and Nagraj, T.S., (2006), Phenomenological model for assessment of strength development in Class F fly ash based geopolymer mortars, *ICI Journal,* April – June, 23 – 27.

Roy, D., (2005), Fly ash transportation by optimizing the usage of the disposal area and minimizing the environmental consequences through high concentration slurry system, *Journal of CAII*, Coal Ash Institute of India, Vol. VII, Mareh, 45 – 49.

Saha, S. and Pal, S.K. (2012), Compressibility behavior of soil and fly ash used in successive layers, *EJGE*, Vol. 17, Bund. T, 2659 – 2670.

Senapati, M.R., (2011), Fly ash from thermal power plants – Waste management and overview, *Current Science*, 100 (2).

Singh, G.B., (2011), Cellular lightweight concrete produced under ambient conditions, The Masterbuilder, December, 88 – 94.

Sivasundaram, V. and Malhotra, V.M., (2004), High-performance high-volume fly ash concrete, *The Indian Concrete Journal*, 78(11), 13 – 21.

Sivasundaram, V., (2004), Evaluation of Indian fly ashes for the HVFA concrete, Part II : Behaviour in concrete, *The Indian Concrete Journal*, 78(11), 41 – 50.

Sridharan, A., Pandian, N.S. and Rajasekhar, G., (1996), Geotechnical characterization of pond ash, Ash ponds and ash disposal systems (V.S. Raju et al, eds.), Narosa Publishing House, New Delhi, 97 – 110.

Taylor, H.F.W., (1990), Cement Chemistry, Academic Press, London.

USBR (1948), Construction of Hungry Horse Dam in Mantana, No. CH – 95, USA.

Venkatesh, D.S. and Harish, (1997), Process the fly ash for beneficial utilization, *Indian Cement Review*, December, 25 – 28.

Venkatesh, D.S. and Srikanth, D.V., (2004), Co-production of sintered fly ash nodules in cement plants, *Indian Cement Review*, May, 9 – 13.

Zachariasen, W.H., (1932), The atomic arrangement in glass, *J. Am. Ceram. Soc.*, 54, 384 – 395.

PART C: HEALTH AND ENVIRONMENTAL IMPACTS OF FLY ASH

In: Fly Ash
Editor: Prabir Kumar Sarker
ISBN: 978-1-62948-044-2

Chapter 14

Oxygen Metabolism and Mitochondrial Function in the Lung and Heart after Exposure to Residual Oil Fly Ash

Pablo Evelson*[*], *Natalia Magnani, Timoteo Marchini and Silvia Alvarez
Instituto de Bioquímica y Medicina Molecular (IBIMOL UBA-CONICET),
Facultad de Farmacia y Bioquímica. Universidad de Buenos Aires,
Buenos Aires, Argentina

Abstract

There is strong evidence that ambient air particulate matter (PM) currently presents a serious risk to human health. Epidemiological studies have established a direct correlation between the levels of these particles and acute morbidity and mortality from cardiopulmonary diseases. Human and animal models have shown a pulmonary and systemic inflammatory response and oxidative stress associated with the exposure to environmental particles which could, in turn, alter heart oxygen metabolism and cardiovascular function. Increased production of reactive oxygen species leading to oxidative damage have been suggested to significantly contribute to the cardiopulmonary toxicity of PM inhalation. Because of the unique composition of Residual Oil Fly Ashes (ROFA), especially rich in metals, it is the most frequently PM surrogate to evaluate their role in many biological studies. Tissue injury after ROFA inhalation has been suggested to be triggered by local reactive oxygen species production which, in turn, could be generated not only from the particles themselves, but from chemicals coated on their surface as well. Moreover, oxygen-derived free radicals may also be generated by the interaction of particulate pollutants and their components, with cellular enzymes and organelles. It is well established that mitochondria represent an important modulator of the cellular response to a wide variety of metabolic and environmental stressors. Moreover, superoxide anion is produced by this organelle under normal physiological conditions as a by-product of redox reactions during electron transport in the respiratory chain. Therefore, in addition to serving as an oxidant production source, mitochondria are

[*] Corresponding Author address: Email: pevelson@ffyb.uba.ar.

highly susceptible to reactive oxygen species, which could also indirectly perturb the organelle function. Considering that tissue injury mediated by reactive oxygen species appears to be a key mechanism in the observed damage triggered by exposure to ROFA, oxygen metabolism and mitochondrial function are important areas of study for clarifying the molecular mechanisms involved in this process.

Keywords: Residual oil fly ash, lung, heart, reactive oxygen species, mitochondria, NAPDH oxidase

1. Composition and Characteristics of Air Pollution

The pollutant mix is complex and this complexity is magnified because of the variation in its components, between places and over time. The major primary pollutants in ambient air which are of concern regarding health effects, are sulfur dioxide (SO_2), nitrogen oxides (NOx), carbon monoxide (CO), volatile organic compounds (VOCs), and particualte matter (PM) mostly coming from a single source, which is the combustion of fossil fuels. The burning of fuels for road transport and electricity generation, is of particular importance.

On the one hand, the pollutants which are directly emitted into the atmosphere are known as primary pollutants. On the other hand, secondary pollutants are those formed in the air as a result of chemical reactions with other pollutants and atmospheric gases. Examples of primary pollutants are CO and SO_2, while ozone is an example of a secondary pollutant. Others can be considered as both primary and secondary pollutants, such as nitrogen dioxide (NO_2) and some PM. They are both emitted directly into the atmosphere, and formed from other pollutants. For example, some of the NO_2 is emitted directly from power stations and vehicle exhaust, while most is formed by the oxidation of nitric oxide (NO) in the air. Fine PM is emitted directly from a number of natural and anthropogenic sources as well as being formed within the atmosphere. Secondary PM is formed mainly from the oxidation of SO_2 and NO_2. The reactions of hydrocarbons in the air to produce organic particles observed in some areas are an important source of secondary PM.

Thus, it is accepted that air pollution exists as a complex mixture and that effects attributed to different pollutants such as ozone (O_3), NOx, or PM, may be influenced by the underlying toxicity of the full mixture of all air pollutants.

Suspended PM is composed of solid and liquid particles that come not only from natural sources, such as forest fires, volcanic emissions and sea spray, but from anthropogenic sources as vehicle exhaust and fossil combustion as well (Nel, 2005). Of the different air pollutants, PM seems to be the major concern from a health perspective (WHO, 2005). It is now well established that PM cause respiratory diseases and increased lung cancer, as well as death due to cardiovascular diseases (Pope et al, 2006).

PM consists of a mixture of particle components, including traffic and combustion-derived carbon-centered particles, secondary particles (nitrates and sulphates), wind-blown dust of geological origin and potentially containing endotoxin, and biological particles (spores, pollen) with their associated allergens. Based on the size, PM is often classified into three main groups: coarse, also known as "thoracic" particles (with an aerodynamic diameter between 2.5–10 microns, PM_{10}), fine (aerodynamic diameter between 0.1-2.5 microns, $PM_{2.5}$) and ultrafine (aerodynamic diameter less than 0.1 micron). PM_{10} deposit predominantly in the lower airways (nose and throat) and are cleared by exhalation or mucociliary and swallowing. Fine and ultrafine particles can penetrate into the lower airways and alveoli and are more closely related with the adverse health effects of PM than coarser particles (Schins et al., 2004). The fine fraction or "respirable" particles, smaller than 2.5μm ($PM_{2.5}$), can penetrate into the gas-exchange region of the lung and the ultrafine fraction (sometimes called nanoparticles) which contribute little to particle mass but which are most abundant in terms of number and offer a very large surface area, with increasing degrees of lung penetration.

2. Residual Oil Fly Ash (ROFA)

Anthropogenic emissions are the main contributors to environmental PM burden and consist mainly of motor vehicle emissions and fossil fuel combustion during power generation and industrial processes (Nel, 2004). The inorganic residue that remains after the incomplete oxidation of such carbonaceous materials contributes to PM in urban air and is termed Residual Oil Fly Ash (ROFA) (Ghio et al., 2002). Diverse PM surrogates have been assayed in different animal models in order to study the biological effects of PM exposure. Among them, ROFA has been particularly useful given that it is especially rich in soluble transition metals (namely iron, nickel and vanadium), and because of its low concentration of organic compounds (Schroeder et al, 1987). Therefore, ROFA is the most frequently used combustion-derived particle in order to evaluate the contribution of transitions metals in the biological effects of PM inhalation (Chen and Lipmann, 2009). Moreover, ROFA particles often present an aerodynamic diameter smaller than 2.5 μm ($PM_{2.5}$), a size that have been shown to be more closely associated with PM adverse health effects than coarser particles ($PM_{10\text{-}2.5}$) (Brook et al, 2004).

3. PM and ROFA Adverse Health Effects

Respiratory and cardiovascular events are those more strongly associated with particle pollution (Donaldson et al., 2001; Zareba et al., 2001). Inhalation of ambient particles can provoke a significant amount of pulmonary inflammation (Pope, 2000; Souza, 1998), alterations of the heart autonomic function, and changes in blood parameters (Dockery, 1994; Schwartz, 2001). Although the epidemiologic evidence is convincing, the biologic mechanisms by which ROFA evokes systemic effects, remain to be defined.

3.1. Effects on Lung Tissue

3.1.1. Pulmonary Inflammation

Pulmonary inflammation after PM exposure is well documented in animals as well as humans (Dreher et al. 1997; Nurkiewicz et al. 2004; Schaumann et al. 2004). Immune cells such as neutrophils, eosinophils, and macrophages contain a membrane bound enzyme, the NADPH oxidase, which is induced during the inflammatory response. NADPH oxidase produces superoxide anion ($O_2^{\bullet-}$) using O_2. These reactive oxygen species (ROS) are removed by superoxide dismutase, which is present at the surface of the epithelial cells of the lung (47). The product of this reaction is hydrogen peroxide (H_2O_2), which can easily diffuse through plasma membranes. H_2O_2 can therefore initiate intracellular signaling pathways, or contribute to the generation of other ROS. Alternatively, another enzyme, mieloperoxidase, which is also released by neutrophils, uses H_2O_2 as a substrate and produces hypoclorous acid (HClO), another powerful oxidant.

A critical component of the inflammatory response to particles in the lung, is the release of cytokines from activated macrophages and lung epithelial cells, resulting in neutrophil recruitment. This response may be caused by the deposition of ROFA into the alveolar space in the lung, inducing the release of cytokines from alveolar macrophages. Cultured human alveolar macrophages produce TNF-α and proinflammatory cytokines such as granulocyte-macrophage colony-stimulating factor, IL-6, and IL-1 after phagocytosing PM (Oury et al, 1994). The release of proinflammatory mediators from PM-exposed macrophages appears to be important in stimulating cytokine release from lung epithelial cells, thus amplifying the inflammatory response (Jimenez et al., 2000).

Attention has been focused on transition metals and ultrafine particles as components of ROFA that may be responsible of the inflammatory response. Each of these individual components of PM has been shown to produce an inflammatory response after exposure in animal models (Dye et al., 1994; Brown et al., 2000).

The inflammatory genes induced after exposure to ROFA (TNF-α and IL-8, for example) are regulated by redox sensitive transcription factors such as NF-κβ, AP-1 (activator protein-1) and C/EBP (CCAAT enhancer binding protein) family. NF-κβ is one of key transcription factors that are involved in the inflammatory responses to the PM in the lungs. The result of NF-κβ activation is the generation of proinflammatory cytokines such as TNF-α, IL-6, IL-8 by airway and alveolar epithelial cells (Shcukla et al, 2000). Quay et al reported that ROFA activated NF-κβ via a process thought to be related to the presence of vanadium in the particles; this effect could be inhibited by the addition of the antioxidants deferoxamine and N-acetylcysteine (Quay et al., 1998).

Taken all this data into consideration, the probable sequence of events for ROFA-induced lung inflammation involves the following: (a) injury to epithelial cells by ROS, possibly enhanced in the presence of metals via Haber-Weiss and Fenton chemistry, accompanied by activation of nuclear regulatory factors, leading to the production of proinflammatory cytokines, including IL-8 and IL-6, and increased expression of nitric oxide synthase (NOS), with increased levels of nitric oxide (NO) in exhaled air; (b) activation of vascular endothelium and circulating leukocytes; emigration of inflammatory cells from blood to tissue sites involves up-regulation of adhesion molecules and other markers on vascular endothelium and on circulating leukocytes; and (c) the process of leukocyte-endothelial binding includes increased expression of adhesion molecules followed by their shedding as cells "tether and roll", leukocyte activation, stable adhesion, and transmigration through the epithelium.

3.1.2. Oxidative Stress

The mechanisms by which ROFA causes oxidative stress are still unknown but the phenomenon may be attributed to different causes, which include: (a) generation of oxidants at the particle surface, (b) release of metals from the particle, and (c) triggering of an inflammatory response.

The direct generation of ROS at the surface of the particles is supported by the concept that the particle surface offers a unique physicochemical interface to catalyze reactions resulting in oxidant production. The interaction of PM with membrane components was recognized by the presence of free radicals and oxidants on the particle surface (Fubini et al., 2004).

ROFA contains a large number of soluble metals that have the ability of redox cycling. The involvement of transition metals, such as Fe, Va, Cr, Mn, Co, Ni and Cu, which are able to catalyze Fenton-type reactions and generate hydroxyl radicals, has been proposed. Indeed, the production of oxidative damage, as measured by TBARS levels, showed significant correlations with the overall metal content and with the content of individual metals such as Fe and Va. This effect was inhibited by the presence of antioxidants such as dimethylthiourea and deferoxamine (Sorensen et al., 2005; Dreher et al., 1997). These and another works have led to the suggestion that the dose of bioavailable transition metal, rather than particulate mass, may be the principal determinant of the acute inflammatory response (Samet et al, 2004).

Activation of inflammatory cells capable of generating reactive oxygen and nitrogen species is other proposed mechanism: chronic inflammatory lung diseases are characterized by activation of epithelial cells and resident macrophages and the recruitment and activation of neutrophils, eosinophils, monocytes, and lymphocytes. It is understood that the oxidative stress caused by the activation of the inflammatory system, comprised by alveolar macrophages and neutrophils, plays an important role in the deleterious effects of PM in multicellular organisms. ROS are generated during phagocytosis of the particles, leading to enhancement of oxidative stress and triggering the inflammatory response (Tao et al., 2003).

3.2. Cardiac Effects Caused by Exposure to PM

It has been postulated that pulmonary PM exposure may trigger a cardiovascular response through three different mechanisms (Brook et al. 2004; Nemmar et al. 2002; Oberdorster et al. 2004). The first hypothesis proposes that ROFA deposited in the lung acts through a neural mechanism to alter central nervous system function. In the lung, nociceptive neurons are stimulated by ROFA (Veronesi et al. 2000). Cardiac autonomic function is also altered by PM exposure, suggesting that central input to the heart is altered (Chen and Hwang 2005). Acute electrocardiographic changes after PM exposure are also suggestive of an activated neural mechanism (Wichers et al. 2004). The second proposed mechanism is that ROFA deposited in the lung may enter in the systemic circulation and directly interact with different organs. It has been reported that after lung exposure, PM deposits has been found in several non-pulmonary tissues, including the blood, ventricular microvascular walls, liver, spleen, heart, and brain (Calderon-Garciduenas et al. 2001; Kreyling et al. 2002; Nemmar et al. 2002; Oberdorster et al. 2002). PM deposition was found in the cardiac arteriolar wall of dogs where polymorphonuclear leukocyte (PMNL) margination and microthrombi were also observed (Calderon-Garciduenas et al. 2001). When PM is exposed to cells or tissues *in vitro*, cytokine and tumor necrosis factor-alpha (TNF-α) production, cytotoxicity via endotoxins, oxidative stress, and smooth muscle relaxation was observed (Bagate et al. 2004; Dye et al.

1997; Li et al. 2003; van Eeden et al. 2001). The third postulated mechanism is that PM deposited in the lung initiates a local inflammatory response that develops into a systemic inflammatory response, characterized by alterations in circulating factors and cells associated with inflammation. Circulating interleukin (IL)-1 and IL-6 are elevated in humans exposed to PM (van Eeden et al. 2001). IL-1, TNF-α, and the immune-related transcription factor nuclear factor κB are elevated in the brain tissue of mice exposed to PM (Campbell et al. 2005). Furthermore, blood samples from healthy humans exposed to PM reveal elevations in immature PMNL, neutrophils, and platelets (Salvi et al. 1999; Tan et al. 2000).

Although the evidence for these three proposed hypothesis is significant, the basic mechanisms responsible for the biological effects in the heart remain unclear.

4. Mitochondrial Function after Inhalation of Ambient Air Particles

It has been shown that mitochondrial damage is an important event in PM-induced cytotoxicity (Hiura et al, 1999), but there is little information available about the specific effects of the exposure to ROFA. The first response to PM observed in isolated mitochondria is a decrease in the mitochondrial membrane potential and an augmented generation of O_2^- production, that is followed by cytochrome c release and inner mitochondrial membrane damage (Upadhyay et al, 2003). It is also important to point out that ultrafine particles, the smallest and probably the most toxic component of PM, were found deposited inside damaged mitochondria (Li et al, 2003). However, it is still controversial if PM can damage mitochondria in a direct way or enter to the organelle as a consequence of oxidative damage. PM induced mitochondrial disruption has important biological effects, which include the initiation of apoptosis and decreased ATP production (Hiura et al, 1999; Li et al, 2003). A key mitochondrial target for oxidizing chemicals is the permeability transition pore (PTP) (Jajte et al, 1997). PTP is permeable to molecules of < 1.5 kDa and opens in the mitochondrial inner membrane when matrix Ca^{2+} levels are increased, especially when this is accompanied by oxidative stress (Bernardi, 1999). The opening of PTP produces, in vitro, mitochondrial swelling, outer membrane rupture, and the release of proapoptotic factors such as cytochrome c. In addition, mitochondria become depolarized, causing inhibition of oxidative phosphorylation and stimulation of ATP hydrolysis (Susin et al, 1998).

5. Oxygen Metabolism and Mitochondrial Function in the Lung and Heart After Exposure to ROFA

5.1. Experimental Model

Female Swiss mice weighing 20-25 g were anesthetized by an intraperitoneal (i.p.) injection of ketamine (10 mg/kg) and xylazine (0.1 mg/kg), and exposed to ROFA particles (1.0 mg/kg body weight) or saline solution (control group) by intranasal instillation in a single dose. Mice were immobilized in a 60° inclined supine position while 50 μL of the ROFA suspension was delivered dropwise to the nares by the use of an automatic pipette. After 1, 3 or 5 h of exposure, animals were sacrificed and lung and heart samples were collected. Control mice were handled in parallel, instilled with 50 μL of sterile saline solution, and sacrificed at the same time points. Due to the presence of fluid in the mouse nasal cavity, a respiratory reflex is triggered which ensures that the maximum delivered volume reaches the lung (19). The amount of particles administered in this study was 25 μg per mouse. The selected dose falls within the range of concentrations consistently used in several animal studies (Ghio et al., 2002, Nurkiewicz et al., 2004) and it can be considered as a medium to low dose when compared with pollution levels attained in large cities worldwide (Yosgizaki Akinaga et al., 2009).

5.2. Time Course of Oxidative Stress in Mice Lung

The first approach to study the effects of the acute exposure to ROFA was to evaluate the oxidative mechanisms occurring in lung mice after an acute ROFA exposure, and the temporal sequence of oxidative damage after this exposure. Instillation with ROFA increased oxygen consumption in lung by 55% after 1 h of exposure. A 23% increase in comparison with the control was observed 3 h after the exposure (control value: 225 ± 7 ng-at O/min mg tissue; 1 h: 407 ± 33 ng-at O/min mg; 3 h: 277 ± 12 ng-at O/min mg). In order to evaluate the occurrence of oxidative after the exposure to ROFA, several markers were measured. The macromolecular oxidative damage showed different induction time, since there was a statistically significant increase in phospholipid oxidation after 1 h of exposure, while oxidative damage to proteins was significant only after 3 h exposure. After 1 h of exposure, a 29% increase (control value: 249 ± 5 pmol/mg prot) in phospholipid oxidation, measured as TARS content, was observed. After 3 h of exposure this marker of oxidative damage to lipids returned to control values. When the oxidative damage to proteins (as protein carbonyl levels) was determined, no statistical changes were observed after 1 h of exposure, a 30% increase was found after 3 h (control value: 2.2 ± 0.1 nmol/mg prot). The different time course is consistent with different susceptibility to oxidative damage of these macromolecules, as suggested by previous works (Gurgueira et al., 2002). The antioxidant status was measured as Total Antioxidant Reactive Potential (TRAP), a methodology that estimates the hydrosoluble antioxidants levels present in biological samples. TRAP values after 3 h of exposure were 40% lower than the corresponding control value (14.8 ± 0.99). This observation shows a moderate consumption of antioxidants, since TRAP values are an indication of the levels of low molecular weight antioxidants (mainly glutathione, ascorbic acid and other hydrosoluble molecules capable of trapping free radicals) (Evelson et al., 2001).

With the aim of clarifying the underlying mechanisms of this observation, NADPH oxidase activity and mitochondrial oxygen consumption were measured as the main sources of oxygen consumption after ROFA exposure. The activation of NADPH oxidase 1 h after the instillation, measured by lucigenin chemiluminescence (a significant 40 % increase was observed; control value: control value: 0.76 ± 0.02 UA/mg prot) may be a relevant non-mitochondrial source of reactive oxygen species in this model. No significant changes were observed 3 h after the treatment. The inflammatory process triggered by the ROFA exposure, includes leukocyte recruitment, activation (respiratory burst) and increased alveolar macrophages count (Nurkiewicz et al. 2004; Schaumann et al. 2004). It has been shown that the NADPH oxidase homolog NOX2 expressed in macrophages is an important source of reactive oxygen species under airway inflammation elicited by environmental factors, including PM exposure (van der Vliet, 2005). However, through the use of NOX2 knockout mice exposed to PM, it was suggested that monocyte infiltration could also contribute to the augmented $O_2^{\bullet-}$ production (Kampfrath et al, 2011). Tissue oxygen consumption is related to mitochondrial energy production and reactive oxygen species release. Consequently, the increased tissue oxygen consumption observed after 1 h may also be originated by an elevated mitochondrial subpopulation in state 3 (active respiration), in order to respond to high ATP demand (Vanasco et al., 2012). ROFA treatment after 1 h of exposure significantly increased mitochondrial state 3 oxygen consumption by 34% (control value: 127 ± 4 ng-at O/min mg prot; p<0.005). No significant changes were observed in state 4 respiration. No changes were observed after 3 h of exposure. This observations are supported by the positive correlation found between phospholipid oxidation and NADPH dependent $O_2^{\bullet-}$ production, measured by lucigenin chemiluminescence, and mitochondrial oxygen consumption in state 3 as it is shown in Figure 1.

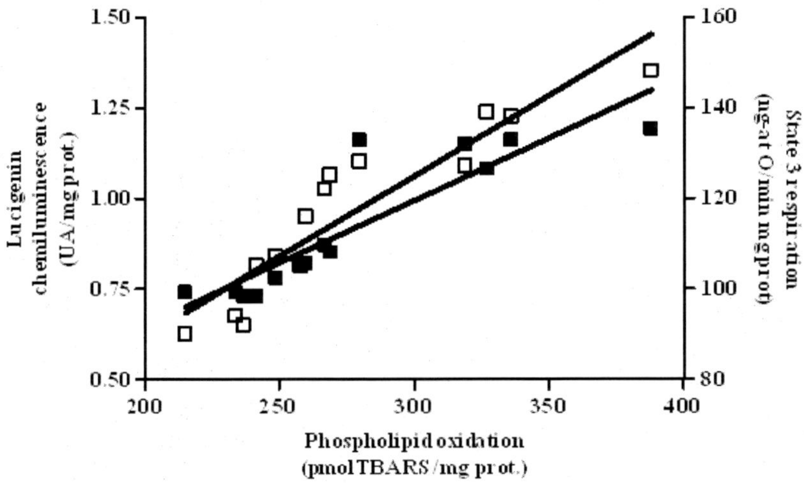

Figure 1. Linear correlation between phospholipid oxidation assayed as TBARS content and NADPH dependent $O_2^{\bullet-}$ production, assayed by lucigenin chemiluminiscence in lung homogenate (□, $p < 0.0001$, $r^2 = 0.80$), and mitochondrial oxygen consumption in state 3 (■, $p < 0.0001$, $r^2 = 0.85$), from lung of control and ROFA treated animals. Taken with permission from Magnani et al, 2011.

5.3. Analysis of Reactive Oxygen Species Production Sources Involved in Lung Oxygen Metabolism

Two different approaches were used to deeply analyse O_2 consumption after the exposure to ROFA particles: a) pretreatment with the Nox inhibitor gp91ds-tat and, b) measurements in the presence of 4 mM KCN. Pretreatment with gp91ds-tat was used to establish the relative contribution of Nox to the observed increase in tissue O_2 consumption. It is well known that the activation of Nox is triggered via protein kinase C (PKC)-mediated phosphorylation of cytosolic p47phox, which then binds to membrane-associated gp91phox (Berard et al., 2007). The gp91ds-tat peptide inhibits the interaction between gp91phox and p47phox, necessary for fully activation of the enzyme (Rey et al., 2001). Results obtained when animals were previously injected with gp91ds-tat showed that Nox is an important source of the observed raise in O_2 consumption.

Taking into account the data presented in Table 1 and the results described in the previous section, an analysis of the contribution of the different sources of tissue O_2 consumption rates values can be performed. Our first finding was a 55% increase in O_2 consumption for the ROFA exposed group. When animals exposed to ROFA were pretreated with the Nox inhibitor gp91 ds-tat, a significant 10% increase was still observed, which may account for an increase in mitochondrial respiration (other sources may not be relevant, since the values from animals pretreated with gp91 ds-tat together with the presence of KCN showed no significant differences when compared to the control group). When O_2 consumption rates were assessed in isolated mitochondria, a 34% increase in state 3 was found for ROFA-instilled mice, while no changes were observed in state 4 respiration. In the lung, besides Nox and other O_2-consuming enzymes, the whole tissue O_2 consumption accounts for the sum of the O_2 uptakes of mitochondria respiring in state 3 and of mitochondria respiring in state 4. Under physiological conditions, it can be estimated that the mitochondrial fraction in state 3 is approximately 40% for non-muscular tissues (Boveris and Boveris, 2007). As shown in Table 1, the increase in mitochondrial state 3 respiration may explain the significant 10% increase in animals pretreated with the Nox inhibitor.

Table 1. Oxygen consumption by tissue cubes in lung after exposure to ROFA in the presence or absence of KCN 4 mM and gp91 ds-tat. Data is presented as mean ± SEM, n = 6. Taken with permission from Magnani et al, 2013

	Control	ROFA	ROFA + gp91 ds tat
Oxygen consumption (ng-at O/min g tissue)	228 ± 7	$354 \pm 8^{*}$	$249 \pm 13^{*\#}$
Oxygen consumption + KCN (ng-at O/min g tissue)	48 ± 2	$61 \pm 2^{*}$	$38 \pm 5^{\#}$

$^{*}p < 0.001$ vs control group, $^{\#}p < 0.001$ vs ROFA, (unpaired t-test).

Table 2. Oxygen consumption rates by isolated mice heart mitochondria. Measurements were performed 3 h after an acute exposure to ROFA. Succinate (8 mM) was used as substrate to establish state 4 respiration rate. State 3 was triggered by the addition of ADP (125 μM) to the reaction chamber. RCR was calculated as state 3/state 4 respiration rates. Results are expressed as mean ± SEM of at least 6 independent experiments. (Taken with permission from Marchini et al., 2013)

Metabolic state	Control	ROFA
	(ng-at O/min mg protein)	
State 4	88 ± 5	61 ± 4 **
State 3	240 ± 20	180 ± 20 *
RCR	2.8	3.0

*p<0.05 and **p<0.01 compared with the control group.

Furthermore, in order to evaluate the contribution of non-mitochondrial sources to the observed values of lung O_2 consumption, the same measurements were carried out in the presence of KCN. Under these conditions, a significant 27% increase was found in ROFA-exposed mice, which may be attributed to NOX activation since it was abolished in animals pretreated with gp91 ds-tat. Consistently, when NOX activity was estimated, an average 28% increase was found. In conclusion, and in time with the results, the main contributors to the observed increase in lung O_2 consumption appear to be Nox activation, and mitochondrial state 3 respiration increase.

5.4. Oxygen Metabolism and Mitochondrial Function in Mice Heart

As a first approach in order to evaluate cardiac O_2 metabolism after an acute ROFA exposure, heart O_2 consumption was assessed in tissue cubes. In these conditions, heart O_2 consumption was found to be significantly decreased by 23% as early as 1 h after the ROFA instillation, with a minimum at 3 h after the treatment, indicating deficient tissue O_2 uptake. The observed decrease seemed to be not only time-dependent but also reversible, since O_2 consumption rates of ROFA-exposed mice were not significantly different from control animals at 5 h after the treatment.

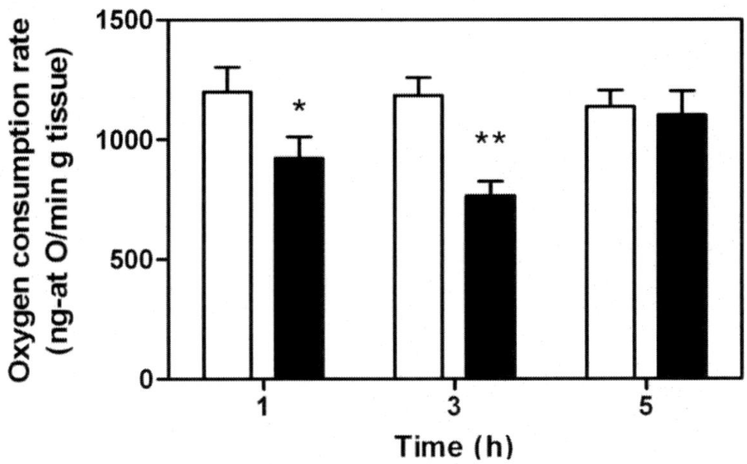

Figure 2. Oxygen consumption rates by 1 mm^3 heart tissue cubes. Mice were intranasally instilled with a single dose of either the ROFA suspension (black bars) or saline solution (white bars). Results are expressed as mean ± SEM of at least 6 animals per group. *$p<0.05$ compared with control 1 h and **$p<0.01$ compared with control 3 h. (Taken with permission from Marchini et al., 2013).

Under physiological conditions, 85-90% of tissue O_2 uptake is consumed by mitochondria in the oxidative phosphorylation process (Hill et al., 2009). Therefore, the evaluation of mitochondrial respiration was carried out in order to clarify the observed alteration in tissue O_2 consumption. Results presented in Table 2 suggest that the altered mitochondrial function accounts for the decrease in tissue O_2 consumption, and indicates a decreased mitochondrial function due to an inhibition of the electron transport chain with maintenance of mitochondrial membrane integrity. This was confirmed by *in vitro* spetrophometric determination of respiratory chain complexes activities in mitochondrial membranes, where complex II activity was significantly decreased by 25% in ROFA-exposed mice in comparison with the control group (control value: 65 ± 5 nmol/min mg protein). In agreement with these results, impaired mitochondrial function (resulting in a substantial reduction of cardiac O_2 consumption) is a common finding in the early steps of cardiovascular diseases which are associated with PM exposure, such as congestive heart disease and heart failure (Neubauer, 2007). Several studies have shown a decreased mitochondrial membrane potential triggered by PM-associated transition metals in human alveolar epithelial A549 cells (Di Pietro et al., 2011).

Conclusion and Perspectives

Nox activation and mitochondrial respiration are involved in the altered lung O_2 metabolism observed after ROFA inhalation. Nox activation seems to be a relevant non-mitochondrial source of reactive oxygen species in this model. The exposure to ROFA also impacts on mitochondrial function. In the heart, an acute ROFA exposure triggers a cardiac O_2 metabolism imbalance induced by mitochondrial dysfunction and decreased mitochondrial complex II activity. Considering that tissue injury mediated by reactive oxygen species appears to be a key mechanism in the observed damage triggered by exposure to ROFA, oxygen metabolism and mitochondrial function are important areas of study for clarifying the molecular mechanisms involved in this process.

Acknowledgments

This study was supported by grants from the University of Buenos Aires, CONICET, and the Agencia Nacional de Promoción Científica y Tecnológica (ANPCyT). P.E and S.A. are career investigators from CONICET.

References

Bagate, K., Meiring, J.J., Cassee, F.R., Borm, P.J. (2004). The effect of particulate matter on resistance and conductance vessels in the rat. *Inhal. Toxicol.* 16, 431–436.

Bedard, K., Krause, K.H. (2007). The NOX family of ROS-generating NADPH oxidases: physiology and pathophysiology. *Physiol. Rev.* 87, 245-313.

Bernardi, P. (1999). Mitochondrial transport of cations: channels, exchangers, and permeability transition. *Physiol. Rev.* 79, 1127-1155.

Brook, R.D., Franklin, B., Cascio, W., Hong, Y., Howard, G., Lipsett, M., Luepker, R., Mittleman, M., Samet, J., Smith, S.C., Tager, I. (2004). Air pollution and cardiovascular disease: a statement for healthcare professionals from the Expert Panel on Population and Prevention Science of the American Heart Association. *Circulation* 109, 2655–2671.

Brown, D., Stone, V., Findlay, P., MacNee, W., Donaldson, K. (2000). Increased inflammation and intracellular calcium caused by ultrafine carbon black is independent of transition metals or other soluble components.*Occup. Environ. Med.* 57, 685-691.

Calderon-Garciduenas, L., Gambling, T.M., Acuna, H., Garcia, R., Osnaya, N., Monroy, S., Villarreal-Calderón, A., Carson, J., Koren, H.S., Devlin, R.B. (2001). Canines as sentinel species for assessing chronic exposures to air pollutants: part 2. Cardiac pathology. *Toxicol. Sci.* 61, 356–367.

Campbell, A., Oldham, M., Becaria, A., Bondy, S.C., Meacher, D., Sioutas, C., Misra, C., Mendez, L.B., Kleinman, M. (2005). Particulate matter in polluted air may increase biomarkers of inflammation in mouse brain. *Neurotoxicology* 26, 133–140.

Chen, L.C., Hwang, J.S. (2005). Effects of subchronic exposures to concentrated ambient particles (CAPs) in mice. IV. Characterization of acute and chronic effects of ambient air fine particulate matter exposures on heart-rate variability. *Inhal. Toxicol.* 17:209–216.

Chen, L.C., Lippmann, M. (2009) Effects of metals within ambient air particulate matter (PM) on human health, *Inhal. Toxicol.* 21, 1-31.

Di Pietro, A., Visalli, G. Baluce, B., Micale, R.T., La Maestra, S., Spataro, P., De Flora, S. (2011) Multigenerational mitochondrial alterations in pneumocytes exposed to oil fly ash metals. *Int. J. Hyg. Environ. Health.* 214, 138-144.

Dockery, D.W. (1994). Epidemiologic evidence ofcardiovascular effects of particulate air pollution. Environ. Health Perspect. 109, 483–486.

Donaldson, K., Stone, V., Seaton, A., MacNee, W. (2001). Ambient particle inhalation and the cardiovascular system: potential mechanisms. *Environ. Health Perspect.* 109, 523–527.

Dreher, K.L., Jaskot, R.H., Lehmann, J.R., Richards, J.H., McGee, J.K., Ghio, A.J., Costa, D.L., (1997). Soluble transition metals mediate residual oil fly ash-induced acute lung injury. *J. Toxicol. Environ. Health* 50, 285–305.

Dye JA, Adler KB, Richards JH, Dreher KL. (1997). Epithelial injury induced by exposure to residual oil fly-ash particles: role of reactive oxygen species? *Am. J. Respir. Cell. Mol. Biol.* 17, 625–633.

Evelson, P., Travacio, M., Repetto, M., Escobar, J., Llesuy, S., Lissi, E. (2001). Evaluation of total reactive antioxidant potential (TRAP) of tissue homogenates and their cytosols, *Arch. Biochem. Biophys.* 388, 261-266.

Fubini, B., Fenoglio, I., Ceschino, R., Ghiazza, M., Martra, G., Tomatis, M., Borm, P., Schins, R., Bruch, J. (2004). Relationship between the state of the surface of four commercial quartz flours and their biological activity in vitro and in vivo. *Int. J. Hyg. Environ. Health* 207, 89-104.

Ghio, A.J., Silbajoris, R., Carson, J.L., Samet, J.M. (2002) Biologic effects of oil fly ash, *Environ Health Perspect.* 110, 89-94.

Gurgueira, S.A., Lawrence, J., Coull, B., Krisha Murthy, G.G., González-Flecha, B. (2002) Rapid increases in the steady-state concentration of reactive oxygen species in the lungs and heart after particulate air pollution inhalation. *Environ. Health Perspect.* 110, 749-755.

Hill, B.G., Dranka, B.P., Zou, L., Chatham, J.C., Darley-Usmar, V.M. (2009) Importance of the bioenergetic reserve capacity in response to cardiomyocyte stress induced by 4-hydroxynonenal. *Biochem J.* 424, 99-107.

Hiura, T.S., Kaszubowski, M.P., Li, N., Nel, A.E. (1999), Chemicals in diesel exhaust particles generate reactive oxygen radicals and induce apoptosis in macrophages. *J. Immunol.* 163, 5582-5591.

Jajte, J.M., Chemical-induced changes in intracellular redox state and in apoptosis. (1997). *Int. J. Occup. Med. Environ.Health* 10, 203.

Jimenez, L., Drost, E., Gilmour, P., Rahman, I., Antnicelli, F., Ritchie H., MacNee, W., Donaldson, K. (2000) PM(10)-exposed macrophages stimulate a proinflammatory response in lung epithelial cells via TNF-alpha. *Am. J. Physiol.* 282, L237-L248.

Kampfrath, T., Maiseyeu, A., Ying, Z., Shah, Z., Deiulis, J.A., Xu, X. Kherada, N., Brook, R.D., Reddy, K.M., Padture, N.P., Parthasarathy, S., Chen, L.C., Moffatt-Bruce, S., Sun, Q., Morawietz, H., Rajagopalan, S. (2011) Chronic fine particulate matter ecposure induces systemic vascular dysfunction via NADPH oxidase and TLR4 pathways. *Circ. Res.* 108, 716-726.

Kreyling, W.G., Semmler, M., Erbe, F., Mayer, P., Takenaka, S., Schulz, H., Oberdörster, G., Ziesenis, A. (2002). Translocation of ultrafine insoluble iridium particles from lung epithelium to extrapulmonary organs is size dependent but very low. *J. Toxicol. Environ. Health A* 65, 1513–1530.

Li, N., Sioutas, C., Cho, A., Schmitz, D., Misra, C., Sempf, J., et al. (2003). Ultrafine particulate pollutants induce oxidative stress and mitochondrial damage. *Environ. Health Perspect.* 111, 455–460.

Magnani, N., Marchini, T., Tasat, D., Alvarez, S., Evelson, P. (2011). Time course of lung oxidative metabolism after exposure to ambient particles. *Biochem. Byophys. Res. Comm.* 412, 667–672.

Magnani, N., Marchini, T., Vanasco, V., Tasat, D., Alvarez, S., Evelson, P. (2013). Reactive oxygen species produced by NADPH oxidase and mitochondrial dysfunction in lung after an acute exposure to Residual Oil Fly Ashes. *Toxicol. Appl. Pharmacol.* 270, 31–38.

Marchini, T., Magnani, N., D'Annunzio, V., Tasat, D., Gelpi, R., Alvarez, S., Evelson, P. (2012). Impaired cardiac mitochondrial function and contractile reserve following an acute exposure to environmental particulate matter. *Biochim. Biophys. Acta* 1830, 2545-2552.

Nel, A. (2005). Atmosphere. Air pollution related illness: effects of particles. *Science* 308, 804-806.

Nemmar, A., Hoet, P.H., Vanquickenborne, B., Dinsdale, D., Thomeer, M., Hoylaerts M.F, Vanbilloen, H., Mortelmans, L., Nemery, B. (2002). Passage of inhaled particles into the blood circulation in humans. *Circulation* 105,411–414.

Neubauer, S. (2007). The failing heart-an engine out of fuel. *N. Engl. J. Med.* 356, 1140-1151.

Nurkiewicz, T.R., Porter, D.W., Barger, M., Castranova, V., Boegehold, M.A. (2004). Particulate matter exposure impairs systemic microvascular endothelium-dependent dilation. *Environ. Health Perspect.* 112, 1299–1306.

Oberdörster, G., Sharp, Z., Atudorei, V., Elder, A., Gelein, R., Lunts, A., Kreyling, W., Cox, C. (2002). Extrapulmonary translocation of ultrafine carbon particles following whole-body inhalation exposure of rats. *J. Toxicol. Environ. Health. A* 65, 1531–1543.

Oury, T., Chang, L., Marklund, S., Day, B.J., Crapo, J.D. (1994) Immunocytochemical localization of extracellular superoxide dismutase in human lung. *Lab. Invest.* 70, 889-898.

Pope, III, C.A. (2000). Epidemiology offine particulate air pollution and human health: biologic mechanisms and who's at risk? *Environ. Health Perspect.* 108, 713–723.

Pope, C. A., Muhlestein, J. B., May, H. T., Rendlund, D. G., Anderson, J. L., Horne, B. D. (2006) Ischemic heart disease events triggered by short-term exposure to fine particulate air pollution. *Circulation* 114, 2443-2448.

Quay, J., Reed, W., Samet, J., Devlin, R. (1998). Air pollution particles induce IL-6 gene expression in human airway epithelial cells via NF-kappaB activation. *Am. J. Respirat. Cell Mol. Biol.* 19, 98-106.

Rey, F., Cifuentes, M.E., Kiarash, A., Quinn, M.T., Pagano, P.J. (2001). Novel competitive inhibitor of NAD(P)H oxidase assembly attenuates vascular O_2 and systolic blood pressure in mice. *Circ. Res.* 89, 408-414.

Salvi, S., Blomberg, A., Rudell, B., Kelly, F., Sandstrom, T., Holgate, S.T., Frew, A. (1999). Acute inflammatory responses in the airways and peripheral blood after short-term exposure to diesel exhaust in healthy human volunteers. *Am. J. Respir. Crit. Care Med.* 159, 702–709.

Samet, J., DeMarini, D., Malling, H. (2004). Biomedicine. Do airborne particles induce heritable mutations? *Science* 304, 971-972.

Schaumann, F., Borm, P.J., Herbrich, A., Knoch, J., Pitz, M., Schins, R.P., Luettig, B., Hohlfeld, J.M., Heinrich, J., Krug, N. (2004). Metal-rich ambient particles (particulate matter 2.5) cause airway inflammation in healthy subjects. *Am. J. Respir. Crit. Care Med.* 170, 898–903.

Schins, R.J., Lightbody, P., Borm, T., Shi, K., Donaldson, T., Stone, V. (2004). Inflammatory effects of coarse and fine particulate matter in relation to chemical and biological constituents. *Toxicol. Appl. Pharmacol.* 184, 1-11.

Schroeder, W.H., Dobson, M., Kane, D.M, Johnson, N.D. (1987). Toxic trace elements associated with airborne particulate matter: a review. *JAPCA* 37 1267-1285.

Schwartz, J. (2001). Air pollution and blood markers ofcardiovascular risk. *Environ. Health Perspect.* 109, 405–409.

Shukla, A., Timblin, C., BeruBe, K., Gordon, T., McKinney, W., Driscoll, K., Vacek, P., Mossman, B. (2000). Inhaled particulate matter causes expression of nuclear factor (NF)-kappaB-related genes and oxidant-dependent NF-kappaB activation in vitro.*Am. J. Respirat. Cell Mol. Biol.* 23, 182-187.

Sorensen, M., Schins, R., Hertel, O., Loft, S. (2005). Transition metals in personal samples of PM2.5 and oxidative stress in human volunteers. *Cancer Epidemiol. Biomarkers Prev* 14, 1340-1343.

Southam, D.S., Dolovich, M., O'Byrne, P.M., Inman, M.D. (2002) Distribution of intranasal instillations in mice: effects of volume, time, body position, and anesthesia. *Am J Physiol Lung Cell Mol Physiol.* 282, 833-839.

Souza, M.B. (1998). Respiratory changes due to long-term exposure to urban levels ofair pollution. *Chest* 113, 1312–1318.

Susin, S., Zamzami, N., Kroemer, G. (1998). Mitochondria as regulators of apoptosis: doubt no more. *Biochim. Biophys. Acta* 1366, 151-165.

Tan, W.C., Qiu, D., Liam, B.L., Ng, T.P., Lee, S.H., van Eeden, S.F., D'Yachkova, Y., Hogg, J.C. (2000). The human bone marrow response to acute air pollution caused by forest fires. *Am. J. Respir. Crit. Care Med.* 161, 1213–1217.

Tao, F., Gonzalez-Flecha, B., Kobzik, L. (2003) Reactive oxygen species in pulmonary inflammation by ambient particulates. *Free Radic. Biol. Med.* 35, 327-340.

Upadhyay, D., Panduri, V., Ghio, A., Kamp, D. (2003). Particulate matter induces alveolar epithelial cell DNA damage and apoptosis: role of free radicals and the mitochondria. *Am. J. Respir. Cell. Mol. Biol.* 29, 180-187.

van der Vliet, A. (2005) NADPH oxidases in lung biology and pathology: Host defenses enzymes and more *Free Radic. Biol. Med.* 44, 938-955.

van Eeden, S.F., Tan, W.C., Suwa, T., Mukae, H., Terashima, T., Fujii, T., Qui, D., Vincent, R., Hogg, J.C. (2001). Cytokines involved in the systemic inflammatory response induced by exposure to particulate matter air pollutants (PM10). *Am. J. Respir. Crit. Care Med.* 164, 826–830.

Vanasco, V., Magnani, N., Cimolai, M.C., Valdez, L., Evelson, P., Boveris, A., Alvarez, S. (2012). Endotoxemia impairs heart mitochondrial function by decreasing electron transfer, ATP synthesis and ATP content without affecting membrane potential. J. *Bioenerg. Biomembr.* 44, 243-252.

Veronesi, B., Oortgiesen, M., Roy, J., Carter, J.D., Simon, S.A., Gavett, S.H. (2000). Vanilloid (capsaicin) receptors influence inflammatory sensitivity in response to particulate matter. *Toxicol. Appl. Pharmacol.* 169, 66–76.

World Health Organization (WHO), 2005, WHO Air quality guidelines for particulate matter, ozone, nitrogen dioxide and sulfur dioxide. Global update 2005, World Health Organization (Ed.), Copenhagen, World Health Organization.

Wichers, L.B., Nolan, J.P., Winsett, D.W., Ledbetter, A.D., Kodavanti, U.P., Schladweiler, M.C., Costa, D.L., Watkinson, W.P. (2004). Effects of instilled combustion-derived particles in spontaneously hypertensive rats. Part I: Cardiovascular responses. *Inhal. Toxicol.* 16, 391–405.

Yosgizaki Akinaga, L.M., Lichtenfels A.J., Crvalho-Oliveira, R., Garcia Caldini, E., Dolhnikoff, M., Ferraz Siva, L.F., De Siqueira Bueno, M., Pereira, A., Saldiva, P.H.N., Bueno Garcia M.L. (2009) Effects of chronic exposure to air pollution from Sao Paulo city on coronart of swiss mice, from birth to adulthood. *Toxicol. Pathol.* 37, 306-314.

Zareba, W., Nomura, A., Couderc, J.P. (2001). Cardiovascular effects of air pollution: what to measure in ECG? *Environ. Health Perspect.* 109, 533–538.

In: Fly Ash
Editor: Prabir Kumar Sarker

ISBN: 978-1-62948-044-2

Chapter 15

INCINERATED SEWAGE SLUDGE ASHES – POTENTIAL AND CHALLENGES

Paula Guedes[1,2,*], Nazaré Couto[1,2], Lisbeth M. Ottosen[2] and Alexandra B. Ribeiro[1]

[1]CENSE, Departamento de Ciências e Engenharia do Ambiente, Faculdade de Ciências e Tecnologia, Universidade Nova de Lisboa, Caparica, Portugal
[2]Department of Civil Engineering, Technical University of Denmark, Lyngby, Denmark

ABSTRACT

Incineration is a "growing" technology to treat wastes and is highly representative in countries of Northern Europe. Literature reports the potential to upgrade incinerated sewage sludge waste incineration ashes in order to remove heavy metals with different technologies, like the electrodialytic process. Consequently, and according to specifications, ashes could be reused as adsorbents, construction materials or in agriculture. Ashes from sewage sludge generally contain high concentrations of phosphorus (P) and can be regarded as secondary P resources. For this, a viable option is to recover phosphorus as a clean product so it can be reused. The conducted review presented here, showed that the electrodialytic process has proven to be a suitable technique for phosphorus recovery and heavy metals removal. The finally obtained clean phosphorus product can be reused in agriculture and/or industries, the removed heavy metals can also be recycled, and the cleaned ash can be further reused as a raw material, saving expensive landfill volume.

Keywords: Incinerated sewage sludge ash; electrodialytic process; phosphorus; heavy metal

[*] CENSE, Departamento de Ciências e Engenharia do Ambiente, Faculdade de Ciências e Tecnologia, Universidade Nova de Lisboa, 2829-516 Caparica, Portugal; Email:p.guedes@campus.fct.unl.pt.

1. BACKGROUND

Treatment of municipal wastewater inevitably leads to the generation of large volumes of sewage sludge. Currently over 6.5 millions tonnes of dry solid sludge are produced in the European Union (EU) *per* year.

European Union introduced the Directive (91/271/EEC) to improve the aquatic environment through the regulation of the municipal wastewater treatments, prior to its release, and eliminating the possible adverse effects where such discharges occur. The basic idea of the Directive is to encourage all cities above 2000 person equivalent (PE) to implement secondary wastewater treatment leading to a significant increase in sludge production. This is the main drawback, in the implementation of the directive, which takes further the issue of sustainable handling of sludge defined as a *method that meets requirements of efficient recycling of resources without supply of harmful substances to humans or the environment* (Commission of European Communities, 1998).

The amount of sludge produced is affected in a limited scale by the treatment efficiency while the sludge quality is strongly dependent on the original pollution load of the treated effluent and also, on the technical and design features of the wastewater treatment process.

The outlets for their disposal/utilisation were in 2004 agriculture (37%), landfill (40%), incineration (11%), others (12%) (Fytili & Zabaniotou, 2008).

The use of sludge is considered a viable option to maintain/enhance the recycling of elements with socio-economic value such as nutrients and organic matter. Nevertheless, it is necessary to increase the scope of the existing regulations and manage it in outlets such as green areas, reclaimed land and silviculture (European Commission, 2000).

1.1. Biosolids

Land application is an attractive option for biosolids (treated sewage sludge) management (Peters & Rowley, 2009) as it is a cheap and easy solution that promotes the recycling of valuable nutrients, like phosphorus (P), while improving soil organic content.

The use of biosolids is approved by Member States in several situations (e.g. short-rotation plantations, plantations for growing energy crops (European Commission, 2000). Biosolids can be used when the legislation values both for organics as well as heavy metals are met (Table 1) as there is an agronomic interest for nutrients or for the improvement of the content of organic matter in soil; the quantity of nutrients introduced is adapted to the needs of the crops or the soil, according to best practice, and its use does not cause unreasonable odour nuisance to the nearest dwellings.

But biosolids should not be used on water saturated, flooded, frozen or snow-covered ground or soils with pH lower than 5.0. Sludge should also be treated in order to prevent the spreading of pathogens. The use of sludge should minimize the risk of negative effects to humans, animals and plants, quality of waters as well soil quality and biodiversity (European Commission, 2000).

Table 1. Limit values proposed for concentration of organic compounds and heavy metals in sludge for use on land (European Commission, 2000)

Organic compound	Limit values (mg/kg DM)
AOX[1]	500
LAS[2]	2600
DEHP[3]	100
NP/NPE[4]	50
PAH[5]	6
PCB[6]	0.8
Dioxins	Limit values (ng TE/kgDM)
PCDD[7]/F	100
Heavy metal	Limit values (mg/kg DM)
Cd	10
Cr	1000
Cu	1000
Hg	10
Ni	300
Pb	750
Zn	2500

[1] Sum of halogenated organic compounds;
[2] Linear alkylbenzene sulphonates;
[3] Di(2-ethylhexyl)phthalate;
[4] It comprises the substances nonylphenol and nonylphenolethoxylates with 1 or 2 ethoxy groups;
[5] Sum of the following polycyclic aromatic hydrocarbons: acenapthene, phenanthrene, fluorene, flouranthene, pyrene, benzo(b+j+k)fluoranthene, benzo(a)pyrene, benzo(ghi)perylene, indeno(1, 2, 3-c,d)pyren;
[6] Sum of the polychlorinated byphenils components number 28, 52, 101, 118, 138, 153, 180;
[7] Polychlorinated dibenzodioxins/ dibenzofuranes.

However, concerns have been risen about municipal biosolids applications, composed of waste streams from residential, commercial and industrial sources, which increase the probability of including potential chemicals, solvents and pharmaceuticals. The presence of pathogens can also be problematic in terms of land contamination and it was already shown that their concentrations increase over time in biosolids, even in samples that indicated no measureable pathogens after the application of destruction methods (Gibbs, Hu, Ho, & Unkovich, 1997).

High concentrations of heavy metals in biosolids are also of concern, as they can contaminate and reduce the productivity of land used for disposal by reducing the bacterial diversity within the soil (Moffett, et al., 2003)

Some of the pollutants that can be found in biosolids have potential to bioaccumulate in plant tissues, namely in vegetables. For example, triclocarban (3,4,4′-trichlorocarbanilide) and triclosan (5-chloro-2-(2,4-dichlorophenoxy)-phenol), antimicrobial products usually used in personal care products such as hand soap, are toxic, lipophilic and persistent and were found in vegetables tissues of pumpkin and zucchini (Aryal & Reinhold, 2011).

Besides the above mentioned factors, direct use of biosolids in agriculture may be difficult due to public perception, odours or difficulties of transport and storage.

1.2. Incineration Process

The technology of incineration in terms of process engineering, energy efficiency and compactness of plant has greatly improved (Roy, Dutta, Corscadden, Havard, & Dickie, 2011).

The incineration process costs about 210-310 euros *per* ton of dry solids but some controversy arises from this technique as it promotes a potentially hazardous ash (Smith, Fowler, Pullket, & Graham, 2009) due to the concentration of inorganic contaminants and residual amounts of hazardous organics like dioxins. Public concern about possible harmful emissions is also a constraint to the widespread of incineration. These problems can be minimized by introducing new technologies for controlling gaseous emissions diminishing the adverse effects, while the reduction in the corresponding cost gives incineration considerable advantages in the future (Roy, Dutta, Corscadden, Havard, & Dickie, 2011)

Incineration has several advantages when compared to other available disposal routes as it allows to reduce the volume and mass of the waste by 90% and 70%, respectively (Chen & Yan, 2012; Hjelmar, 1996; Kirby & Rimstidt, 1993; Medici, Piga, & Rinaldi, 2000) and also as it destroys pathogenic agents, oxidizes organic compounds and generates thermal energy that can be reused (Marani, Braguglia, Mininni, & Maccioni, 2003; Porteous, 2005)

The increase of incineration process is a consequence of the increasing quantity of wastes but also EU directives from the last two decades (Saner, Blumer, Lang, & Koehler, 2011).

In the EU-27, between 1995 and 2009 municipal waste rates changed: landfilled between 286 to 191 kg *per* capita, incinerated between 65 and 101 kg *per* capita, recycled between 46 and 118 kg *per* capita, and composted between 28 and 89 kg *per* capita (Eurostat, 2011). Incineration is playing a significant role in municipal solid waste management of European countries (Saner, et al., 2011), as in 2008 there were 405 treatment facilities in the EU-27 member states and 453 all over the Europe with 43 new plants planned in the next years (2020) (Eurostat, 2011), with 2 plants in Portugal (Ministério do Ambiente, do Ordenamento do Território e do Desenvolvimento Regional, 2007). In total, the share of landfilling in the EU-27 dropped from 68 % in 1995 to 38 % in 2008. Although national implementation of waste policies has different levels of incineration percentages, there is a tendency to increase the use of the process in the scope of waste management strategies. Denmark is one of the countries that presently incinerates a substantial share of their municipal solid waste (MSW) (around 50–55%) and has developed recycling system (recycling rates 35–40%) (Saner, et al., 2011). But still, there are Member States without incineration facilities (Eurostat, 2011). Summing up: the municipal solid waste treated in 2009, sorted by percentage was 40-50% in Switzerland, Sweden, Denmark, Norway; 30-40% in France, Netherlands, Germany, Belgium, Luxembourg; 20-30% Austria, Finland; 10-20% Italy, United kingdom, Portugal; 0-10% Spain, Ireland, Slovenia and 0% Estonia, Malta, Greece, Cyprus, Latvia. Nevertheless there is still a share of 40% for the landfill of municipal solid waste.

Incineration of sewage sludge from urban wastewater treatment plants, meaning all sludge that is disposed by direct incineration or after mixing with other waste, was in 2009: 7 kg *per* capita in Denmark, 8 kg *per* capita in Slovenia, 4 kg *per* capita in Greece and 0 kg *per* capita in countries such as Bulgaria, Poland, Lithuania and Estonia (Eurostat, 2012). For Portugal there was no available data.

From 1997 till now, Portugal has been applying a strategic plan allowing the country to reflect fundamental steps in wastes policies by the Strategic Plan of Urban Solid Wastes

(*Plano Estratégico para os Resíduos Sólidos Urbanos* - PERSU). In this perspective, the country aims to "intensify reduction policies, recycling and reutilization, as well as to assure the necessary infrastructures of treatment and disposal, creating PERSU II, covering a period between 2007 and 2016.

PERSU II aims to decrease landfill disposal through recycling maximization and, in complement, other valorisation ways, and additionally improve pollution prevention, recycling maximization and minimization of waste disposal. In 2005, 21% of Portuguese solid wastes were incinerated, very near to the established value of 22% (Ministério do Ambiente, do Ordenamento do Território e do Desenvolvimento Regional, 2007).

1.3. ISSA Potential Use

As shown before, incineration is already a well-established technology in some countries (Werther & Ogada, 1999) being estimated that 300-400 kg of incinerated sewage sludge ash (ISSA) is produced *per* ton of dried sewage sludge (Cyr, Coutand, & Clastres, 2007).

Ashes are generally considered a waste material to be disposed into landfill (Donatello, Tyrer, & Cheeseman, 2010) but it can also be reused as adsorbents (Pan, Lin, & Tseng, 2003), in geotechnical applications (Ferreira, Ribeiro, & Ottosen, 2003) or in construction materials (Al Sayed, Madany, & Buali, 1995; Anderson, Elliott, & Hickson, 2002; Lin, Luo, & Sheen, 2005). However, none of the above applications make use of the valuable P resource (discussed below) in the ashes. Still, each potential application for ashes results in three main advantages: first, the use of a zero-cost raw material, secondly, the conservation of natural resources, and thirdly, the elimination of waste (Ferreira, et al., 2003).

Agricultural application of the ash offers several advantages as it reduces the need for commercial fertilizers, as they enclose sodium, potassium, calcium, magnesium, boron, sulphates and other nutrients (Wong & Wong, 1989), except nitrogen that is lost during the combustion. Ashes also counteract the ongoing acidification, as typically have alkaline pH values (Moliner & Street, 1982), and a solution to the potential problem of ash disposal. For these reasons, their positive effects by enclosing available nutrients has been shown by increasing crop yields (Ayeni, Adetunji, Ojeniyi, Ewulo, & Adeyemo, 2008; Bierman & Rosen, 1994; Jakobsen & Willett, 1986; Lee, Ha, Lee, & Kim, 2008; Mellbye, Hemphill, & Volk, 1982; Rosen, Olson, & Bierman, 1994; Singh, Thind, Manchanda, & Kansal, 2009). A study performed on plant growth in an amended soil with MSW fly ash showed a positive influence (1.5 to 2 times greater than in non-amended soil) on alfalfa and Swiss chard (Rosen, et al., 1994), but it was also reported that it can also cause potential problems. The nutritional benefits of the ash were also showed in the same work as the growth plant species in ash-amended soils was similar or better than that in soils amended with P and K fertilizers. Despite no phytotoxic effects have been observed, some alfalfa treatments exceeded livestock dietary tolerances for Mo (>5 mg/kg) and Cd (>0.5 mg/kg). The mixture of ash with gypsum (1:1) increased significantly rice yield, and did not result in accumulation of heavy metals in soil or excessive uptake of heavy metals by rice grains (Lee, et al., 2008). The application of ash also increased grain yield of maize by 31% and by 74% on residual basis on a sandy loam soil (Ayeni, et al., 2008) and the productivity of cotton–wheat system (Singh, et al., 2009).

The application of ash in soils is still a controversial subject as it raises questions of toxicity that must be carefully addressed (Ferreira, et al., 2003), prior to its application, due to the high levels of heavy metals and salts present in the ash. Major concerns generally revolve around the metals, which can accumulate in the soil over time, and enter the food chain or groundwater systems. However, the risk for this occurring depends primarily on metal concentrations in ash, their antecedent concentrations in the soil, their mobility from the ash and subsequently from the soil, and their uptake by plants.

1.3.1. ISSA As a P Source

Nowadays, concepts to recover P within the wastewater treatment scheme are particularly attractive, if they promise to obtain a product which is free from contaminants and of a high quality as fertilizer (Blöcher, Niewersch, & Melin, 2012).

In wastewater treatment plants without phosphorus removal, 90–95% of the incoming phosphorus load is contained in the sewage sludge (Blöcher, et al., 2012; Cornel & Schaum, 2009). Its theoretical recovery potential is significantly higher than with separation processes from the aqueous phase (Cornel & Schaum, 2009).

Phosphorus recovery from sewage sludge ash first involves the re-dissolution of the bound phosphorus followed by its separation, e.g., electrokinetic process followed by precipitation processes. The advantage of treating sewage sludge ashes is their exclusively inorganic formation, a fact which, in contrast to sewage sludge, facilitates phosphorus recovery. Furthermore, the incineration of organic matter causes the enrichment of phosphorus in the ashes (Cornel & Schaum, 2009).

Ashes from incineration of sewage sludge generally contain high concentrations of phosphorus and can be regarded as secondary P resources but they do not have direct value as fertilizer as P is not plant available (Ottosen, Kirkelund, & Jensen, 2013). So, it becomes important to recover P from the ashes in order to recycle a nutrient that has high socio-economical value and is also indispensable to the ecosystem. For this reason, it is worth to briefly discuss important matters of this subject such as P scarcity and cycle.

1.3.2. Phosphorus Cycle

The P-cycle is similar to several other mineral nutrient cycles as it exists in soils, minerals, living organisms and water but, although P is widely distributed in nature, it is not found by itself in its elemental form.

The global P cycle has four major components (Ruttenberg, 2003): i) tectonic uplift and exposure of P-bearing rocks to the forces of weathering; ii) physical erosion and chemical weathering of rocks producing soils and providing dissolved and particulate P to rivers; iii) riverine transport of P to flood plains, lakes and the ocean; and iv) sedimentation of P associated with organic and mineral matter and buried sediments. The cycle is continuously beginning with the uplift of sediments into the weathering regime (Figure 1).

Human activities have increased the global P-cycle by an average of 400% (Falkowski et al., 2000) and, as a result of anthropogenic distortion of the macronutrient balance, profound ecosystem-level changes are now evident. The P inputs from non-point and point sources in the freshwater bodies may lead into an excessive algal growth (eutrophication) in lakes and rivers (Carpenter and Bennett, 2011), decreasing water quality through undesirable colour, odour and taste, and possible toxicity to humans, livestock and wildlife.

The continuous application of 40–120 kg/ha of mineral P_2O_5-fertilizer [ca. 18–52 kgP/ha] *per* year is necessary to supply agricultural systems nowadays (Ott & Rechberger, 2012). Net *per* capita consumption in the EU-15 is 4.7 kgP/yr of which only 1.2 kgP/yr reach the consumer. The main losses are a net accumulation in agricultural soils (2.9 kgP/yr), followed by losses to landfills (1.4 kgP/yr) and to the hydrosphere (0.55 kgP/yr). Only 0.77 kgP/yr are recycled (Ott & Rechberger, 2012). Optimizing P fertilization, collecting and recycling P-rich wastes, increasing the connection of households to sewer systems, and implementing tertiary wastewater treatment comprehensively could significantly reduce Europe's import dependence on phosphorus (Ott & Rechberger, 2012).

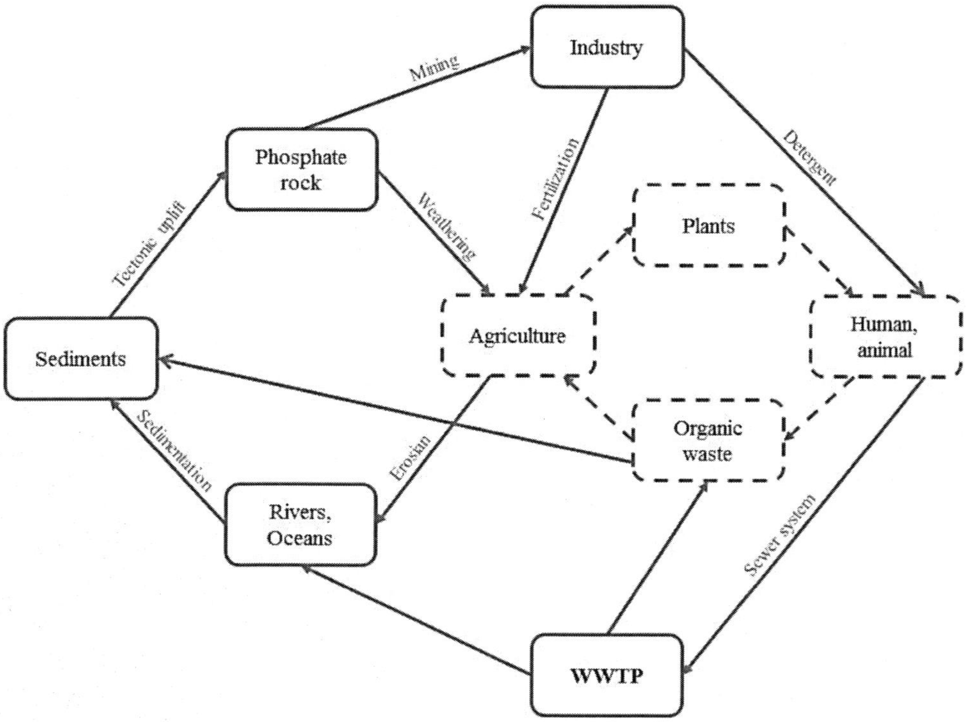

Figure 1. Geological and organic phosphorus cycles including human impacts (adapted from Cornel & Schaum, 2009).

1.3.2.1. Phosphorus Scarcity

One element indispensable for plant growth is phosphorus, a non-renewable resource, dominant application of which is as soil fertilizer (Heffer, Prud'homme, Muirheid, & Isherwood, 2006).

Current population growth rates require an increased supply of staple foods and to guarantee it, a sufficient nutrient level of agricultural soils needs to be maintained by application of soil fertilisers. Unfortunately, phosphate rock, the primary source of P, is becoming increasingly scarce and more expensive. It was reported that phosphate rock reserves are expected to last ca. one century (Cordell, Drangert, & White, 2009; IFDC, 2010; Survey, 2012) and that P peak has been estimated to occur by 2035.

P has no substitute in food production and the European Union is almost entirely dependent upon imports, with China, Jordan, Morocco, South Africa and USA controlling 85% of global phosphate reserves (Smit, Bindraban, Schröder, Conjin, & Meer, 2009). Therefore there is a need to find ways to recover phosphorus for reuse.

There is a large quantity of phosphate in the effluent of wastewater treatment plants (Carpenter & Bennett, 2011). Therefore, to limit eutrophication potential of wastewaters, P removal has been in the toolbox of wastewater-treatment engineers for several decades (Rittmann, Mayer, Westerhoff, & Edwards, 2011) and several methods where developed in order to remove P from wastewater. But these approaches do not recycle it as a truly sustainable product as it is removed from liquid phase together with various other waste products like organic chemicals (Bright & Healey, 2003; Clarke & Smith, 2011), metals (Basta, Ryan, & Chaney, 2005) or even pathogens (Elliott & O'Connor, 2007; Gerba & Smith, 2005; Sidhu & Toze, 2009).

Nowadays, and taking into account that phosphate rock is becoming increasingly scarce, inaccessible or unaffordable, the new challenge is to recover P for further reuse instead of only removing it due to pollution concerns.

2. ELECTROKINETIC PROCESS

2.1. Process Overview

An electrokinetic (EK) setup consists of external hardware including a power supply, electrodes and the connection between both. For optimum current conduction the electrodes are typically inserted in through wells where electrolytes are circulated.

The EK process uses a low level direct current (DC) density, in the order of a few mA/cm^2 and low potential gradient, in the order of V/cm (Acar & Alshawabkeh, 1993), between suitably locate electrodes, which induce physicochemical changes in the applied media. These changes lead to species transport by coupled mechanisms, and acts as the "cleaning agent" by which the contaminants are moved out of the matrix towards one of the electrode compartments. The transport processes include:

i) electroosmosis, the movement of water which generally takes place from the anode to the cathode and is due to the existence of a space-charge on the solution side of the particle/solution interface;
ii) electromigration, the transport of ions and ion complexes toward the electrode of opposite charge;
iii) electrophoresis, the movement of charged particles or colloids under the influence of an DC electric field, with contaminants bound to mobile particulate matter also being transported.

Electrokinetic treatment relies on several interacting mechanisms but the dominant and most important electron transfer reactions that occur at electrodes during the process is the electrolysis of water (equations 1.1 and 1.2):

$$H_2O \rightarrow 2\,H^+ + \tfrac{1}{2}\,O_2\,(g) + 2\,e^-\ (\text{anode}) \qquad (1.1)$$

$$2\,H_2O + 2\,e^- \rightarrow 2\,OH^- + H_2\,(g)\ (\text{cathode}) \qquad (1.2)$$

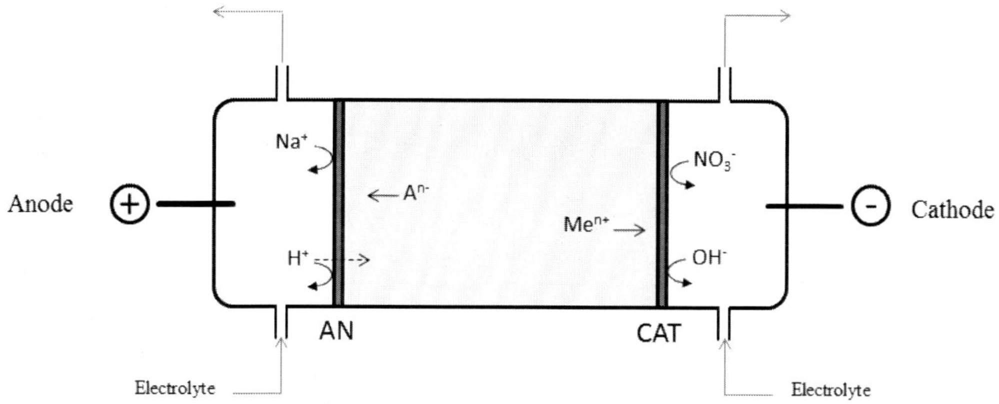

Figure 2. Schematic representation of the electrodialytic laboratory (AN - anion exchange membrane; CAT- cation exchange membrane).

Due to the formation of hydrogen ions in the anode an acid front is carried towards the cathode by electrical migration, diffusion and advection. Consequently there is a pH decrease near the anode and, at the same time, an increase in the pH near the cathode due to the formation of hydroxide ions (Virkutyte, Sillanpää, & Latostenmaa, 2002).

This change in pH in the central cell compartment is observed with the use of passive membranes between compartments, but ion exchange membranes can also be used, nominated as electrodialytic (ED) process (Figure 2).

The anion exchange membrane allows only passage of anions and is placed between the sludge and the electrolyte solution at the anode end, and between the sludge and the electrolyte solution at the cathode end a cation exchange membrane is placed, that allows only the passage of cations.

However, the anion exchange membrane is not a 100% perfect rectifier and some hydrogen ions are still transported from the anode end towards the cathode end, passing through the middle compartment and consequently lowering the pH.

The ion exchange membranes prevent the waste of current in transporting ions from one electrode compartment through the sludge into the second electrode compartment. Furthermore, the ion exchange membranes make the conditions in the central compartment less dependent on the choice of electrolyte solution than if passive membranes were used. This is beneficial especially when further concentration of the removed compounds for reuse is planned. ED remediation presents several advantages (Velizarova, Ribeiro, & Ottosen, 2002):

i) application of a low-level direct current to move the charged species;
ii) use of ion-exchange membranes allowing to regulate the direction and magnitude of the ion fluxes;
iii) possibility to control the degree of elements removal by adjusting the flow rates of the solutions contacting with membranes;
iv) the elements removed might be recycled;
v) the “cleaned” matrix might be further used.

2.2. Electrodialytic Process Applied to Ashes

2.2.1. Cell Design

Electrodialytic process has shown promising results in extracting organic and/or inorganic contaminants from several matrices like soil (Ribeiro, Mateus, & Rodríguez-Maroto, 2011), mine tailings (Hansen, Rojo, Pino, Ottosen, & Ribeiro, 2008), timber waste (Mateus, Zrostlíková, Gomes da Silva, Ribeiro, & Marriott, 2010), harbour sediment (Nystroem, Ottosen, & Villumsen, 2005) and wastewater sludge (Ottosen, Pedersen, Hansen, & Ribeiro, 2007).

Regarding ED remediation of heavy metals from ashes, it has also been applied to, e.g., municipal solid waste incineration fly ash (Ferreira, Jensen, Ottosen, & Ribeiro, 2005; Ottosen, Lima, Pedersen, & Ribeiro, 2006; Pedersen, Ottosen, & Villumsen, 2003, 2005), bio-ashes (Hansen, Ottosen, & Villumsen, 2004; Lima, Ottosen, & Ribeiro, 2009; Ottosen, et al., 2007) and, more recently, ISSA (Guedes, Couto, Ottosen & Ribeiro, (submitted); Pazos, Kirkelund, & Ottosen, 2010).

In order to remove the heavy metals from the porous media, there is a need to transform the sorbed heavy metals into their ionic forms in solution or in exchangeable sites on the surface of the particles. To aid desorption, it may be necessary to add an enhancement solution to the polluted matrix prior to the ED treatment (Ferreira, et al., 2005; Ottosen, et al., 2003). Assisting agents are usually defined as solubilisation enhancement compounds, which form charged stable complexes with the substances to be removed, so that they can be mobilized by the electric field.

Electrodialytic cells can also have varying numbers of compartments as well as use different membrane types. This flexibility allows the development of process-specific designs. A cell in which the sample is continuously stirred (Pedersen, 2003) can also be used, and showed improved cell efficiencies when comparing to an unstirred cell (Ottosen, Hansen, & Hansen, 2000). This is due to the high percentage of easily soluble particles that compose the ash, and for that reason a consequent decrease of mass and volume along the process is expected (Hansen, et al., 2004; Pedersen, 2003) which can be a problem if a stationary cell is used. An appropriate density and electric field strength is also an important factor. However, it depends on the electrochemical properties of the sample to be treated: the higher the conductivity, the higher the electric current density must be used to maintain the electric field strength required (Alshawabkeh, Yeung, & Bricka, 1993)

In case the ED treatment succeeds, it could be possible to replace other contaminants containment methods such as current stabilization/landfilling procedure by a new approach based on the reuse/recycling of fly ash, now out of reach due to the high loads of heavy metals (Ferreira, et al., 2005).

The successful ED application may avoid the costs of controlled deposition and the residue obtained could be possibly reused in new applications such as: production of cement, concrete, ceramic and glass; road pavement and embankments; agriculture as a soil amendment; sorbent in industry; and also for sludge conditioning (Ferreira, et al., 2003).

2.2.2. Heavy Metal Removal from ISSA

Pazos et al. (2010) showed that the application of ED to a ISSA mixed with distilled water, in a stirred cell, reduced the Cd concentration, and after 14 days of treatment to values below the Danish legislation threshold, allowing ash direct application into agricultural soil. Using the same cell design, but mixing ISSA with sulfuric acid also achieved good removal efficiencies after 14 days, being the highest for Cu (ca. 80 %) and the lowest for Pb (ca. 4%) (Guedes, et al., (submitted)).

Ferreira et al. (2005) applied a broad range of ED experimental conditions (studied including current densities, remediation times, use of assisting agents and cell design) to municipal solid waste incineration (MSWI) fly ash. And, although low remediation efficiencies obtained, several operational problems were identified during the ED, among which are formation of precipitates, dryness of sample and partial dissolution of sample, creating preferential pathways for the electric current. However, a comparison between experiments showed that generally the use of Na-gluconate as assisting agent leads to better results than distilled water and the concentration increase of the assisting agent also results in higher removals.

Pedersen, et al. (2005) showed that suspending MSWI fly ashes in ammonium citrate in the central compartment gives higher removals for Pb but for a remediation period of 70 days, showing Pb removal is not an easy task.

Ottosen, et al. (2006) treated MSWI suspended in water and were able to decrease the Cu concentration of the ash from 2200 to 860 mg/kg but the Pb concentration increased from 8560 to 16 800 mg/kg, showing that Pb is mainly found in the ash fraction that is least soluble. Hence ED treatment of the ash suspended in water is not a solution to improve the ash quality in terms of Pb. The water-soluble Cl content per unit weight of the original ash was 12.4%. The removal of water-soluble Cl was efficient and >98% of Cl was removed (calculated on the basis of mean initial and final concentrations). This result indicates that ED extraction may be a method that can be used for the removal of Cl from ash prior to its utilization in concrete.

Lima, et al. (2009) applied ED to straw ashes (pre-washed) and distilled water in the central compartment, and achieved high removal rates for Ca, Cu, Cr and particularly for Cd, which has been reduced to a level below the established by Danish regulations. However, acidification also induced a high dissolution, lower pH and agricultural value of the ash (K was mainly removed during the pre-washing) reducing its application in agriculture but it can still be applied as an amendment. Hansen, et al. (2004) made experiments with Cd removal from straw ash in a stationary ED cell and more than 97% Cd was removed in 36 days with a current density of 5.6 mA/cm^2. However, as a high percentage of the fly ash are easily soluble particles, mass and volume decreased during the experiments being an operational problem during treatment in the stationary cell.

Ottosen, et al. (2007) applied ED in bio-ashes (straw and wood) using an ED method and after the treatment the Cd removals were of 45 and 75% from wood ash and straw ash respectively, and the concentration was bellow Danish limiting value allowing the spread in agricultural land of the ashes. Pedersen, et al., (2003) also tested ED for Cd removal from wood ash in a stirred suspension. Citric acid was used with different concentrations and pH and ammonia was tested as enhancement solution and more than 70% Cd could be removed.

Kirkelund, Damoe, & Ottosen (2013) recently conducted a series of laboratory scale ED experiments for 14 days with different ashes (combustion of straw, wood chips, and co-firing of wood pellets and fuel oil). In spite of such large variations between the ashes, the ED method showed to be sufficiently robust to treat the ashes so the final Cd concentration was below 2.0 mg Cd/kg DM in at least one experiment done with each ash.

Pedersen & Ottosen (2006) were able to reduce by 86% for As, 15% for Cu and 33% for Cr the amounts present in the ash that was suspended in sulfuric acid in a stirred cell.

The results obtained in several works that use the ED technique showed that it can be used to reduce heavy metals concentration in several types of ashes, including from ISSA. In some cases, no extracting reagents were necessary to accomplish a sufficient reduction of some heavy metals concentration in the ash samples. This fact along with the simplicity of the used equipment reduces the cost of the possible implementation of this treatment in industrial scale (Pazos, et al., 2010). However, together with heavy metals, alkalis are also removed, which may decrease the fertilizing value of the remaining solid residue for agricultural applications (Lima, Ottosen, Ribeiro, & Hansen, 2008), although some amount is still present after the treatment (Pazos, et al., 2010).

2.2.3. Influence of ISSA Characteristics

In the short combustion process, thermodynamic equilibrium of minerals in fly ash is not achieved; consequently, the primary mineral association formed at high temperature is generally unstable under room temperature and humidity conditions at the disposal site. Hence fly ash will undergo chemical reactions such as hydration, carbonation or oxidation/reduction processes, which produce secondary minerals in the landfill (Zevenbergen & Comans, 1994). However, before the final destination of ISSA (landfill or reuse), there is often a storage step in the facilities of the incineration plant where weathering can change the mineralogical characteristics of the material. When stored in an open air structure, the ash will be exposed to environmental conditions such as precipitation, solar radiation or wind, which can alter the physico-chemical properties (Bayuseno & Schmahl, 2011; Meima & Comans, 1997; 1999).

Three major stages in weathering were identified for MSWI, with each stage presenting a characteristic pH that is largely controlled by Ca minerals and pCO_2, but also by soluble Al and SO_4 (Meima & Comans, 1997):

i) unweathered bottom ash with pH > 12 (grate siftings and unquenched samples);
ii) quenched/noncarbonated bottom ash with pH between 10 and 10.5 (freshly quenched and 6-week-old samples);
iii) carbonated bottom ash with pH between 8 and 8.5 (1.5- and 12-year-old samples).

The leaching mechanism identified for these three stages revealed major differences for Ca, Al, Fe, Si, and SO_4 but not for Na, K, and Cl (Meima & Comans, 1997)

Literature reported the effects of weathering in bottom ashes from MSWI (Rendek, Ducom, & Germain, 2006) or in the treatment methods of fly ash from MSWI prior to its reuse or final disposal (Bayuseno & Schmahl, 2011). But there was a lack of knowledge regarding the influence of disposal conditions on ISSA.

Aiming to study the potential of acid extraction (Couto, Guedes, Ottosen, & Ribeiro, (submitted)) and ED remediation ((Guedes, et al., (submitted)) for P recovery and heavy metals removal from ISSA, a parallel a study was carried out to test the effect of ashes natural aging (due to open air disposal of ashes in the incineration facility) in the extraction / removal efficiencies. Results were discussed under the light of ashes aging.

Two types of ashes were collected at *Lynneten* incineration plant, Copenhagen, Denmark, a "reference" country in terms of incineration ratio. One ash sample was collected immediately after incineration and the other after being stored in a deposit. Differences were found in the pH, conductivity, water content, loss on ignition (LOI) and solubility, and in heavy metal concentration, namely for Al, Cd, Cu and Zn (Couto, et al., (submitted)). With a LOI reflecting sufficient incineration conditions, concentrations of different elements were, in general, in the same range for both ashes, except for Ca concentration that was slightly higher in the ash collected immediately after incineration. Regarding nutritional potential, ISSA presented ca. 13% of P per weight, which is higher than the average value (4-9%) that can be majorly attributed to population diet (Mihelcic, Fry, & Shaw, 2011), as well as to the type of industries located in the area. *Deposited* ashes presented higher solubility, conductivity and water content, but lower pH, due to carbonation (Rendek, et al., 2006), promoting changes in ash buffering capacity. Water content was related with the exposure to environmental factors such as precipitation and carbonation occurring when moist ashes are in contact with atmospheric CO_2 promoting pH decrease, precipitation of calcite (Meima & Comans, 1997; Rendek, et al., 2006).

In fact, it was reported that ISSA properties can vary significantly dependent on the characteristics of the catchment area (e.g. population diet, type of industries) and the applied process in the wastewater treatment plant (Donatello, Tong, & Cheeseman, 2010). In this last case, one of the steps directly influencing the concentration of some elements is related to addition of precipitating agents (e.g. Ca, Al, Fe) to the wastewater in order to form phosphate complexes for retaining P, avoiding contamination of natural waters.

Recently, ED experiments were carried out with ashes, before and after being deposited, *new* and *deposited* ash (Guedes, et al., (submitted)). The ashes were ED treated as stirred suspensions in sulfuric acid for 3, 7 and 14 days and the percentage of the analysed elements present in the ISSA after the treatment, can be seen in Figure 3.

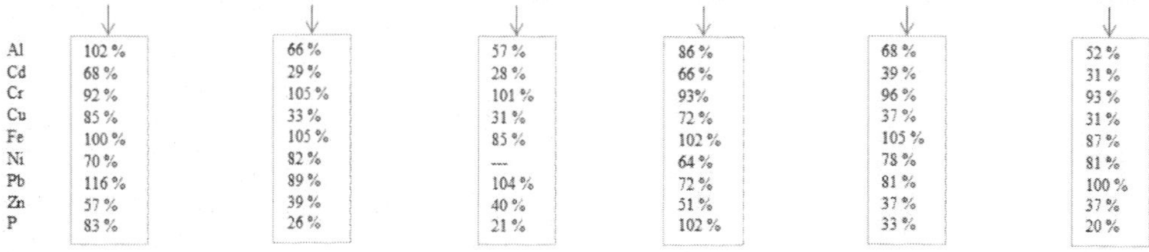

Figure 3. Percentage of elements present in the new and deposited ISSA samples after 3, 7 and 14 days of ED treatment.

There was a tendency for higher concentrations in the *new* ash after ED treatment, and there were statistically significant differences (p<0.05) for Al, Cr and Zn that presented higher concentrations in the *new* ash (Figure 3). Although, the ash collected after the incineration already presented higher concentration of Al and Cr ((Couto, et al., (submitted)). This shows that the removal efficiencies were affected by the ISSA characteristics as, for example, before ED Cu concentration was higher in the *new* ash that presented more 3% of this element and were statistically different, and after ED the concentrations were similar. This means that more 3% of Cu was extracted from the *new* ash during ED, probably due to the presence of less stable and less bounded Cu forms. The same happened to Cd that also presented slightly higher concentration in the beginning for the *new* ash and consequently higher removal efficiency in the *new* ash.

When ISSA is exposed to the weather conditions, leaching may occur and the concentration of some heavy metal decrease over time. When ED is performed, those elements are less bounded to the ash and are more easily removed, resulting the in final similar concentrations obtained for Cd and Cu.

Conducted studies showed that changes in the physico-chemical properties of the ashes after natural aging, namely metal speciation (Guedes, et al., (submitted)), and the extraction solvent used (Ottosen, et al., 2013), need to be taken into account when delineating their treatment.

2.2.4. Potential for P Recovery

Potential for P recovery from *new* and *deposited* ISSA has been studied by Couto, et al., (submitted), performing acid extraction with HNO_3. In total, 5% of P was mobilized from ISSA, representing 53 g/kg for *new* ash and 51 g/kg for *deposited* ash, indicating that even though ashes characteristics changed due to natural aging, P is still available for extraction with no significant influence in the amount extracted. Ottosen, et al., (2013) tested acid extraction with H_2SO_4 and HNO_3 aiming P recovery in two different ashes from mono-incineration of sewage sludge, where for P precipitation from the wastewater both Al or Fe was used. In both studies, simultaneously to the release of P, a fraction of heavy metals also occurred (Couto, et al., (submitted); Ottosen, et al., 2013) and, consequently, a separation step is needed. It was also accessed that in order to obtain the cleanest liquid in relation to heavy metals and a maximum P concentration, the duration of the extraction process should not be too long (Ottosen, et al., 2013). It has been suggested that 2 h is the optimum time for extraction of P in H_2SO_4 as the major part of P was extracted during this time (Donatello, Tong, et al., 2010; Donatello, Tyrer, et al., 2010).

The application of ED treatment for P removal (Guedes, et al., (submitted)) resulted in ca. two times higher removal percentages than the ones obtained with acid wash (Couto, et al., (submitted)) for the same ISSA. Only 12 to 17% of P was found in the cathode end. This shows that, despite simultaneously P removal with heavy metals during the ED process, P was concentrated in the anode end, where heavy metals concentration was very low. Final P concentration in the anolyte was 79314 mgP/kg for the *new* ash and 89057 mgP/kg for *deposited* ash, representing 59% and 69% of ashes initial content. In this compartment the presence of heavy metals is low, meaning that the obtained P product is cleaner comparing to the initial ash. As ion exchange membranes were used, they physically separated the dissolved charged particles that migrated to the electrolytes from the ash compartment, and the filtration step, that is difficult and expensive, can be avoided (Pedersen, et al., 2005).

The concentration of the metals in the electrolytes also results in a more concentrated solution, with less volume, making possible their recovery by commercial methods (e.g. electro-deposition or ion-exchange resin). The remaining liquid could be reused as electrolyte for a new treatment as it still contains high amount of salts (Pazos, et al., 2010). In another way, the anolyte contains P that could be directly used in agriculture (after e.g. pH adjustment) or be recovered through different methods like precipitation for further processing or use in industry. The cleaned ash could also be used as a raw material. However, the use of H_2SO_4 causes formation of a high quantity of gypsum crystals in the remaining ash and this must be taken into account when handling the remaining ash in e.g. construction materials or as an increased volume to be deposited (Ottosen, et al., 2013).

2.2.5. Electrodialytic Remediation vs Acid Wash

When an electric direct current is applied to the system, the heavy metals in solution and in exchangeable forms are removed continuously as charge-bearing ions in the electric field, what should enhance metal desorption compared to traditional batch extraction experiments (Pedersen, et al., 2005). By comparing the results obtained by ED experiments for the 14 days period (Guedes, et al., (submitted)) with acid extraction experiments (Couto, et al., (submitted)) for the ISSA from Lynetten samples, this theory is supported (Table 2).

In general, ED presented higher removal efficiencies than acid wash except for Zn and Ni. The removal of Fe, Pb was also low in both techniques, although higher in the ED.

For Al, Cr, Cu, the ED removals were between 1.4 to 5.2 times higher than in the batch experiments.

In ED experiments, the final Pb concentration in the ash was 4% higher than in the beginning for the *new* ash, and this was not observed in the batch experiments probably due the a higher mass loss in ED experiments.

Comparing both methods, ED process presents the advantage of removing charged particles from the sample as soon as they are solubilized, due to electromigration, whereas in leaching/washing they remain in contact with the sample, hindering further solubilisation due to saturation of the sample (Ferreira, et al., 2005). Also, the processes based on leaching/washing require longer treatment times and higher liquid:solid ratios in order to be effective, making the process of phosphorus and heavy metals recovery more difficult, as well as creating a "new problem" related to the waste generated (Ferreira, et al., 2005).

Table 2. Comparison of removal efficiencies between ED (14 days) and acid wash for the two ashes tested

Element	Removal efficiencies (%)			
	ED experiments [a]		Batch experiments [b]	
	New ash	*Deposited* ash	*New* ash	*Deposited* ash
Al	54.4	52.3	10.4	19.2
Cd	55.9	53.1	59.3	58.3
Cr	14.7	12.3	4.39	6.30
Cu	78.6	82.4	51.2	55.8
Fe	6.24	4.90	0.80	1.67
Ni	-	27.1	32.3	33.1
Pb	3.81	4.32	0.24	0.51
Zn	50.2	53.1	65.3	65.9
P	76.2	81.3	36.3	36.3

[a] (Guedes, et al., (submitted)).
[b] (Couto, et al., (submitted)).

Conclusion and Final Remarks

Incineration of urban wastes is already a well-established technology in some countries. It is expected to growth in the scope of the European Union waste management strategy, as it presents several advantages. Incineration reduces the volume and mass of the waste, destroys pathogenic agents, oxidizes organic compounds and generates thermal energy that can be reused.

The ash can be potentially hazardous due to the presence of leachable heavy metals, high concentrations of soluble salts and residual amounts of hazardous organics (e.g. dioxins) (Ferreira, et al., 2005). But they can also contain high phosphorus concentrations, but they do not have a direct value as fertilizer, as P is in a non-plant available form (Ottosen, et al., 2013).

Therefore, it becomes important to recover P from the ashes in order to recycle a nutrient that has a high socio-economic value and is also indispensable to the ecosystem.

Several works with ED have shown promising results in extracting organic and inorganic contaminants from a few types of ashes. In some cases, extraction was performed with water with significant reduction of some heavy metals concentration in the ash samples. More recently, ED process was applied for P recovery and the results were very satisfactory. The application of ED to ISSA samples allowed recovering in the anolyte between 59 and 69% of the P initial ash content. As P was removed to the anolyte, contrary to the heavy metals that were removed towards the catholyte, the obtained P product is clean, avoiding a filtration step that is difficult and expensive.

In future, the design of an optimized ED process should take into account ISSA characteristics as both natural aging and previous precipitation of P with Al, influence extraction efficiencies. The enhancement solution to be used should also be carefully chosen according to ash future use.

Knowing that sewage sludge is a matrix with high socio-economical values even after incineration, future works can also take advantage of it. The suitability of ED process with stirring conditions for the removal of P, heavy metals and the differences of a common "disposal" conditions were here discussed. It can be concluded that ED is a viable process for P recovery as a clean product while promoting a cleaner ash through heavy metals removal.

Still, research on this field should increase as waste streams are continuously raising and it becomes urgent to adopt sustainable options to the future.

References

Acar, Y. B., & Alshawabkeh, A. N. (1993). Principles of electrokinetic remediation. *Environmental Science & Technology*, 27 (13), 2638-2647.

Al Sayed, M. H., Madany, I. M., & Buali, A. R. M. (1995). Use of sewage sludge ash in asphaltic paving mixes in hot regions. *Construction and Building Materials*, 9(1), 19-23.

Alshawabkeh, A. N., Yeung, A. T., & Bricka, M. R. (1993). Practical aspects of in-situ electrokinetic extraction. *Environmental Science and Technology*, 23, 26-38.

Anderson, M., Elliott, M., & Hickson, C. (2002). Factory-scale proving trials using combined mixtures of three by-product wastes (including incinerated sewage sludge ash) in clay building bricks. *Journal of Chemical Technology & Biotechnology*, 77(3), 345-351.

Aryal, N., & Reinhold, D. M. (2011). Phytoaccumulation of antimicrobials from biosolids: Impacts on environmental fate and relevance to human exposure. *Water Research*, 45(17), 5545-5552.

Ayeni, L. S., Adetunji, M. T., Ojeniyi, S. O., Ewulo, B. S., & Adeyemo, A. J. (2008). Comparative and cumulative effect of cocoa pod husk ash and poultry manure on soil and maize nutrient contents and yield. *American-Eurasian Journal of Sustainable Agriculture*, 2(1), 92-97.

Basta, N. T., Ryan, J. A., & Chaney, R. L. (2005). Trace element chemistry in residual-treated soil: key concepts and metal bioavailability. *Journal of Environmental Quality*, (34), 49-63.

Bayuseno, A. P., & Schmahl, W. W. (2011). Characterization of MSWI fly ash through mineralogy and water extraction. Resources, *Conservation and Recycling*, 55(5), 524-534.

Bierman, P. M., & Rosen, C. J. (1994). Phosphate and Trace Metal Availability from Sewage-Sludge Incinerator Ash. *Journal of Environmental Quality*, 23(4), 822-830.

Blöcher, C., Niewersch, C., & Melin, T. (2012). Phosphorus recovery from sewage sludge with a hybrid process of low pressure wet oxidation and nanofiltration. *Water Research*, 46(6), 2009-2019.

Bright, D. A., & Healey, N. (2003). Contaminant risks from biosolids land application: Contemporary organic contaminant levels in digested sewage sludge from five treatment plants in Greater Vancouver, British Columbia. *Environmental Pollution*, 126(1), 39-49.

Carpenter, S. R., & Bennett, E. M. (2011). Reconsideration of the planetary boundary for phosphorus. *Environmental Research Letters*, 6(1).

Chen, T., & Yan, B. (2012). Fixation and partitioning of heavy metals in slag after incineration of sewage sludge. *Waste Management*, 32(5), 957-964.

Clarke, B. O., & Smith, S. R. (2011). Review of 'emerging' organic contaminants in biosolids and assessment of international research priorities for the agricultural use of biosolids. *Environment International*, 37(1), 226-247.

Commission, E. (27 April 2000). Working document on sludge. 3rd draft. Brussels.

Commission of European Communities (1998). Council Directive 91/271/EEC of 21 March 1991 concerning urban waste-water treatment (amended by the 98/15/EC of 27 February 1998).

Cordell, D., Drangert, J.-O., & White, S. (2009). The story of phosphorus: Global food security and food for thought. *Global Environmental Change*, 19(2), 292-305.

Cornel, P., & Schaum, C. (2009) Phosphorus recovery from wastewater: Needs, technologies and costs. *Water Science and Technology*, 59(6), 1069-1076.

Couto, N., Guedes, P., Ottosen, L. M., Kirkelund, G. M., & Ribeiro, A. B. (submitted). Natural weathering of incinerated sewage sludge: insights from characterization and acid extraction.

Cyr, M., Coutand, M., & Clastres, P. (2007). Technological and environmental behavior of sewage sludge ash (SSA) in cement-based materials. *Cement and Concrete Research*, 37(8), 1278-1289.

Donatello, S., Tong, D., & Cheeseman, C. R. (2010). Production of technical grade phosphoric acid from incinerator sewage sludge ash (ISSA). *Waste Management*, 30(8–9), 1634-1642.

Donatello, S., Tyrer, M., & Cheeseman, C. R. (2010). EU landfill waste acceptance criteria and EU Hazardous Waste Directive compliance testing of incinerated sewage sludge ash. *Waste Management*, 30(1), 63-71.

Elliott, H. A., & O'Connor, G. A. (2007). Phosphorus management for sustainable biosolids recycling in the United States. *Soil Biology and Biochemistry*, 39(6), 1318-1327.

Eurostat. (2011). Municipal waste statistics - Statistics Explained, March 2011, accessed in 2013/4/4, from http://epp.eurostat.ec.europa.eu/statistics_explained/index.php/Municipal_waste_statistics#

Eurostat. (2012). "Water statistics" - Statistics Explained, September 2012, accessed in 2013/4/4, from http://epp.eurostat.ec.europa.eu/statistics_explained/index.php/Water_statistics

Ferreira, C., Jensen, P., Ottosen, L., & Ribeiro, A. (2005). Removal of selected heavy metals from MSW fly ash by the electrodialytic process. *Engineering Geology*, 77(3–4), 339-347.

Ferreira, C., Ribeiro, A., & Ottosen, L. (2003). Possible applications for municipal solid waste fly ash. *Journal of Hazardous Materials*, 96(2–3), 201-216.

Gerba, C. P., & Smith, J. E. (2005). Sources of Pathogenic Microorganisms and Their Fate during Land Application of Wastes. *Journal of Environmental Quality*., 34(1), 42-48.

Gibbs, R. A., Hu, C. J., Ho, G. E., & Unkovich, I. (1997). Regrowth of faecal coliforms and salmonellae in stored biosolids and soil amended with biosolids. *Water Science and Technolo*gy, 35(11–12), 269-275.

Guedes, P., Couto, N., Ottosen, L. M., & Ribeiro, A. B. (submitted). Electrodialytic recovery of phosphorus from new and deposited sewage sludge ashes

Hansen, H. K., Ottosen, L. M., & Villumsen, A. (2004). Electrodialytic removal of cadmium from straw combustion fly ash. *Journal of Chemical Technology & Biotechnology*, 79(7), 789-794.

Hansen, H. K., Rojo, A., Pino, D., Ottosen, L. M., & Ribeiro, A. B. (2008). Electrodialytic remediation of suspended mine tailings. *Journal of Environmental Science and Health, Part A*, 43(8), 832-836.

Heffer, P., Prud'homme, M., Muirheid, B., & Isherwood, K. F. (2006). Phosphorus Fertilization Issues and Outlook, Proceedings no 586. International Fertilizer Society, York UK, 32.

Hjelmar, O. (1996). Disposal strategies for municipal solid waste incineration residues. *Journal of Hazardous Materials*, 47(1–3), 345-368.

IFDC- International Fertilizer Development Center. (2010). World Phosphate Rock Reserves and Resources (Press Release), accessed in January 2013, from http://www.ifdc.org/

Jakobsen, P., & Willett, I. R. (1986). Comparisons of the fertilizing and liming properties of lime-treated sewage sludge with its incinerated ash. *Fertilizer research*, 9(3), 187-197.

Kirby, C. S., & Rimstidt, J. D. (1993). Mineralogy and surface properties of municipal solid waste ash. *Environmental Science & Technology*, 27(4), 652-660.

Kirkelund, G. M., Damoe, A. J., & Ottosen, L. M. (2013). Electrodialytic removal of Cd from biomass combustion fly ash suspensions. *Journal of Hazardous Materials*, 250–251(0), 212-219.

Lee, Y. B., Ha, H. S., Lee, C. H., & Kim, P. J. (2008). Coal Fly Ash and Phospho-gypsum Mixture as an Amendment to Improve Rice Paddy Soil Fertility. *Communications in Soil Science and Plant Analysis*, 39(7-8), 1041-1055.

Lima, A. T., Ottosen, L. M., & Ribeiro, A. B. (2009). Electroremediation of straw and co-combustion ash under acidic conditions. *Journal of Hazardous Materials*, 161(2–3), 1003-1009.

Lima, A. T., Ottosen, L. M., Ribeiro, A. B., & Hansen, H. K. (2008). Electrodialytic removal of Cd from straw ash in a pilot plant. *Journal of Environmental Science and Health - Part A Toxic/Hazardous Substances and Environmental Engineering*, 43(8), 844-851.

Lin, D.-F., Luo, H.-L., & Sheen, Y.-N. (2005). Glazed Tiles Manufactured from Incinerated Sewage Sludge Ash and Clay. *Journal of the Air & Waste Management Association*, 55(2), 163-172.

Marani, D., Braguglia, C. M., Mininni, G., & Maccioni, F. (2003). Behaviour of Cd, Cr, Mn, Ni, Pb, and Zn in sewage sludge incineration by fluidised bed furnace. *Waste Management*, 23(2), 117-124.

Mateus, E., Zrostlíková, J., Gomes da Silva, M., Ribeiro, A., & Marriott, P. (2010). Electrokinetic removal of creosote from treated timber waste: a comprehensive gas chromatographic view. *Journal of Applied Electrochemistry*, 40(6), 1183-1193.

Medici, F., Piga, L., & Rinaldi, G. (2000). Behaviour of polyaminophenolic additives in the granulation of lime and fly-ash. *Waste Management*, 20(7), 491-498.

Meima, J. A., & Comans, R. N. J. (1997). Geochemical Modeling of Weathering Reactions in Municipal Solid Waste Incinerator Bottom Ash. *Environmental Science & Technology*, 31(5), 1269-1276.

Meima, J. A., & Comans, R. N. J. (1999). The leaching of trace elements from municipal solid waste incinerator bottom ash at different stages of weathering. *Applied Geochemistry*, 14(2), 159-171.

Mellbye, M. E., Hemphill, D. D., & Volk, V. V. (1982). Sweet Corn Growth on Incinerated Sewage Sludge-Amended Soil1. *Journal of Environmental Quality*, 11(2), 160-163.

Mihelcic, J. R., Fry, L. M., & Shaw, R. (2011). Global potential of phosphorus recovery from human urine and feces. Chemosphere, 84(6), 832-839.

Ministério do Ambiente, do Ordenamento do Território e do Desenvolvimento Regional. (2007). PERSU II, Plano Estratégico para os Resíduos Sólidos Urbanos 2007-2016, 1ª edição, ISBN 978-989-8097-01-9 (in Portuguese).

Moffett, B. F., Nicholson, F. A., Uwakwe, N. C., Chambers, B. J., Harris, J. A., & Hill, T. C. J. (2003). Zinc contamination decreases the bacterial diversity of agricultural soil. FEMS *Microbiology Ecology*, 43(1), 13-19.

Moliner, A. M., & Street, J. J. (1982). Effect of fly ash and lime on growth and composition of corn (Zea mays L.) on acid sandy soils. *Soil and Crop Science Society of Florida*, 41, 217–220.

Nystroem, G. M., Ottosen, L. M., & Villumsen, A. (2005). Electrodialytic removal of Cu, Zn, Pb, and Cd from harbor sediment: influence of changing experimental conditions. *Environmental Science & Technology*, 39(8), 2906-2911.

Ott, C., & Rechberger, H. (2012). The European phosphorus balance. Resources, *Conservation and Recycling*, 60(0), 159-172.

Ottosen, L. M., Hansen, H. K., & Hansen, C. B. (2000). Water splitting at ion-exchange membranes and potential differences in soil during electrodialytic soil remediation. *Journal of Applied Electrochemistry*, 30(11), 1199-1207.

Ottosen, L. M., Kirkelund, G. M., & Jensen, P. E. (2013). Extracting phosphorous from incinerated sewage sludge ash rich in iron or aluminum. *Chemosphere*, 91(7), 963-969.

Ottosen, L. M., Kristensen, I. V., Pedersen, A. J., Hansen, H. K., Villumsen, A., & Ribeiro, A. B. (2003). Electrodialytic Removal of Heavy Metals from Different Solid Waste Products. Separation Science and Technology, 38(6), 1269-1289.

Ottosen, L. M., Lima, A. T., Pedersen, A. J., & Ribeiro, A. B. (2006). Electrodialytic extraction of Cu, Pb and Cl from municipal solid waste incineration fly ash suspended in water. *Journal of Chemical Technology & Biotechnology*, 81(4), 553-559.

Ottosen, L. M., Pedersen, A. J., Hansen, H. K., & Ribeiro, A. B. (2007). Screening the possibility for removing cadmium and other heavy metals from wastewater sludge and bio-ashes by an electrodialytic method. *Electrochimica Acta*, 52(10), 3420-3426.

Pan, S.-C., Lin, C.-C., & Tseng, D.-H. (2003). Reusing sewage sludge ash as adsorbent for copper removal from wastewater. *Resources, Conservation and Recycling*, 39(1), 79-90.

Pazos, M., Kirkelund, G. M., & Ottosen, L. M. (2010). Electrodialytic treatment for metal removal from sewage sludge ash from fluidized bed combustion. *Journal of Hazardous Materials*, 176(1–3), 1073-1078.

Pedersen, A. J. (2003). Characterization and electrodialytic treatment of wood combustion fly ash for the removal of cadmium. *Biomass and Bioenergy*, 25(4), 447-458.

Pedersen, A. J., & Ottosen, L. M. (2006). Elemental analysis of ash residue from combustion of CCA treated wood waste before and after electrodialytic extraction. *Chemosphere*, 65(1), 110-116.

Pedersen, A. J., Ottosen, L. M., & Villumsen, A. (2003). Electrodialytic removal of heavy metals from different fly ashes: Influence of heavy metal speciation in the ashes. *Journal of Hazardous Materials*, 100(1–3), 65-78.

Pedersen, A. J., Ottosen, L. M., & Villumsen, A. (2005). Electrodialytic removal of heavy metals from municipal solid waste incineration fly ash using ammonium citrate as assisting agent. *Journal of Hazardous Materials*, 122(1–2), 103-109.

Peters, G. M., & Rowley, H. V. (2009). Environmental Comparison of Biosolids Management Systems Using Life Cycle Assessment. *Environmental Science & Technology*, 43(8), 2674-2679.

Porteous, A. (2005). Why energy from waste incineration is an essential component of environmentally responsible waste management. *Waste Management*, 25(4), 451-459.

Rendek, E., Ducom, G., & Germain, P. (2006). Carbon dioxide sequestration in municipal solid waste incinerator (MSWI) bottom ash. *Journal of Hazardous Materials*, 128(1), 73-79.

Ribeiro, A. B., Mateus, E. P., & Rodríguez-Maroto, J.-M. (2011). Removal of organic contaminants from soils by an electrokinetic process: The case of molinate and bentazone. Experimental and modeling. *Separation and Purification Technology*, 79(2), 193-203.

Rittmann, B. E., Mayer, B., Westerhoff, P., & Edwards, M. (2011). Capturing the lost phosphorus. *Chemosphere*, 84(6), 846-853.

Rosen, C. J., Olson, D., & Bierman, P. M. (1994). Swiss Chard and Alfalfa Responses to Soils Amended with Municipal Solid Waste Incinerator Ash: Growth and Elemental Composition. *Journal of Agricultural and Food Chemistry*, 42(6), 1361-1368.

Roy, M. M., Dutta, A., Corscadden, K., Havard, P., & Dickie, L. (2011). Review of biosolids management options and co-incineration of a biosolid-derived fuel. *Waste Management*, 31(11), 2228-2235.

Ruttenberg, K. C. (2003). *The Global Phosphorus Cycle*. Editors-in-Chief: In D. H. Heinrich & K. T. Karl (Eds.), Treatise on Geochemistry (pp. 585-643). Oxford: Pergamon.

Saner, D., Blumer, Y. B., Lang, D. J., & Koehler, A. (2011). Scenarios for the implementation of EU waste legislation at national level and their consequences for emissions from municipal waste incineration. *Resources, Conservation and Recycling*, 57(0), 67-77.

Sidhu, J. P. S., & Toze, S. G. (2009). Human pathogens and their indicators in biosolids: A literature review. *Environment International*, 35(1), 187-201.

Singh, C. J., Thind, H. S., Manchanda, J. S., & Kansal, B. D. (2009). Effect of coal fly ash on crop yield and soil health under cotton–wheat cropping sequence. *Environmental and Ecology*, 27 519–523.

Smit, A. L., Bindraban, P. S., Schröder, J. J., Conjin, J. G., & Meer, H. G. v. d. (2009). Phosphorus in agriculture: global resources, trends and developments. *Plant Research International B.V.*, Wageningen, Report 282.

Smith, K. M., Fowler, G. D., Pullket, S., & Graham, N. J. D. (2009). Sewage sludge-based adsorbents: A review of their production, properties and use in water treatment applications. *Water Research*, 43(10), 2569-2594.

Survey, U. S. G. (2012). Mineral Commodity Summaries - Phosphate rock, accessed in February 2012 from http://minerals.usgs.gov/minerals/pubs/commodity/phosphate_rock/mcs-2012-phosp.pdf

Velizarova, E., Ribeiro, A. B., & Ottosen, L. M. (2002). A comparative study on Cu, Cr and As removal from CCA-treated wood waste by dialytic and electrodialytic processes. *Journal of Hazardous Materials*, 94(2), 147-160.

Virkutyte, J., Sillanpää, M., & Latostenmaa, P. (2002). Electrokinetic soil remediation — critical overview. *Science of the Total Environment*, 289(1 3), 97 121.

Werther, J., & Ogada, T. (1999). Sewage sludge combustion. *Progress in Energy and Combustion Science*, 25(1), 55-116.

Wong, M. H., & Wong, J. W. C. (1989). Germination and seedling growth of vegetable crops in fly ash-amended soils. *Agriculture, Ecosystems & Environment*, 26(1), 23-35.

Zevenbergen, C., & Comans, R. N. J. (1994). *Geochemical factors controlling the mobilization of major elements during weathering of MSWI bottom ash*. In Environmental aspects of construction with waste materials, H. A. van de Sloot, J.J.J.M. Goumans & Th. G. Aalbers (Eds.), Studies in Environmental Science, 60, 179-194, Elsevier Science B. V., Amsterdam.

In: Fly Ash
Editor: Prabir Kumar Sarker

ISBN: 978-1-62948-044-2

Chapter 16

FLY ASH AND THE USE OF BIOLEACHING FOR FLY ASH DETOXIFICATION

Thulasya Ramanathan and Yen-Peng Ting*
Department of Chemical and Biomolecular Engineering,
National University of Singapore, Singapore

ABSTRACT

Fly ash refers to the finely divided residue which rises with flue gas during the combustion of coal or any other solid waste. The two major types of fly ash include coal fly ash and municipal solid waste incineration fly ash. Increasing global consumption of energy and the generation of solid wastes has led to large quantities of fly ash production which require huge landfill area for disposal. Only about one-third of the fly ash generated worldwide is utilized in some applications while the rest is discarded in landfills. This has not only increased the demand for space but has also proved to be an environmental issue of global concern.

Fly ash is heavily loaded with toxic metals (such as aluminium, zinc, cadmium, copper and lead) and is generally regarded as a hazardous waste. The potential leaching of metals in landfill sites through groundwater and rainwater poses grave environmental threat. However, with the presence of high concentration of some of these metals, the ash may be considered an artificial ore from which metal values may be derived. Although chemically treating and detoxifying fly ash may improve the utilization of the ash, the traditional approach such as chemical treatment using organic and inorganic acids or chloride evaporation, and thermal treatment has become obsolete due to high cost, increased energy consumption as well as negative environmental impact. Bioleaching is a well-known technology that utilizes microorganisms for the extraction of metals from low-grade ores. While bioleaching may offer an economical and less energy intensive alternative, this approach is unfortunately fraught with challenges such as poor moisture retention, nitrogen and organic matter deficiency, as well as the alkaline nature of the ash, and toxic levels of heavy metals in the ash which are detrimental to microbial growth and leaching activity. This review focuses on environmental issues related to fly ash and the possible use of bioleaching for the removal or recovery of metals from fly ash.

* Corresponding author address: Email: chetyp@nus.edu.sg.

Keywords: Fly ash, Bacteria, Bioleaching, Metal recovery, Detoxification

1. Introduction

Fly ash is defined as “the finely divided residue resulting from the combustion of ground or powdered coal, which is transported from the firebox through the boiler by flue gases” [Cement and Concrete Terminology (ACI Committee 116)]. The major source of fly ash is the coal combustion process (Sustainable Sources, 2013); the quantity of ash produced is about 10% of the original volume of coal burnt, and forms about 74% of the total amount of ash generated from coal combustion (USEPA, 2012). The composition of coal fly ash greatly depends on the coal burned.

On the basis of composition, coal fly ash is classified as Class F fly ash if it originates from the combustion of anthracite and bituminous coal, and Class C fly ash if it arises from the burning of lignite or sub-bituminous coal (Sustainable Sources, 2013). These two types vary in their CaO (lime) content; Class F fly ash usually has less than 15% CaO while Class C fly ash has CaO between 15% and 30%. Besides, Class F fly ash has higher amounts of silica, aluminium and iron (around 70%); Class C has only 50%-70% (Headwaters resources, 2008).

Another major source of fly ash is the municipal soild waste incineration process. Disposal of municipal solid waste is of global concern due to several factors: diminishing availability of landfill capacity, generation of toxic leachate because of rainwater infiltration, and production of noxious methane gas during decomposition of organic matter in the waste (Clapp, Magee II, Ahlert, & Kosson, 1988). Incineration is widely used to reduce the volume and potential toxicity of hazardous chemical and biological waste, as well as the potential infectious properties of medical waste (Committee on Health Effects of Waste Incineration, 2000).

Metals in coal or municipal solid waste are not destroyed or broken down during combustion but are converted into metal oxides or organometallic complexes or into vapour or fine particulates during the process of burning. These metal compounds and vapours formed are more injurious than the metallic complexes present before incineration. Moreover, incineration converts solid waste into fine ash which is detrimental to health if inhaled or ingested. In addition, the process of incineration results in increased surface area which enhances metal leaching when in contact with rainwater (Montague, 1990). Table 1 lists the general composition of coal and municipal solid waste (MSW) incineration fly ashes.

Besides coal and MSW incineration fly ash, several other types of fly ash such as medical waste fly ash and petroleum fly ash are also generated from industrial activities. Indeed, the alarming rate of production of fly ash calls for the development of effective disposal techniques and the allocation of landfill for ash disposal. Apart from the quantity of fly ash produced, the heavy metal content in fly ash also poses a major problem. Although waste minimization (i.e. ash generation) is always the preferred approach, an increase in the utilization of the generated ash helps to alleviate the ash disposal problem.

Table 1. Typical elemental composition of coal and MSW fly ashes determined by many investigators (Theis & Gardner, 1990)

Element	Coal fly ash (%)	MSW fly ash (%)	Element	Coal fly ash (%)	MSW fly ash (%)
Al	0.1-17.3	0.53-17.6	Cl	0.005-0.1	0.12-1.12
Ca	0.11-22.2	3.3-27	Co	0.00063-0.0035	0.0025-0.0054
Fe	4.9-23.5	0.09-8.7	Cr	0.0015-0.09	0.0021-0.19
K	0.15-3.5	1.1-6.58	Cu	0.0016-0.04	0.0069-0.2
Mg	0.04-7.6	0.5-2.1	F	0.012-0.0671	0.15-0.31
Na	0.01-2.03	0.97-4.95	Hg	0.000004-0.00001	0.000009-0.0035
P	0.04-0.8	n.r.	Mn	0.01-0.1	0.017-0.85
S	0.1-1.5	1.9-3.6	Mo	0.00084-0.0033	n.r.
Si	1.41-28.6	0.18-26.6	Ni	0.00018-0.0013	0.0039-0.096
As	0.0006-0.12	0.00094-0.075	Pb	0.0011-0.08	0.02-2.66
B	0.001-0.06	n.r.	Se	0.00069-0.076	0.000048-0.0016
Ba	0.01-0.107	0.0088-0.9	V	0.0073-0.0256	0.011-0.017
Cd	0.000029-0.0051	<0.0001-0.112	Zn	0.005-0.9	0.08-15.2

n.r.- not reported . Note: These values are only representative and vary based on the source of coal/ waste, location, incineration process etc.

2. Statistics on Fly Ash Production and Use

A recent survey by American Coal Ash Association (ACAA) showed that in US alone, coal fly ash production increased from 49 million tons in 1990 to 60 million tons in 2011. Of this, only 38.36% (around 23 million tons) is used and the rest is discarded in landfills (ACAA, 2011). Similarly, in India, about 175 million tonnes of coal fly ash was produced in 2012, which required about 40,000 hectares of land for the construction of ash ponds. It has been estimated that the figure would rise to 600 million tonnes by 2031-2032. Although fly ash utilization has increased from 1 million tonne in 1993 to around 60 million tonnes in 2006-2007, chemically treating and detoxifying fly ash will further improve utilization (Singh, n.d.-b). China, one of the leading producers of coal fly ash, reported a similar trend, with increasing production and increasing utilization yearly, despite strict measures taken by the Government in decreasing the former and increasing the latter (Wei, 1990). The situation in Europe is different; the production of coal fly ash remained relatively constant at around 60 million tonnes annually from 1993 to 2008 but the utilization rose significantly from 2.3 million tonnes in 1993 to 5.9 million tonnes in 2007(Berg, n.d.). Although a significant rise in the use of fly ash has been reported, the narrow scope of commercial utilization limits its usage. The dearth of strategies for utilization has fuelled research laboratories around the world to identify and widen the possibility of exploiting the use of fly ash in diverse fields. Some of the well-established and recently evolved–techniques for the use of fly ash are discussed below.

3. Use of Fly Ash

Apart from pollution prevention and waste minimization, the use of fly ash is one of the best ways to reduce the need for landfill sites for disposal, as well as generate valuable products from a waste material. Table 2 describes some of the possible ways by which fly ash can be utilized, and the characteristics of fly ash necessary for its application. (This chapter briefly introduces various utilization routes of fly ash. Only limited information is presented since a large number of journal articles and reports are published annually.)

Table 2. Properties of fly ash and its uses

Application	Properties of fly ash	Remarks
Manufacture of concrete	Abundance, size range, pozzolanic nature etc.	Fly ash helps not only in reducing cement manufacturing cost but also enhances durability, flowability and strength of cement. Class F fly ash replaces about 30-75% of total cement usage (Singh, n.d.-b). Several improvisations and optimizations have been tried on fly ash incorporated into cement (e.g. using fly ash microspheres (Kwan & Chen, 2012), and using finer fly ash, etc.) for superior properties.
Adjunct to Portland cement	Abundance, size range, pozzolanic nature etc.	1 tonne of Portland cement production releases about 0.87 tonnes of carbon dioxide. This emission can be reduced by using fly ash as an adjunct to Portland cement, minimizing cement production. Hydration of Portland cement releases free lime which, on reaction with fly ash silicates, improves the cement strength and durability (Singh, n.d.-b).
Manufacturing bricks, pavement tiles and interlocking blocks	Abundance, mechanical strength, chemical composition	Fly ash bricks are stronger and offer greater thermal insulation compared to normal baked clay bricks (Singh, n.d.-a). Compared to the traditional brick manufacturing process, more economical and eco-friendly techniques have been developed to make fly ash bricks. These do not employ high pressure molding or high temperature kiln firing and hence reduce greenhouse gas emissions and energy consumption (Shakir, Naganathan, & Mustapha, 2013).
Synthesis of geopolymers	Si and Al content	Geopolymers are synthetic aluminosilicate inorganic polymers with amorphous three-dimensional networks (Li et al., 2012). The structure of geopolymers renders them as good cement binders with superior mechanical, chemical and thermal properties compared to Portland cement (OPC), making them a promising alternative for a variety of applications, such as removal of metal ions and dyes (Al-Zboon, Al-Harahsheh, & Hani, 2011), and stabilization of wastes (Ogundiran, Nugteren, & Witkamp, 2012).
Agriculture	pH, chemical composition	Fly ash when added to soil acts as an ameliorant, pH buffering agent and stabilizing agent. It neutralizes acidic soils, thus favouring plant growth. In addition, the lime in fly ash reacts with acidic components in the soil and releases micro- and macronutrients such as S, Mo, B etc. which in their available form fortify plants and boost their growth. Fly ash also stabilizes soil by preventing soil erosion. The high Ca concentration is also useful in the flocculation/ aggregation of soil particles brought about by the ions and promotes water and root penetration into other hard and compacted soil. Although some have noted improved yield in crops such as legumes, undesirable effects of fly ash on soil salinity and metal toxicity have also been reported. Fly ash addition affects soil microbiota, contaminates ground water and earth-lined lagoons with its heavy metals and radionuclides, and the cementing properties of certain types of fly ash negatively impact root development in some cases (Basu, Pande, Bhadoria, & Mahapatra, 2009; Jala & Goyal, 2006; Pandey & Singh, 2010).

Application	Properties of fly ash	Remarks
Synthesis of calcium phosphate hydrogel	Ca content	Calcium phosphate hydrogels are promising candidates for use as electrolytes in fuel cells because of their high proton conductivities and good thermal resistance. Fly ash hydrogels have been successfully used for electricity generation in fuel cells (Fukui, Arimitsu, Kidoguchi, Yamamoto, & Yoshida, 2009).
Synthesis of coagulant for wastewater treatment	Al^{3+} and Fe^{3+} content	Fly ash may be used for the treatment of industrial waste water such as tanning wastewater, papermaking wastewater and coal washing wastewater, (Yan et al., 2011). Fly ash coagulants display more superior performance than single $Al_2(SO_4)_3$ and $FeCl_3$ because the higher electric charge in fly ash favours greater adsorption of suspended colloidal impurities in wastewater (Liu and Zhang, 2011).
Synthesis of fly ash based composites for electromagnetic shielding	Non-conductivity and dielectric nature, Fe_2O_3 content	Electromagnetic interference shielding is in critical demand due to the interference of wireless digital devices and the increasing sensitivity and importance of electronic devices. Fe_2O_3 content in fly ash enhances shielding by absorption (Cao & Chung, 2004). Fly ash is insulating in nature and thus helps in regulating the conductivity of the polymer composite (Singh, Kumar, Chandra, & Dhawan, 2011).
Enzyme immobilization on fly ash based support	Low cost, adsorption capability, non-interference with enzyme activity, and mechanical strength	Fly ash (Karim & Husain, 2010), modified/activated fly ash (Karim & Husain, 2011a), and glass-ceramic support (Albertini et al., 2012b) formed from fly ash have been successfully employed in the immobilization of enzymes such as peroxidase (Karim & Husain, 2010, 2011a) and invertase (Albertini, et al., 2012b). Fly ash-immobilized enzymes result in reduction in activation energy of reaction (Albertini et al., 2012a), increase in reaction rate and conversion (Karim & Husain, 2010) and greater tolerance to external factors (Karim & Husain, 2011b).
Synthesis of catalysts from fly ash	Low cost, thermal stability, chemical composition	Aluminosilicates, Fe_2O_3, Cu, Ni, V, Mo^{2+}, K^+ in the ash confer higher stability and catalyze several reactions such as pyrolysis of polyethylene (Na, Jeong, Chung, & Kim, 2006), oxidation of alcohols (Khatri, 2011), trans-esterification of triglycerides (Babajide, Petrik, Musyoka, Amigun, & Ameer, 2010), and conversion of NO to NH_3 (Xuan, Yue, Li, & Yao, 2003). Advantages of utilizing fly ash as a catalyst support include its availability at low cost, its high thermal stability, and suitability for use in industrial fluidized bed reactors. Carbon nanotubes produced from fly ash derived catalysts can be employed in composite materials (Dunens, MacKenzie, & Harris, 2009; Nath & Sahajwalla, 2011).
Preparation of nanoparticles	Chemical composition	Oil shale ash is a cheap and promising source for silica nanoparticle synthesis because of the rich silica content in the ash (Gao et al., 2009). Metal ferrite nanoparticles (e.g. nickel ferrites) which can be extensively used as gas sensors, and microwave absorbents have also been synthesized with fly ash (Fouad, Halim, & Rashad, 2008). Apart from this, nano-structured fly ash has also been prepared for use in metal-matrix nanocomposites. Nanostructured fly ash is added to reinforce the metal matrix to enhance ductility and strength (Swami, Raju, Rao, & Rao, 2009).

Table 1. (Continued)

Application	Properties of fly ash	Remarks
Synthesis of zeolites	Si and Al content	Zeolites are crystalline aluminum–silicates, with Group I or II elements as counter ions (Querol et al., 2002). The advantage of zeolite compared with other materials is that its structural and chemical properties can be controlled by crystallization. These properties can be modified by substituting the aluminum atoms in the SiO_2–Al_2O_3 structure with Si, P, B, Ge, Co, Fe, etc. as well. Moreover, zeolites exist with low, medium, and high pore size. Nano-porosity in zeolites gives rise to a high specific surface area and many active sites (Medina et al., 2010). Zeolites have been used for the adsorption of metal ions such as As(V) (Medina, et al., 2010), Cd and Zn (Izidoro, Fungaro, Abbott, & Wang, 2012).
Synthesis of adsorbents	Chemical composition, low cost and properties such as porosity, large surface area and small particle size (Wang & Wu, 2006), in-situ adsorption and increased mechanical strength (Wee, 2013).	Fly ash is comparatively inactive in its native state while chemical activation significantly improves its adsorption capabilities (Visa, Isac, & Duta, 2012; Wajima & Munakata, 2011; Xu, Wang, Zhu, Kuang, & Jing, 2013). The crystalline structure, polar surface and SiO_2/Al_2O_3 ratio are the major factors determining the adsorptive nature of fly ash. Fly ash can be used to remove metal cations (Visa, et al., 2012; Wajima & Munakata, 2011), organic compounds such as phenol (Srivastava, Swamy, Mall, Prasad, & Mishra, 2006), dyes such as methyl orange (Visa & Duta, 2012), inorganic anions such as boron (Polowczyk, Ulatowska, Koźlecki, Bastrzyk, & Sawiński, 2012), sulphate (Madzivire, Gitari, Vadapalli, Ojumu, & Petrik, 2011), fluoride and phosphorus (Wang & Wu, 2006) from wastewater. Desorption from fly ash pellets was observed in acidic media and studies have proposed the stabilization of pellets in concrete blocks to reduce metal desorption (Papandreou, Stournaras, & Panias, 2007). Fly ash can be used to adsorb CO_2 from CO_2/N_2 mixture because of the presence of alkali components (CaO and MgO) which are essential for mineral carbonation. Adsorption of SO_2 with fly ash has also been reported (Christopher-Chia, Lee, Fernando, Bhatia, & Mohamed, 2005).
Manufacture of glass and ceramics	Chemical composition	Fly ash has high amounts of CaO, SiO_2 and Al_2O_3 which are glass-network formers and transition metal oxides which aid in nucleation in crystal formation. The varied chemical composition of fly ash reduces the melting point and if fly ash is melted or vitrified above 1300°C, inert glass will be formed (Erol et al., 2000). Glass or glass-ceramic production from fly ash is useful in stabilizing heavy metals in fly ash during vitrification (Barbieri et al., 1999). Glass-ceramics from fly ash are sometimes accompanied by undesirable colouration due to Fe_2O_3 and MnO present in the ash. Formation of foam ceramics (with the addition of a foaming agent) avoids this and also improves porosity which increases thermal and sound insulation properties of the ceramic (Chen, Lu, & Qu, 2012).
Manufacture of fly ash-reinforced polymers	Mechanical strength	In order to improve the adhesion between natural fibers and polymeric binders in fiber-polymer composites, fly ash is introduced at the interface between the fiber and the binder (Acharya, Mishra, & Mishra, 2008). Several composites have been successfully reinforced with fly ash such as jute (Acharya, et al., 2008), coir dust (Tripathy, 2011), and polyetheretherketone (Parvaiz, Mohanty, Nayak, & Mahanwar, 2010).

4. The Need for Fly Ash Treatment

Although the myriad applications of fly ash are appealing, only about one-third of fly ash is used while the rest is landfilled. Indeed, even if fly ash is landfilled, disposed in surface impoundments, lagooned or recycled into ash based-products, the toxic effects of fly ash are not completely eradicated. The 2008 coal fly ash spill at Tennessee valley is a case in point. The spill occurred when an earthen wall covering a fly ash pond in Kingston, Tennessee cracked and more than 1 billion gallons of fly ash was released into the adjacent river valley. Water samples from the affected area analyzed by the U.S. Environmental Protection Agency showed elevated levels of Cd, Cr, Ba, Pb, Tl, Hg, Ni and As. Arsenic was almost 149 times the allowable standard for drinking water (Gottlieb, 2010).

Apart from accidental spills, leachate generated by rainwater infiltration through fly ash landfill sites pose greater concerns. Although landfills sites in countries with stringent environmental regulations are lined with impervious layers and equipped with leachate treatment, leachate infiltration into the surrounding environment often occurs. Studies have reported an increase in groundwater pH and metal concentration in areas adjoining landfill sites. Investigations aimed at mimicking field conditions to determine the metals most susceptible to leaching showed that pure water could extract substantial amounts of arsenic, chromium, cobalt, molybdenum, copper, nickel, strontium, and lead from coal fly ash (Llorens, Fernandez-Turiel, & Querol, 2001). Simulated rainwater leaching of coal fly ash (Praharaj, Powell, Hart, & Tripathy, 2002) revealed the total cumulative concentrations of As, Mn and Mo to be higher than World Health Organization (WHO) recommendation for drinking water, and the concentrations of Fe, Mn and As to be higher than the maximum allowable concentrations set by the United States Environmental Protection Agency (USEPA).

Another alarming issue is windblown transport of fly ash from uncovered landfills. Windblown ash can lead to exposure via dermal contact, inhalation, and incidental ingestion (Electric Power Research Institute, 2009). Due to the fine particle size of fly ash, long-term inhalation of fly ash may cause severe respiratory problems. Indeed, coal fly ash use provides an additional pathway of exposure to ash components (Gottlieb, 2010).

The radioactive content of fly ash has also been investigated. Studies have shown that combustion of coal results in fly ash with uranium and thorium being concentrated up to 10 times their original concentration. Leaching of uranium into soil and water pose a potential risk to people living in the "stack shadow" (the area within a half- to one-mile radius of a coal plant's smokestacks) (Hvistendahl, 2007).

Clearly, fly ash has to be treated appropriately in order to mitigate the above effects before disposal or use.

5. Treatment of Fly Ash Using Conventional Metal Removal Techniques and Their Drawbacks

Infiltration of rainwater into fly ash landfills results in the release of many trace elements. These mobilized elements cannot be easily captured or immobilized by traditional methods. Instead, expensive techniques such as chemical precipitation, ion exchange, reverse osmosis and solvent extraction need to be employed which are difficult to be applied for fly ash leachate due to its large quantity and seepage into groundwater or surrounding land/water body (Bhattacharyya, Donahoe, & Patel, 2009). Hence, preventive measures need to be taken before landfilling the fly ash.

The pretreatment methods commonly used to prevent the release of hazardous elements from fly ash into the environment include

a) Thermal treatment,
b) Chloride evaporation,
c) Chemical treatment, for e.g. extraction with acids, alkalis, salts or chelating agents,
d) Cement solidification, and
e) Stabilization with polymers/binding agents

Thermal treatment involves melting the ash at temperatures > 700°C. This vaporizes metal species present in the ash and minimizes the metal content of the ash (Kirk, Chan, & Marsh, 2002). Thus, thermal treatment stabilizes heavy metals in the ash and at the same time reduces fly ash volume (Hu et al., 2012). However, ash melting is expensive, energy intensive, and is operated at very high temperature making construction and maintenance of the furnace a burden. Added to this, thermal treatment produces a secondary ash. Moreover, metal recovery from the stabilized ash at a later stage is difficult and uneconomical (Nowak et al., 2010). Few reports state that Cr leachability increases 20 times after thermal treatment due to change in speciation of Cr during the treatment (Hu, et al., 2012; Kirk, et al., 2002). Al leachability also follows a similar profile as Cr (Chan, Kirk, & Marsh, 2000). In addition, thermal treatment uses arc plasma to melt fly ash which results in hazardous secondary emission. The use of radio-frequency induction thermal plasma has been suggested as an alternative (Sakano, Tanaka, & Watanabe, 2001).

Another common method for treating fly ash is chloride evaporation. Metal chlorides have lower boiling points than metal oxides, which makes chloride evaporation a more preferred process for fly ash detoxification. Chlorine used for the process can be supplied as gaseous chorine, gaseous HCl or solid chlorides such as $CaCl_2$, $MgCl_2$, NaCl, and $AlCl_3$. Solid chlorides are advantageous over the gaseous forms as they are easy to handle and selectively chlorinate heavy metals such as Cd, Cu, Pb and Zn. $CaCl_2$, $FeCl_2$ and $MgCl_2$ are more effective than $AlCl_3$ and NaCl at temperatures of 500 to 1050°C as $AlCl_3$ produces only HCl over this temperature range and NaCl evaporates due to high vapour pressure in this range (Nowak, Rocha, Aschenbrenner, Rechberger, & Winter, 2011). $CaCl_2$ reacts with water vapour or oxygen to give HCl and Cl_2 wherein the gaseous chloride reacts with the metals to form metal chlorides. The reaction takes place according the following equations (Nowak, et al., 2010):

With oxygen: $$CaCl_2 + 12O_2 \rightarrow CaO + Cl_2$$

With water: $$CaCl_2 + H_2O \rightarrow CaO + 2HCl$$

CaO formed may react with metal oxides such as Cr_2O_3 to form $CaCr_2O_4$ as an insoluble complex.

$$2Cr_2O_3 + 4CaO + 3O_2 \rightarrow 4CaCrO_4$$

$MgCl_2$ reacts similar to $CaCl_2$ (Nowak, et al., 2010). The temperature range used in chloride evaporation is much lower than the melting point of ash and hence, energy consumption is greatly minimized in this process. When the metal chlorides form and evaporate, a substantially decontaminated product remains which can be disposed safely. The rising heavy metal chlorides are captured in the flue gas cleaning system (e.g. a scrubber) and can be recovered in an ensuing treatment step.

The process can be significantly improved by modifications such as

1. Wetting the mixture (fly ash + $CaCl_2$) with water and drying it increases the contact area maximizing the volatilization of metals (Chan, et al., 2000), and
2. Pelletizing the mixture in order to decrease entrained particles thus mitigating storage and handling issues and also greater losses due to the flue gas formation (Nowak, et al., 2010).

Although chloride evaporation consumes lower energy than thermal treatment, the cost and severe conditions of operation still remain an issue.

A popular alternative to the above processes is chemical treatment. Studies have reported the use of chemicals such as calcium aluminates (Auer, Kuzel, Pöllmann, & Sorrentino, 1995), phosphoric acid (Bournonville, Nzihou, Sharrock, & Depelsenaire, 2004), sodium hydroxide (Nagib & Inoue, 2000; Phair, Deventer, & Smith, 2004), ammonium nitrate, ammonium chloride, organic acids (Fedje, Ekberg, Skarnemark, & Steenari, 2010), chelating agents like ED3A (Qian, Zhang, Zhang, & Chui, 2005), concentrated acids (Nagib & Inoue, 2000) etc. for mobilizing hazardous elements from the solid phase. Ferrous sulphate with or without calcium carbonate can also be used (Bhattacharyya, et al., 2009). Heavy metals immobilized in chemically treated fly ash may leach when there is a change in chemical environment, in particular pH (Shimaoka & Hanashima, 1996). Moreover, large amounts of acid are required to treat fly ash, resulting in an acidic residue which poses an environmental problem. An easier and more eco-friendly way is to use chelating agents, for example, aminopolyacetates such as nitrilotriacetate, ethylenediaminetetraacetate and diethylenetriaminepentaacetate, which form complexes with the heavy metals and thus leach the metals. If the purpose is to stabilize the fly ash, then insoluble complex forming chelating agents containing thiol –SH or dithiocarbamate –NH–CSSH group need to be employed (Hong, Tokunaga, & Kajiuchi, 2000). Natural chelating agents such as canine serum (Harris & David, 1988), gluconic acid, biomass residue molasses hydrolysate (Bipp, Wunsch, Fischer, Bieniek, & Kettrup, 1998), saponins (Hong, Tokunaga, Ishigami, & Kajiuchi, 2000) etc. have also been examined for leaching fly ash. Although chelating agents are environmentally friendly, large scale use of these agents is uneconomical.

In order to recover metals from the leachate obtained after chemical treatment processes (as discussed above), precipitation techniques are employed where the metals are usually precipitated as metal sulphides, hydroxides, carbonates or chelates. Metals in fly ash, instead, can directly be converted to sulphides and removed by flotation using Na_2S. This avoids the chemical treatment step, thus reducing the generation of acidic residue and production of H_2S gas during metal precipitation after chemical treatment (Kuchar, Fukuta, Onyango, &

Matsuda, 2007). The feasibility of this approach on a large scale, however, has not been investigated.

Cement solidification is the most attractive method for ash treatment and disposal because of its simple operation and cost effectiveness. This process decreases the hydraulic conductivity and porosity of fly ash but increases its durability, strength, volume and tortuosity (Shimaoka & Hanashima, 1996). In addition, cement solidification is ineffective for Pb-rich fly ash as Pb leaches from the solidified mass due to high pH of the cemented fly ash (Shimaoka & Hanashima, 1996). Moreover, immobilization makes recovery of heavy metals at a later stage more difficult and expensive (Nowak, et al., 2010). Above all this, a substantial increase in volume raises the issue of space availability.

Stabilization in polymers such as polyethylene, polystyrene and polyvinyl chloride has also been reported (Shimaoka & Hanashima, 1996). Stabilization using binding agents such as calcium phosphate (Nzihou & Sharrock, 2002), and powdered blast furnace slag (Conner, 1990) has also been studied. However, polymer- and binding agent- employed stabilization is more expensive than cementation and hence is not considered for large scale. In addition, a high binding agent-ash ratio increases the volume of resulting ash by two- or three-fold (Mangialardi, 2003). Both cementation and stabilization are not ultimate solutions as continuous monitoring is necessary even after the process (Ferreira, Jensen, Ottosen, & Ribeiro, 2005).

The metals-laden fly ash may also be treated in a biological process termed bioleaching. This approach is gaining popularity as the process is considered clean, environmentally friendly and sustainable, and is economically attractive with lower reagent and operating cost and energy demand as compared to conventional methods such as pyrometallurgical and hydrometallurgical treatments (Bosecker, 1997; Cui & Zhang, 2008; Krebs, Brombacher, Bosshard, Bachofen, & Brandl, 1997). The use of bioleaching and the prospects for fly ash bioleaching is discussed in detail in the next section.

6. Bioleaching

6.1. Introduction to Bioleaching

Bioleaching is a biomining process; it relies on the use of micro-organisms to recover base and precious metals from mineral ores and concentrates. Traditional mining techniques are unable to economically extract metals from ores which have been worked and are lean. In biomining, microorganisms act as biocatalysts in extracting metals from metal laden sources and can be used to exploit secondary sources. The process of biomining includes biooxidation and bioleaching (Brandl, 2008). In biooxidation, the microbes selectively dissolve undesired metals from the solid matrix leaving the metal values of interest enriched in the solid phase to be recovered by other processes. The obtained supernatant is thereby discarded. Bioleaching on the other hand refers to the solubilization of metals of interest from their unavailable solid state by microorganisms such as bacteria and fungi. Here, the metal of interest is recovered from the solution (Brandl, 2008). The discussion that follows focuses on bioleaching rather than biooxidation.

Bioleaching processes are based on the ability of microorganisms to transform solid compounds resulting in soluble and extractable elements which can be recovered (Krebs, Brombacher, Bosshard, Bachofen, & Brandl, 2006). Earliest reports on bioleaching date back to the late 19th century. Rio Tinto mines in South-west Spain with low-grade Cu ores are regarded as the birthplace for commercial bioleaching practices (Brandl, 2008). Initially, bioleaching was employed only for ores. With relatively lower operating temperatures and lower consumption of reagents, bioleaching has been shown to be an economic and cost-efficient alternative to conventional pyrometallurgical and hydrometallurgical methods of metal extraction. The biological process enables the recycling of metals in a process close to natural biogeochemical cycles of waste materials, thus reducing the demand for natural resources such as ores, energy and landfill space (Krebs, et al., 1997).

Bioleaching has emerged as an economical and simple technology for leaching of metals such copper, gold, silver, uranium etc. from their respective ores. With increasing demand for metals, industrial wastes such fly ash, electronic scrap, and spent catalysts have also been considered for metal recovery. Bioleaching has increased the possibility of extracting metals from natural ores and industrial residues not recoverable by conventional methods (Krebs, et al., 1997). Furthermore, with increasing focus on environmentally friendly and sustainable processes, bioleaching has proven to be a clean and green technology for metal extraction (Brierley, 2008).

Interests in bioleaching of fly ash intensified when a study compared hydrochloric acid leaching, chloride evaporation, acetic acid leaching, and bioleaching by *Aspergillus niger* (Tateda, Ike, & Fujita, 1998). The work listed the advantages of traditional techniques over the emerging methods as well as their shortcomings. Chloride evaporation showed excellent extraction efficiency and short reaction time but the process was uneconomical with high input cost and required the use of harsh conditions and high energy consumption. Inorganic and organic acid extractions consumed less energy and were more cost efficient but less process efficient than chloride evaporation. The study considered bioleaching inferior to the other technologies due to extended reaction rate. However, the dearth of knowledge but versatile, cost-efficient and eco-friendly nature of bioleaching process encouraged research in this area. The following sections discuss how bioleaching is carried out in the laboratory and commercial scale, the mechanisms involved in fly ash bioleaching, and how research has progressed.

6.2. Types of Bioleaching

Leaching operations, based on the volume of operations, are usually classified into three domains: laboratory scale (up to 10 dm^3), pilot-plant scale (<10 m^3) and industrial scale (>10 m^3) (Krebs, et al., 1997). Figure 1 shows the different types of bioleaching techniques.

6.2.1. Industrial Scale Techniques

Industrial scale techniques were first used in the extraction of copper from copper mines. The techniques used include dump leaching, heap leaching, vat leaching, in situ leaching and tank leaching (Figure 1). Major factors under consideration in industrial scale techniques include capital, operational and labor costs.

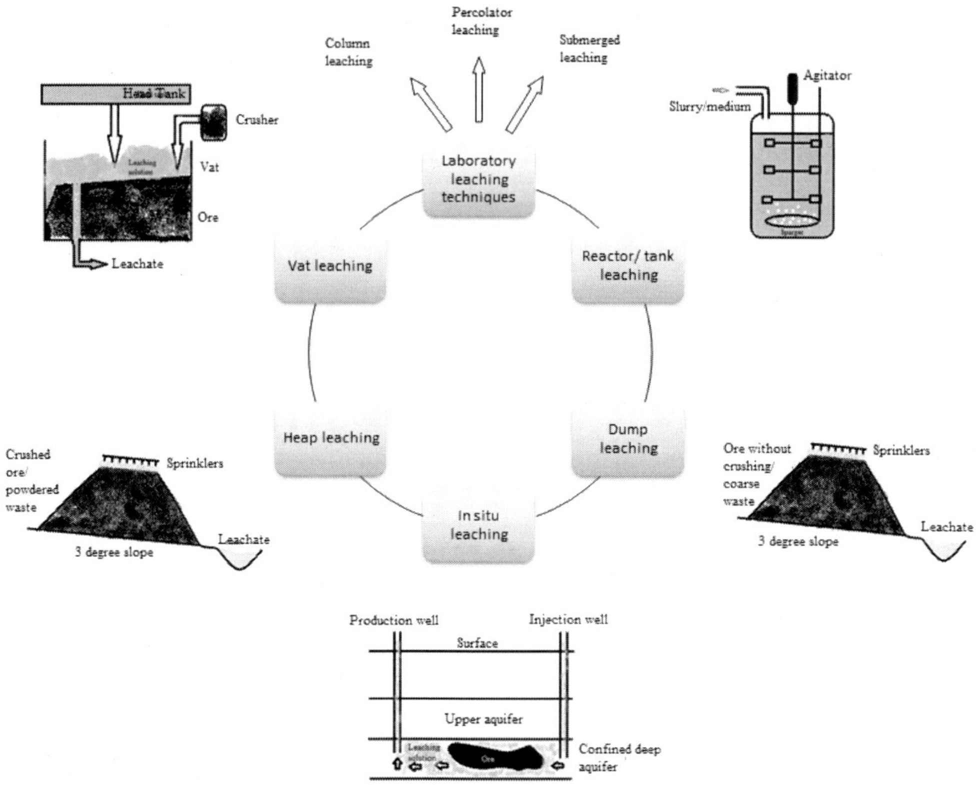

Figure 1. Types of bioleaching (Adapted from Bosecker, 1997; Burgess, 2008; Coloradoans Against Resource Destruction (CARD), n.d.; Mackie, n.d.; Peninsula Energy Limited, n.d.).

a. Dump Leaching

In dump leaching, submarginal ores (i.e. ore that has been mined, but is not of sufficient value to warrant treatment) are removed from mines to a hydrologically isolated area and stacked up to a depth of 300 to 750 ft and a lixiviant (a liquid medium which selectively extracts the desired metal from the ores) is sprayed onto the dump from the top (Hartman, 1992). The lixiviant usually contains media with microorganisms. Sometimes, only media is supplied, encouraging native microbes in the ore to grow and leach the metals. The sprayed lixiviant trickles down the dump and is collected in a reservoir where it is concentrated with metals (AMPREP Metal Recovery Systems, n.d.; Bosecker, 1997; Thakur, 2006). Modern day dumps are lined with an impervious lining at the bottom, made generally of clay to prevent percolation of the leachate into groundwater. The microbial population dynamics varics grcatly within the dump. As the temperature of the interior of the dump is higher than that outside due to the exothermic nature of bioleaching, it promotes the growth of thermophiles.

The nature of the ore or waste treated defines the bacterial population present within the dump. An example of dump leaching is copper mining carried out by Kennecott Copper Bingham Mine, USA (Brierley & Brierley, 2001).

b. Heap Leaching

Heap leaching is very similar to dump leaching, except that heaps are much smaller than dumps. In heap leaching, the ore is crushed or the waste to be treated is ground into smaller particles, and bioleaching is carried out in a way similar to dump leaching (Bosecker, 1997; Thakur, 2006). As heaps are usually only 10 to 35 ft in height, percolation of the lixiviant is faster and more uniform, which enhances metal recovery (Hartman, 1992).

Both dump and heap leaching face problems such as lack of close monitoring systems to keep a check on the toxic leachate infiltration into groundwater. Although environmental awareness has heightened in the present era, no monitoring systems are available on already constructed and used dumps.

Other problems faced are related to the type of ores. Low permeability of ore leading to saturation and stagnation of liquid, channeling of the liquid, blocking of pipes and filter zones, collapsing of pipes and fracturing of lining, salt and silt deposits, erosion, etc. are problems that commonly occur (Smith, 2002). Less of such problems are encountered in modern heap leaching practices.

One area where heap leaching is employed is the mining of precious metal ores such as silver and gold (Kappes, 2002).

Heap and dump leaching suffer from problems related to climatic conditions, with evaporation losses and freezing being common in the heap in dry and cold conditions respectively.

In regions with heavy rainfall, heaps require shelter which increases the operation costs. More importantly, heaps occupy large land areas (INNOVAT Limited, n.d.). Vat and tank leaching are alternatives to these operations.

c. Vat Leaching

Vat leaching employs large containers usually made of wood or concrete and are used for treating coarse ores. Sequential operation of vats is common and is performed to enhance the contact time between the ores and the leaching solution (media with bacteria). In this type of leaching, solids are first loaded into the vats and flooded with the leaching solution. The leachate is collected and then transferred to another vat and the operation is repeated until the leachate is saturated. Counter-current extraction wherein the leachate and the ore move counter-currently is exploited to maximize yield in some cases (Brombacher, Bachofen, & Brandl, 1997). Mantos Blancos Mine in Chile uses vat leaching to extract copper from Copper mines (INNOVAT Limited, n.d.).

d. *In Situ* Leaching

In situ leaching, also known as solution mining, is more common for ores than wastes. The technique is usually applied for low grade ores and ores present in small quantities. Holes are drilled into the ore deposit present in the deep aquifer below the ground. A leaching solution (media containing bacteria) is then passed through the holes for dissolution of metals in the ore and the pregnant solution is pumped to the surface where the metals are recovered (Bosecker, 1997; World Nuclear Association, 2012). Such an operation results in minimal surface disturbance and tailings or waste rock generation. The only limitations are the permeability of the ore, performance of bacteria underground and contamination that may occur to the groundwater adjacent to the leaching site (World Information Service on Energy,

2012; World Nuclear Association, 2012). *In situ* leaching is extensively employed for uranium ore (World Nuclear Association, 2012).

e. Reactor/Tank Leaching

In tank leaching, crushed ores are placed in large tanks with built-in agitators which keep the solids in suspension and improve nutrient supply to bacteria. Baffles are also introduced to prevent the formation of dead zones and to improve mixing within the reactor. The most important factors affecting bioleaching in tanks are retention times within the tanks, agitation and shear stress experienced by the cells (AMPREP Metal Recovery Systems, n.d.). Tank leaching is the best amongst the industrial scale techniques as it gives good extraction yield within shorter time periods but is also the most expensive technique and is used exclusively for materials of high value (Bosecker, 1997).

6.2.2. Laboratory Scale Techniques

As shown in Figure 1, laboratory scale bioleaching techniques include percolator leaching, column leaching and submerged leaching. These techniques are mainly performed to mimic large scale operations and to determine the optimal parameters for operation in large scale.

a. Percolator Leaching

Percolator leaching is carried out in glass columns with a sieve plate at the bottom and with ore/waste particles as the packing. A leaching solution (media with bacteria) is passed from the top. The solution trickles down and is recirculated using sterile compressed air at the bottom which also serves as a source of aeration. Percolator leaching is a bench-scale operation (Bosecker, 1997).

b. Column Leaching

Column leaching is similar to percolator leaching but on a larger scale, and is operated as a pilot plant scale leaching treating few kilograms to tonnes of sample. Leaching is performed to mimic heap and dump leaching conditions (Bosecker, 1997) where monitoring of pH and temperature fluctuations, oxygen supply, fluid flow etc. in the column gives an approximation of conditions in large scale operations such as heap leaching. Column leaching is also used to obtain kinetic information about the ore/waste and to evaluate the effect of various parameters in large scale bioleaching processes (Leach Inc., 2007).

c. Submerged Leaching

Some of the limitations of percolator and column leaching include poor oxygen supply within the interior of the column, and low surface to volume ratio. These limitations are overcome in submerged leaching where fine solid particles and microorganisms are held in suspension or agitation in shake flasks or bioreactors. With improved oxygen and nutrient availability, the microorganisms grow well in submerged leaching which leads to shorter reaction time and enhanced metal recovery (Bosecker, 1997). Bioleaching in laboratory scale is usually performed using submerged leaching.

6.3. Modes of Bioleaching

Where bioleaching is performed in shake flasks or bioreactors, three modes of leaching are usually adopted: one-step, two-step and spent medium leaching. These modes differ in the way the bioleaching operation is carried out and are applied in all types of leaching, although one-step bioleaching is usually used in large scale operations.

a. One-Step Bioleaching

In one-step bioleaching, the microorganisms are inoculated together with the solid waste in the medium where the growth of the microorganism and leaching takes place together. Although this mode is the easiest to perform, the dissolved metal ions arising from the leaching may affect the metabolism and growth of the microorganisms and ultimately limit the bioleaching process efficiency (Yang, Wang, & Wu, 2008). Nonetheless, one-step bioleaching may sometimes be advantageous if the waste contains metals that enhance rather than inhibit bacterial growth and lixiviant production.

b. Two-Step Bioleaching

In two-step bioleaching, the microorganisms are first grown in the medium until it reaches exponential phase (i.e. actively growing state) before the addition of the solid waste. Two-step is usually more effective than one-step bioleaching (Bosshard, Bachofen, & Brandl, 1996) due to possible inhibition of microbial growth (leading to reduced metabolite secretion) in the latter. Metal tolerance is better in two-step bioleaching compared to one-step bioleaching (Yang, et al., 2008).

c. Spent Medium Leaching

In spent medium leaching, the microorganisms are cultured in the growth medium until the maximum production of metabolites occurs. The suspension and microorganisms are subsequently filtered to obtain the cell-free spent medium containing only the biogenically-produced metabolites which is used for bioleaching (Bharadwaj & Ting, 2012). This approach decouples the growth of the microorganisms (and the production of metabolites) and the leaching process, and hence allows different optimum operating conditions for bacterial growth and leaching.

The absence of microorganisms in spent medium leaching permits the use of more aggressive leaching conditions (e.g. low pH, high temperature, etc.) and hence, a higher metal removal. In addition, as the bacteria are not in contact with the toxic metals present in waste, they may be recycled for further growth and production of the metabolites which are separated (e.g. filtered) and used in spent medium leaching. However, since leaching by direct mechanisms such as extracellular polymeric substances production cannot occur in spent medium leaching, poor metal recoveries have been observed in some cases (Wu & Ting, 2006).

6.4. Mechanisms Involved in Bioleaching

Microorganisms leach metals through three major principles, namely

(i) Redoxolysis. Redox reactions, which are based either on electron transfer from minerals to microorganisms in the case of physical contact between organisms and solids, or on bacterial oxidation of Fe^{2+} to Fe^{3+} where ferric iron subsequently catalyzes metal solubilization as an oxidizing agent (Islam & Ting, 2009; Krebs, et al., 1997),

(ii) Acidolysis. Formation of organic or inorganic acids (protons), for instance, in the case of *Acidithiobacillus thioxidans* which produces extraneous sulphuric acid in the presence of sulphur. The sulphuric acid produced leaches the metals from the solid waste (Bosecker, 1987), and

(iii) Complexolysis. Secretion of complexing agents which chelate metal ions by forming stable co-ordinate complexes. For example, cyanogenic bacteria such as *Chromobacterium violaceum* produce extracellular cyanide which forms water-soluble complexes with many metals (Campbell, Olson, Clark, & McFeters, 2001).

Leaching mechanisms have also been classified as direct leaching and indirect leaching.

a. Direct Leaching

Direct leaching refers to leaching mediated through physical contact of the microorganisms with the solid substrate. It occurs through oxidation/ reduction reactions where electrons are transferred from metals in fly ash to microbial cells. Within the cell, these electrons are involved in energy generation through biochemical reactions and oxidative phosphorylation. In the process, metals in fly ash are oxidized and oxygen in the cell is reduced to water.

Generally, prokaryotic cells carry out direct leaching because of the presence of the catalytic system for electron transfer in their cell envelope (Ehrlich, 1992). Another mechanism through which direct leaching occurs is the production of extracellular polymeric substances (Kinzler, Gehrke, Telegdi, & Sand, 2003). These exopolymers mediate attachment to a (metal) sulfide surface, and concentrate iron(III) ions by complexation through uronic acids or other residues at the mineral surface, thus allowing an oxidative attack on the sulfide. Consequently, dissolution of the metal sulfide is significantly enhanced (Kinzler, et al., 2003).

b. Indirect Leaching

Indirect leaching refers to the leaching caused by biogenically secreted metabolic products. These products act as chemical oxidants or reductants to leach metals from fly ash. The metabolites usually involved are organic compounds such as organic acids and ligands, or inorganic compounds like carbonate, phosphate, acids etc. Indirect leaching includes acidolysis (leaching promoted by production of citrate, gluconate etc), complexolysis (leaching by production of cyanide etc.), redoxolysis (leaching by oxidation/ reduction reactions carried out by metabolites usually produced by chemolithotrophic bacteria), alkalolysis (leaching by ammonia produced during urea hydrolysis or amino acid deamination), or a combination of the above listed mechanisms (Bosshard, et al., 1996; Burgstaller & Schinner, 1993; Ehrlich, 1992, 1997).

6.5. Microorganisms Commonly Used in Fly Ash Bioleaching

a. *Aspergillus niger*

Aspergillus niger is a fungus belonging to the genus *Aspergillus,* subsection *Circumdati,* section *Nigri.* Fungi are heterotrophic eukaryotes which require pre-formed organic carbon and an energy source for growth. These organisms are mesophilic (i.e. moderate temperature loving) and grow over a wide pH range. A greater pH tolerance (compared to bacteria) renders fungi more suitable for fly ash bioleaching. Fungi produce extracellular metabolites which serve as complexing agents for metal ions in fly ash and thus reduce toxicity and effect bioleaching (Burgstaller & Schinner, 1993). The growth of fungi in sugars-containing media results in production of various organic acids such as oxalic acid, citric acid, gluconic acid and fumaric acid. Fungi generally have a tendency to produce large quantities of metabolites which increase the availability of complexing agents for bioleaching (Burgstaller & Schinner, 1993). Moreover, fungi oxidize the substrate only partially; this incomplete oxidation causes accumulation of organic acids, thereby improving bioleaching efficiency. The metabolites produced depends largely on the fungal strain and medium composition.

A. niger produces citrate in the absence of fly ash, and gluconate in its presence. Detailed mechanism for citrate production is illustrated in Figure 2:

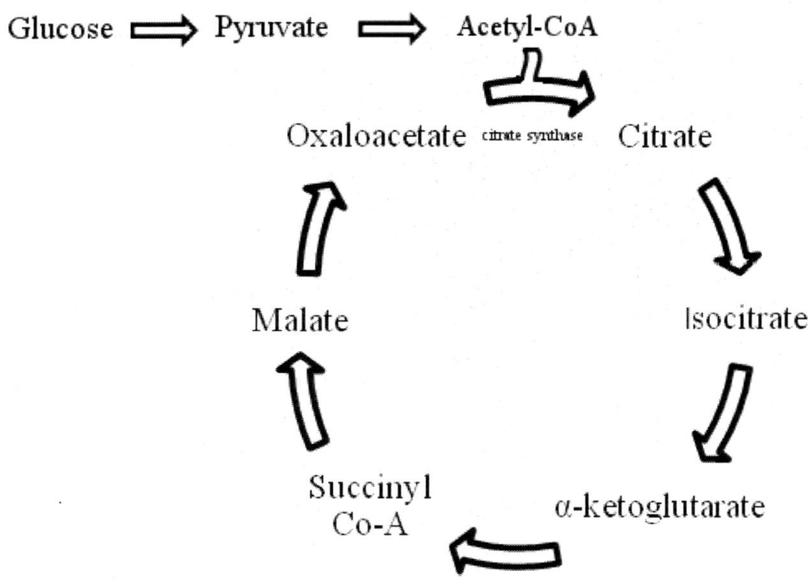

Figure 2. Synthesis of citrate by TCA cycle in the absence of fly ash (Nelson, & Cox, 2005).

A.s niger has always been the paramount and highly valued organism for the bioleaching fly ash. Developments in *A. niger* bioleaching hastened after a report by Singer et al., in 1982, who witnessed the fungi as a successful bioleaching strain.

Brief descriptions on the findings are given in Table 3

Table 3. Research publications on *A.niger* bioleaching of fly ash

Reference	Metals recovered	Highlights of the work
Singer, Navrot, & Shapira, 1982	Al	Microbially produced and commercial citric acid gave similar extraction yields.
Torma & Singh, 1993	Al	Calcination of fly ash increased extraction rate of aluminium.
Bosshard, et al., 1996	Al, Cd, Cr, Cu, Fe, Mn, Ni, Pb, Zn	Two-step bioleaching showed greater extraction yields than one-step bioleaching. Gluconate was produced in the presence of fly ash and citrate in the absence of fly ash as leaching agents.
Xu & Ting, 2004	Al, Cu, Fe, Pb Zn	Factors affecting bioleaching were studied using central composite design and co-optimal conditions were determined using Taylor's series approximation. Sucrose concentration and pulp density were found to be important factors. Spore concentration and time of addition of fly ash did not significantly affect the process.
Wu & Ting, 2006	Al, Cu, Fe, Mn, Pb, Zn	Lower pulp densities showed higher leaching efficiencies. Spent medium leaching was less efficient than two-step bioleaching.
Yang, et al., 2008	Cd, Cr, Cu, Fe, Mn, Pb, Zn	*A.niger* could tolerate 5% fly ash in two-step bioleaching but not in one-step bioleaching
Xu & Ting, 2009	Al, Cu, Fe, Mn, Pb, Zn	Fungal growth kinetics and toxic metals inhibition kinetics were determined in one-step bioleaching. Modified Gompertz model fit the experimental data for biomass concentration and citric acid production well. Han and Levenspiel model was found to be the best fit for inhibitory effects on fungi by fly ash. Addition of fly ash after two days resulted in diminished fungal growth and extended lag phase.
Wang, Yang, & Wu, 2009	Cd, Cu, Pb, Zn	Deionized water washing of fly ash improved percentage recovery
Yang, Wang, & Wu, 2009; Yang, Wang, Luo, Wang, & Wu, 2009	Al, Fe, Pb, Zn	Biosorption of metals leached by fungal biomass was observed. Al, Fe and Zn leaching followed pseudo-first order reaction kinetics while Pb leaching followed pseudo-second order reaction kinetics due to chemisorption being the rate-controlling step. Sorption of Pb, Zn and Fe followed Langmuir model, indicating that monolayer biosorption took place at the functional groups/binding sites on the surface of the biomass, but sorption of Al followed Freundlich isotherm suggesting non-ideal sorption on heterogeneous surfaces and multilayer sorption.

Acidithiobacillus species:

Acidithiobacillus belongs to the phylum Proteobacteria (UNIPROT, n.d.). Organisms from this genus in particular *Acidithiobacillus thiooxidans* and *Acidithiobacillus ferroxidans* (formerly called *Thiobacillus thiooxidans* and *Thiobacillus ferroxidans* respectively and renamed (Kelly & Wood, 2000)) have been extensively used as catalysts in bioleaching metals from a variety of sources ranging from ores to wastes such as fly ash, electronic waste, etc. These organisms are acidophilic and exist in low pH conditions with optimal growth at pH 2-3. The pH range of these microbes is more suitable for ores than for alkaline industrial wastes such as municipal solid waste incineration fly ash (White, Sayer, & Gadd, 2006). These organisms are chemolithoautotrophic and metabolize sulphur as their name suggests. Chemolithoautotrophic bacteria obtain energy by oxidizing inorganic compounds, and carbon by fixing carbon dioxide (Ehrlich, 1997). Organisms of this genus are usually mesophilic. Most chemolithoautotrophs are aerobic. Species of the genus *Acidithiobacillus* are Gram-negative, motile and rod-shaped.

Both direct and indirect leaching are exhibited by *Acidithiobacillus* species. *A. ferrooxidans* obtains energy by oxidising Fe^{2+} to Fe^{3+}. The ferric ion generated is a powerful oxidizing agent; it oxidizes metals from their insoluble form to the soluble form. During the process, ferric ion is reduced to ferrous ion and sulphur is precipitated. The released sulphur reacts with water and oxygen in the presence of bacterial catalysis to form sulphuric acid. The mechanism of leaching by *A. ferrooxidans* is illustrated below (Ehrlich, 1992, 1997):

$$2Fe^{2+} + 2H^{+} + 1/2\ O_2 \xrightarrow{\text{Bacteria}} 2Fe^{3+} + H_2O$$

$$MS + 2Fe^{3+} \longrightarrow M^{2+} + 2Fe^{2+} + S^0$$

$$S^0 + H_2O + 3/2\ O_2 \xrightarrow{\text{Bacteria}} H_2SO_4$$

This is the indirect form of leaching performed by the bacteria. Direct leaching occurs by a similar mechanism when iron sulphide is the substrate (Bosecker, 1997).

A.s thioxidans produces sulphuric acid extraneously by catalyzing the following reaction:

$$S^0 + H_2O + 3/2\ O_2 \xrightarrow{\text{Bacteria}} H_2SO_4$$

Microbially produced sulphuric acid acts as the leaching agent. In systems wherein metals are not present as metal sulphides or when the amount of sulphur present in the waste is minimal, sulphur is added extraneously to promote the formation of sulphuric acid.

Brief descriptions on the findings are given in Table 4 below.

Table 4. Research publications on *Acidithiobacilli* bioleaching

Reference	Organism	Metals recovered	Highlights of the work
Bosecker, 1987	*Acidithiobacillus strains*	Zn	Indirect bioleaching with extraneously produced sulphuric acid was reported. Higher pulp densities were toxic to bacteria.
Fass, Geva, Shalita, White, & Fleming, 1994	*Acidithiobacillus thiooxidans* ZYR1	Al, Ti	Salt and metal tolerant bacteria was developed and used for bioleaching to improve extraction yield.

Table 4. (Continued)

Reference	Organism	Metals recovered	Highlights of the work
Brombacher, Bachofen, & Brandl, 1998	*Acidithiobacillus thiooxidans*	Al, Fe	Enhanced extracellular polymeric substances production in the presence of fly ash was observed; this promoted cell-solid surface interactions and thus improved direct bioleaching.
Seidel, Zimmels, & Armon, 2001	Mixed culture: *Acidithiobacillus thiooxidans* and *Acidithiobacillus ferrooxidans*	Al, Cd, Cr, Cu, Fe, Mn, Ni, Pb, Zn	Semi-continuous laboratory scale leaching unit was used. The degree of bioleaching followed the order: inoculated medium > filtered cell-free filter medium > autoclaved sterile spent medium > uninoculated medium.
Krebs, Bachofen, & Brandl, 2001	Sewage suspensions with four *Acidithiobacilli* strains, *A.neopolitanus, A.acidophilus* and two strains of *A.thioxidans*	Al, Cd, Cr, Cu, Fe, Mn, Ni, Pb, Zn	To minimize cost and to use native microbes, co-cultures of *Acidithiobacilli* and anaerobic sewage sludge were used. The combination of bacteria and sludge required 50% less time to reduce the suspension pH to 1 compared to pure bacterial culture. The stimulation of acidification by anoxic sewage sludge was due to the presence of additional nutrients for microbial growth and because of the wide variety of indigenous bacteria. The time for metal solubilization was reduced 50%.
Ishigaki, Nakanishi, Tateda, Ike, & Fujita, 2005	Mixed culture: *Acidithiobacillus thiooxidans* and *Acidithiobacillus ferrooxidans*	As, Cd, Cr, Cu, Zn	IOB (Iron oxidizing bacteria/ *Acidithiobacillus ferroxidans*) showed higher metal leaching ability but poor tolerance to heavy metals than SOB (Sulphur oxidizing bacteria/ *Acidithiobacillus thiooxidans*). IOB used contact leaching/ redox mechanisms for metal leaching whereas SOB employed acid/lixiviant leaching.

Acidianus brierleyi:

A. brierleyi is an extremely acidophilic, thermophilic archaea. It is a chemolithotroph, and is unique in that it can both oxidize and reduce sulfur depending on the availability of oxygen. A facultative anaerobe which appears yellow-orange under aerobic condition and grayish-black under anaerobic condition, *A. brierleyi* has an irregular coccoid shape and is 1 to 1.5 μm in diameter. The organism is halotolerant and grows between 45 and 75°C and pH 1 to 6; optimum growth, however, occurs at 70°C and pH 1.5 to 2.0. Another unique feature of *A. brierleyi* is its ability to grow either heterotrophically on yeast extract or autotrophically on carbon dioxide. High temperature promotes growth of microbes as well as bioleaching. *A. brierleyi*, like *A. thioxidans*, oxidizes sulphur to form sulpuric acid extraneously. Microbial oxidation of sulphur is very fast compared to chemical oxidation. The sulphuric acid produced acts as the leaching agent (Barclay, 2010).

A. brierleyi reportedly extracted 90% zinc from municipal waste incinerator fly ash in 9 days in batch operation (shake flasks) at 65°C and pH 2 with a pulp density of 1 kg/m^3 and sulphur-liquid loading ratio of 5 kg/m^3 (Konishi, Matsui, Fujiwara, Nomura, & Nakahara, 2003). Leaching rate by *A.brierleyi* showed temperature dependence confirming that microbially-produced acids leached fly ash indirectly at higher temperatures. Sulphur was essential for the production of acids. The ash had particles smaller than 20 μm and 1.34% Zn,

0.568% Pb, 0.091% Cr, 0.068% Cd, 26.4% SiO_2. A comparison of chemical leaching (with dilute sulphuric acid) and bioleaching at an ash-liquid loading ratio of 0.2 kg/m^3 and sulphur-liquid loading ratio of 5 kg/m^3 showed that biogenic acid and metabolites were more effective than commercial acid. At pH 1.3, 22% of Zn was leached in 2.4 hours, after which no leaching occurred and at pH 2.5, no leaching was seen. However, at pH 1.3, 52% of Zn was leached in 6 days and 90% in 9 days by *A. brierleyi*, when all other conditions were similar. Bioleaching rate greatly depended on the amount of elemental sulphur added as solid substrate. Optimal sulphur-liquid loading ratio was 5 kg/m^3 beyond which sulphur adversely affected the rate. Ash-liquid loading ratio also showed significant impacts; adding ash increased pH of the solution rapidly due to release of components such as $Ca(OH)_2$ thus repressing bacterial growth and bioleaching. This was overcome by adding fly ash after the production of microbially produced acids Addition time of 2 days was found to be optimal. When fly ash was added after 2 days with an ash-liquid loading ratio of 5 kg/m^3, the amount of zinc leached exceeded 40% in 9 days. For ash-liquid loading ratio of 1 kg/m^3 and addition time 2 days, the zinc leached was 90% in 9 days after addition of fly ash.

Yarrowia lipolytica:

Yarrowia lipolytica is a hemiascomycetous dimorphic fungus of family Dipodascaceae (Coelho, Amaral, & Belo, 2010; EMBL-EBI, n.d.). The organism is generally employed to produce methane from lipids (EMBL-EBI, n.d.). *Y. lipolytica* is an aerobic and mesophilic yeast, and grows in the pH range of 2 to 6 (Bankar et al., 2012; Coelho, et al., 2010). The yeast has an average length of 6 μm (Bankar, et al., 2012), and assimilates hydrocarbons and produces citric acid from n-alkanes, vegetable oils or glucose under aerobic conditions (EMBL-EBI, n.d.). Citric acid is produced via tricarboxylic acid cycle similar to *A. niger*. The only advantage is its substrate flexibility. The organism can oxidize fatty acids to acetyl-coA, which is then converted to citrate via TCA cycle (Nelson, & Cox, 2005). The ability of the organism to utilize hydrocarbons more efficiently than A.niger renders it of industrial interest (Coelho, et al., 2010). *Y. lipolytica* has been used extensively in a wide range of industrial as well as research applications, and serves as a model organism for genetic engineering research (Beckerich, Boisramé-Baudevin, & Gaillardin, 2010). Industrially, it is used for the production of lipase, organic acids, single cell oils etc. It is also used for environmental remediation such as the degradation of n-paraffins and oils (Coelho, et al., 2010).

A recent study on the bioleaching of fly ash by *Y. lipolytica* at 8% pulp density showed that the main mechanism of leaching was via the production of citric acid, and that fly ash had no inhibitory effect on citric acid production (Bankar, et al., 2012). Furthermore, the organism did not produce a different metabolite in the presence of fly ash, unlike *A.niger.* Apart from citric acid production, two extracellular proteins were found to be excreted exclusively in the presence of fly ash and were suspected to take part in the bioleaching. Accumulation of metal ions at the cell wall, cell membrane and cytoplasm was also observed. The yeast cells were found to leach Cu more effectively than commercial citric acid. The study also reported the attachment of *Y .lipolytica* cells to fly ash, due to be the hydrophobic nature of fly ash and the cells. An increase in average cell length, cell volume and surface area of the yeast cells, and a decrease in surface-to-volume ratio were observed with increasing concentrations of fly ash.

6.6. Comparison of Two-Step Leaching Efficiencies of Reported Organisms

In two-step bioleaching, the organisms are grown in media, and fly ash is added only after the organisms have attained exuberant growth or produced the necessary metabolites for bioleaching. In one-step leaching where the organism and the fly ash are added at the same time, the organism has to withstand the pH and metal toxicity posed by fly ash. Not unexpectedly, two-step bioleaching shows better leaching efficiency than one-step bioleaching. Table 5 compares the leaching efficiencies reported in two-step leaching at 1% pulp density of fly ash with different organisms.

Table 5. Comparison of two-step leaching efficiencies of reported organisms

Reference	Two-step leaching with 1% fly ash pulp density								
	Organism	Al	Cu	Pb	Mn	Zn	Fe	Cd	Cr
(Wu & Ting, 2006)	*Aspergillus niger*	Almost 100	60-70	55-70	Almost 100	Almost 100	30-40	-	-
(Yang, Wang, Wang, & Wu, 2008)	*Aspergillus niger*	-	65	negligible	70	40	25	20	20
(Ishigaki, et al., 2005)	*Acidithiobacillus thioxidans*	-	70	-	-	60	-	100	-
	Acidithiobacillus ferroxidans	-	77	-	-	100	-	100	100
	Mixed culture	-	67	-	-	78	-	100	100
(Konishi, et al., 2003)	*Acidianus brierleyi* (0.5% pulp density)	-	-	-	-	40	-	-	-

Note: These values though reported in literature, may vary based on the strain and the bioleaching conditions.

Table 5 shows that *A. thiooxidans* is more effective in leaching Cd, Cr and Cu. The recoveries of Al, Mn, Pb and Fe cannot be compared due to insufficient data. Variation in bioleaching efficiency exhibited by different strains of the same species under different experimental conditions is shown in the first two references in the table which showed different extraction efficiencies for Pb, Mn and Zn by *A.niger*. *Y. lipolytica* is not included in the comparison due to unavailability of data at 1% pulp density. The bioleaching of a particular metal from a particular waste is dependent not only on the organism but several other factors. The next section briefly discusses the various factors which influence the bioleaching process.

6.7. Factors Influencing Bioleaching

Bioleaching is dependent on operational conditions which are influenced by the leaching environment, and microbial activity (such as microbial growth and nutritional requirements) and properties of the waste (Figure 5). These may be classified as physiochemical, microbiological and mineralogical factors (Talvivaara Mining Company Plc, 010).

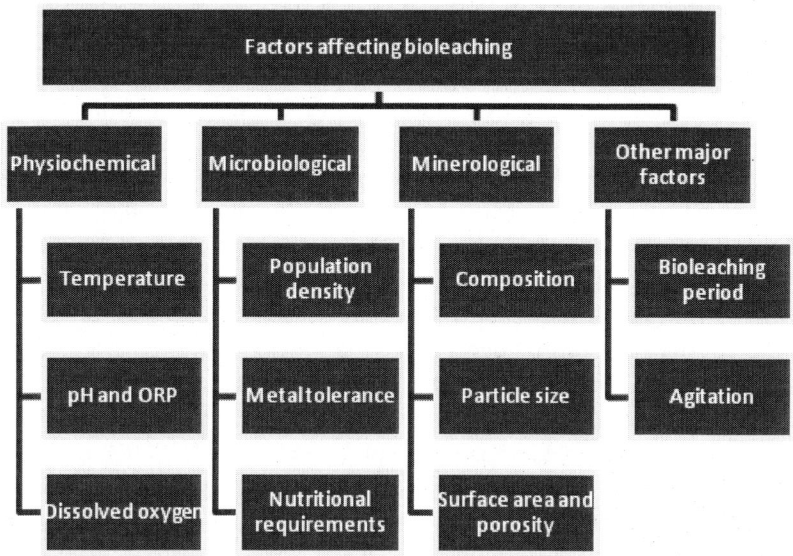

Figure 3. Factors affecting bioleaching (Chen & Lin, 2000; Das & Ayyappan, 1999; Talvivaara Mining Company Plc, 2010).

a. Effect of Physiochemical Factors on Bioleaching

Temperature is a very important factor in determining the growth of microorganisms. Depending on their temperature preference, microorganisms are classified as psychrophiles, mesophiles and thermophiles(Willey, Sherwood, & Woolverton, 2008). Bioleaching microorganisms such as *Acidithiobacillus sp.* and *A. niger* are mesophiles and usually grow optimally at 30°C (Astoreca, Magnoli, Ramirez, Combina, & Dalcero, 2007; Lin, Huang, & Huang, 2010) and bioleaching carried out at 30°C showed better results than at 37 or 28°C (Lin, et al., 2010). Optimal growth conditions alone do not determine the leaching efficiency (Vilcáez, Suto, & Inoue, 2008), since the interdependency of factors plays a critical role. While temperature control can be implemented for laboratory-scale bioleaching experiments, large scale bioleaching such as heap leaching suffers from fluctuations in ambient temperature and exothermic oxidation reactions (Sissing & Harrison, 2003).

pH of the solution varies throughout the bioleaching process due to acid-production and acid-consumption reactions, and has a significant impact on microbial growth and its activity. The pH range of growth of *Acidithiobacillus sp.* is most suitable for leaching operation and pH enhances solubility of most metals (White, Sayer, & Gadd, 1997). Since fly ash is alkaline, acidification of fly ash is necessary to support the growth of acidophiles (Acevedo, 2000). However, heterotrophic fungi can tolerate a wider range of pH and hence is employed for treating alkaline wastes (White, et al., 1997).

The oxidation-reduction potential is an important factor when the leaching agent is a chemolithoautotroph. The microbes need an ORP of +300mV for the production of Fe^{3+} ions. The ORP increases to +600mV during the leaching process (Bosecker, 1987) because of the ferric ion leached from the ash. Increasing the ORP facilitates growth of the microbes as well as improves bioleaching yield.

Oxygen and carbon dioxide supply are other important factors. Sufficient supply of oxygen is required to support the growth of aerobic bacteria, and inadequate supply disrupts the production of metabolites. Carbon dioxide is particularly important in the case of chemolithoautotrophic bacteria which use CO_2 as a carbon source for growth. O_2 is the final electron acceptor in all oxidation reactions and hence O_2 is also important to the chemolithoautotrophic organisms (Acevedo, 2000; Burgstaller & Schinner, 1993). In many large scale operations such as heap leaching, dump leaching and tank leaching, insufficient supply of O_2 and CO_2 is a common problem which inhibits the leaching process (Pradhan, Nathsarma, Srinivasa Rao, Sukla, & Mishra, 2008).

b. Effect of Microbiological Factors on Bioleaching

Microbiological factors greatly depend on the organism and the waste employed. Microbial growth depends on various physiochemical factors discussed above as well as the inoculum concentration, nutritional source, and absence of inhibitory substances in the media. Inoculum size determines the time required to achieve a certain population density; a denser microbial population will be achieved with a large inoculum size within a shorter time. However, a higher population density does not necessarily lead to higher leaching. Several studies have shown a decrease in metabolite production with increased inoculum size and hence reduced leaching efficiency (Crolla & Kennedy, 2001; Xu & Ting, 2004). Therefore, an optimal inoculum size should be determined.

The nutritional requirements of the organism must be satisfied for growth and metabolite production. Chemolithoautotrophic bacteria require CO_2 as the carbon source, and some trace amounts of salts as electron donors (Willey, et al., 2008). These trace salts are sometimes supplied by the mineral leached (Bosecker, 1987). Heterotrophic bacteria require an organic carbon source, a nitrogen source as well as salts (Willey, et al., 2008). The carbon source determines the production of metabolites.

For *A. niger*, sucrose as carbon source enhanced metabolite production when compared to fructose, lactose or galactose (Hossain, Brooks, & Maddox, 1984). In addition, some bacteria have special requirements. For instance *A. thiooxidans* and *A. ferrooxidans* require elemental sulphur and iron for growth (Rawlings, 2005). Fly ash does not usually promote growth of microorganisms in the absence of media because of the lack of sufficient nutrition.

Another major concern is the presence of various inhibitory substances in the waste to be treated. Heavy metal in the waste is detrimental to the growth of most microbes. High concentrations of metal ions inhibit bacterial growth by various mechanisms. For instance, functional groups of enzymes/biomolecules are blocked by metal ions. Membrane integrity and transport of molecules across the membrane are also disrupted by metal ions, and important metal ions in the biomolecules are substituted/displaced by non-essential metal ions in the waste etc. (Burgstaller & Schinner, 1993). Thus, microbes are generally incapable of handling higher pulp densities of the waste (Wu & Ting, 2006; Yang, et al., 2008). The concentration and range of heavy metals present in fly ash causes it to be a very poor medium for bacterial growth.

c. Effect of Mineralogical Factors on Bioleaching

The type of waste to be treated and its properties significantly affect bioleaching efficiency. As discussed earlier, chemical composition of fly ash plays an important role in defining the bioleaching yield. Particle size, surface area and porosity of the particles are other influencing parameters. Generally, smaller size particle have larger surface areas providing a larger contact area between the microbes/lixiviant and waste particles. Although small particle size is advantageous, very fine particles may damage the cell structure thereby inhibiting cellular activities (Nemati, Lowenadler, & Harrison, 2000). Therefore, the optimal particle size has to be determined. A higher porosity of the particles is advantageous. For non-porous particles, efforts have been taken to increase the porosity before bioleaching. As fly ash is generally very fine and non-porous, pre-treatment of fly ash by calcination would be preferable before bioleaching (Torma & Singh, 1993).

d. Effect of Bioleaching Duration on Bioleaching

Bioleaching is performed for a sufficiently long time in order to obtain a good percentage recovery. The duration of bioleaching greatly depends on the microorganism employed for bioleaching. Bacteria such *Acidithiobacilli* are slow-growing and have a long lag phase before the commencement of an active growth phase. This is further extended when the bacteria are exposed to stress such as high metal concentrations in the waste (Bosshard, et al., 1996). Therefore, providing a sufficiently long period for microbial growth, metabolite production and bioleaching is important.

However, a longer leaching period does not always lead to a higher recovery. There have been instances where decrease in bioleaching yield of certain metals occurs with increase in duration of bioleaching. This could be due either to precipitation of metals over time or uptake of metals by the microorganisms (Xu & Ting, 2009; Yang, et al., 2009). Hence, an optimal period for bioleaching has to be established.

e. Effect of Agitation on Bioleaching

Agitation increases the rate of gas transfer, heat transfer and nutrient transfer within the system. In the absence of proper agitation, dead zones may be formed within the system wherein nutrient supply is limited and insufficient for bacterial growth. In addition, as bioleaching is an exothermal process, proper agitation may help in maintaining a constant temperature within the system. The supply of gases such as oxygen in the case of aerobes and CO_2 in the case of autotrophic bacteria is aided by agitation (Acevedo, 2000).

Despite the advantages of agitation, very strong agitation induces shear stress on the cells which results in cell disruption, or damage to cell envelope, and interferes with mechanisms such as electron transfer in bioleaching, thereby reducing bioleaching recovery (Loi, Rossi, Trois, & Rossi, 2006). Adverse effects on fungal growth and activity have also been reported (Bailey, & Ollis, 1986). Hence, an optimal agitation speed has to be determined.

6.8. Comparison of Certain Characteristics of Reported Organisms

From the preceding discussion, it is evident that bioleaching efficiency is highly dependent on certain characteristics of the microorganisms. pH and fly ash tolerance are most important amongst them, and are particularly critical when the process needs to be extrapolated to large scale and with high pulp density. pH of alkaline fly ash is usually around 11-13 and use of acidophilic organisms requires considerable acidification of fly ash. Thus, an organism with good pH and fly ash tolerance is sought. Temperature tolerance, although not critical, may be important during heap leaching where the temperature in the interior of the heap rises during the process of bioleaching. This is because bioleaching is an exothermic process. However, thermophilic organisms are usually not preferred due to their effect on carbon footprint. Organisms with extensive nutritional requirements increase the cost of the process and hence are not favoured. Table 6 compares the fly ash, pH and temperature tolerance as well as nutrient requirements of the most commonly reported bioleaching organisms.

Compared to acidophiles, heterotrophic fungi tolerate a wider range of pH and hence are employed for treating moderately alkaline wastes. As fly ash is extremely alkaline with a pH range varying from 11-13, alkali-tolerant or alkaliphilic bacteria are preferred in order to minimize the need for acidification. In addition, all the microbes reported for fly ash bioleaching tolerate no more than 10% (w/v) of fly ash. Clearly microorganisms with greater metal tolerance are required for bioleaching in large scale processes (Ramanathan & Ting, 2013).

Table 6. Comparison of pH and metal tolerance of reported organisms

Organism	Fly ash tolerance	pH tolerance	Temperature tolerance	Nutrient requirements
Acidithiobacillus thiooxidans	Maximum reported value 10% (Seidel, et al., 2001)	0.5-6 (Krebs, et al., 1997)	10-37°C (Krebs, et al., 1997)	Chemolitho-autotrophic (Krebs, et al., 1997)
Acidithiobacillus ferrooxidans	Less than 10% (Ishigaki, et al., 2005)	1.4-6 (Krebs, et al., 1997)	28-35°C (Krebs, et al., 1997)	Chemolitho-autotrophic (Krebs, et al., 1997)
Aspergillus niger	Maximum reported value 10% (Bosshard, et al., 1996)	2-8(Andersen, Lehmann, & Nielsen, 2009)	Mesophilic(Clinical Microbiology Proficiency Testing, 2008)	Heterotrophic (Krebs, et al., 1997)
Acidianus brierleyi	Maximum reported value 0.5% (Konishi, et al., 2003)	1-6(Barclay, 2010)	45-75°C (Krebs, et al., 1997)	Facultative heterotroph (Krebs, et al., 1997)
Yarrowia lipolytica	Maximum reported value 8% (Bankar, et al., 2012)	2-6(Bankar, et al., 2012)	Mesophilic(Bankar, et al., 2012)	Heterotrophic(Coelho, et al., 2010)

6.9. Challenges in Bioleaching

Although there has been considerable research in bioleaching of fly ash, the technology still remains at its infancy. Indeed, even though bioleaching of mineral ores has been applied commercially in various parts of the world, many challenges remain which also apply to fly ash bioleaching.

The main challenge in bioleaching is the long leaching time required to recover the same amount of metals as compared to chemical leaching, since the lixiviants are biogenically produced at lower concentration. Slow reaction rate and poor process efficiency pose significant challenges to commercialization. Several studies are currently focused on overcoming these limitations. To hasten the process, techniques such as catalyst-driven bioleaching (Zeng, Deng, Luo, Luo, & Zou, 2012), electrochemical bioleaching (Yong, 2011) and sonobioleaching (Swamy, Narayana, & Misra, 2005) are being investigated.

The microorganisms employed in most of the cases are unable to tolerate the high concentration of toxic metals thereby making operation at high pulp densities extremely difficult. Higher efficiency can be obtained with more tolerant strains of bacteria. Indigenous microbial population isolated from heaps and dump sites have been examined extensively in order to identify waste-resistant strains and better performers. Further studies need to be done on using genetically modified strains or strains selected by mutation in order to accelerate and enhance the efficiency of the bioleaching process.

Laboratory scale and pilot scale studies which simulate commercial scale scenarios are carried out to obtain information on problems faced in large scale systems. Column bioleaching which mimics large scale heap bioleaching has been studied extensively in order to determine the concerns with commercial scale leaching, with aeration and temperature control being the major issues. To overcome the problem of aeration and high temperature at the interior of the heaps, the combined use of aerobes and anaerobes as well as mesophilic and thermophilic bacteria, and the efficient supply of air and water to create a conducive growth environment for the microbes have been attempted. Hydrodynamic problems are being solved by regulating the distribution of media and inoculum in the heap. Despite the efforts, agglomeration and compaction of waste particles remain as a hindrance to scaling up (Brierley, 2008).

6.10. Future of Bioleaching of Fly Ash

Bioleaching has been referred to as the "mining of the future". It is simple in operation, is more economical and has low environmental impact, compared to traditional mining techniques. Release of CO_2 and toxic chemicals are much lower in bioleaching compared to conventional processes (Moskvitch, 2012). Several commercial plants have been set up over the last two decades on the mining of ores using microbes. As has been rightly pointed out a decade ago, "Today bioleaching is no longer a promising technology but an actual economical alternative for treating specific mineral ores" (Acevedo, 2002).

With fly ash considered as a secondary ore, owing to its high concentration of heavy metals, bioleaching of fly ash on an industrial scale may be applied in the near future, possibly as an alternative, or adjunct to current (chemical) technologies. Better bioleaching techniques are expected to evolve and result in more optimal controlled operating conditions and higher metal removal and recoveries. Indeed, over time, the enforcement of more stringent legislation to limit environmental pollution would make bioleaching of fly ash even more attractive. With increasing concern over the impact of fly ash disposal on the environment as the driving force, as well as recognizing the need for resource conservation in the modern era of environmental sustainability, industry may consider adopting bioleaching of fly ash for detoxification and resource recovery.

REFERENCES

ACAA. (2011). Coal Ash Recycling Rates Remain Stalled As Regulatory Uncertainty Continues. Retrieved 18-09-2012, from http://www.acaa-usa.org/associations/8003/files/News_Release- Coal_Ash_Production_and_Use_2011.PDF

Acevedo, F. (2002). Present and future of bioleaching in developing countries. *Electronic journal of biotechnology, 5*(2), 18-19.

Acevedo, F. (2000). The use of reactors in biomining processes. *Electronic journal of biotechnology, 3*(3), 10-11.

Acharya, S.K., Mishra, P., & Mishra, S.C. (2008). Effect of environment on the mechanical properties of fly ash-jute-polymer composite. *Indian Journal of Engineering & Materials Sciences, 15*, 483-488.

Al-Zboon, K., Al-Harahsheh, M.S., & Hani, F.B. (2011). Fly ash-based geopolymer for Pb removal from aqueous solution. *J. Hazard Mater, 188*(1-3), 414-421. doi: 10.1016/j.jhazmat.2011.01.133

Albertini, A.V.P., Silva, J.L., Freire, V.N., Santos, R.P., Martins, J.L., Cavada, B.S., . . . Martínez, C.R. (2012a). Immobilized invertase studies on glass–ceramic support from coal fly aches. *Chemical Engineering Journal.*

Albertini, A.V.P., Silva, J.L., Freire, V.N., Santos, R.P., Martins, J.L., Cavada, B.S., . . . Martínez, C.R. (2012b). Immobilized invertase studies on glass–ceramic support from coal fly ashes. *Chemical Engineering Journal, 214*, 91-96.

AMPREP Metal Recovery Systems. (n.d.). Precious Metal Leaching - General Information Retrieved 12/4/2013, from http://www.americhem.biz/_amprep/leaching.htm

Andersen, M.R., Lehmann, L., & Nielsen, J. (2009). Systemic analysis of the response of Aspergillus niger to ambient pH. *Genome Biol., 10*(5), R47.

Astoreca, A., Magnoli, C., Ramirez, M.L., Combina, M., & Dalcero, A. (2007). Water activity and temperature effects on growth of *Aspergillus niger, A. awamori and A. carbonarius* isolated from different substrates in Argentina. *International journal of food microbiology, 119*(3), 314-318.

Auer, S., Kuzel, H-J., Pöllmann, H., & Sorrentino, F. (1995). Investigation on MSW fly ash treatment by reactive calcium aluminates and phases formed. *Cement and Concrete Research, 25*(6), 1347-1359.

Babajide, O., Petrik, L., Musyoka, N., Amigun, B., & Ameer, F. (2010). Use of coal fly ash as a catalyst in the production of biodiesel. *Petroleum & Coal, 52*(4), 261-272.

Bailey, J.E., & Ollis, D.F. (1986). Biochemical engineering fundamentals.

Bankar, A., Winey, M., Prakash, D., Kumar, A.R., Gosavi, S., Kapadnis, B., & Zinjarde, S. (2012). Bioleaching of fly ash by the tropical marine yeast, Yarrowia lipolytica NCIM 3589. *Applied biochemistry and biotechnology, 168*(8), 2205-2217.

Barbieri, L., Lancellotti, I., Manfredini, T., Queralt, I., Rincon, J., & Romero, M.. (1999). Design, obtainment and properties of glasses and glass–ceramics from coal fly ash. *Fuel, 78*(2), 271-276.

Barclay, J. (2010). *Acidianus brierleyi. Microbes of the Week* Retrieved 14-2-2013, from http://web.mst.edu/~microbio/BIO221_2010/A_brierleyi.html

Basu, M., Pande, M., Bhadoria, P.B.S, & Mahapatra, S.C. (2009). Potential fly-ash utilization in agriculture: A global review. *Progress in Natural Science, 19*(10), 1173-1186.

Beckerich, J.M., Boisramé-Baudevin, A., & Gaillardin, C. (2010). *Yarrowia lipolytica:* a model organism for protein secretion studies. *International Microbiology, 1*(2), 123-130.

Berg, J.V.D., and Feuerborn, H.J., (nd). Production and Utilisation of CCPs in Europe - Fly ash in cement and concrete Retrieved 16-2-2013, from http://www.cementdistribution.com/industryinfo/articles/production_and_utilization_ccp_europe.pdf

Bharadwaj, A., & Ting, Y.P. (2012). Bioleaching of spent hydrotreating catalyst by acidophilic thermophile *Acidianus brierleyi*: leaching mechanism and effect of decoking. *Bioresource technology, 130*, 673-680. doi: 10.1016/j.biortech.2012.12.047

Bhattacharyya, S., Donahoe, R.J, & Patel, D. (2009). Experimental study of chemical treatment of coal fly ash to reduce the mobility of priority trace elements. *Fuel, 88*(7), 1173-1184.

Bipp, H.P., Wunsch, P., Fischer, K., Bieniek, D., & Kettrup, A. (1998). Heavy metal leaching of fly ash from waste incineration with gluconic acid and a molasses hydrolysate. *Chemosphere, 36*(11), 2523-2533.

Bosecker, K. (1987). Microbial recycling of mineral waste products. *Acta biotechnologica, 7*(6), 487-497.

Bosecker, K. (1997). Bioleaching: metal solubilization by microorganisms. *FEMS Microbiology Reviews, 20*(3-4), 591-604.

Bosshard, P.P., Bachofen, R., & Brandl, H. (1996). Metal Leaching of Fly Ash from Municipal Waste Incineration by *Aspergillus niger. Environmental Science & Technology, 30*(10), 3066-3070. doi: 10.1021/es960151v

Bournonville, B., Nzihou, A., Sharrock, P., & Depelsenaire, G.. (2004). Stabilisation of heavy metal containing dusts by reaction with phosphoric acid: study of the reactivity of fly ash. *J Hazard Mater, 116*(1), 65-74.

Brandl, H. (2008). Microbial leaching of metals. *Biotechnology Set, Second Edition*, 191-224.

Brierley, C.L. (2008). How will biomining be applied in future? *Transactions of nonferrous metals society of China, 18*(6), 1302-1310.

Brierley, J.A., & Brierley, C.L. (2001). Present and future commercial applications of biohydrometallurgy. *Hydrometallurgy, 59*(2), 233-239.

Brombacher, C., Bachofen, R., & Brandl, H. (1998). Development of a Laboratory-Scale Leaching Plant for Metal Extraction from Fly Ash by Thiobacillus Strains. *Applied and environmental microbiology, 64*(4), 1237-1241.

Brombacher, C., Bachofen, R., & Brandl, H. (1997). Biohydrometallurgical processing of solids: a patent review. *Applied microbiology and biotechnology, 48*(5), 577-587.

Burgess, L.. (2008). A Copper Mining Company To Watch, A Big Surprise in the Copper Market. *Wealth daily* Retrieved 15-2-2013, from http://www.wealthdaily.com/articles/copper-mining-company/2059

Burgstaller, W., & Schinner, F. (1993). Leaching of metals with fungi. *Journal of biotechnology, 27*(2), 91-116.

Campbell, S.C., Olson, G.J., Clark, T.R., & McFeters, G. (2001). Biogenic production of cyanide and its application to gold recovery. *Journal of Industrial Microbiology and Biotechnology, 26*(3), 134-139.

Cao, J., & Chung, D.D.L. (2004). Use of fly ash as an admixture for electromagnetic interference shielding. *Cement and concrete research, 34*(10), 1889-1892.

Chan, C.C., Kirk, D.W., & Marsh, H. (2000). The behaviour of Al in MSW incinerator fly ash during thermal treatment. *J Hazard Mater, 76*(1), 103-111.

Chen, S.Y., & Lin, J.G. (2000). Factors affecting bioleaching of metal contaminated sediment with sulfur-oxidizing bacteria. *Water science and technology, 41(12)*, 263-270.

Chen, X., Lu, A., & Qu, G. (2012). Preparation and characterization of foam ceramics from Red mud and Fly ash using sodium silicate as foaming agents. *Ceramics International, 39(2)*, 1923-1929

Christopher-Chia, K.C., Lee, K.T., Fernando, W., James, N., Bhatia, S., & Mohamed, A.R. (2005). Modeling and simulation of flue gas desulfurization using $CaO/CaSO_4$/coal fly ash sorbent. *Journal of chemical engineering of Japan, 38*(6), 391-396.

Clapp, T.L., Magee II, J.F., Ahlert, R.C., & Kosson, D.S. (1988). Municipal solid waste composition and the behavior of metals in incinerator ashes. *Environmental progress, 7*(1), 22-30.

Clinical Microbiology Proficiency Testing. (2008). *Aspergillus niger. Mycology plus* Retrieved 18-4-2013, from http://www.cmpt.ca/pdf_mycology_2008/my_plus_0809_3_aniger.pdf

Coelho, M.A.Z., Amaral, P.F.F., & Belo, I. (2010). *Yarrowia lipolytica*: an industrial workhorse. In Méndez-Vilas (Ed.), *Current Research, Technology and Education Topics in Applied Microbiology and Microbial Biotechnology* (pp. 930–944): Formatex.

Coloradoans Against Resource Destruction(CARD). (n.d.). In situ leaching impacts Retrieved 18-2-2013, from http://www.nunnglow.com/impacts/in-situ-leaching-isl-impacts.html

Committee on Health Effects of Waste Incineration, Board on Environmental Studies and Toxicology, National Research Council (2000). Waste Incineration and Public Health Retrieved from http://www.nap.edu/catalog.php?record_id=5803

Conner, J.R. (1990). *Chemical fixation and solidification of hazardous wastes*. New York: Van Nostrand Reinhold

Crolla, A., & Kennedy, K.J. (2001). Optimization of citric acid production from *Candida lipolytica Y-1095* using *n*-paraffin. *Journal of Biotechnology, 89*(1), 27-40.

Cui, J., & Zhang, L. (2008). Metallurgical recovery of metals from electronic waste: a review. [Research Support, Non-U.S. Gov't Review]. *J Hazard Mater, 158*(2-3), 228-256. doi: 10.1016/j.jhazmat.2008.02.001.

Das, T., & Ayyappan, S. (1999). Factors affecting bioleaching kinetics of sulfide ores using acidophilic micro-organisms. *BioMetals, 12*(1), 1-10.

Dunens, O.M., MacKenzie, K.J., & Harris, A.T. (2009). Synthesis of multiwalled carbon nanotubes on fly ash derived catalysts. *Environmental science & technology, 43*(20), 7889-7894.

Ehrlich, H.L. (1992). Metal extraction and ore discovery. *Encyclopedia of microbiology, 3*, 75-80.

Ehrlich, H.L. (1997). Microbes and metals. *Applied Microbiology and Biotechnology, 48*(6), 687-692.

Electric Power Research Institute. (2009). Is coal fly ash toxic? Retrieved 17-2-2013, from http://www.ameren.com/Environment/Documents/EPRI-IsCoalAshToxic.pdf

EMBL-EBI. (n.d.). Yarrowia lipolytica is a "non-conventional" species of yeast, often used in genetic research because it differs from other well-studied species Retrieved 13-2-2013, from http://www.ebi.ac.uk/2can/genomes/eukaryotes/Yarrowia_lipolytica.html

Erol, M., Genç, A., Öveçoğlu, M.L., Yücelen, E., Küçükbayrak, S., & Taptık, Y. (2000). Characterization of a glass-ceramic produced from thermal power plant fly ashes. *Journal of the European Ceramic Society, 20*(12), 2209-2214.

Espinosa, D.C.R., Bernardes, A.M., & Tenório, J.A.S. (2004). An overview on the current processes for the recycling of batteries. *Journal of Power Sources, 135*(1), 311-319.

Fass, R., Geva, J., Shalita, Z.P., White, M.D., & Fleming, J.C. (1994). Bioleaching method for the extraction of metals from coal fly ash using Thiobacillus: The Israel Electric Corporation Ltd.

Fedje, K.K., Ekberg, C., Skarnemark, G., & Steenari, B.M. (2010). Removal of hazardous metals from MSW fly ash—An evaluation of ash leaching methods. *J. Hazard Mater, 173*(1), 310-317.

Ferreira, C., Jensen, P., Ottosen, L., & Ribeiro, A. (2005). Removal of selected heavy metals from MSW fly ash by the electrodialytic process. *Engineering Geology, 77*(3), 339-347.

Fouad, O.A., Abdel Halim, K.S., & Rashad, M.M. (2008). Catalytic Oxidation of CO Over Synthesized Nickel Ferrite Nanoparticles from Fly Ash. *Topics in Catalysis, 47*(1), 61-65.

Fukui, K., Arimitsu, N., Kidoguchi, S., Yamamoto, T., & Yoshida, H. (2009). Synthesis of calcium phosphate hydrogel from waste incineration fly ash and bone powder. [Research Support, Non-U.S. Gov't]. *J Hazard Mater, 163*(1), 391-395. doi: 10.1016/j.jhazmat.2008.06.103.

Fukui, K., Arimitsu, N., Jikihara, K., Yamamoto, T., & Yoshida, H.. (2009). Performance of fuel cell using calcium phosphate hydrogel membrane prepared from waste incineration fly ash and chicken bone powder. *J Hazard Mater, 168*(2), 1617-1621.

Gao, G.M., Zou, H.F., Gan, S.C., Liu, Z.J., An, B.C., Xu, J.J., & Li, G.H. (2009). Preparation and properties of silica nanoparticles from oil shale ash. *Powder technology, 191*(1), 47-51.

Gottlieb, B., Gilbert, S.G., Evans, L.G., (2010). Coal ash: A toxic threat to our health and environment Retrieved 12 2 2013, from http://www.psr.org/assets/pdfs/coal-ash.pdf

Harris, W.R., & David, S. (1988). Leaching of metal ions from fly ash by canine serum. *Environmental Science & Technology, 22*(1), 109-112.

Hartman, H.L. (1992). *SME mining engineering handbook* (Vol. 1): Society for Mining Metallurgy.

Headwaters resources. (2008). Fly ash - Types and Benefits Retrieved 12-2-2013, from http://www.flyash.com/data/upimages/press/tb.1%20fly%20ash%20-%20types%20&%20benefits.pdf

Hong, K.J., Tokunaga, S., Ishigami, Y., & Kajiuchi, T. (2000). Extraction of heavy metals from MSW incinerator fly ash using saponins. *Chemosphere, 41*(3), 345-352.

Hong, K.J., Tokunaga, S., & Kajiuchi, T. (2000). Extraction of heavy metals from MSW incinerator fly ashes by chelating agents. *J Hazard Mater, 75*(1), 57-73.

Hossain, M., Brooks, J.D., & Maddox, I.S. (1984). The effect of the sugar source on citric acid production by *Aspergillus niger. Applied Microbiology and Biotechnology, 19*(6), 393-397.

Hu, H., Luo, G., Liu, H., Qiao, Y., Xu, M., & Yao, H. (2012). Fate of chromium during thermal treatment of municipal solid waste incineration (MSWI) fly ash. *Proceedings of the Combustion Institute*.

Hvistendahl, M. (2007). Coal ash is more radioactive than nuclear waste. *Scientific American* Retrieved 17-2-2013, from http://www.scientificamerican.com/article.cfm?id=coal-ash-is-more-radioactive-than-nuclear-waste

INNOVAT Limited. (n.d.). Continuous vat leaching -Gold/silver leaching Retrieved 12/4/2013, from http://66.165.158.187/innovat/PDF%20Resources%5CGoldLeach.pdf

Ishigaki, T., Nakanishi, A., Tateda, M., Ike, M., & Fujita, M. (2005). Bioleaching of metal from municipal waste incineration fly ash using a mixed culture of sulfur-oxidizing and iron-oxidizing bacteria. [Research Support, Non-U.S. Gov't]. *Chemosphere, 60*(8), 1087-1094. doi: 10.1016/j.chemosphere.2004.12.060.

Islam, M., & Ting, Y.P. (2009). Fungal Bioleaching of Spent Hydroprocessing Catalyst: Effect of Decoking and Particle Size. *Advanced Materials Research, 71*, 665-668.

Izidoro, J.C, Fungaro, D.A, Abbott, J.E, & Wang, S. (2012). Synthesis of zeolites X and A from fly ashes for cadmium and zinc removal from aqueous solutions in single and binary ion systems. *Fuel, 103*, 827-834.

Jala, S., & Goyal, D. (2006). Fly ash as a soil ameliorant for improving crop production—a review. *Bioresource technology, 97*(9), 1136-1147.

Kappes, D.W. (2002). Precious metal heap leach design and practice. *Proceedings of the Mineral Processing Plant Design, Practice, and Control, 1*, 1606-1630.

Karim, Z., & Husain, Q. (2010). Application of fly ash adsorbed peroxidase for the removal of bisphenol A in batch process and continuous reactor: Assessment of genotoxicity of its product. *Food and Chemical Toxicology, 48*(12), 3385-3390.

Karim, Z., & Husain, Q. (2011a). Immobilization of peroxidase on a cost free support from salt fractionated proteins of bitter gourd. *Journal of Scientific and Industrial Research, 70(05)*, 352-359.

Karim, Z, & Husain, Q. (2011b). Immobilization of peroxidase on a cost free support from salt fractionated proteins of bitter gourd.

Kelly, D.P., & Wood, A.P. (2000). Reclassification of some species of Thiobacillus to the newly designated genera Acidithiobacillus gen. nov., Halothiobacillus gen. nov. and Thermithiobacillus gen. nov. *International Journal of Systematic and Evolutionary Microbiology, 50*(2), 511-516.

Khatri, C., Mishra, M.K. (2011). *Fly ash supported heteropolyacid as green and reusable solid acid catalyst for oxidation of alcohols*. Paper presented at the World of Coal Ash, Denver, CO, USA, May 2011.

Kinzler, K., Gehrke, T., Telegdi, J., & Sand, W. (2003). Bioleaching—a result of interfacial processes caused by extracellular polymeric substances (EPS). *Hydrometallurgy, 71*(1), 83-88.

Kirk, D.W., Chan, C.C.Y., & Marsh, H. (2002). Chromium behavior during thermal treatment of MSW fly ash. *J Hazard Mater, 90*(1), 39-49.

Konishi, Y., Matsui, M., Fujiwara, H., Nomura, T., & Nakahara, K. (2003). Zinc Leaching from Fly Ash in Municipal Waste Incineration by Thermophilic Archaean *Acidianus brierleyi* Growing on Elemental Sulfur. *Separation Science and Technology, 38*(16), 4117-4130. doi: 10.1081/ss-120024722.

Krebs, W., Bachofen, R., & Brandl, H. (2001). Growth stimulation of sulfur oxidizing bacteria for optimization of metal leaching efficiency of fly ash from municipal solid waste incineration. *Hydrometallurgy, 59*(2-3), 283-290.

Krebs, W., Brombacher, C., Bosshard, P.P., Bachofen, R., & Brandl, H. (1997). Microbial recovery of metals from solids. *FEMS Microbiology Reviews, 20*(3-4), 605-617. doi: 10.1016/s0168-6445(97)00037-5

Kuchar, D., Fukuta, T., Onyango, M.S., & Matsuda, H. (2007). Sulfidation treatment of molten incineration fly ashes with Na_2S for zinc, lead and copper resource recovery. *Chemosphere, 67*(8), 1518-1525.

Kwan, A.K.H., & Chen, J.J. (2012). Adding fly ash microsphere to improve packing density, flowability and strength of cement paste. *Powder technology, 234,* 19-25.

Leach, Inc. (2007). Column Leach Test Procedure - Copper Ores Retrieved 12-4-2013, from http://www.leachinc.net/Pages/default.aspx

Li, Q., Xu, H., Li, F., Li, P., Shen, L. & Zhai, J. (2012). Synthesis of geopolymer composites from blends of CFBC fly and bottom ashes. *Fuel, 97,* 366-372. doi: 10.1016/j.fuel.2012.02.059

Lin, H., Huang, M., & Huang, H.. (2010). Effect of Temperature on Bioleaching Heavy Metals from Sewage Sludge. Paper presented at the 4th International Conference on Bioinformatics and Biomedical Engineering (iCBBE), 2010

Liu, B., & Zhang, L. (2011). Experimental study on wastewater treatment by fly ash coagulant. Paper presented at the International Conference on Electric Technology and Civil Engineering (ICETCE), 2011.

Llorens, J.F., Fernandez-Turiel, J.L., & Querol, X. (2001). The fate of trace elements in a large coal-fired power plant. *Environmental Geology, 40*(4), 409-416.

Loi, G., Rossi, A., Trois, P., & Rossi, G. (2006). Continuous revolving barrel bioreactor tailored to the bioleaching microorganisms. *Minerals and Metallurgical Processing, 23*(4), 196-202.

Mackie, D., Trask, F. (n.d.). Continuous vat leaching- First copper pilot trials Retrieved 26/2/2013, from http://66.165.158.187/innovat/PDF%20Resources/ALTA2009.pdf

Madzivire, G., Gitari, W.M., Vadapalli, V.R.K, Ojumu, T.V., & Petrik, L.F. (2011). Fate of sulphate removed during the treatment of circumneutral mine water and acid mine drainage with coal fly ash: Modelling and experimental approach. *Minerals Engineering, 24*(13), 1467-1477.

Mangialardi, T. (2003). Disposal of MSWI fly ash through a combined washing-immobilisation process. *J. Hazard Mater, 98*(1), 225-240.

Medina, A., Gamero, P., Almanza, J.M., Vargas, A., Montoya, A., Vargas, G., & Izquierdo, M. (2010). Fly ash from a Mexican mineral coal. II. Source of W zeolite and its effectiveness in arsenic (V) adsorption. *J. Hazard Mater, 181*(1), 91-104.

Montague, P. (1990). Incineration ash - Part 2: All wastes must go somewhere forever. *Rachel's Environment and health weekly* Retrieved 12-2-2013, from http://www.ejnet.org/rachel/rhwn190.htm

Moskvitch, K. (2012). Biomining: How microbes help to mine copper Retrieved 3-5-2013, from http://www.bbc.co.uk/news/technology-17406375.

Na, J.G., Jeong, B.H., Chung, S.H., & Kim, S.S. (2006). Pyrolysis of low-density polyethylene using synthetic catalysts produced from fly ash. *Journal of Material Cycles and Waste Management, 8*(2), 126-132.

Nagib, S., & Inoue, K. (2000). Recovery of lead and zinc from fly ash generated from municipal incineration plants by means of acid and/or alkaline leaching. *Hydrometallurgy, 56*(3), 269-292.

Nath, D.C.D., & Sahajwalla, V.. (2011). Application of fly ash as a catalyst for synthesis of carbon nanotube ribbons. *Journal of hazardous materials, 192*(2), 691-697.

Nelson, D.L., Cox, M.M. (2005). *Lehninger Principles of Biochemistry* (4 ed.): W.H.Freeman and Company.

Nemati, M., Lowenadler, J., & Harrison, S.T.L. (2000). Particle size effects in bioleaching of pyrite by acidophilic thermophile Sulfolobus metallicus (BC). *Applied Microbiology and Biotechnology, 53*(2), 173-179.

Nowak, B., Rocha, S.F, Aschenbrenner, P., Rechberger, H., & Winter, F. (2011). Heavy metal removal from MSW fly ash by means of chlorination and thermal treatment: Influence of the chloride type. *Chemical Engineering Journal, 179*(1), 178-185.

Nowak, B., Pessl, A., Aschenbrenner, P., Szentannai, P., Mattenberger, H., Rechberger, H., Winter, F. (2010). Heavy metal removal from municipal solid waste fly ash by chlorination and thermal treatment. *J Hazard Mater, 179*(1), 323-331.

Nzihou, A., & Sharrock, P. (2002). Calcium phosphate stabilization of fly ash with chloride extraction. *Waste Management, 22*(2), 235-239.

Ogundiran, M.B., Nugteren, H.W., & Witkamp, G.J. (2012). Immobilisation of lead smelting slag within spent aluminate-fly ash based geopolymers. *J Hazard Mater, 248-249C*, 29-36. doi: 10.1016/j.jhazmat.2012.12.040.

Pandey, V.C., & Singh, N. (2010). Impact of fly ash incorporation in soil systems. *Agriculture, Ecosystems & Environment, 136*(1), 16-27.

Papandreou, A., Stournaras, C.J., & Panias, D. (2007). Copper and cadmium adsorption on pellets made from fired coal fly ash. *J Hazard Mater, 148*(3), 538-547.

Parvaiz, M.R., Mohanty, S., Nayak, S.K., & Mahanwar, P.A. (2010). Polyetheretherketone (PEEK) Composites Reinforced with Fly Ash and Mica. *Journal of Minerals & Materials Characterization & Engineering, 9*(1), 25-41.

Peninsula Energy Limited. (nd). What is in situ leach or in situ recovery? Retrieved 22-2-2013, from http://www.pel.net.au/investor_info/uranium_outlook/what_is_in_situ_leach_or_in_situ_recovery_.phtml

Phair, J.W., Van Deventer, J.S.J., & Smith, J.D. (2004). Effect of Al source and alkali activation on Pb and Cu immobilisation in fly-ash based "geopolymers". *Applied geochemistry, 19*(3), 423-434.

Plumb, J.J., Muddle, R., & Franzmann, P.D. (2008). Effect of pH on rates of iron and sulfur oxidation by bioleaching organisms. *Minerals Engineering, 21*(1), 76-82.

Polowczyk, I., Ulatowska, J., Koźlecki, T., Bastrzyk, A., & Sawiński, W. (2012). Studies on removal of boron from aqueous solution by fly ash agglomerates. *Desalination, 310(1)*, 93-101.

Pradhan, N., Nathsarma, K.C., Srinivasa Rao, K., Sukla, L.B., & Mishra, B.K. (2008). Heap bioleaching of chalcopyrite: a review. *Minerals Engineering, 21*(5), 355-365.

Praharaj, T., Powell, M.A., Hart, B.R., & Tripathy, S. (2002). Leachability of elements from sub-bituminous coal fly ash from India. *Environment International, 27*(8), 609-615.

Qian, G., Zhang, H., Zhang, X., & Chui, P.C. (2005). Modification of MSW fly ash by anionic chelating surfactant. *J Hazard Mater, 121*(1), 251-258.

Querol, X., Moreno, N., Umana, J.C., Alastuey, A., Hernández, E., López-Soler, A., & Plana, F. (2002). Synthesis of zeolites from coal fly ash: an overview. *International Journal of Coal Geology, 50*(1), 413-423.

Ramanathan, T., & Ting, Y.P. (2013). Isolation and characterization of an alkaliphilic and metal-tolerant microorganism from a landfill site for fly ash bioleaching. Paper presented at the International Conference on Environmental and Hazardous Substance Management, Bangkok, Thailand, 21-23 May 2013.

Rawlings, D.E. (2005). Characteristics and adaptability of iron-and sulfur-oxidizing microorganisms used for the recovery of metals from minerals and their concentrates. *Microbial cell factories, 4*(1), 13.

Remonsellez, F., Galleguillos, F., Moreno-Paz, M., Parro, V., Acosta, M., & Demergasso, C. (2009). Dynamic of active microorganisms inhabiting a bioleaching industrial heap of low-grade copper sulfide ore monitored by real-time PCR and oligonucleotide prokaryotic acidophile microarray. *Microbial Biotechnology, 2*(6), 613-624.

Rohwerder, T., Gehrke, T., Kinzler, K., & Sand, W. (2003). Bioleaching review part A. *Applied Microbiology and Biotechnology, 63*(3), 239-248.

Sakano, M., Tanaka, M., & Watanabe, T. (2001). Application of radio-frequency thermal plasmas to treatment of fly ash. *Thin Solid Films, 386*(2), 189-194.

Seidel, A., Zimmels, Y., & Armon, R. (2001). Mechanism of bioleaching of coal fly ash by Thiobacillus thiooxidans. *Chemical Engineering Journal, 83*(2), 123-130.

Shakir, A. A., Naganathan, S., & Mustapha, K.N. (2013). Properties of bricks made using fly ash, quarry dust and billet scale. *Construction and Building Materials, 41*, 131-138. doi: 10.1016/j.conbuildmat.2012.11.077.

Shimaoka, T., & Hanashima, M. (1996). Behavior of stabilized fly ashes in solid waste landfills. *Waste Management, 16*(5), 545-554.

Singer, A., Navrot, J., & Shapira, R. (1982). Extraction of aluminum from fly-ash by commercial and microbiologically-produced citric acid. *Applied Microbiology and Biotechnology, 16*(4), 228-230.

Singh, A.P., Kumar, A.S., Chandra, A., & Dhawan, S.K. (2011). Conduction mechanism in Polyaniline-flyash composite material for shielding against electromagnetic radiation in X-band & Ku band. *AIP Advances, 1*(2), 022147.

Singh, Y. (nd-a). Fly ash reuse. *Wealthy waste : waste management* Retrieved 17-1-2013, from http://www.wealthywaste.com/fly-ash-reuse

Singh, Y. (nd-b). Fly Ash utilization in India. *Wealthy Waste- Waste management* Retrieved 3-12-2012, from http://www.wealthywaste.com/fly-ash-utilization-in-india

Smith, M.E. (2002). Potential Problems in Copper Dump Leaching Retrieved 12/4/2013, 2013, from http://www.ausenco.com/uploads/papers/64065_Potential_Problems_in_Copper_Dump_Leaching.pdf

Srivastava, V.C, Swamy, M.M, Mall, I.D., Prasad, B., & Mishra, I.M. (2006). Adsorptive removal of phenol by bagasse fly ash and activated carbon: Equilibrium, kinetics and thermodynamics. *Colloids and Surfaces A: Physicochemical and Engineering Aspects, 272*(1), 89-104.

Sustainable sources. (2013). Flyash concrete Retrieved 25/2/2013, from http://flyash.sustainablesources.com/

Swami, P.N., Raju, B.N., Rao, D.V., & Rao, J.B. (2009). Synthesis and Characterization of Nano-Structured Fly Ash: A Waste from Thermal Power Plant. *Proceedings of the*

Institution of Mechanical Engineers, Part N: Journal of Nanoengineering and Nanosystems, 223(2), 35-44.

Swamy, K.M., Narayana, K.L., & Misra, V.N. (2005). Bioleaching with ultrasound. *Ultrasonics sonochemistry, 12*(4), 301-306.

Talvivaara Mining Company Plc. (2010). Bioleaching Retrieved 10/4/2013, from http://www.talvivaara.com/files/talvivaara/Presentations/Talvivaara_Technical_Seminar_London_May_Presentation.pdf

Tateda, M., Ike, M., & Fujita, M. (1998). Comparative evaluation of processes for heavy metal removal from municipal solid waste incineration fly ash. *Journal of Environmental Sciences China-English Edition-, 10*, 458-465.

Thakur, I.S. (2006). *Environmental Biotechnology: Basic Concepts and Applications*: IK International Pvt Ltd.

Theis, T.L., & Gardner, K.H. (1990). Environmental assessment of ash disposal. *Critical Reviews in Environmental Science and Technology, 20*(1), 21-42.

Torma, A.E., & Singh, A.K. (1993). Acidolysis of coal fly ash by Aspergillus niger. *Fuel, 72*(12), 1625-1630.

Tripathy, S. (2011). Development of a novel hybrid polymer composite with coir dust and fly ash reinforcements and investigation of its dielectric nature. *The National High School Journal of Science.*

UNIPROT. (nd). Thiobacillus ferroxidans Retrieved 1-3-2013, from http://www.uniprot.org /taxonomy/920

USEPA. (2012). Coal ash Retrieved 10-2-2013, from http://www.epa.gov/radiation/tenorm/ coalandcoalash.html

Vilcáez, J., Suto, K., & Inoue, C. (2008). Bioleaching of chalcopyrite with thermophiles: Temperature–pH–ORP dependence. *International Journal of Mineral Processing, 88*(1), 37-44.

Visa, M, & Duta, A. (2012). Methyl Orange and Cadmium Simultaneous Removal Using Fly-Ash and Photo-Fenton Systems. *J Hazard Mater, 244-245*, 773-779.

Visa, M., Isac, L., & Duta, A. (2012). Fly ash adsorbents for multi-cation wastewater treatment. *Applied Surface Science, 258*(17), 6345-6352.

Wajima, T., & Munakata, K. (2011). Material conversion from paper sludge ash in NaOH solution to synthesize adsorbent for removal of Pb^{2+}, $NH_4^+$and PO_4^{3-} from aqueous solution. *Journal of Environmental Sciences, 23*(5), 718-724.

Wang, Q., Yang, J., & Wu, T. (2009). Effects of water-washing pretreatment on bioleaching of heavy metals from municipal solid waste incinerator fly ash. [Research Support, Non-U.S. Gov't]. *Journal of hazardous materials, 162*(2-3), 812-818. doi: 10.1016/j.jhazmat.2008.05.125.

Wang, S., & Wu, H.. (2006). Environmental-benign utilisation of fly ash as low-cost adsorbents. *J Hazard Mater, 136*(3), 482-501.

Wee, J.H. (2013). A review on carbon dioxide capture and storage technology using coal fly ash. *Applied Energy, 106*, 143-151.

Wei, L., Golden, D., and Naik. T.R. (1990). Utilization of fly ash as a construction material in China Retrieved 24-2-2013, from http://www4.uwm.edu/cbu/Papers/1990% 20CBU%20 Reports/CBU%201990-09.pdf

White, C., Sayer, J.A., & Gadd, G.M. (1997). Microbial solubilization and immobilization of toxic metals: key biogeochemical processes for treatment of contamination. *FEMS Microbiology Reviews, 20*(3-4), 503-516.

White, C., Sayer, J.A., & Gadd, G.M. (2006). Microbial solubilization and immobilization of toxic metals: key biogeochemical processes for treatment of contamination. *FEMS Microbiology Reviews, 20*(3-4), 503-516.

Willey, J.M., Sherwood, L.M., Woolverton.C.J. (2008). *Prescott, Harley and Klein's Microbiology* (7 ed.): The McGraw-Hill Companies.

World Information Service on Energy. (2012). Impacts of Uranium In-Situ Leaching Retrieved 12/4/2013, from http://www.wise-uranium.org/uisl.html

World Nuclear Association. (2012). In Situ Leach (ISL) Mining of Uranium Retrieved 12/4/2013, 2013, from http://www.world-nuclear.org/info/Nuclear-Fuel-Cycle/Mining-of-Uranium/In-Situ-Leach-Mining-of-Uranium/#.UWkHDrV-Z6-.

Wu, H.Y, & Ting, Y.P. (2006). Metal extraction from municipal solid waste (MSW) incinerator fly ash—Chemical leaching and fungal bioleaching. *Enzyme and Microbial Technology, 38*(6), 839-847. doi: 10.1016/j.enzmictec.2005.08.012.

Xu, J., Thomas, H.R., Francis, R.W., Lum, K.R., Wang, J., & Liang, B. (2008). A review of processes and technologies for the recycling of lithium-ion secondary batteries. *Journal of Power Sources, 177*(2), 512-527. doi: 10.1016/j.jpowsour.2007.11.074.

Xu, T.J, & Ting, Y.P. (2004). Optimisation on bioleaching of incinerator fly ash by Aspergillus niger – use of central composite design. *Enzyme and Microbial Technology, 35*(5), 444-454. doi: 10.1016/j.enzmictec.2004.07.003.

Xu, T.J, & Ting, Y.P. (2009). Fungal bioleaching of incineration fly ash: Metal extraction and modeling growth kinetics. *Enzyme and Microbial Technology, 44*(5), 323-328. doi: 10.1016/j.enzmictec.2009.01.006.

Xu, W., Wang, H., Zhu, T., Kuang, J., & Jing, P. (2013). Mercury removal from coal combustion flue gas by modified fly ash. *Journal of Environmental Sciences, 25*(2), 393-398.

Xuan, X., Yue, C., Li, S., & Yao, Q. (2003). Selective catalytic reduction of NO by ammonia with fly ash catalyst☆. *Fuel, 82*(5), 575-579.

Yan, L., Wang, Y., Ma, H., Han, Z., Zhang, Q., & Chen, Y. (2011). Feasibility of fly ash-based composite coagulant for coal washing wastewater treatment. *J Hazard Mater, 203-204*, 221-228.

Yang, J., Wang, Q., & Wu, T. (2008). Comparisons of one-step and two-step bioleaching for heavy metals removed from municipal solid waste incineration fly ash. *Environmental Engineering Science, 25*(5), 783-789.

Yang, J., Wang, Q., & Wu, T. (2009). Heavy metals extraction from municipal solid waste incineration fly ash using adapted metal tolerant *Aspergillus niger*. [Research Support, Non-U.S. Gov't]. *Bioresource technology, 100*(1), 254-260. doi: 10.1016/j.biortech.2008.05.026.

Yang, J., Wang, Q., Luo, Q., Wang, Q., & Wu, T. (2009). Biosorption behavior of heavy metals in bioleaching process of MSWI fly ash by *Aspergillus niger*. *Biochemical Engineering Journal, 46*(3), 294-299. doi: 10.1016/j.bej.2009.05.022.

Yang, J., Wang, Q., Wang, Q., & Wu, T. (2008). Comparisons of One-Step and Two-Step Bioleaching for Heavy Metals Removed from Municipal Solid Waste Incineration Fly Ash. *Environmental Engineering Science, 25*(5), 783-789. doi: 10.1089/ees.2007.0211.

Yong, H.K. (2011). Electro-bioleaching of Spent Hydroprocessing Ni-Mo Catalysts. Master of Engineering, National University of Singapore.

Zeng, G., Deng, X., Luo, S., Luo, X., & Zou, J. (2012). A copper-catalyzed bioleaching process for enhancement of cobalt dissolution from spent lithium-ion batteries. [Research Support, Non-U.S. Gov't]. *Journal of hazardous materials, 199-200*, 164-169. doi: 10.1016/j.jhazmat.2011.10.063.

Zhao, L., Wang, L., Yang, D., & Zhu, N. (2007). Bioleaching of spent Ni-Cd batteries and phylogenetic analysis of an acidophilic strain in acidified sludge. *Frontiers of Environmental Science & Engineering in China, 1*(4), 459-465.

In: Fly Ash
Editor: Prabir Kumar Sarker
ISBN: 978-1-62948-044-2

Chapter 17

ASSESSMENT OF POTENTIAL IMPACTS OF FLY ASH UTILIZATION IN THE SLOVAK ROAD ENGINEERING

Marcela Ondova*[*] *and Nadezda Stevulova
Technical University of Kosice, Faculty of Civil Engineering,
Institute of Environmental Engineering, Kosice, Slovak Republic

ABSTRACT

The prognosticators emphasize the dangers of natural resources exhaustion, environmentalists emphasize the necessity of changes in the consumption of limited resources in order to ensure the energy saving, and material recycling. In order to follow and fulfill the defined aims and criteria, our society is increasingly oriented on the construction industry, which causes the decline of the natural resources, disturbs chain sustainability and has major impacts on the environment. On the other hand, it has predisposition to be one of the biggest "consumers" of the produced industrial waste containing recyclable components as secondary raw material. The secondary raw material is used in still higher replacement of the mineral wealth. Its appropriate characteristics and usability without prior treatment in large volumes offer not only material benefits but also great economic benefits to the constructors, the building companies. Several aspects are important to consider the complete benefits of the industrial waste management: economic benefits, savings of natural resources, environmental aspect (toxic metals, radioactivity) and regional "the quality of life" aspect as well. The coal combustion processes and exhaust cleanings in the energy industry are associated with the production of the large quantity of the solid waste. Solid residues from the coal combustion and flue gas desulphurization have major environmental impacts. The fly ash from gas retained in the separation plants has the largest share on generation of solid waste in the energy production, usually around 80% of the total produced waste amount. Fly ash can be used in different areas of the building engineering thanks to its appropriate characteristics. This chapter describes the possibilities and assesses impacts of Slovak brown coal fly ash utilization in the area of Slovak road engineering.

Keywords: Fly ash, cement, concrete, concrete road, secondary raw material, environment

[*] Vysokoskolska 4, 042 00 Kosice, Slovakia; Email: marcela.ondova@tuke.sk.

1. INTRODUCTION

Coal is used as a major source of energy throughout the world. It is an important raw material, which is naturally found in large quantities. In 2008, the global European production represented 579 million tonnes of black coal and 965 million tonnes of brown coal. Coal covers 27% of global primary energy needs and produces 41% of global electricity. Coal used for electricity generation remains important, as it is suggested it will produce 44% of world electricity in 2030. At current production levels proven coal reserves will last about 119 years (report of the International Energy Agency (IEA), 2008; World Coal Association).

1.1. Fly Ash

Combustion of pulverized coal is necessary in order to produce energy. During the combustion process, the volatile matter and carbon burn off, and the coal impurities such as clays, shale, quartz, feldspar (Mehta, P. K., 1983). These fused particles are carried along with the flue gas. As the flue gas approaches the low temperature zones, the fused substances solidify to form predominately spherical particles called fly ash. The remaining matters, which agglomerate and settle down at the bottom of the furnace, are called bottom ash. Fly ash is captured by mechanical separators, electrostatic precipitators, or bag filters. It is a mixture of particles varying in shape, size, and composition. These particles can be classified as carbon from unburned coal, thin-walled hollow spheres, and their fragments, magnetic iron containing spherical particles, and spherical particles (Berry and Malhotra, 1980; Berry and Malhotra, 1987; Berry et al., 1989; Mehta, 1989). Size of spherical fly ash particles to lie in the range of 1-150μm (Berry et al., 1989; Mehta, 1989) and their morphology ilustrated in Figure 1.

In the European Union (EU) of the EU 15 (15 countries) the production was about 52 million tonnes of coal combustions products (CCPs) in 2009 (Figure 2) and in the larger EU of 27 member states the total production is estimated to be about 100 million tonnes of CCPs. Fly ash represents the greatest proportion of total CCP production.

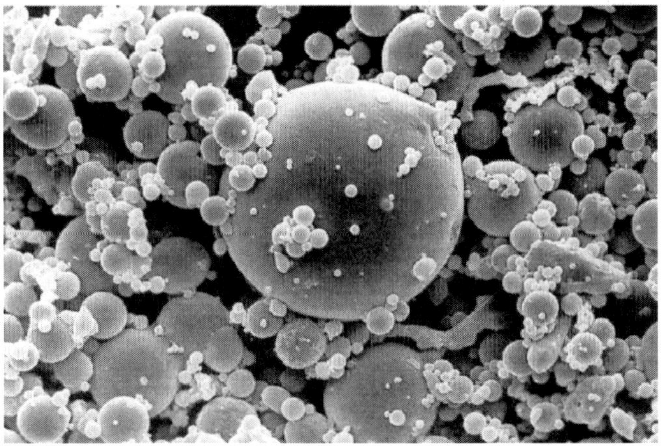

Figure 1. Alumosilicate spherical particles of fly ash.

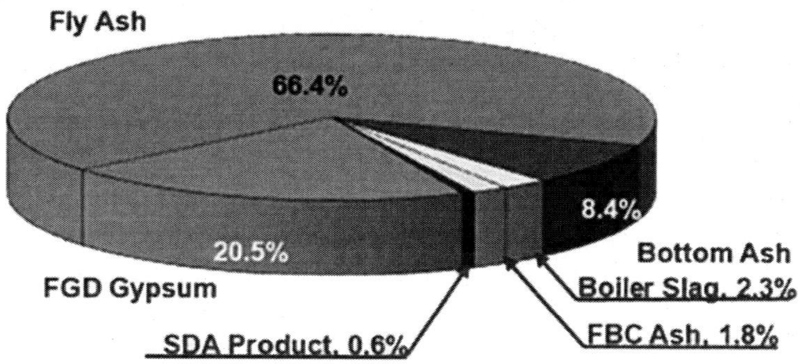

Figure 2. Productions of CCPs in Europe (EU 15) in 2009 total production: 52 million tonnes (source: European Coal Combustion Products Association).

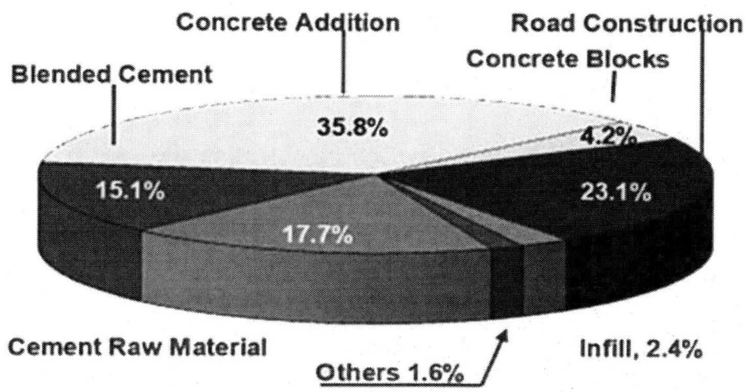

Figure 3. Utilization of fly ash in the construction industry in Europe (EU 15)in 2009; total utilization: 15.4 million tonnes (source: European Coal Combustion Products Association).

Within the EU, the utilization of fly ash in the construction industry is currently around 48% and of bottom ash around 45%, while the utilization for boiler slag is 100%. In the majority of cases CCPs are used as a replacement for naturally occurring resources and therefore offer environmental benefits by avoiding the need to quarry or mine these resources. CCPs also help to reduce energy demand as well as emissions to atmosphere, for example CO_2, which result from the manufacturing process of the replaced products. CCPs are utilised in a wide range of applications in the building and construction industry. Applications for CCPs include their use as an addition in concrete as a cement replacement material and as an aggregate or binder in road construction. They can also be utilized as mineral fillers and as fertilisers. Where necessary, CCPs meet any relevant national and European building materials standard and regulation. A graph of the fly ash utilization in the construction industry in 2009 is presented Figure 3 (ECOBA: European Coal Combustion Products Association).

Physical, Chemical and Mineralogical Properties

It is necessary to say that the type of combusted coal has the significant impact on the shape, size, composition as well as on the chemical, physical and mineralogical characteristics of the fly ash and its further utilization has the type of combusted coal (it is known that fly ash generated from the combustion of Slovak black coal is not suitable for further use due to its unsuitable properties, especially high loss on ignition). The basic physical properties are granularity, respectively the specific surface area, morphology, density, bulk and apparent density, hardness, compatibility, frost resistance, fusibility and others. The content of combustible substances Ca content, the amount of harmful (S, Cd, As, Pb ...) and radioactive (Ra, Th, K) elements, as well as the loss on ignition and content of the mineral neologisms created during thermal processes belongs to the basic chemical properties (Noskievic, 1993; Saman, 1971). Fly ash contains some heavy metals (Cu, Zn, Pb, etc.). Presence of heavy metals in the fly ash (As, Cr, V, Mo, Pb) causes problems for industrial use. Heavy metals present in the fly ash can be removed by leaching, but the creation of wastewater and their subsequent disposal is not environmentally acceptable alternative (Kolousek, D. et al, 2002).

One of the most important indicators of the fly ash quality is the content of SiO_2 and Al_2O_3. Very important fly ash characteristic is its ability to react with Ca $(OH)_2$ and to create hydrated calcium silicates and aluminates with typical binding property (Michalikova, F. et al., 2005). Fly ash itself is non-hydraulic; it is not able to react with water. However, fly ash mixed with the calcium hydroxide (for example with cement) reacts. The reaction of silica oxide with the calcium hydroxide generates same products as the reaction of cement with water (hydrated calcium silicates) (Tkacova, K., Stevulova, N., 1992). This reaction is known as pozzolanic; its process depends on the fly ash type.

Pozzolana is fine natural or artificial material containing active silicon dioxide and aluminum oxide without the ability of self-coupling reactions with calcium hydroxide and water react to insoluble compounds with strength properties (Keil, F., 1971).

Fly Ash is a pozzolan, which in finely divided form and in the presence of moisture, chemically reacts with the calcium hydroxide released by the hydration of Portland cement to form additional calcium silicate hydrate and other cementitious compounds. The hydration reactions are similar to the reactions occurring during the hydration of Portland cement (Betontechnische Empfehlungen, 2002; Helmuth, R., 1987). However, this reaction is slower than the hydration of clinker minerals. The products of pozzolanic reaction seal up the pore structure of the cement stone gradually and ensure the high strength of concrete, its water resistance and resistance to aggressive elements compared to straight Portland cement concrete mixtures. At the same time fly ash application leads to the reduction of cement consumption and heat hydration (by reduction of cement dose).

Considering fly ash particles shape is a heterogeneous material, because of unburned coal and vitrified alumosilicate compounds of different morphology. Particles of spherical and oval shape, called microspheres, are dominant. Depending on melting processes spherical particles occur as hollow (cenospheres) or/and full. Bigger particle sized fly ash can be fulfilled with cenospheres of smaller size (plerospheres) or formed of metallic iron or silicon.

The fly ash produced from the combustion of black coal in wet bottom boilers is typical by its particles of spherical shape. This shape is a result of ash melting and cooling process. The fly ash from brown coal burned in dusty granulation boilers is characterized by particles of original coal grain shape, but the grains are rounded in consequence of melting. Fluid fly ash from black and brown coal has morphology of coal grain, spherical particles are rare.

Mineralogical composition of fly ash, as well as share and composition of glassy phase depends on origin, chemical composition and mineralogy of burned coal. Other fly ash characteristics are affected by conditions and burning process as well. It is assumed that mineralogical composition of fly ash affects significantly its application characteristics regarding its influence on hydration processes with cement by temperatures and conditions corresponding regimes by increasing temperature of concrete (Stevulova, N. et al., 2008).

Relevant mineral components of fly ash are mullite ($3Al_2O_3.2SiO_2$), free silica (SiO_2) thermal changed into α-cristobalit and partly reacted with calcium oxide, and a relatively large amount of glassy phase of texture SiO_2. Silica dioxide occurs mostly in a form of silica or glassy phase, until a part of Al_2O_3 is in mullite and the rest is contained in glassy phase. Iron occurs in a form of hematite (Fe_2O_3) and magnetite (Fe_2O_3, FeO) (Asokan, P., et al., 2005). Based on literature (Grier, D.G., et al., 1997), content of silica and glassy phase determines pozzolanic activity of fly ash. The higher content of silica and unburned components is in the fly ash, the lower is its pozzolanic activity.

Amorphous phases with high content of Al_2O_3 and crystalline phases (metakaolin) belong to the hydraulic active mineralogical components. CaO and MgO belong to the active non-hydraulic components. Particles of unburned coal slag with high content of silicon dioxide and crystalline phase of silica and mullite are to the hydraulic inactive components. Alumosilicate glass and a different number of iron and crystalline phases are dominant for fly ash with low content of Ca (silica, mullite, magnetite, ferrit-spinel), which are considered as chemically inactive in concrete. The presence of relevant content of Ca effects nature of a glass and crystalline phases in fly ash. Amorphous phase in fly ash with high Ca content can be described as calcium-alumosilicate glass, although different glass types coexist in wide range of composition. Influence of Ca in fly ash mineralogy is proven in the chemical reactivity in different ratio $Si0_2$: Al_2O_3. Alumosilicate glass in fly ash with low content of Ca reacts slowly and requests alkali or lime resource, for example a lime formed by hydration of Portland cement. Fly ash with low Ca content is considered pozzolanic with no or small concrete characteristics. Ca substitution in glassy phase increases glass reactivity, and in fly ash with high Ca content ensures forming of Ca silicates or hydrates of calcium aluminate in the absence of a separate lime resource. Other concrete characteristics of fly ash with high Ca content are attributed to hydration and reactions of crystalline Ca - bearing phases (Brylska, E. et al., 2001).

In technical literature we find several internationally used approaches in fly ash classification. Their basic goal is to standardize assumptions for utilization of different types of fly ash in the industry based on knowledge of their physical, chemical characteristics, and mineralogy. It is known seven current classifications (Rafalowski, 2003; McCarthy et al., 1990).

Effects of Fly Ash on Fresh and Hardened Concrete Performances

The utilization of fly ash in concrete as partial cement replacement leads to increase of a binder´s amount and improves workability (pumpability, keepability and others) of fresh concrete, as fly ash behaves as a solid fluidifier. Application of fly ash in combination with a fluidifier is very effective, because it improves the workability of fresh concrete significantly. It enables to extend the time of workability of fresh concrete. Regarding characteristics of lower class fresh concrete the main benefit of fly ash utilization with fluidifier is complement of fine binder, especially particles up to 0.25mm. At the same time, fly ash hinders segregation and flowing of very soft and liquid fresh concrete (cone consolidation 100 mm and more) contributes to the higher structure homogeneity and higher quality of concrete construction surfaces (Slanicka, 1982, Pavlik et al., 1973). The phases of hydrate created in pozzolanic reaction cause higher strength of fly ash mixtures, structure concretion, and finer porosity (Sybertz, F., 2004). These pore fulfilling in concrete structure with products of this reaction and increasing of concrete density by fly ash application leads to decrease of cement consumption as well as:

- increase of concrete strength
- increase of concrete water-tightness
- reduction of hydration heat (by a lower concrete amount)
- increase of concrete resistance against some aggressive influences, for example sulphated waters (by a lower concrete amount, it leads to lower concentration C_3A in concrete).
- increase of concrete durability.

The utilization of fly ash in concrete reduces bulk density of concrete, as fly ash density is a lower than concrete density (Sicakova, A., 2000). The fact that by fly ash application the binder´s amount increases and gravel amount in $1m^3$ of concrete decreases, contributes to the reduction of bulk density. The freezing and hardening of concrete is slower with fly ash application. It is related to a fact that pozzolanic reaction is slower than concrete hydration. Therefore fly ash utilization is not proper for concrete production at low temperatures, eventually if rapid increase of its early initial strengths is required. However, by utilization of fly ash the higher concrete strength after 28 days of hardening can be calculated (BetonRacio Company).

Utilizations of Fly Ash in Concrete Roads

Road Engineering offers a several ways of waste utilization (Junak, J., 2011). Thanks to proper concrete parameters the real value of the fly ash within the building practice is expressed in the road engineering, in the road concrete production, where the investment projects, considered from the long term and complex view, have to serve for decades. With regard to the excellent characteristics of the road concrete they belong to the most progressive road construction and to the world trends in the highway constructions. The biggest share of the roads with the cement-concrete surface based on high amount of the fly ash as cement replacement have Belgium, Germany, France, and the USA mainly. The road infrastructure belongs to the key factors influencing the economy and the organization of the state area significantly. The rising living standards cause the road automobile traffic growing, the individual traffic mainly, and lead to the necessary modernization and road net spreading.

The Slovak Republic makes efforts to connect its eastern and western side not only due to the requirement of the society but also due to the world trends of concrete road. However, in comparison to the western European norms the Slovak strict guidelines determine the maximal share of admixtures about 5 wt.%. Except this fact, the fly ash use has to fulfill other specifications ralated to its chemical and physical characteristics defined by the Slovak Technical Standards. The fly ash generated from the Slovak brown coal combustion fulfills the strength quality standards of concrete as the results of researches prove. Higher share of fly ash in concrete (cement replacement) means significant financial savings for the construction companies engaged in road construction.

However, only meeting the specific standard criteria i.e. mechanical (strength characteristics), chemical (frost and chemical resistance) and economic indicators are not sufficient for the concluding statement of fly ash utilization. In terms of comprehensive assessment, it is necessary to draw attention to the analysis of its life cycle as well.

Life Cycle Assessment (LCA)

Increased public awareness of the threats posed by global warming has led to greater concern over the impact of anthropogenic carbon emissions on the global climate. The current level of CO_2 in the atmosphere is approaching 380 ppm (Feely et al., 2004; Sabine et al., 2004). According to the International Energy Agency's (IEA) Greenhouse Gas R&D Program (Hendriks et al., 2000), Portland cement production generates an average world carbon emission of 1 ton CO_2 / 1 ton cement produced (Lippiatt, B., Ahmad, S., 2004). Coal fly ash as a supplementary cementations material in the concrete is preferable for the applications presented in the Green Building Guide and the main type offered for transportation engineering because of its performance enhancing and cost-saving benefits. It has also been praised for its environmental benefits as a "green" building material — putting to use an energy production byproduct that reduces demand for carbon-intensive Portland cement, requires less water in the setting process, and would otherwise wind up in a landfill.

For globally significant products such as cement, environmental life cycle assessment (LCA) is a valuable tool for improving our understanding of the environmental hazards posed by a product's life stages. In addition it allows cement producers to optimize the manufacturing process by reducing adverse environmental impacts (Palaščáková, L. et al., 2011; Eštoková, A., Palaščáková, L., 2013). Life cycle assessment is a method of evaluation used to assess the environmental impacts of technologies from "cradle to grave" and may be performed on both products and processes. This method belongs to a group of analytical instruments for environmental management that evaluate potential environmental impacts independent on the place of manufacture or product use (Vilcekova, S., Kridlova B., E., 2010). There are several benefits of LCA including the ability to evaluate the material and energy efficiency of a system, identifying pollution shifts between operations, and providing benchmarks for improvement (Ross, Evans, 2002; Owens, 1997). LCA results from different areas of building industry (assessment of building materials and structures as well as whole buildings) are presented in the works (Estokova, Porhincak, 2012; Estokova et al., 2012; Kridlova B., Vilcekova, 2012; Kridlova B., Vilcekova, 2013).

2. Material and Methods

2.1. Material

Experiments were based on European Standards (EN 13877-1;-2), which determine rather general conditions for materials and methods of properties evaluation, but in many ways they refer to national standards. The national standards for concrete pavements are in place. ES´s do not cancel but only partly modify them. The materials selection for experimental works, as well as testing the contribution of fly ash to quality of road concrete was performed in the terms of national standard requirements for roads of classes I – II (motorways, international roadways, parking areas). A lot of expert studies were taken into consideration; however national conditions should be respected not only in the terms of technical requirements, but also due to specific parameters of fly ash. It is well known, that properties of fly ash vary significantly and strongly depend on coal quality, conditions of combustion etc. (Stevulova et al., 2012; Ondova et al., 2012).

Slovak Requirements for Properties of Fly Ash in Concrete

The Slovak Republic is governed in assessing the suitability of fly ash chemical and physical properties in accordance with current European legislation (European Technical Standards). Specific requirements of chemical and physical properties of fly ash for its use in concrete are given in Table 1, 2.

Table 1. European requirement for chemical properties of fly ash for its use in concrete

Properties	Test methods	Characteristic value	
LOI (loss on ignition)	STN EN 196-2 STN EN 450-1	category A	≤ 5.0 % wt.
		category B	from 2.0 to 7.0 % wt
		category C	from 4.0 to 9.0 % wt.
Cl^-	STN EN 196-2	≤ 0.10 % wt.	
SO_3	EN 196-2	≤ 3.0 % wt.	
free CaO content	EN 451-1	≤ 2.5*) % wt.	
active CaO	STN EN 196-2	≤ 10 % wt.	
active SiO_2	STN EN 196-2 EN 197-1	≤ 25 % wt.	
SiO_2, Al_2O_3, Fe_2O_3	STN EN 196-2	sum of the contents ≥ 70 % wt.	
alkali content (Na_2O equivalent)	STN EN 196-2	≤ 5.0 % wt.	
MgO	STN EN 196-2	≤ 4.0 % wt.	
soluble phosphates (P_2O_5)	STN EN 450-1	≤ 100 mg/kg	

Note: *) fly ash with the greater content of free CaO than 1.0% by weight, but less than 2.5% by weight meets the requirement of the bulk stability.

Table 2. European requirements for physical properties of fly ash for its use in concrete

Properties	Test methods	Characteristic value	
Fineness (overflow of the sieve 0.045 mm)	STN EN 450-2	category N	≤ 40.0 % wt. D*± 10 % wt.
		category S	≤ 12.0 % wt.
Activity index	STN EN 196-1	after 28 day	≥ 75 %
		after 90 day	≥ 85 %
Bulk stability	STN EN 196-3	tension ≤ 10.0 mm	
Specific weight of grains	STN EN 196-6	D ± 200 kg/m^3	
Start of curing	STN EN 196-3	≤ 120 min compared to the reference test	
Water consumption	STN EN 450-1 STN EN 1015-3	category S ≤ 95.0 % compared to the reference test	
Hazardous substances and radioactivity	It may not contain substances that are hazardous to health, hygiene and the environment, when they release from concrete		

Note: *) D – the value declared producer.

Table 3. Chemical composition of fly ash from the coal combustion

Component [%]	SiO_2	Al_2O_3	Fe_2O_3	TiO_2	CaO	MgO	K_2O	Na_2O	MnO
FA (black)	36.57	16.06	10.82	0.65	3.30	1.22	1.41	0.18	0.27
FA (brown)	37.50	15.60	7.67	1.30	22.94	2.77	1.21	0.63	0.11
Component [%]	P_2O_5	SO_3 (S)	S_{total}*	S_{pyrite}	Loss on ignition	Combustible substances by 830 °C	Residual of organic Carbon	Loss on drying	
FA (black)	0.56	0.17	0.09	0.02	28.53	28.53	27.55	–	
FA (brown)	0.18	7.29 (2.91)	2.91	< 0.01	2.59	2.14	0.28	0.16	

Note: * $S_{total} = S\ (SO_3) + S_{pyrite}$.

The chapter discusses about the strict conditions of the Slovak Technical Standards in the field of admixture replacement (5%). However, in the field of road construction is also an increase of 10% replacement visible to economic savings. Therefore, the 0-15% fly ash cement replacement was studied. To study the possibility of fly ash utulization in concrete road (max. 15% cement replacement) were chosen two types of Slovak fly ash. The fly ash (FA) from the black and the brown coal combustion (chemical properties of these selected types are presented in Table 3) was tested in cooperation with the building praxis (Central Building Laboratory of Slovak a corporation). Fly ashes used in experiments must fulfill the requirements according to the STN EN 450 – Fly ash into the concrete. However, the black fly ash has not verified these requirements because of high value of loss on ignition; therefore its chemical properties were modified by thermal treatment. This modification has reduced the value of loss on ignition from primary 28.53% to 16.2%. Side effect of this modification was change in particle size distribution of fly ash. Figure 4 illustrates the change of mean particle size before and after thermal treatment from the value x = 30.538μm to x = 60.94 μm. This fact was pointed at the next physical properties modification. The black fly ash tests were rejected, because of high energy costs to modify of fly ash.

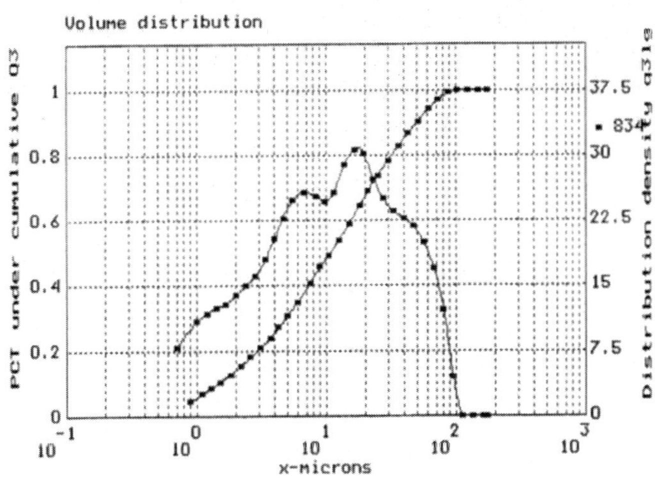

Figure 4. (Continued).

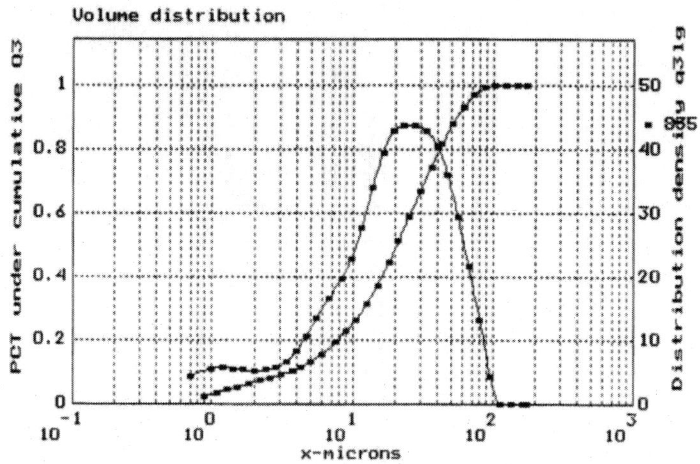

Figure 4. Particle size of black fly ash before (a) and after (b) thermal treatment.

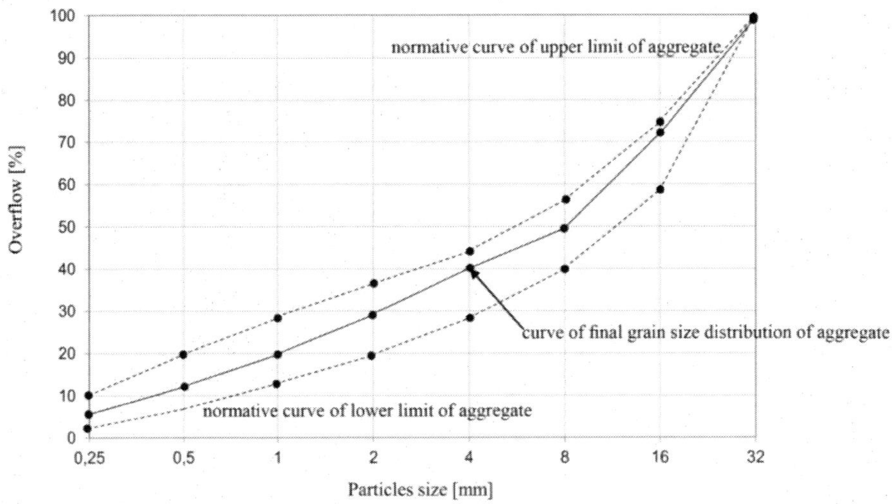

Figure 5. Final grain size distribution of aggregate used in concrete.

In accordance to the proposed prescription, the C30/37 grade concrete was prepared with 0 - 15% fly ash replacement of special kind of Portland cement CEM I 42.5 N. Water cementations material ratio was 0.36 and natural gravel aggregate from stone – pit Soporna and Hanisberk in specific ratio of the fine to coarse aggregate 40 (0/4): 10 (4/8): 50 (8/16, 16/32) was used in the mixture (final grain size distribution of aggregate is shown in Figure 5) (Ondova et al., 2012[a]).

2.2. Methods

The mixtures were prepared in the laboratory mixer type ZZ 150 SH with horizontal rotary drum of 150 l capacity. Four based properties of fresh concrete: consistence, air content, density and temperature (EN 206-1: 2002) were tested according to EN 12350 (parts 3 and 7) before the forms filing. The forms filling (the cube form with size 150 mm and the prism form with size 100 × 100 × 400 mm) were performed in two layers. Each layer was compacted on the vibration plate VSB 40 for 8 seconds. Next day the composites were taken out of the forms and saved in the water bath. 144 pieces of testing composite were cured at temperature 20 ± 2 °C. After 7, 28 and 90 days of hardening the composites were taken out of the water bath and tested for the required properties for concrete class C 30/37 (Table 4). The composites were tested on compressive strength (CS), flexural strength (FS), determination of concrete chemical substance resistance (Slovak Technical Standards: STN 73 1326, STN 73 6123) as well as on freezing and thawing. For the comparative study the reference sample (RS) concrete class (C 30 / 37) was prepared in accordance with requirements of Slovak Technical Standard STN 73 6123 – Road Construction. Cement Concrete Pavements. Wearing Courses, (Ondova et al., 2012[a]; Ondova et al., 2012[b]).

Table 4. Tested parameters and values required for road concrete (class I- I roadsI)

Properties	Method	Required value
Consistency of fresh concrete	Slump test	S1 (10-40 mm)
Air content in fresh concrete	Pressure method	4 - 8%
Fresh concrete temperature	Just after mixing	+ 5 °C ≤ T ≤ + 30 °C
Compressive strength	7, 28 and 90 days	32 MPa (in 28 days)
Flexural strength	7, 28 and 90 days	4.5 MPa (in 28 days)
De-icing salts resistance	150 freezing-thawing cycles	max. 300 g/m^2 /100 [1)]
Frost resistance	300 freezing-thawing cycles	min 0.85 [2)]

Note: [1]Rate of destruction/minimal number of cycles. Rate of destruction is expressed by scaling of tested concrete [g/m^2]; [2]Frost index - rate of flexural strength before and after freezing-thawing cycles.

Testing Methods – Fresh Concrete

In accordance with the Slovak Standards the individual tests of fresh concrete mixtures were prepared as follows:

The Slump Test

The concrete slump test is an empirical test that measures the workability of fresh concrete. The behavior of a self-compacted inverted cone of concrete (concrete's workability or the dampness of concrete) was monitored under the action of gravity. (Ondova, M., Stevulova, N., 2012[c]).

Air Volume Test

It is important to notice the air content changes during the concrete transport from the concrete plant to the paver and also through the paver. A specifier has to decide how much air is acceptable, and a contractor has to understand the implications in terms of yield loss and strength. The pressure method is based on Boyle's law, which states that the volume occupied by air is proportional to the applied pressure. Two meters types, A and B, are covered by the standard. The Type A meter is used rarely. With the Type B meter a separate air chamber is connected through a valve to the test bowl that is filled with concrete. With the valve closed, the separate chamber is pressurized to a predetermined operating pressure. When the valve is open, the air expands into the test chamber, and the pressure drops in proportion to the air contained within the concrete sample. The pressure gauge is read in units of air content (Ondova, M., Stevulova, N., 2012[c]).

Temperature and Fresh Concrete Density

Other key parameters to consider are temperature and density of fresh mixture. A concrete mixture design is composed of several ingredients: Portland cement, supplementary cementitious materials, fine aggregate, coarse aggregate, admixtures, air, and water. All of these materials have different specific gravities. A variation in the unit weight of a mixture indicates a change in proportioning of the mixture, often in the water or air content. The density of fresh concrete was determined in following steps: determine the level-full volume of the air pot, determine weigh of the empty air pot, consolidate a sample of fresh concrete in the air pot using the tamping rod or vibrator, using the strike-off plate, finish the concrete so that it is level-full with the top rim of the air pot, clean off all excess concrete from the exterior of the air pot, determine weigh of the air pot full of concrete, record the empty mass, full mass, and volume of the air pot. The density of the concrete mixture is reported in kg per cubic meter (kg/m^3). The density of the mixture should be compared with the density of the mixture design to identify potential problems in the batching process or changes in raw materials.

Testing Methods – Concrete Composites

Among many concrete tests, the compressive strength of concrete is the most important parameter, which gives an overview of its character. Flexural strength is other measure of the tensile strength of concrete. It is a measure of an unreinforced concrete beam or slab to resist failure in bending. Composites were tested on both: compressive strength (CS) and flexural strength (FS) after 28 days of hardening. For the comparative study the reference composite (RC) concrete class (C 30 / 37) was prepared in accordance with requirements of Technical Standards (STN 73 6123). Determination of concrete chemical substance resistance (STN 73 1326, STN 73 6123) as well as on freezing and thawing were also performed (Ondova et al., 2013).

Compressive Strengths

This simple analysis proves whether concreting has been done properly or not. Two different types of cubes, depending on aggregate size, are used for the cube test. Cubical moulds with size of 150 mm x 150 mm x 150 mm are used in most experiments. Prepared specimens were tested with compression testing machine after 7 and 28 days of curing. Load was applied gradually until the specimens were destroyed. Load at the failure divided by area of specimen give us the compressive strength of concrete (Ondova, M., Stevulova, N., 2012[c]). Satisfying and unsatisfying failure modes of test cubes are presented in Figures 6, 7. Followed, the compressive strength of concrete f_c (MPa) is determined according to the formula:

$$f_c = \frac{F}{A_c} \tag{1}$$

where F - maximal load breaking strength [N] A_c - sectional area of the test specimen, which acting the load force [mm^2] (Svoboda et al., 2005).

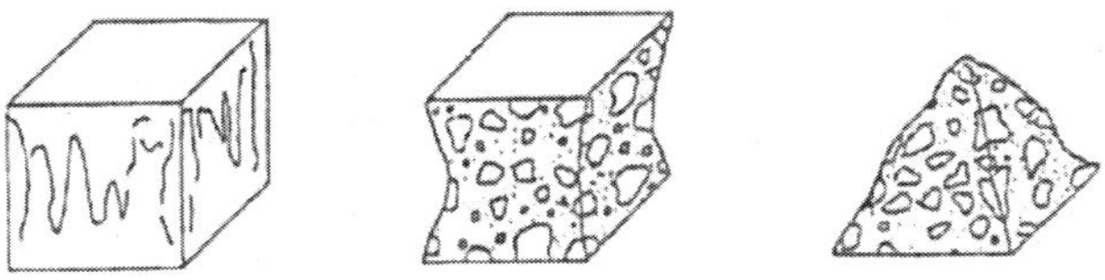

Figure 6. Satisfying failure modes of test cubes (source: EN 12390-3).

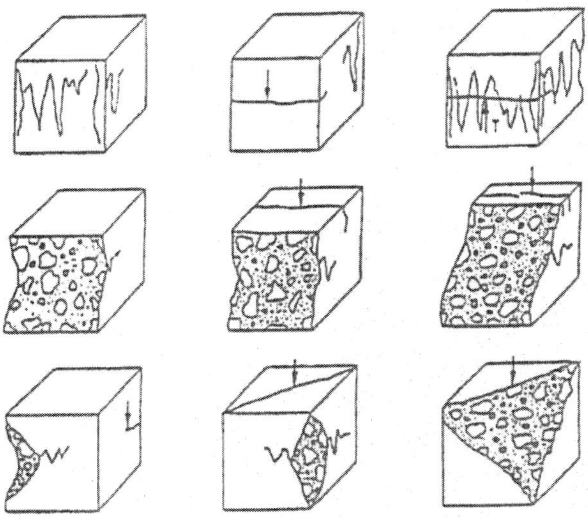

Figure 7. Some unsatisfying failure modes of test cubes (source: EN 12390-3) Flexural strengths.

Flexural strength is measure of the tensile strength of concrete. The tests were provided with the third point loading of 100 x 100 mm x 400 mm concrete beams (Ondova, M., Stevulova, N., 2012[c]). 4 point load test is illustrated in Figure 8. The result of measurement is the maximum value of the strength (F) obtained in breach of the body (specimen).

Strength of the concrete tensile strength f_{cf} (MPa) is determined according to the formula:

$$f_{cf} = \frac{F.l}{d_1.d_2.d_2} \tag{2}$$

where F – maximal load strength; l – the distance between the support rollers (mm). The value is rounded to the nearest; 0.1 MPa.

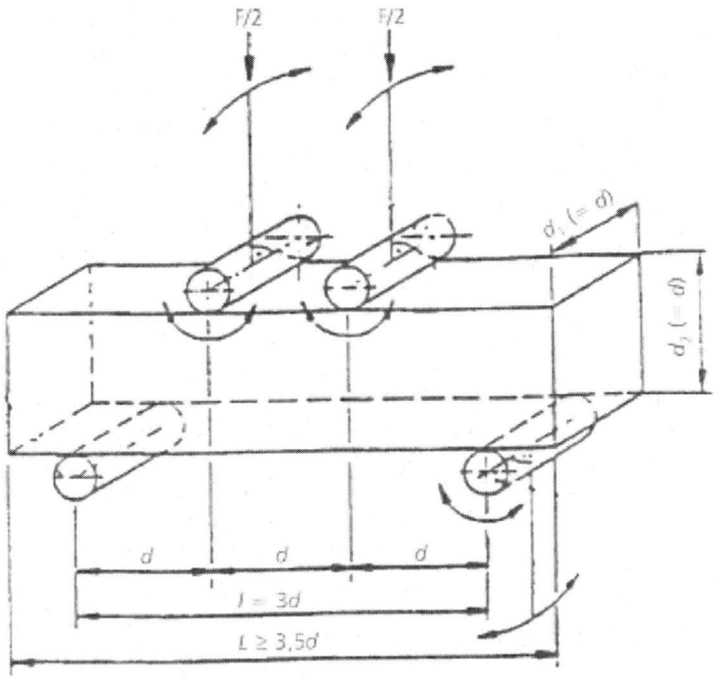

Figure 8. Flexural strengths - 4 point load test.

De-Icing Salts Resistance; Frost Resistance

Chemicals used in pavements include sodium chloride, calcium chloride, magnesium chloride, and potassium chloride. These chemicals reduce the freezing point of the precipitation, as it falls on pavements whose high concentration reduces the number of freezing and thawing cycle exposures to the pavement. Maximal values of wastes after components testing (method A- automatic cycling) is 1000 $g.m^{-2}$. It expands about 9 % by water freezing. As the water freezes in moist concrete freezes, it produces pressure in the pores of the concrete. If the pressure developed exceeds the tensile strength of the concrete, the cavities dilate and rupture. The accumulative effect of successive freeze thaw cycles and disruption of paste and aggregate can cause expansion and cracking, scaling, and crumbling of the concrete. Tests were provided by the Freeze Thaw Resistance Tester CDF / CIF with specified minimal frost resistance coefficient of 0.85 (Ondova, M., Stevulova, N., 2012[c]).

Analytical Methods

Repeated research of mechanical properties of fly ash - concrete composites by changing the chemical (plasticizers and aeration) additives showed slight deviations in the mechanical strength results. Specifications of tested additives are given below.

Admixture 1: Plasticizer 1 is superplasticizer based on polycarboxylates modified especially for the production of road concrete. It is presented by a minimal decrease of the consistency of the concrete in the summer. Aeration admixture 1 based on synthetic surfactants, which must be used wherever concrete comes into contact with weathering, respectively aggressive chemical. Substantially it contributes to the increased resistance of concrete to frost and de-icing substances.

Admixture 2: Plasticizer 2 is a plasticizer based on sodium salt of naphthalene sulphate formaldehyde condensate. It has a strong plasticizing effect and significantly improves the consistency of fresh concrete. Aeration admixture 2 is an aeration additive that allows air dispersion of large quantities of stable and separated microscopic bubbles in fresh concrete. It can be used for long-term construction suffer adverse climatic conditions. Increases frost resistance, durability and resistance to road salt (cement-concrete roads, construction of bridges, concrete structures, water works, concrete pavers, concrete for roads).

Admixture 3: Plasticizer 3 is high-performance special superplasticizer and concrete-plasticizer for ready-mix concrete and the prefab industry. Increase of the early and final strengths beacuse of the possible reduction of the w/c-value. Aeration admixture 3 is a synthetic aeration ingredient that entrains very finely distributed air micro-pores (< 0.3 mm) into the concrete. Typical areas of application are the construction of concrete road surface, bridge curbs, parking lots or runways or concretes with a high resistance against frost and de-icing salts.

Through analytical method (X-ray fluorescence method - XRF) were:

X-Ray Fluorescence Method

XRF was used for the chemical compositions investigation of concrete samples and admixtures. The samples of concrete were pulverised by using planetary ball mill SFM (MTI corp., USA) and prepared as tablets of diameter 32 mm by mixing of 5 g of concrete powder and 1 g of dilution material and pressing at pressure of 0.1 MPa/m^2. The chemical composition was determined by using SPECTRO iQ II (Ametek, Germany) with SDD silicon drift detector with resolution of 145 eV at 10 000 pulses. The primary beam was polarized by Bragg crystal and Highly Ordered Pyrolytic Graphite - HOPG target. The prepared samples were measured during 300 sec. at voltage 25 and 50 kV, at current 0.5 and 1.0 mA, respectively. The standardized cement method of fundamental parameters was used for the measurements. The admixtures were measured during 180 sec. the same way by using the method of fundamental parameters for liquid samples.

Thermal Analysis

Thermogravimetry continuously measures the mass of a sample subjected to a steady increase of temperature in order to quantify reactions involving gaseous emissions. Thermogravimetric analysis (TG) is thus a convenient method to determine concrete carbonation, because it quantifies the calciumcarbonates and portlandite contents in a sample of mortar extracted from concrete ("concrete mortar") and reduced into powder (Parrott, L.J., Killoh, D.C., 1989; Platret, G., Deloye, F.-X., 1994).

Determination of the thermal behavior of the concrete samples prepared with different chemical admixtures was studied. Thermal properties of prepared concrete composite were studied by using STA 449F3 (Netzsch, Germany) in the temperature range from 25 to 1000 °C with the heating rate of 10 K/min under nitrogen atmosphere using DSC/TG mode.

Economic Assessment

The Slovak cost calculation software CENKROS was used to demonstrate the economic benefit of fly ash utilization. CENKROS is software for measurement and quality control of building production. This system covers all activities associated with the preparation and execution of the contract. It allows to prepare a quality bid quickly, to calculate costs of using the effective utilization and billing of the work performed, and to prepare price estimates by financial indicators (Ondova, M., Stevulova, N., 2012[c]).

Costs were calculated with no production and administrative expenses and without any profit as well. Current inputs of materials, transport costs, team and machine utilization was specified for Dx highway space.

The expected cost saving was calculated within the study of Slovak brown coal fly ash utilization in concrete cements with regard to application of raw materials in the road and highway constructions. Concrete mixture with 15 % FA compensation in a road construction was selected for our calculation. The road construction study of two coat concrete cement (CC) pavement of T1 tunnel on Dx highway was used as a basis for this project. Its structure and dimension are in compliance with requirements of Slovak building company no. 0803/2003 – design of concrete-cement carriageway at road communication. The T1 tunnel is designed as a double-pipe highway tunnel in Slovak rural area with 80 km/h maximum speed in one way traffic, or with 60 km/h maximum speed in two way traffic (in case of closure of one tunnel pipe). The pipe length is 698 m including concrete cement pavement in front of the portal with optimum length of 50 m. Width configuration of the tunnel is specified in classification T1= 9.0 m. The engineering characteristic of the T1 tunnel on Dx highway is presented in Table 5 (Ondova et al., 2011; Zelenakova, Ondova, 2010).

Table 5. Engineering characteristic of T1tunnel on Dx highway

Two - layer concrete cover *	
Two layer concrete cover	170/80 mm
Coated intermediate aggregate	50 mm
Infiltration road spray	1.0 kg/m^2
Cement stabilization I	180 mm
Aggregate 0-32, 0-63 (120+150mm) - Macadam	270 mm

Note: *minimal thickness according to Slovak standard requirement is 50 mm.

Environmental and Life Cycle Assessment

As part of the environmental assessment of harmful elements in human health, the research goals are focused on determining the contens of Cr^{VI} and ^{40}K, ^{226}Ra, ^{232}Th of cement CEM I 42.5 N and fly ash itself.

Hexavalent Water-Soluble Chromium

Chromium is an indelible non-volatile trace element of raw materials (clay, limestone and iron additives in particular) used in cement clinker production in the form of chromium (III). Naturally occurring chromium (III) is not initially harmful, since it is chemically stable. Only at high temperatures found in cement rotary kilns, inert trivalent chromium oxidizes to form reactive hexavalent chromium. The measured concentration is suggested to be the hexavalent water-soluble chromium which is harmful and allergenic.

The content of soluble hexavalent chromium was determined in fly ash leachates, according to STN EN 196-10 (STN EN, 2007). The same amount of fly ash and of re-destilled water (Rodem 6) with conductivity of 5.72 μS/cm and pH of 6.81 was mixed during

15 min at laboratory temperature. The prepared fly ash paste was separated by vacuum filtration through the glass filter with porosity 4 (Morton). The obtained filtrate was adjusted to final volume of 250 mL and the concentration of soluble hexavalent chromium was measured in fly ash leachates by the DR 2800 spectrophotometer (Hach Lange, Germany) at 540 nm.

Radioactivity

Measurement of radioactivity was carried out using a detection system EMS-1A SH (Empos) with a multichannel analyzer MC4K with optimized resolutions of 818 V, 4096 channel and with 0.9 cm of lead shielding and internal lining of 2 mm tinned copper.

The emphasis was on the determination of specific activity concentration of ^{226}Ra, ^{232}Th and ^{40}K. The radioactivity of ^{40}K was measured directly through its gamma ray energy peak at 1460.8 keV, while activities of ^{226}Ra and ^{232}Th were calculated based on the mean value of their respective decay products. Activity of ^{226}Ra was measured using the 351.9 keV gamma rays from ^{214}Pb and the activity of ^{232}Th was measured using the 238.6 keV gamma rays of ^{212}Pb (Estokova, A., Palascakova, L., 2013).

To assess whether the safety requirements for building materials are being fulfilled, a gamma index proposed by the European Commission (European Commission, 1999) was used. It is defined as:

$$I\gamma = \frac{ARa}{300} + \frac{ATh}{200} + \frac{AK}{3000} \qquad (3)$$

where ARa, ATh and AK are the specific activities of isotopes ^{226}Ra, ^{232}Th and ^{40}K in the building material.

The gamma index should also take into account typical amounts and ways in which the material is used. The limit values depend on the dose criteria, the way and amount of the material and the manner in which it was used. For material used in bulk amounts (concrete) $I\gamma \leq 1$ as well as superficial and other materials with restricted use (tiles, boards, etc.) $I\gamma \leq 6$ corresponds to an absorbed gamma dose rate of 1 mSv y-1 (European Commission, 1990). Thus the material with $I\gamma > 6$ should be avoided to use as building material, since these values correspond to the dose rates higher than 1 mSv y-1, which is higher than recommended values (UNSCEAR, 2000).

Life Cycle Assessment (Cradle-to-Gate)

Many works studied problems of life cycle assessment of concrete (Sjunnesson, 2005; Sleeswijk, 2011; Nisbet et al., 2002) through a variety of environmental indicators. The basic step of the multi-criteria analysis is the determination of the functional unit, which is defined as a key element of life cycle assessment. The functional unit is a measure of the function of the studied system and it provides a reference to which the inputs and outputs can be related. This enables comparison of two essential different systems. For example, the functional unit for a paint system may be defined as the unit surface protected for 10 years. A comparison of the environmental impact of two different paint systems with the same functional unit is therefore possible.

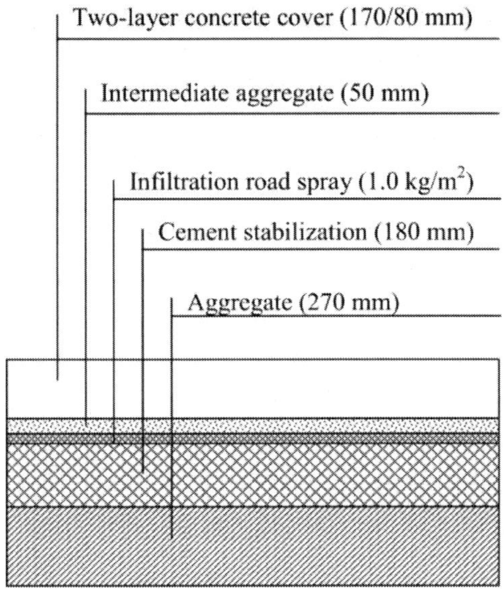

Figure 9. Structure of concrete road.

Life cycle assessment of building materials is a complicated evaluation process of materials. Complexity of the process is exactly in a suitable selection of evaluated areas (cradle-to-grave). Because the life cycle assessment calculations in chapter are very general and approximate, the boundaries were established only in the interval "cradle to gate". The task was to compare the environmental profile of building materials of concrete roads (Figure 9) constructed with and without 15 % fly ash replacement in cement. Functional unit 1m^2 was determined. The calculations of the environmental profiles used input data from the catalog of an international database of building materials Ecoinvent (Czech database ENVIMAT). We chose the following indicators out of the multitude of generally, currently used environmental categories: PEI, GWP and AP for our assessment. Their values were calculated for 1m^2 of road.

- PEI (Primary energy intensity)

The energy consumed from "cradle to gate" of a product or service is called primary energy content. It is derived accounting for both renewable and non-renewable sources of energy. Oil, natural gas, coal and uranium are defined as non-renewable resources, whereas renewable resources are wood, hydroelectric power, solar and wind energy. The primary, non-renewable energy content - PEI ne - is calculated by adding the gross calorific value of all the non-renewable resources used in the process (such as manufacturing of a product). Similarly the primary, renewable energy content - PEI e - sums all renewable energy sources used (Kridlova et al., 2012).

- GWP (Global Warming Potential)

The most common substances contributing to global warming are also defined using the term Global Warming Potential (GWP). This parameter is expressed in relation to carbon dioxide (CO_2) emissions. The Global Warming Potential defines the contribution of a given substance to global warming in comparison to the contribution of an equal amount of CO_2 (Kridlova et al., 2012; Sjunnesson, 2005).

- AP (Acidification Potential)

Acidification occurs primarily through the reaction of nitrogen oxides (NO_x) and sulphur dioxide with other components in the air such as hydroxyl radicals. The Acidification Potential (AP) is a measure of the tendency of a component to become acidified (Kridlova et al., 2012; Sjunnesson, 2005).

3. Results and Discussion

The use of fly ash in concrete is in accordance with principles of the sustainable development of the cement and concrete industries. Our results showed that type of concrete is for Slovak Road industry more economic, environmental friendly, reduces the Portland cement use and CO_2 emissions.

Results of Fresh Concrete and Concrete Composites

Summary results of tested parameters of fresh concrete as well as of concrete composites in comparison with the reference sample are given in Tables 6, 7.

Table 6. Results of tested parameters

Parameter	Unit	Time of testing	RS	5 wt. % of FA	10 wt. % of FA	15 wt. % of FA
Consistency	[mm]	-	30	30	40	40
Air content	[%]	-	6.0	6.0	6.4	6.5
Temperature	[°C]	-	23.5	22.5	19.5	19.5

The Slump Test (Consistency)

All samples comply with requirement S1, while improvement in consistency is also slightly visible with increasing amount of fly ash.

Air Volume Test

Air content: All samples comply with requirement for 4-8% amount of fly ash does not influence the air content significantly.

Temperature and Fresh Concrete Density

All samples comply with requirement for temperature range, while increasing portion of fly ash is causes slight decrease in temperature. Decrease of hydration heat is caused by smaller amount of clinker (Ondova et al., 2012[a]).

Flexural Strength and Compressive Strength of Concrete Composites

Both of tested strengths fulfill requirements for 28 day values (4.5 and 32 MPa respectively). With increasing amount of fly ash, slight decrease in both strengths is visible, including early strengths and ultimate strengths.

De-Icing Salts Resistance

Requirement for maximum 300 g/m^2 of scaling in 100 cycles is not fulfilled only when replacing 15% of cement with fly ash. However, it is necessary to say that testing was done in strict conditions – with 150 cycles. In publication (Brandes, Chr., Schiebl, P., 2006), the value of scaling 1500 g/m^2 is classified as small damage, while according to laboratory testing such samples did not show any degradation even at real exposure.

Frost Resistance - Frost Index, Compressive Strength

All samples comply with requirement for min. 0.85. Increase in amount of fly ash causes decrease of frost index. Frost resistance - Compressive strength: Standard requirement for value of compressive strength after freezing cycles does not defined; however it was included this in our testing. As per results, compressive strengths after freezing cycles increased slightly. Increasing amount of fly ash basically doesn't influence values of strength after freezing (its decrease is only slightly visible after 15% replacement) (Ondova et al., 2012[a]).

Table 7. Results of tested parameters of concrete composites

Parameter	Unit	Time of testing	RS	5 wt. % of FA	10 wt. % of FA	15 wt. % of FA
Flexural strength	[MPa]	7	5.8	6.1	5.7	5.5
		28	6.9	6.6	6.4	6.0
		90	8.2	8.1	7.8	7.6
Compressive strength	[MPa]	7	44.2	40.0	38.7	37.0
		28	48.4	44.2	42.4	40.2
		90	57.2	53.7	52.6	50.1
De-icing salts resistance - Scaling	[g/m^2]	*freezing	47.6	90.7	209.1	557.0
Frost resistance - Flexural strength	[MPa]	**freezing	7.1	6.5	6.2	5.8
		*freezing	6.5	5.9	5.6	5.1
Frost resistance - Frost index	[%]	-	0.92	0.91	0.90	0.88
Frost resistance - Compressive strength	[MPa]	**freezing	48.6	45.3	42.7	38.2
		*freezing	50.6	50.6	50.6	42.7

Note: * after, ** before freezing.

Analytical Methods

X-Ray Fluorescence Method

The results of first three experiments show significant differences in the measured values of the strength for concrete composites with the same portion of fly ash. The differences in strength values result from various admixture constitutions probably. The results of XRF chemical analysis of admixtures used in tested concrete samples are summarized in Table 8.

Obviously, the chemical composition of admixtures 1 and 3 was very similar. The concentrations of elements measured in admixture 2 were quite different. Silicon, phosphorous, manganese and iron contents were measured to be much higher in admixture 2. On the contrary, the calcium and sulphur concentrations reached lower values when comparing to the admixtures 1 and 3. The high sulphur content in admixtures 1 and 3 can influence the surface of particles negatively as well as hydration reactions. Along with the Si content seems to be crucial in terms of the strength parameters evaluation.

Based on obtained outputs, the study of potential chemical admixtures influence on concrete properties has been started. The experiments were repeated and new composites were prepared according to the identical recipe, same value of fly ash replacement and with a variation of chemical admixture to confirm our previous results (Bacikova, Stevulova, 2009; Sedlak, T., 2011). The measured compressive and flexural strength of new tested sample are shown in Table 9 (the reference sample represents always the best results).

Table 8. XRF chemical analysis of tested admixtures

Chemical composition	Admixture 1 [mg/kg]	Admixture 2 [mg/kg]	Admixture 3 [mg/kg]
Mg	<101	2140	<101
Si	<8.3	113	<5.1
P	<3.0	473.4	<3.0
S	93 160	47 280	99 680
Cl	188.9	294.4	195
K	222	324.2	224
Ca	5 799	2 476	5 130
Mn	<5.1	184.1	39.7
Fe	30.4	129.3	57.4
Co	15.5	8.1	14.7

Table 9. Results of mechanical properties of composites based on various admixtures

Strength [MPa]/ Mixtures based on various admixtures	Admixture 1	Admixture 2	Admixture 3
Compressive strength – ref. sample	19.3	28.2	19.9
Compressive strength 15% FA	13.0	23.0	15.7
Flexural strength – ref. sample	7.2	8.0	7.2
Flexural strength 15 % FA	5.8	7.4	4.5

Considering the results of chemical composition of prepared composites, it can be stated that composites based on various admixtures were similar to the chemical composition of the assessed Portland cement (Lam et al., 2010). Differences in composition depend on the structure of the composites as well as on the processes that take place in composite materials during hardening. Mixtures 1 - 3 are presented by the different value of CaO / SiO_2 ratio (0.50 – mixture 1; 0.43 – mixture 2; 0.6 – mixture 3). The lowest value of CaO / SiO_2 ratio corresponded with the highest value of flexural strength (Table 9 – admixture 2). As consequence, the strength differences by the thermal analysis were monitored (Ondova et al., 2012[d]).

Thermal Analysis

The prepared concrete samples with 15 % of cement replacement by fly ash and concrete samples without fly ash replacement, respectively with various chemical admixtures were chosen for the thermal properties testing. As an example, the DSC/TG spectrum of concrete sample with 15 % of cement replacement by fly ash containing admixture 2 is illustrated in Figure 10.

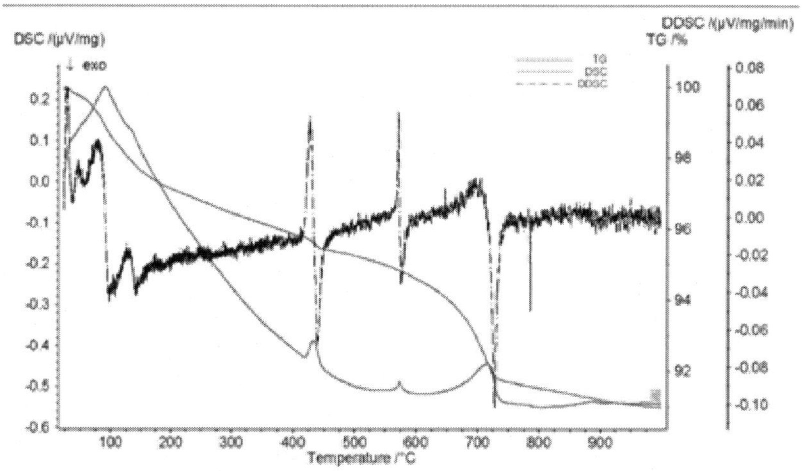

Figure 10. DSC/TG spectrum of concrete sample with 15 % wt. of FA.

The comparison DSC curves (endothermic processes courses) and TG curves (mass of losses) of the concrete samples with fly ash and admixture addition and the reference concrete is presented in Figures 11 and 12.

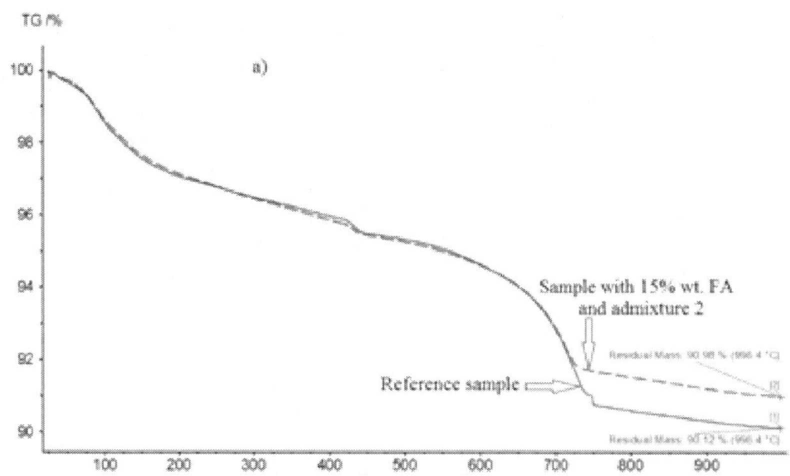

Figure 11. (Continued).

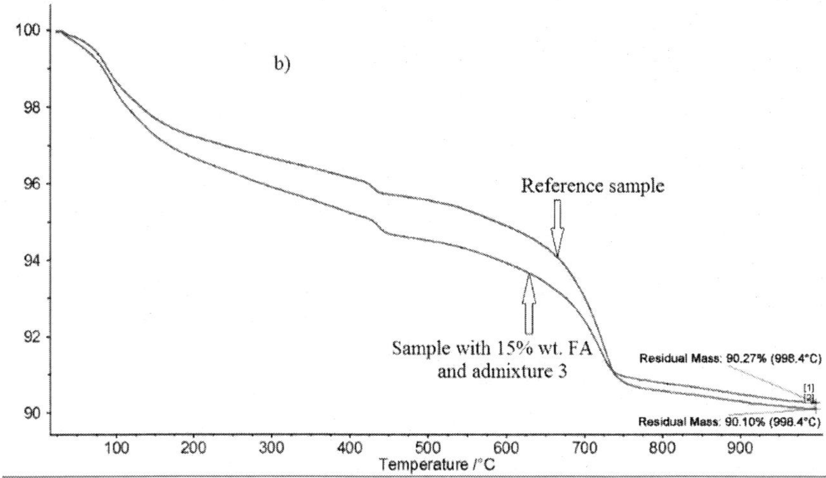

Figure 11. Comparison of the TG curves of the concrete samples with 15 % of cement replacement by fly ash to the reference samples without fly ash.

Several endothermic processes in the range 25 to 1000°C were detected during heating of all measured concrete samples. The peaks and onset temperature values as well as the related mass changes for studied concrete composites with 15 % of cement replacement by fly ash are summarized in Table 10. Similar DSC curves with four principal peaks were observed for all studied concrete samples. The most significant shift in peak temperature towards to the higher temperature by 6°C was detected in process 4 in case of sample with admixture 2 (Figure 12b) for reference sample. The other related peak temperatures for all processes and each measured sample were very close to each other (Ondova et al., 2012[d]).

No significant differences in peak temperatures were detected for the investigated samples evaluating the endothermic processes (Table 10). The TG measurements expressed in mass of lost gases for tested powdered samples show to be similar; the higher percentage difference was detected only in case of sample with admixture 3 at process 4. Residual mass after the heating process ranged in close interval from 90.1 to 92.02 %. Admixtures are likely not to be the factor of importance in terms of thermal behavior of concrete composites.

Table 10. Characterisation of the thermal processes of studied concrete samples

		Admixture 1	Admixture 2	Admixture 3
Process 1	Peak [°C]	87.6	90.6	91.4
	Onset [°C]	71.9	75.6	70.8
	Mass loss [%]	1.32	1.87	1.87
Process 2	Peak [°C]	431.2	432.8	430.0
	Onset [°C]	419.0	420.9	418.3
	Mass loss [%]	0.33	0.47	0.42
Process 3	Peak [°C]	573.0	573.0	572.8
	Onset [°C]	568.7	568.7	569.3
	Mass loss [%]	0.44	0.4	0.29
Process 4	Peak [°C]	720.8	715.4	728.0
	Onset [°C]	683.4	680.0	690.0
	Mass loss [%]	2.67	2.4	3.62
	Residual mass [%]	92.02	90.98	90.1

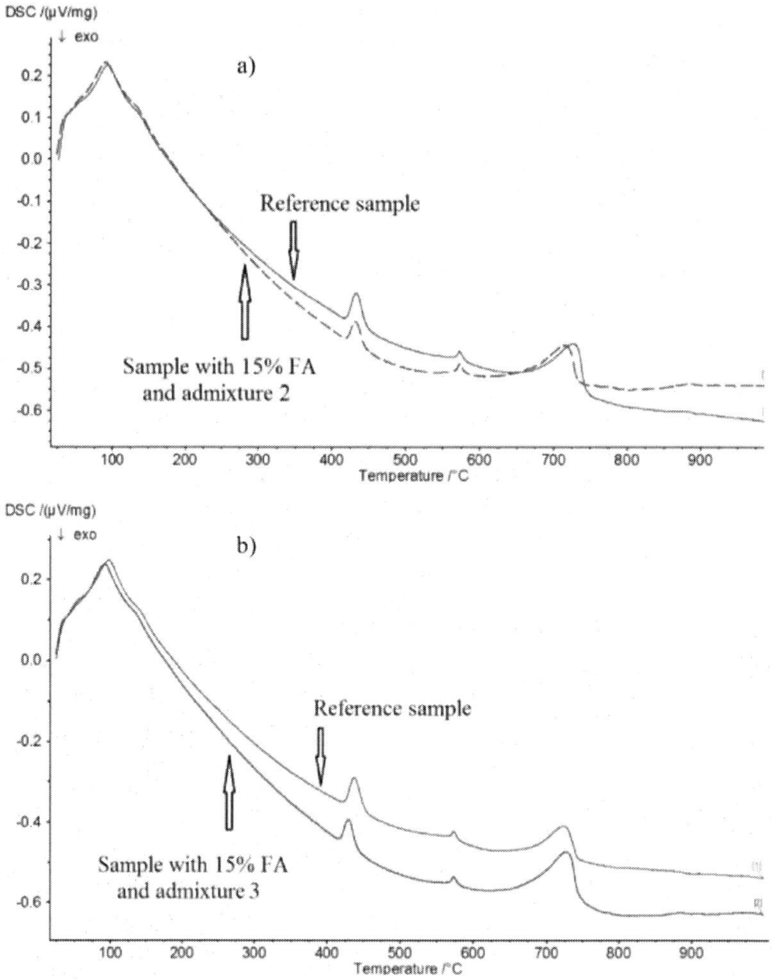

Figure 12. Comparison of the DSC curves of the concrete samples with 15 % of cement replacement by fly ash to the reference samples without fly ash.

When comparing the TG curves of the samples with and without fly ash addition, the results are very similar for all studied samples. The total mass losses after heating process were measured 9.35, 9.88 and 9.73 % for reference samples with admixture 1, 2 and 3, respectively; 7.98, 9.02 and 9.9 % for fly ash samples with admixture 1, 2 and 3, respectively. In summary, the mass loses of reference samples were measured to be higher (except for sample with admixture 3).

Courses of TG curves (Figure 11) are consistent with the curves (Figure 13) published in work (Villain et al., 2007). TG method supplement by chemical analysis, its comparison to gamma-ray absorption method and using FT IR spectroscopy as well as X-ray diffraction analysis was applied to identify of dehydration and decarbonation processes occuring during thermal behaviour study of an ordinary Portland cement paste, and three concrete mixes, containing siliceous or calcareous aggregates. It can be seen from Figure 13 that the mass of a carbonated concrete powder sample containing siliceous aggregates subjected to a steady increase of temperature in order to quantify reactions during its thermal decomposition. Each phase is characterized by its own temperature range of decomposition and by a specific mass loss. In the first stage of the thermal processes (Figure 11), the release of physical bounded water from concrete sample occures at low temperatures (up to 100°C). In the temperature range of ≈430-570°C (process 2), the dehydration of the residual portlandite ($Ca(OH)_2$) occures. Decomposition of the calcium carbonates measured between 570°C and 730°C (process 3) results mainly from the dissociation of the poorly crystallized and thermally unstable forms of $CaCO_3$ such as vaterite and aragonite resulting from carbonation of calcium silicate hydrates (C-S-H phase) which dissociate between 500 and 700 °C according to literature sources (Sauman, Z., 1971).

The more stable calcite ensuing from portlandite carbonation decomposes between 730 and 1000°C (process 4). The obtained decomposition temperatures of carbonated products are a little shifted in comparison to the literature data (Villain, 2007), where calcite formed from portlandite by carbonation reaction dissociates between roughly 760 and 950°C (or 650-950°C). It is difficult to determine a precise temperature range of decarbonation processes because they depend on the kinetics of the carbonation process of hydrated products (portlandite and C-S-H) (Thiery et al., 2007).

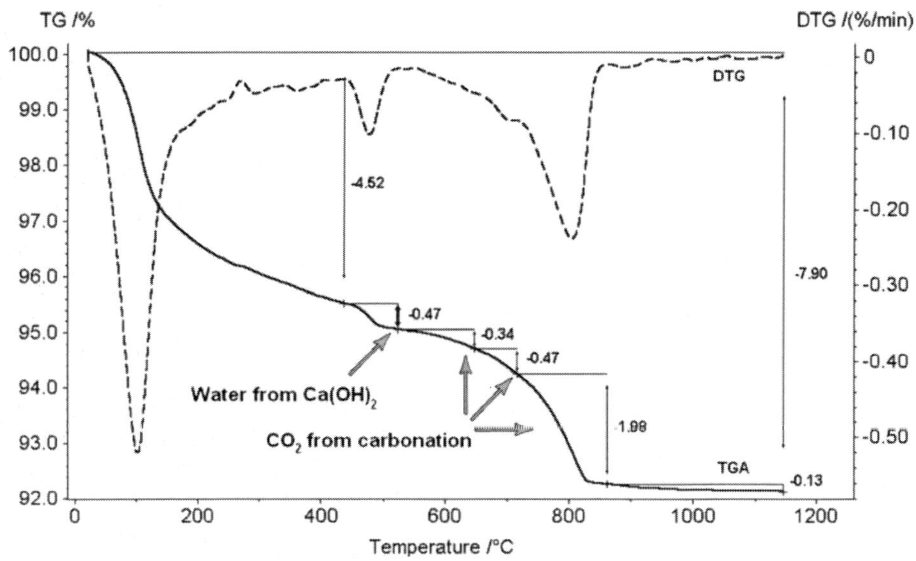

Figure 13. Dissociation of a carbonated sample of crushed concrete mortar during a TGA test: concrete containing siliceous aggregates (source: Villain et al., 2007, with permission).

Economic Assessment

Respecting the specific technology finaly economic results presented in Table 11 came out in this calculation. The most effective alternative seems to be the variant III with 15% fly ash compensation. It represents 21 261.08 € cost savings per kilometer of cement concrete pavement (Zelenakova, E., Ondova, M., 2010).

Table 11. Final calculation of CC two layer cover of pavement

Description (calculation according to the Slovak database - CENKROS)	Total price
Variant I. Concrete two-layer reinforced cover of pavement class I. thickness to 250mm – calculation according to the Slovak CENEKON 2009 database)	553 555.88 €
Variant II. Concrete two layer reinforced cover of pavement class I. thickness to 250mm – calculation based on our own recipe with 100 % Portland cement using)	549 758.76 €
Variant III. (Concrete two layer reinforced cover of pavement class I. thickness to 250mm - calculation based on our own prescriptions with 85 % Portland cement using and 15% Slovak fly ash using)	532 294.80 €
Cost saving	21 261.08 €

Environmental and Life Cycle Assessment

Result of determination of hexavalent water-soluble chromium as well as of radioactivity (^{226}Ra, ^{228}Th, ^{40}K) of cement CEM I 42.5 N and fly ash are summarized in Table 12.

Table 12. Summary of toxicity tests

Tested hazardous elements	Unit	Fly ash	CEM I 42.5 N
Cr^{VI}	mg/kg Cr^{6+}	0.01	2.46
Radioactivity: ^{226}Ra	[Bq/kg]	39.259	13.081
^{232}Th	[Bq/kg]	10.144	19.839
^{40}K	[Bq/kg]	698.167	169.232
Gamma index Iγ	-	0.411	0.198

Hexavalent Water-Soluble Chromium

The total content of Cr^{VI} was measured. The value of CEM I 42.5 N is 2.46 mg/kg and 0.01 mg/kg of fly ash. Taking into account the maximum value for packed cements 2.0 mg per kg of cement (2.0 ppm = 0.0002%) stated by Slovak legislative (Notice no. 275/2004 Collection of Laws), CEM I samples had higher content of chromium (VI). That means this cement is not too dangerous for human health, but it may be distributed only in the fully automated operation without cement contact with the skin of workers.

Radioactivity

The activity concentration index should be used as a screening tool for identifying materials which might be of concern. It has been referred that the criteria of gamma index of material used in bulk amounts is set to the value $I\gamma \leq 1$ as well as superficial and other materials with restricted use (tiles, boards, etc.) $I\gamma \leq 6$. The material with $I\gamma > 6$ should be avoided to use as building material because it doesn´t meet specified requirements. The results of measured gamma index of CEM I 42.5 N (0.198) and fly ash (0.411) presented in

Table 12 showed that both of tested materials have value of Iγ less than 1 and are suitable for use in building construction.

Life Cycle Assessment (Cradle-to-Gate)

The results of calculated environmental profile of two different concrete roads are showed in Table 13. The calculated profiles showed a reduction of emissions PEI, GWP as well as AP. Emission reduction per m^2 was: PEI 35.1 MJ/m^2 (2.7%), GWP 6.54 $kg.CO_2.eq/m^2$ (2.6%), AP 0.011 $kg.SO_2.eq/m^2$ (2.8%). In considering the conditions used in the economic assessment (Tunnel with length: 698m and width 9m) the result of calculation is following: PEI 217663.36 MJ/m^2, GWP 41103.04 $kg.CO_2.eq/m^2$, AP 70.52 $kg.SO_2.eq/m^2$. By converting to km we obtain a real emissions reductions produced by road construction. The results indicate that the construction of many kilometers of highways, the numbers in terms of the impact on the environment are not insignificant. It is important to say that the environmental profiles and with it connected emission reductions were calculated only within the boundaries of cradle to gate.

Table 13. Comparison of LCA of concrete roads with and without fly ash

Layer	Density kg/m^3	Thickness mm	Mass kg/m^2	PEI MJ/m^2	GWP $kg\ CO_2eq/m^2$	AP $kg\ SO_2eq/m^2$
Concrete road						
2nd layer concrete cover	2445	170	415.7	262.34	48.90	0.084
1st layer concrete cover	2445	80	195.6	123.45	23.01	0.039
Coated intermediate aggregate - gravel	1650	50	82.5	5.65	0.23	0.001
Infiltration road spray	1160	1	1.2	57.54	1.28	0.007
Cement stabilization	1200	180	216.0	777.26	177.55	0.251
Aggregate 2nd layer - macadame	1650	150	247.5	30.76	1.09	0.006
Aggregate 1st layer - macadame	1650	120	198.0	24.61	0.87	0.005
Sum		**751**	**1356.41**	**1281.62**	**252.94**	**0.395**
Concrete road based on 15% FA cement replacement in concrete cover						
2nd layer concrete cover	2280	170	387.6	238.47	44.45	0.076
1st layer concrete cover	2280	80	182.4	112.22	20.92	0.036
Coated intermediate aggregate - gravel	1650	50	82.5	5.65	0.23	0.001
Infiltration road spray	1160	1	1.2	57.54	1.28	0.007
Cement stabilization	1200	180	216.0	777.26	177.55	0.251
Aggregate 2nd layer - macadame	1650	150	247.5	30.76	1.09	0.006
Aggregate 1st layer - macadame	1650	120	198.0	24.61	0.87	0.005
Sum		**751**	**1315.16**	**1246.51**	**246.39**	**0.383**
Difference (reduction)		0	41.25	35.10	6.54	0.011
Difference (reduction) [%]				2.7%	2.6%	2.8%

Summary of obtained results are presented in Table 14. Determined test results were presented through a simple assessment method in this chapter. The brown fly ash produced heat and power plant is useful in road construction. Thanks to fly ash properties as well as thanks to the our obtained results is the fly ash suitable partial replacement (minimal 15 wt.%) for cement in concrete road cover in Slovakia. In regards to the subjective ranking of assessed areas it could be possible to reconsider and modify the strict criteria of Slovak Technical Standards related to Slovak Highway Engineering.

Table 14. Final summary of results

Type of assessed area	Fly ash	CEM I 42.5 N
Technical parameters	*/** [1]	**
Economical	**	*
Environmental	**	**
Life cycle "cradle to gate"	**	*
Sum	7/8	6

Note: [1] the results of mechanical tests showed lower values than concrete without FA, but met the requirements of Technical standard.

CONCLUSION

Over a period of last ten years, the image of fly ash has completely been changed from a "polluting waste" to "resource material". Fly ash if segregated, collected and used properly can solve the major problems of cement use, which consumes a lot of energy and natural resources.

Rapid industrialization has resulted in increased use of natural resources, which has also brought along serious ecological and environmental imbalance due to the dumping of industrial wastes. Principles of sustainable construction, especially in the building construction, have to be accepted with regards to the consumption of natural resources and the production of harmful emissions. Concerning raw materials cost and produced CO_2 emission the replacment of cement in concrete mixtures with more sustainable materials is necessary. Fly ash concretes are proposed as a potential approach for achieving substantial reductions in cement.

Based on this assessment it can be stated that fly ash improves the workability of concrete, extends the life cycle of concrete roads, and reduces energy use and greenhouse gas as well as amount of coal combustion products that must be disposed in landfills.

The results of our study confirm the Slovak fly ash as appropriate resource material. Mechanical characteristics of concrete composites with fly ash fulfill the Slovak and European Technical Standards. Pointed economical benefits present fly ash as gainful alternative for building industry. Values of hazardous substances in fly ash as well as results of environmental profiles show other advantages and potential of fly ash utilization in the field of construction industry as well as in the field of construction materials production.

ACKNOWLEDGMENTS

This research has been carried out in terms of the project NFP 26220120037 Centre of excellent research of the progressive building structures, materials and technologies, supported from the European Union Structural funds and project no. 1/0481/13 Study of selected environmental impacts of building materials.

REFERENCES

Asokan, P., et al., (2005). Coal combustion residues–environmental implications and racycling potentials. *Resourse Conservation and Recycling* 43, 239-262.

Bacikova, M., Stevulova, N., (2009). Examination of fly ash utilization suitability for the production of cement – concrete cover of pavement. *Chemine Technologija*, 24 –29.

Berry, E. E., Hemmings, R. T., Langley, W. S., Carette, G. G., (1989). Beneficiated Fly Ash: Hydration, Microstructure, and Strength Development in Portland Cement Systems. Proceedings of the Third International Conference on the Use of Fly Ash, Silica Fume, Slag and Natural Pozzolans in Concrete, ACI Special Publication, 241-273.

Berry, E. E., Malhotra, V. M., (1980). Fly Ash Use in Concrete - A Critical Review. *ACI Journal* 77, 59-73.

Berry, E. E., Malhotra, V. M., (1987). Supplementary Cementing Materials for Concrete. Canadian Government Publishing Center, p. 428.

Brandes, Chr., Schiebl, P., (2006). Effect of aging related to freeze/thaw and de-icing salt resistance of concretes. Concrete repair, rehabilitation and retrofitting. Taylor and Francis Group, London, p. 187.

Brylska E., Dyczek J., Gawlicki M., Roszczynialski W., (2001). Application of wastes from thermal power stations in building materials. SourcePopioły z energetyki : VIII międzynarodowa konferencja, 191–200.

BVK – Betontechnische Empfehlungen, Neuausgabe 2002.

Cudic, V. et al., (2007). Ash from thermal power plants as Secondary raw material. *Thermal power plant ash as secondary raw material* 58, 233-238.Estokova, A., Palascakova, L., (2013). Study of Natural Radioactivity of Slovak Cements, *Chemical Engineering Transactions* 32, in press.

Estokova, A., Porhincak, M., (2012). Reduction of Primary Energy and CO_2 Emissions Through Selection and Environmental Evaluation of Building. *Theoretical Foundations of Chemical Engineering* 46, 704–712.

Estokova, A., Porhincak, M., Paulikova, A., (2012). Sustainability evaluation of masonry walls using a multi criteria assessment tool. Building Sustainability Assessment 2012, 117-124.

Estokova, A., Palascakova, L., 2013. Study of Natural Radioactivity of Slovak Cements. *Chemical Engineering Transactions*, 32, in press.

European Commission (EC), (1999). Radiation Protection 112-radiological protection principles concerning the natural radioactivity of building materials Directorate-General Environment. Nuclear safety and civil Protection.

European Coal Combustion Products Association (www.ecoba.com, Accessed March, 14, 2013).

Feely, R. A. et al., (2004). Impact of anthropogenic CO_2 on the $CaCO_3$ system in the oceans. Science, 362.

Grier, D.G., at al., (1997). Coal combustio product diagenesis II. Proc. of International Ash Utilization Symposium, 385-394.

Helmuth , R., (1987). *Fly ash in Cement and Conrete*. Illinois : Portland Cement Association.

Hendriks, C.A. et al., (2000). Emission reduction of greenhouse gases from the cement industry. IEA Greenhouse gas R&D Programme.

Junak, J., (2011). Glass waste and it utilization in civil engineering. Construmat 2011. 1-4.

Lam, H. K., Barford, J. P., McKay, G., (2010). Utilization of incineration waste ash residues as Portland cement clinker. *Chemical Engineering Transactions* 21, 757-762.

Lippiatt, B., Ahmad, S., (2004). Measuring the lifecycle environmental and economic performance of concrete the BEES approach, Inter. workshop on sustainable development and concrete technology.

Keil, F., (1971). *Zement, Herstellung und Eigenschaften.* Verlag : Springer.

Kolousek, D. et al., (2002). Biological - chemical interactions accelerated the removal of impurities from fly ash. *Acta Montanistica Slovaca*, 60-63.

Kridlova B., E., Vilcekova, S., (2012). Environmental assessment of building materials and constructions. *Applied Mechanics and Materials* 174-176, 3161-3165.

Kridlova B., E., Ondova, M., Vilcekova, S., (2012). Environmental assessment of building materials. Proc. of the 9. International Scientific Conference FCE TUKE. Technical University of Košice, Slovakia.

Kridlova B., E., Vilcekova, S., (2013). *Building environmental assessment system in Slovakia*. Saarbrücken : LAP LAMBERT Academic Publishing.

McCarthy et al., (1990). Use of a database of chemical, mineralogical and physical properties of North American fly ash to study of nature of fly ash and its utilization as a mineral admixture in concrete. *Material Research Society Symposium Proceedings* 178, 3-34.

Mehta, P. K., (1989). Pozzolanic and Cementitious By-Products in Concrete-Another Look. Proceedings of the Third International Conference on the Use of Fly Ash, Silica Fume, Slag and Natural Pozzolans in Concrete. ACI Special Publication, 1 - 43.

Mehta, P. K., (1983). Pozzolanic and Cementitious By-Products and Mineral Admixtures for concrete- A Critical Review.Proceedings of the First International Symposium on the Use of Fly Ash, Silica Fume and other minerals in Concrete. ACI Special Publication, 1-40.

Michalikova, F., Jacko, V., Sisol M., Kozakova, I., (2005). Modifying technology of obtaining commercial components of solid waste from the coal combustion in power plants. *Acta Montanistica Slovaca* 10, 49-55.

Noskievic, P., (1993). *Coal* combustion. Ostrava: VŠB TU.

Nisbet, M. A., Medgar, M. L., Martha, V. G., (2002). Environmental Life Cycle Inventory of Portland Cement Concrete. Portland Cement Association, no. 2137a.

Ondova, M., Stevulova, N., Zelenakova, E., (2011). Energy Savings and Environmental Benefits of Fly Ash Utilization as Partial Cement Replacement in the Process of Pavement Building. *Chemical Engineering Transactions* 25, 297 - 302.

Ondova, M., Sicakova, A., Stevulova, N., (2012[a]). Assessment of selected indicators of portland cement containing fly ash in road concrete. *Journal of Interdisciplinary* Research 2, 114-116.

Ondova, M., Stevulova, N., Estokova, A., Junak, J., (2012[b]). The influence of chemical composition of mixture on the mechanical properties of fly ash. *SGEM* 2012, 945-950.

Ondova, M., Stevulova, N., (2012[c]). Benefits of Fly Ash Utilization in Concrete Road Cover. *Theoretical Foundations of Chemical Engineering* 46, 713-718.

Ondova, M., Stevulova, N., Estokova, A., (2012[d]). The Study of the Properties of Fly Ash Based Concrete Composites with Various Chemical Admixtures. *Procedia Engineering* 42, 2044-2054.

Ondova, M., Stevulova, N., Palascakova, L., Estokova, A., (2013). The study of concrete properties prepared with a proportion of fly ash. Pollack Periodica, in press.

Owens, J.W., (1997). LCA:constraints on moving from inventory to impact assessment. *Journal of Industrial Ecology* 1, 37.

Palascakova, L., Estokova, A., Balintova, M. and Petrilakova, A., (2011). Assessment of the content of chromium (VI) in selected types of cement as a part of cement eco-labelling, *Pollack Periodica* 6, 123-129.

Parrott, L.J., Killoh, D.C., (1989). Carbonation in a 36 year old in-situ concrete, *Cem. Concr. Res.* 19, pp. 649–656.

Pavlik A., Doležel J., Fiedler K., (1973). Concrete technology. Praha : SNTL.

Platret, G., Deloye, F.-X.,(1994). Thermogravimetry and carbonation of cements and concretes, Actes des Journées des Sciences de l'Ingénieur du réseau des Laboratoires des Ponts et Chaussées, Publication LCPC, Paris, pp. 237–243, (in French).

Rafalowski, M., (2003). Fly Ash Facts for Hightway Engineers. *American Coal Ash Association.*

Report of the International Energy Agency (IEA), 2008;

Ross, S., Evans, D., Webber, M., (2002). How LCA studies deal with uncertainty. *The International Journal of L. C. Assessment* 1, 47.

Ruzickova, Z., Srb, J., Mayerova, M., (1993). Flyash – modification and utiliyation. Praha : SNTL.

Sabine, C.L. et al., (2004). The oceanic sink for anthropogenic CO_2. Science, 367.

Saman, Z., (1971). The importance of mineralogical composition of fly ash in the production of porous concrete. Stavivo 49, 4-8 (in Czech).

Sauman, Z., (1971). Carbonization of porous concrete and its main binding components. *Cement and Concrete Research* 1, 645-662.

Sedlak, T., (2011). Coal fly ash utilization as an active ingredient in road concrete. Košice: Technical University of Košice.

Sicakova, A., (2000). Influence of fly ash on the quality of concrete production technology. TOP 2000, 199-204.

Sjunnesson, J., (2005). Life Cycle Assessment of Concrete. Master thesis.

Slanicka, S., (1982). Possibilities of fly ash utilization in concrete. Bratislava: Research Institute of Civil Engineering.

Sleeswijk, A. W., (2011). Regional LCA in global perspective. A basis for spatially differentiated environmental life cycle assessment. *Int. J. Life Cycle Assess* 16, 112-116.

Sybertz, F., (2004). Beurteilung der Wirksamkeit von Steinkohlenflugaschen als Betonzusatzstoff Schriftenreihe des Deutschen Ausschusses für Hochleistungsbetone. Beton 54, 132 – 133.

Stevulova, N. et al., (2008). Utilization possibilities of selected waste kinds in building materials preparing. SGEM 2008, 193-200.

Stevulova, N., Sicakova, A., Terpakova, E. (2012). Some aspects of fly ash valorisation in road concrete. WasteEng 2012, 1533-1538.

Svoboda, L. et al., (2005). Building Materials. Bratislava.

Thiery, M., Villain, G., Dangla, P., Platret, G, (2007). Investigation of the carbonation front shape on cementitious materials : Analysis of the effects of the chemical kinetics. *Cement and Concrete Research* 37, 1047-1058.

Tkacova, K., Stevulova, N., (1992). Mechanical activation of solid-phase process by milling and its importance to the advanced technology of ceramic materials. Ceramics- Silikaty 36, 109-117 (in Slovak).

UNSCEAR, (2000). Sources and Effects of Ionizing Radiation. United Nations Scientific Committee on the Effects of Atomic Radiation, United Nations Publication, New York, USA.

Vilcekova, S., Kridlova B., E., (2010). Building Environmental Assessment. Kosice : Technical University of Kosice (in Slovak).

Villain, G., Thiery, M., Platret, G., (2007). Measurement methods of carbonation profiles in concrete: Thermogravimetry, chemical analysis and gammadensimetry. *Cement and Concrete Research* 37, pp.1182–1192.

Zelenakova, E., Ondova, M., (2010). The economic balance of the financial costs of construction concrete roadway with a share of fly ash in cement, Innovative approach to modeling intelligent construction elements in building 2010, 202 (In Slovak).

http//: www.betonracio.sk, [Accessed March, 14, 2013].

http//: www.envimat.cz, [Accessed March, 14, 2013].

European Technical standards:

STN EN 450-1, (2005). Fly ash for concrete - Part 1: Definition, specifications and conformity criteria.

EN 450-2, (2005). Fly ash for concrete - Part 2: Conformity evaluation.

EN 451-1, (1995). Method of testing fly ash. Determination of free calcium oxide content.

STN EN 196-1, (2005). Methods of testing cement - Part 1: Determination of strength.

STN EN 196-2, (2005). Methods of testing cement - Part 2: Chemical analysis of cement.

STN EN 196-3, (2009). Methods of testing cement - Part 3: Determination of setting times and soundness.

STN EN 196-6, (2010). Methods of testing cement - Part 6: Determination of fineness.

EN 197-1 Cement - Part 1, (2011). Composition, specifications and conformity criteria for common cements.

STN EN 1015-3, (2000). Methods of test for mortar for masonry - Part 3: Determination of consistence of fresh mortar.

EN 12 390-3, (2001). Testing hardened concrete.Part 3: Compressive strength of test specimens.

EN 13 877-1, (2004). Concrete pavements - Part 1: Materials.

EN 13 877-2, (2004). Concrete pavements - Part 2: Functional requirements for concrete pavements.

EN 206-1, (2002). Concrete - Part 1: Specification, performance, production and conformity.

EN 12 350, (2010). Testing fresh concrete.

Slovak Technical standards:

STN 73 1326, (1984). Resistance of cement concrete surface to water and defrosting chemicals.

STN 73 6123, (2010). Road Construction. Cement Concrete Pavements. Wearing Courses).

INDEX

A

B

C

D

E

F

G

H

I

L

M

N

O

P

R

X

Z